The Handbook of AI for Clean Water

The Handbook of AI for Clean Water: Innovations in Treatment and Monitoring is a comprehensive guide that delves into the transformative role of artificial intelligence (AI) in addressing the global water crisis through advanced treatment, monitoring, and management technologies. Divided into nine thematic parts and spanning 21 chapters, this volume provides a structured and interdisciplinary examination of how AI can revolutionize clean water access, treatment efficiency, contamination control, and policymaking.

This book offers in-depth insights into AI applications across water quality monitoring, predictive maintenance, microbial detection, desalination, and governance. With a technical emphasis on AI–Internet of Things (IoT) integration, machine learning algorithms, data-driven decision-making, and sustainable water management, it explores case studies, technical models, and futuristic solutions aimed at reducing environmental impact and enhancing system efficiency. This handbook empowers readers with knowledge of both foundational principles and cutting-edge innovations.

This is an ideal resource for researchers, environmental scientists, policymakers, engineers, graduate students, and professionals in water resource management. It is also intended for AI developers and decision-makers seeking to implement data-driven strategies for clean water access and sustainability.

The Handbook of AI for Clean Water

Innovations in Treatment and Monitoring

Edited by
Smriti Agarwal and Manoj Chandra Garg

CRC Press
Taylor & Francis Group
Boca Raton London New York Leiden

CRC Press is an imprint of the
Taylor & Francis Group, an **informa** business
A BALKEMA BOOK

Cover: Shutterstock.com

First published 2026
by CRC Press
4 Park Square, Milton Park, Abingdon, Oxon, OX14 4RN

and by CRC Press
2385 NW Executive Center Drive, Suite 320, Boca Raton FL 33431

CRC Press/Balkema is an imprint of the Taylor & Francis Group, an informa business

British Library Cataloguing-in-Publication Data
A catalogue record for this book is available from the British Library

ISBN: 978-1-032-98806-1 (hbk)
ISBN: 978-1-032-98805-4 (pbk)
ISBN: 978-1-003-60071-8 (ebk)

DOI: 10.1201/9781003600718

Typeset in Times
by codeMantra

Contents

About the Editors viii
List of Contributors ix
Preface xiii
Acknowledgments xiv

PART I Understanding the Water Crisis and AI Integration 1

1 Introduction to the Global Water Crisis 3
Manoj Chandra Garg, Sheetal Kumari, Jyoti Chowdhry, and Smriti Agarwal

2 Role of AI in Mitigating the Water Crisis 22
Charu Saxena

PART II AI Applications in Water Quality Monitoring 39

3 Synergy of AI and IoT in Water Monitoring 41
Preeti Karki, Priyanjali Chowdhury, Joel Benny, Alakto Choudhury, Abhishek Sengupta, and Priyanka Narad

4 Detecting Chemical Contaminants in Water Using AI 62
Dibyarupa Pal, Partha Sarathi Mandal, Ritam Paul, Avishek Kumar Kaser, and Pranabi Maji

5 Artificial Intelligence Applications in Water Monitoring and Risk Assessment 80
Minh-Ky Nguyen

6 Advanced Dynamic Water Quality Assessment Using Machine Learning Algorithms 92
Elif Esra Altuner, Hasan Ulus, Halil Burak Kaybal, and Fatih Sen

PART III Predictive Modeling for Effective Water Resource Management 113

7 AI-Driven Predictive Maintenance and Downtime Optimization in Water Treatment Plants 115
Paramasivam Balakrishnan and Marimuthu Subramani

8 Future Prospects: AI Innovations for Enhanced Water Management 138
Sundram Singh, Swati Sangolgi, and Tapas Ranjan Das

9 **AI and the Future of Water Scarcity: Machine Learning for Efficient Resource Management** 155
G. R. Manjunathareddy and H. R. Raveesha

PART IV AI Revolutionizing Water Treatment 175

10 **Automation and Optimization in Water Treatment** 177
Neha Raghuwanshi, Shagun Sharma, and Chaitanya Vijay Kale

11 **Energy Optimization and Environmental Impact Reduction through AI** 191
Ayoub Belcaid and Ayoub Ouarga

PART V AI Combatting Contaminants for Safe Drinking Water 215

12 **Detection and Mitigation of Microbial Threats Using AI** 217
Shubhrajit Sarkar and Ankita Mazumder

13 **AI-Powered Decision Support Systems for Sustainable Wastewater Management: Addressing Nutrient Cycling and Heavy Metal Contamination** 234
Qudrat Ullah, Muhammad Waqar, Sankar Sangeetha Vasanthkumar, and Thanet Khomphet

PART VI AI and Desalination: A Path to Water Security 255

14 **AI-Driven Optimization in Seawater Desalination** 257
Sumeda Bandara and Thilini Bandara

PART VII AI-Integrated Water Policy and Governance 271

15 **AI Enabled Water Policy of Formulation and Decision Making** 273
Kavitha Manikandan and Satthiyaraju Mani

16 **Investigation on Some Aspects of Data-Driven Insights for Effective Water Governance** 282
Abhijit Bora, Lima Hazarika, and Bamdeb Dey

17 **Leveraging Data-Driven Approaches: A Case Study on Enhancing Water Governance in Africa** 293
Basil Osayin Daudu, Salisu Ameh Amodu, Peace Ojochenemi Yakubu, and Esther Besty Anaiye

PART VIII AI and Community Engagement for Water Sustainability **307**

18 Community-Centric Water Conservation Using AI **309**
Surya Pratap Singh, Sudarshana Banerjee, Diwesh Kumar, and Shailesh Kumar Sarangi

PART IX Challenges and Future Perspectives **325**

19 Ethical and Privacy Challenges in AI-Driven Water Management **327**
Naveen Kumar, Devesh Singh, Piyush Sharma, Pankaj Kumar Chaurasia, Sunita Singh, and Ajay

20 Environmental Impacts of Artificial Intelligence in Water Management **341**
Surya Konappan, Cheran Easwaran, Sathya Priya Ramalingam, and Krishnan Ramanujam

21 Understanding Water Scarcity: The Role of AI in Sustainable Solutions **356**
Elif Esra Altuner, Halil Burak Kaybal, Hasan Ulus, and Fatih Sen

Index 371

About the Editors

Dr. Smriti Agarwal is an accomplished academician and researcher with over 17 years of experience in teaching and research. She specializes in machine learning, AI-driven optimization techniques, microwave and millimeter wave radar imaging, antenna design, and microwave absorbers. Currently, she serves as an assistant professor in the Department of Electronics and Communication Engineering at Motilal Nehru National Institute of Technology (MNNIT), Allahabad, a premier Indian institution of national importance.

Dr. Agarwal earned her Ph.D. in Microwave Imaging from the Indian Institute of Technology Roorkee in 2015, following her M.Tech. and B.Tech. in Electronics and Communication Engineering. She has previously held academic positions at G. B. Pant Government Engineering College, New Delhi, and SRMCEM, Lucknow.

Her work has been published in high-impact international journals and presented at prestigious global conferences. She successfully completed a DST-funded research project on "Terahertz Nanoantennas and Its Application in Imaging." She is a recipient of numerous accolades, including the UCOST Young Scientist Award, DST Women Scientist Fellowship, and CSIR Research Intern Fellowship. A senior IEEE member, she also contributes as a technical reviewer for reputed journals across multiple publishers.

Dr. Manoj Chandra Garg is an assistant professor at the Amity Institute of Environmental Sciences, Amity University Uttar Pradesh, with over 10 years of teaching and research experience. He earned his Ph.D. from the Indian Institute of Technology (IIT) Roorkee, focusing on solar photovoltaic-assisted nanofiltration and reverse osmosis systems for water treatment. He also holds an M.Tech. in Environmental Science and Technology from Thapar University, where he was awarded a Gold Medal.

Dr. Garg specializes in water and wastewater treatment, membrane technologies, solar desalination, biosorption, and the application of artificial intelligence and machine learning in environmental systems. He successfully led a SERB-funded research project under the Early Career Research Award.

He has published with several SCI-indexed journals (with a total impact factor exceeding 150), authored two books, and presented his research at numerous national and international conferences. His innovations have resulted in several patent filings. He has also received several accolades, including international travel grants (DST), MHRD, and Ministry of Drinking Water and Sanitation fellowships.

Contributors

Ajay
Department of Commerce
Sanatan Dharm College
Muzaffarnagar, Uttar Pradesh, India

Smriti Agarwal
Department of Electronics and Communication Engineering, MNNIT Allahabad, Prayagraj, Uttar Pradesh, India

Elif Esra Altuner
Program of Medical Laboratory Techniques
Department of Medical Services and Techniques
Vocational School of European
University of Kocaeli Health and Technology
Kocaeli, Turkiye

Salisu Ameh Amodu
Department of Philosophy
Kogi State University
Anyigba, Nigeria

Esther Besty Anaiye
Department of Nursing Science
National Open University of Nigeria
Abuja, Nigeria

Paramasivam Balakrishnan
Department of Mechanical Engineering
SRM Institute of Science and Technology
Kattankulathur, Tamil Nadu, India

Sumeda Bandara
Department of Building Economics
University of Moratuwa
Moratuwa, Sri Lanka

Thilini Bandara
Department of Zoology and Environment Sciences
Faculty of Science
University of Colombo
Colombo, Sri Lanka

Sudarshana Banerjee
Department of Physics
School of Science, Srinath University
Jamshedpur, India

Ayoub Belcaid
High Throughput Multidisciplinary Research Laboratory (HTMR)
Mohammed VI Polytechnic University (UM6P)
Ben Guerir, Morocco

Joel Benny
Center for Computational Biology and Bioinformatics
Amity Institute of Biotechnology
Amity University
Uttar Pradesh, Noida, India

Abhijit Bora
Department of Computer Applications
Assam Don Bosco University
Guwahati, Assam, India

Pankaj Kumar Chaurasia
P.G. Department of Chemistry
L.S. College, B.R. Ambedkar Bihar University
Muzaffarpur, Uttar Pradesh, India

Alakto Choudhury
Center for Computational Biology and Bioinformatics
Amity Institute of Biotechnology
Amity University
Uttar Pradesh, Noida, India

Jyoti Chowdhry
Department of Environmental Science
Maharshi Dayanand University
Rohtak, Haryana, India

Priyanjali Chowdhury
Center for Computational Biology and Bioinformatics
Amity Institute of Biotechnology
Amity University
Uttar Pradesh, Noida, India

Tapas Ranjan Das
Division of Genetics
ICAR-Indian Agricultural Research Institute
New Delhi, India

Basil Osayin Daudu
Department of Philosophy
Kogi State University
Anyigba, Nigeria

Bamdeb Dey
Department of Mathematics
Assam Don Bosco University
Guwahati, Assam, India

Cheran Easwaran
Department of Nano Science and Technology
Centre for Agricultural Nanotechnology
Tamil Nadu Agricultural University
Coimbatore, Tamil Nadu, India

Manoj Chandra Garg
Amity Institute of Environmental Science
Amity University Uttar Pradesh, Noida Gautam Budh Nagar, Uttar Pradesh, India

Lima Hazarika
Department of Zoology
Assam Don Bosco University
Guwahati, Assam, India

Chaitanya Vijay Kale
Department of Industrial Pharmacy
Poona College of Pharmacy
Pune, Maharashtra, India

Preeti Karki
Center for Computational Biology and Bioinformatics
Amity Institute of Biotechnology
Amity University
Uttar Pradesh, Noida, India

Avishek Kumar Kaser
Department of Physics
JIS University
Kolkata, West Bengal, India

Halil Burak Kaybal
Department of Mechanical Engineering
University of Amasya
Amasya, Turkiye

Thanet Khomphet
Food Technology and Innovation Research Center of Excellence
Department of Agro-Industry, School of Agricultural Technology and Food Industry
Walailak University
Tha Sala, Nakhon Si Thammarat, Thailand

Surya Konappan
Department of Agronomy
Tamil Nadu Agricultural University
Coimbatore, Tamil Nadu, India

Diwesh Kumar
Department of Civil Engineering
School of Engineering, Srinath University
Jamshedpur, India

Naveen Kumar
Department of Chemistry
Sanatan Dharm College Muzaffarnagar
Muzaffarnagar, Uttar Pradesh, India

Sheetal Kumari
Amity Institute of Environmental Science (AIES)
Amity University Uttar Pradesh
Gautam Budh Nagar, Uttar Pradesh, India

Pranabi Maji
Department of Physics
JIS University
Kolkata, West Bengal, India

Partha Sarathi Mandal
Department of Electronics and Communication Engineering
Camellia Institute of Engineering
Budbud, West Bengal, India

Satthiyaraju Mani
Department of Mechanical Engineering
SRM Institute of Science and Technology
Chennai, Tamilnadu, India

Kavitha Manikandan
Department of Artificial Intelligence and Data Science
Kathir College of Engineering
Coimbatore, Tamilnadu, India

G. R. Manjunathareddy
Department of Botany
Bangalore University
Bangalore, Karnataka, India

Ankita Mazumder
Department of Chemical Engineering
Haldia Institute of Technology
Haldia, West Bengal, India

Priyanka Narad
Indian Council of Medical Research
New Delhi, India

Minh-Ky Nguyen
Faculty of Environment and Natural Resources
Nong Lam University
Ho Chi Minh City, Viet Nam

Ayoub Ouarga
High Throughput Multidisciplinary Research Laboratory (HTMR)
Mohammed VI Polytechnic University (UM6P)
Ben Guerir, Morocco

Dibyarupa Pal
Department of Biosciences
JIS University
Kolkata, West Bengal, India

Ritam Paul
Department of Biosciences
JIS University
Kolkata, West Bengal, India

Neha Raghuwanshi
Department of Pharmaceutics
Amity Institute of Pharmacy
Amity University
Mumbai, Maharashtra, India

Sathya Priya Ramalingam
Department of Agronomy
Tamil Nadu Agricultural University
Coimbatore, Tamil Nadu, India

Krishnan Ramanujam
Department of Agronomy
Nammazhvar Organic Farming Research Centre
Tamil Nadu Agricultural University
Coimbatore, Tamil Nadu, India

H. R. Raveesha
Department of Botany
Bangalore University
Bangalore, Karnataka, India

Swati Sangolgi
Animal Biochemistry Division
National Dairy Research Institute
Karnal, Haryana, India

Shailesh Kumar Sarangi
Department of Mechanical Engineering
School of Engineering, Srinath University
Jamshedpur, India

Shubhrajit Sarkar
Department of Petrochemical Engineering
Dr. Meghnad Saha Institute of Technology
Haldia, West Bengal, India

Charu Saxena
Department of Pharmaceutics
Kharvel Subharti College of Pharmacy
Swami Vivekanand Subharti University
Meerut, Uttar Pradesh, India

Fatih Sen
Department of Biochemistry
Faculty of Science & Art
University of Kutahya Dumlupinar
Kutahya, Turkiye

Abhishek Sengupta
Center for Computational Biology and Bioinformatics
Amity Institute of Biotechnology
Amity University
Uttar Pradesh, Noida, India

Piyush Sharma
Department of Zoology
Sanatan Dharm College
Muzaffarnagar, Uttar Pradesh, India

Shagun Sharma
Department of Pharmacy
Bombay College of Pharmacy
Mumbai, Maharashtra, India

Devesh Singh
Department of Chemistry
DAV College
Muzaffarnagar, Uttar Pradesh, India

Sundram Singh
Animal Biochemistry Division
National Dairy Research Institute
Karnal, Haryana, India

Sunita Singh
Department of Chemistry
Navyug Kanya Mahavidyalaya Lucknow
University of Lucknow
Lucknow, Uttar Pradesh, India

Surya Pratap Singh
Department of Civil Engineering
School of Engineering, Srinath University
Jamshedpur, India

Marimuthu Subramani
Department of Agronomy
SRM College of Agricultural Sciences
SRM Institute of Science and Technology
Chengalpattu, Tamil Nadu, India

Qudrat Ullah
Food Technology and Innovation Research Center of Excellence
Department of Agro-Industry, School of Agricultural Technology and Food Industry
Walailak University
Tha Sala, Nakhon Si Thammarat, Thailand

Hasan Ulus
Vocational School of Huglu
University of Selcuk
Konya, Turkiye

Sankar Sangeetha Vasanthkumar
Department of Spices and Plantation Crops, Horticulture College and research institute
Tamil Nadu Agricultural University
Coimbatore, Tamil Nadu, India

Muhammad Waqar
Food Technology and Innovation Research Center of Excellence
Department of Agro-Industry, School of Agricultural Technology and Food Industry
Walailak University
Tha Sala, Nakhon Si Thammarat, Thailand

Peace Ojochenemi Yakubu
Department of Philosophy
Kogi State University
Anyigba, Nigeria

Preface

Water is the essence of life, yet access to clean and safe water remains one of the most pressing challenges of the 21st century. As the global water crisis intensifies due to climate change, rapid urbanization, industrial pollution, and growing population demands, traditional water management methods are proving insufficient. In response to these complex challenges, artificial intelligence (AI) has emerged as a transformative force, offering innovative solutions for water monitoring, treatment, conservation, and policy development.

The Handbook of AI for Clean Water: Innovations in Treatment and Monitoring explores the intersection of cutting-edge AI technologies and sustainable water management practices. This book is designed to provide researchers, engineers, policymakers, and practitioners with a comprehensive resource that bridges scientific innovation with real-world water solutions. It brings together interdisciplinary insights from environmental science, data science, engineering, and public policy to address multifaceted water challenges across sectors.

Divided into thematic sections, this book begins with a foundational overview of the global water crisis and the strategic role of AI in mitigating its impacts. Subsequent chapters delve into AI applications in water quality monitoring, predictive modeling for resource management, automation in water treatment, and AI-powered detection of contaminants. A special focus is given to AI integration in desalination technologies, sustainable wastewater management, and the development of intelligent water policies and governance frameworks. The volume also emphasizes community engagement and ethical considerations, ensuring a holistic approach to sustainable water stewardship.

Each chapter offers current research, case studies, and practical insights that highlight how AI technologies—from machine learning and Internet of Things (IoT) integration to decision support systems—are reshaping the landscape of water resource management.

This book aspires not only to advance academic knowledge but also to serve as a practical guide for implementing AI-driven strategies in real-world water systems. By showcasing successful models, innovations, and frameworks, it aims to inspire a new wave of data-informed, resilient, and inclusive water solutions globally.

We hope this volume contributes meaningfully to the ongoing efforts toward achieving the United Nations Sustainable Development Goal 6 (Clean Water and Sanitation) and fosters interdisciplinary collaboration in creating a water-secure future.

Acknowledgments

We would like to express our sincere gratitude to all those who contributed to the development and completion of *The Handbook of AI for Clean Water: Innovations in Treatment and Monitoring.*

First and foremost, we thank the chapter authors and researchers from across the globe who generously shared their expertise, time, and innovative ideas. Their commitment to scientific excellence and interdisciplinary collaboration has laid the foundation for this comprehensive volume.

We are also grateful to the reviewers and editorial advisors whose critical insights and constructive feedback helped refine the content and enhance its scholarly value.

Special thanks go to CRC Press, Taylor & Francis Group, for their support and belief in the importance of this work. Their publishing team provided invaluable assistance throughout the editorial and production process.

We acknowledge the contributions of academic institutions, research centers, and environmental organizations whose ongoing work in water sustainability and artificial intelligence inspired many of the chapters presented in this book.

Our appreciation extends to the global community of water professionals, data scientists, and policymakers who continue to drive innovation in clean water access and management. Your efforts are a constant source of motivation.

Lastly, we thank our families, friends, and colleagues for their patience and encouragement throughout this journey. Their unwavering support made this endeavor possible.

We hope this book serves as a valuable resource and catalyst for collaborative, technology-driven efforts to secure safe and sustainable water for all.

PART I

Understanding the Water Crisis and AI Integration

Introduction to the Global Water Crisis

1

Manoj Chandra Garg, Sheetal Kumari, Jyoti Chowdhry, and Smriti Agarwal

1.1 INTRODUCTION

Ensuring the availability of safe water for all consumers is essential for achieving global sustainability (Kumari et al., 2025). Water scarcity has emerged as a pressing issue, significantly impacting commercial, residential, industrial, and agricultural sectors worldwide (Garg et al., 2024). Despite being a primary objective of the United Nations Sustainable Development Goal 6 (SDG 6) within the UN2030 agenda (United Nations, 2015), the current rate of water depletion is intensifying due to climate change, increased demand, rapid population growth, and poor water resource management (Mishra et al., 2021; Vollmer & Harrison, 2021). The global demand for water is projected to increase by 55%, while nearly 25% of large cities are already experiencing water stress (Biswas et al., 2025). Additionally, approximately 4 billion people face severe water scarcity for at least 1 month annually, exacerbating challenges in water accessibility and quality (Mekonnen & Hoekstra, 2016; Orimoloye et al., 2021). Additionally, 2.2 billion people globally do not have access to clean drinking water (UNICEF and WHO, 2019).

Water shortage has traditionally been assessed in terms of quantity; however, the quality of available water is equally critical. The absence of proper safety measures can compromise water quality, particularly in regions already facing water shortages (Liu et al., 2017a; Liu et al., 2024). Climate change-induced droughts and increasing global temperatures further threaten water security, making efficient water resource management a necessity. The depletion of freshwater reserves, coupled with ageing infrastructure, has placed significant pressure on water utilities to meet rising demands. Strategies such as water reuse, rainwater harvesting (RWH), blending, and intermittent water supply have been implemented to address shortages (Sukanya & Joseph, 2023; Salehi, 2022). Additionally, under extreme conditions, bulk water delivery via tanker trucks has been adopted to provide essential potable water. Households often resort to water storage in building tanks to cope with supply interruptions, yet these practices introduce potential health risks, necessitating continuous monitoring and mitigation strategies.

Excessive water extraction for agricultural, industrial, and municipal purposes further exacerbates depletion, while pollution from fertilizers, pesticides, and industrial waste severely impacts water quality (van Vliet et al., 2021). An integrated, multidisciplinary strategy that maintains ecological health while balancing social and economic demands is needed to address these issues (UN World Development Report, 2022). The main risks connected to water management techniques used by water utilities and

DOI: 10.1201/9781003600718-2

customers are thoroughly reviewed in this study. It further explores monitoring and mitigation measures necessary to minimize public health risks and ensure potable water supply sustainability.

1.1.1 Freshwater Resource and Its Distribution

The availability of renewable water resources across different regions of the world is estimated at approximately 52,580 km^3/year (Musie & Gonfa, 2023). The sustainability of these water resources is influenced by several factors, including population growth, economic development, lifestyle choices, climate conditions, and cultural practices. Proper management of these factors is essential to maintaining water availability for future generations. Global population growth is a key factor impacting water availability. The world's population, which was 6.8 billion in 2009, is projected to reach 8.5 billion by 2025 (Mancosu et al., 2015). This growth is expected to be most significant in Asia, Africa, and South America, leading to a decline in per capita freshwater availability. In 1970, the per capita water availability was approximately 12,900 m^3/year, but it decreased to less than 7,000 m^3 by 2000 and is further projected to drop to 5,100 m^3/year by 2025 (Musie & Gonfa, 2023). This decline has direct implications for global water security, increasing the risk of water scarcity and potential geopolitical tensions.

Another major challenge is the deterioration of water quality due to pollution and improper wastewater (WW) management (Boretti & Rosa, 2019). Countries with high populations, such as India, Pakistan, Argentina, China, Sudan, and several Central Asian countries, face significant issues related to WW discharge (Musie & Gonfa, 2023). Without proper treatment, WW contaminates freshwater bodies, affecting their chemical composition and reducing their usability. Therefore, adopting effective WW treatment and reuse strategies is crucial for mitigating the impact of water pollution and ensuring long-term water security (Das et al., 2023). The following sections will explore water withdrawal trends, scarcity issues, and salinity challenges, along with their associated consequences.

1.1.2 Water Scarcity Level and Its Distribution

The growing global demand for clean water presents significant challenges as freshwater availability continues to decline due to factors such as urbanization, population growth, industrial expansion, climate change, and stricter water quality standards (du Plessis, 2022). Water scarcity arises when water withdrawals become insufficient to meet human and economic needs, leading to increased competition for resources. In 2009, the United Nations estimated that 1.2 billion people across more than 80 countries experienced water shortages, with this number projected to rise to 2.7 billion by 2025. As population numbers grow, per capita water availability decreases, having already fallen from 12,900 m^3/year in 1970 to below 7,000 m^3 in 2000, and it is expected to drop further to 5,100 m^3 by 2025 (Musie & Gonfa, 2023).

Several methods exist to measure water shortage, with two of the most widely used being the International Water Management Institute (IWMI) approach and the Falkenmark indicator (Manungufala, 2021; Rijsberman, 2006). The IWMI method incorporates Sustainable Development Goal (SDG) 6.4.2, which assesses water stress by evaluating the proportion of withdrawn freshwater used across industrial, agricultural, and municipal sectors relative to total renewable freshwater resources, excluding environmental requirements (Vanham et al., 2018). Based on this metric, a country is considered economically water-stressed if withdrawals exceed 25% of available freshwater, approaching physical scarcity at 60% and facing severe scarcity beyond 75%. Physically water-scarce countries lack sufficient resources even with maximum efficiency, whereas economically water-scarce countries have water supplies but require infrastructure improvements to meet the demand (Mancosu et al., 2015).

The Falkenmark classification assesses water scarcity based on per capita freshwater availability. Countries with over 10,000 m^3/person/year, such as Bangladesh and Canada, face no water stress, whereas those with below 500 m^3, including Tunisia, UAE, and Saudi Arabia, experience absolute scarcity (Damkjaer & Taylor, 2017). Many regions, particularly in Central and West Asia, North Africa, and parts of Africa, are approaching critical scarcity levels, with projections indicating further declines in per capita availability across Africa, the Middle East, and South and Southeast Asia by 2050 (Musie & Gonfa, 2023). Addressing these challenges will require strategic water management policies and sustainable resource utilization.

1.2 WATER UTILITIES' RESPONSE TO WATER SCARCITY AND RISKS TO DRINKING WATER QUALITY

1.2.1 Interruption-Based Water Supply

Interruption-based water supply is a response to water scarcity in many regions, affecting over 309 million people, mainly in South America, Africa, and Asia (Carlo Loubser et al., 2021; Li et al., 2020). Intermittent Water Supply (IWS) disrupts continuous water flow, causing pressure changes that lead to pipe damage and water quality deterioration (Farmani et al., 2021). Contaminants can enter through leaks, backflow, and microbial growth in stagnant water (Mohammadi et al., 2020). Biofilm disturbance and reduced disinfectant residuals further compromise water safety (Calero Preciado et al., 2021; Li et al., 2019). Studies link IWS to higher levels of *E. coli* and faecal coliforms, increasing waterborne disease risks (Sharma et al., 2023). Reducing IWS duration, maintaining disinfectant levels, and pressure monitoring can mitigate risks (Erickson et al., 2017). Consumers should avoid first-flush water to minimize exposure (van den Berg et al., 2021).

1.2.2 Blended Water Supply

To combat groundwater depletion, water blending—the process of combining treated groundwater with finished surface water—is becoming more and more popular (Garg et al., 2024). But blending the chemistry of water could upset the microbial biofilm and inorganic scale that have built up in pipelines over many years (Jasechko et al., 2024; Liu et al., 2017b). Changes in pH, alkalinity, and dissolved oxygen can accelerate pipe corrosion, leading to iron release and mobilization of contaminants like arsenic (Zeng et al., 2023). Heavy metals such as cadmium and lead may also leach from pipe materials, posing health risks and causing aesthetic issues (Oladimeji et al., 2024). Water utilities should assess the corrosivity of blended water and monitor distribution systems to detect biofilm destabilization and contaminant release (Zhang et al., 2018). Real-time particle counting could help identify destabilization events (Liu et al., 2017b). Further research is needed to improve monitoring and mitigation strategies.

1.2.3 Bulk Water Supply

Bulk water supply supplies drinking water to isolated rural areas, Arctic regions with permafrost, and emergencies like earthquakes and wildfires (Gora et al., 2020; Nath et al., 2024). In cases of severe water shortages, utilities may rely on tanker trucks to meet potable water demands (Salehi, 2022). However, poor sanitation during filling, transport, and storage can introduce microbial contamination, with tankers identified as sources of coliform bacteria due to improper hygiene and materials (McGuinness et al., 2020). Insufficient disinfectant residuals, extended storage, and high temperatures promote bacterial growth and biofilm formation (Halim, 2021). Tankers used for non-potable purposes must be cleaned before transporting drinking water. Routine water quality monitoring at filling stations and trucks is essential for safety (C. Zhang & Lu, 2021). Research has focused on economic aspects, while microbial risks remain underexplored, requiring better sanitary practices (Ochungo et al., 2019).

1.2.4 Households' Adaptation to Unreliable Water Supply

In regions with unreliable water supply, many households use storage tanks to maintain water availability during supply interruptions (Bose et al., 2024). However, improper design, such as placing pressure booster systems before storage tanks, can create negative pressure, increasing contamination risks through backflow. Storage tanks, made from materials like polyethene and galvanized iron, can promote microbial growth due to leached chemicals, sediments, and corrosion products (Peter & Routledge, 2018). Stagnation and high temperatures accelerate disinfectant decay and biofilm formation, encouraging pathogens like *Legionella pneumophila* and *Pseudomonas* (Slavik et al., 2020). Oversized tanks exacerbate stagnation, reducing disinfectant levels and increasing microbial proliferation (Manga et al., 2021). Regular cleaning reduces sediment accumulation, lowering microbial contamination (Schafer & Mihelcic, 2012). Disinfection of storage tanks further helps prevent microbial regrowth and waterborne diseases (Slavik et al., 2020). Preventing sunlight exposure also minimizes microbial proliferation risks (Miyagi et al., 2017).

1.3 CHALLENGES TOWARDS WATER SCARCITY

1.3.1 Urbanization and Growing Population

The world's population is predicted to grow to around 10 billion people over the next 30 years, with developing countries accounting for the majority of this increase. More than half of the world's population currently lives in cities, and by 2050, that number is expected to increase to more than two-thirds. As freshwater supplies remain constant yet unevenly distributed, rapidly expanding urban centres will experience increased water stress. This will heighten competition for water resources among urban, agricultural, and industrial sectors, exacerbating conflicts over access and allocation (Niva et al., 2020).

Additionally, urbanization will significantly impact water availability and quality, soil health, and overall water demand (Saquib et al., 2022). The intensity and frequency of floods are likely to rise, affecting ecosystem stability and food security. These challenges are further compounded by climatic variations, land use changes, and hydrological shifts, all of which influence the sustainability of water resources (Pizzorni et al., 2024). Given these interconnected issues, long-term water management strategies must be carefully developed to ensure resilience and sustainability.

The implementation of measures to increase water-use efficiency and the exploration of alternative water sources are essential in order to manage the escalating water concerns (Figure 1.1). Sustainable

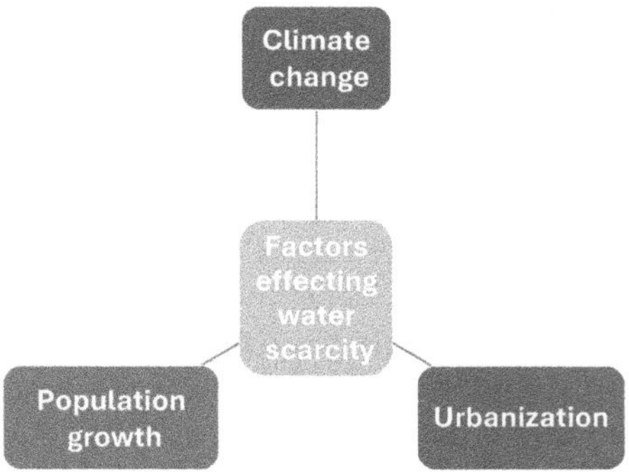

FIGURE 1.1 Various factors affecting water scarcity at the global level.

strategies, such as water reclamation, desalination, and improved WW treatment, can supplement traditional water supplies. Simultaneously, adopting efficient irrigation techniques, reducing industrial water wastage, and promoting conservation at the household level can enhance water sustainability. By integrating innovative water management solutions with proactive policies, urban areas can mitigate water shortage and ensure a stable water supply for future generations.

1.3.2 Climate Change

Global water resources are already being significantly impacted by climate change, which is changing precipitation patterns, temperature trends, and the frequency of extreme weather events including heat waves, droughts, and floods (Sukanya & Joseph, 2023). However, these effects are not uniform, as they vary significantly across regions and timeframes depending on climate variability. For instance, the Mediterranean basin has experienced warming at a much higher rate than the global average, which is expected to alter water availability and increase demand due to more frequent droughts (Lazoglou et al., 2024).

Research indicates that climate change may lead to either intensification of the global hydrological cycle—resulting in higher rates of evaporation and precipitation—or fluctuations between intensification and de-intensification (Tabari, 2020). This could cause unpredictable changes in precipitation and evaporation patterns, ultimately contributing to a decline in global humidity levels. While climate models at both global and regional scales generally align in their projections, translating these projections into actionable data for localized water resource management remains a significant challenge (Tapiador et al., 2019).

Downscaling climate models to watershed or local scales is complex due to methodological limitations, making it difficult to develop precise water management strategies. Furthermore, even at broader scales, uncertainties persist in both global climate models and hydrological models, complicating long-term planning. More developments in climate modelling methods, better regional data gathering, and flexible water resource management plans are needed to address these issues and lessen the uncertainty that climate change will bring about regarding water sustainability and availability.

1.4 ESSENTIAL ACTIONS TO ADDRESS WORLDWIDE WATER SCARCITY

1.4.1 Centralized Administration

A centralized governance system plays a crucial role in overseeing all aspects of water management to ensure a reliable water supply. A dedicated water authority is responsible for key tasks such as licensing well drilling, desalination, WW treatment, infrastructure development, and setting water prices (Di Vaio et al., 2021). While the government must take the lead in implementing water management strategies, it cannot handle this responsibility alone (Singh & Goyal, 2025). To enhance sustainability, equity, and efficiency, a collaborative approach to water governance is essential. This method brings together various stakeholders, NGOs, including government agencies, community organizations, and private sector entities, to jointly manage water resources (Pugel et al., 2022). Collaborative governance offers a comprehensive grasp of water demands at the local and national levels by combining a variety of resources and expertise. Additionally, it guarantees readiness for severe weather conditions like floods and droughts, which can affect the quantity and quality of water. Creating backup plans aids in reducing these risks and preserving the availability of water.

1.4.2 Raising Awareness and Promoting Water Conservation

Addressing water scarcity requires widespread awareness and collective action from individuals, organizations, and governments. Educating children and communities at both national and international levels is essential in fostering an understanding of water's value and the need for its protection (Shemer et al., 2023). Since companies utilize over 22% of the world's water, water conservation initiatives should focus on both residential and commercial use (Kumar et al., 2023). In households, practical conservation methods include installing water-efficient toilets and faucets, harvesting rainwater for irrigation, and planting drought-resistant vegetation. Meanwhile, industries can reduce water usage by implementing water-saving manufacturing techniques and recycling WW. A key strategy is replacing once-through cooling systems with closed-circuit cooling systems, which significantly reduces water waste by replenishing only the evaporated portion (Bauer & Wagner, 2022).

1.4.3 Technologies for Water Catchment and Harvesting

Various water harvesting and catchment technologies are designed to collect and manage water from different sources, including surface water, floodwater, rainwater, and groundwater. These technologies are widely used in domestic, industrial, agricultural, and wetland environments. Common methods include RWH, floodwater harvesting, aquifer recharge, and dam construction (Baba, 2018). RWH is one of the oldest and most sustainable water collection methods. It helps supplement water supply, provides an alternative for non-potable uses, and reduces stormwater runoff (Alim et al., 2020). However, its feasibility depends on water quality, as rainwater can be contaminated by airborne pollutants. Effective treatment is essential for safe use. Furthermore, RWH systems are frequently more expensive than centralized systems and have a smaller supply capacity (Yildirim et al., 2022). Aquifer recharge not only boosts groundwater levels (Basel et al., 2021, 2022) but also prevents soil erosion, ensures a more balanced water distribution throughout the year, and mitigates the effects of droughts and floods (Seddon et al., 2020). It supports vegetation, enhances biodiversity, and regulates water tables. However, the success of community-based aquifer recharge projects depends on a comprehensive understanding of the socio-hydrological context in which they are implemented (Basel et al., 2022).

1.4.4 Irrigation and Agricultural Practices

About 25% of the world's cropland is irrigated by agriculture, which uses 70% of all freshwater withdrawals worldwide. However, water scarcity is becoming worse due to population expansion and climate change, making effective water management essential for sustained agricultural output (FAO, 2020). To address this challenge, precision agriculture and smart irrigation technologies are being adopted to optimize water utilization and prevent plant stress. Smart irrigation systems monitor weather conditions, soil moisture, and plant needs, ensuring water is delivered at the right time, place, and quantity. On the other hand, conventional irrigation techniques frequently result in either excessive or insufficient irrigation, which wastes water, leaches nutrients, and reduces crop output (Bwambale et al., 2022).

1.5 KEY STRATEGIES FOR ENHANCING IRRIGATION EFFICIENCY

1.5.1 Conservation Tillage

Crop residues are left on the field after harvest as part of the conservation tillage (CT) strategy, which keeps soil hydrated and stops erosion. When compared to traditional tillage, this technique can drastically cut water use by up to 50% while simultaneously enhancing soil health and halting the deterioration brought on by excessive water use (Liang et al., 2024). By enhancing soil structure, reducing compaction, and increasing organic matter, CT promotes efficient water use and boosts crop yields (Srivastava et al., 2024). This approach is particularly beneficial for dryland farming, where soil moisture retention is crucial for sustaining yields (Liu et al., 2025). In India, where water resources are scarce, CT has proven effective. For instance, in Haryana, water usage decreased by 40% compared to traditional tillage practices (Tripathi et al., 2017). Similarly, in Gujarat, water use dropped by 30%, while crop yields increased by 20% due to this method (Biswas et al., 2025).

Higher biological and grain yields in maize during the rabi season are the result of CT, which includes decreased tillage and zero tillage. It also improves bio-meteorological indices like the leaf area index and photosynthetically active radiation use efficiency (PARUE) (Pal et al., 2022). Additionally, it is a cost-effective and labour-efficient practice that improves farm productivity. A study in Refugio County, Texas, found that for every 1% increase in CT adoption, the county gained approximately US $16,000 in benefits (Cusser et al., 2023). By reducing water consumption, improving soil quality, and enhancing crop productivity, CT stands as a sustainable and economically viable solution for modern agriculture.

1.5.2 Crop Shifting

Crop shifting is a strategic agricultural practice that involves growing less water-intensive crops during dry seasons or in regions with limited water availability. This approach helps reduce irrigation demand and prevents soil degradation caused by excessive water utilization (Acharjee et al., 2019). In the lower Indo-Gangetic plain, traditional crops like potatoes and Boro rice require intensive irrigation, leading to a steady decline in groundwater levels. To mitigate this issue, farmers are encouraged to cultivate maize instead of rice. Maize requires only one-fifth of the water needed for rice cultivation while offering higher yields and better economic returns for farmers (Kapuria & Banerjee, 2022). By adopting crop shifting, farmers can optimize water resources, maintain soil health, and enhance agricultural sustainability, making it a crucial adaptation strategy in water-scarce regions.

1.5.3 Crop Rotation

Crop rotation involves growing different crops sequentially on the same land within a year, promoting soil fertility and reducing water consumption. Certain crops, such as legumes, improve soil health by naturally fixing nitrogen, decreasing the reliance on synthetic fertilizers, and ultimately conserving water (Zhu et al., 2025). Rice-based cultivation has long been the norm in India's lower Indo-Gangetic plains. But recently, the rotation method has included legumes including cowpea, *Lathyrus*, Bengal gram, green gram, and black gram (Biswas et al., 2025). For example, at B.C.K.V., Kalyani, West Bengal, lentils and chickpeas were successfully cultivated during the rabi seasons of 2018–2019 and 2019–2020 using soil moisture left over from monsoon rice without the need for extra irrigation (Mukherjee et al., 2023). This practice not only conserved water compared to Boro rice cultivation but also generated extra income for farmers. Incorporating crop rotation in agricultural systems can be a valuable solution for enhancing water efficiency and ensuring sustainable farming in water-scarce regions.

1.5.4 Mulching

Mulching is a farming technique in which the soil surrounding plants is covered with organic materials such as leaves, straw, or artificial coverings. This method helps maintain soil moisture, control temperature, and reduce soil evaporation, making it especially useful in dryland environments (Srisawat et al., 2025). Widely used in rain-fed farming systems, mulching effectively conserves water by preventing rapid moisture loss (Zribi et al., 2015). Among different mulching methods, plastic sheet mulching has proven to be more efficient in preserving soil moisture compared to wheat straw mulch (Kader et al., 2017). The key advantage of mulching lies in its ability to reduce soil erosion and surface evaporation, which significantly lowers the irrigation requirements during crop growth (Qin et al., 2016). By improving water retention in the soil, mulching serves as a cost-effective and eco-friendly strategy to enhance agricultural sustainability and optimize water usage (Kader et al., 2017).

1.5.5 Utilization of Drought-Tolerant Crops

Planting drought-tolerant crops is an effective strategy to enhance water efficiency and sustain agricultural productivity in water-limited environments (Das et al., 2022). These crops, developed through advanced breeding techniques, exhibit improved resistance to drought stress while maintaining stable yields. Farmers can reduce the negative consequences of water scarcity and their dependency on irrigation by using drought-tolerant cultivars. Due to their flexibility and high production potential in dry conditions, several water-saving and drought-resistant rice (WDR) varieties, including Huhan-3, Huhan-7, and WDR-48, have become more and more popular in China (Luo et al., 2019). These varieties require less water compared to conventional rice while ensuring food security and economic benefits for farmers. The adoption of drought-resistant crops not only conserves water but also promotes sustainable agriculture, making them a viable option for regions facing recurrent droughts and declining water resources. Expanding research and development of such crops is crucial for future agricultural resilience. By adopting these strategies, farmers can significantly minimize agricultural runoff, lower water consumption, reduce energy demands, and decrease carbon emissions from water extraction and transportation (Bertule, 2018).

By supplying water and nutrients straight to plant roots, drip irrigation is essential to sustainable farming. But it has problems like soil salinization and clogging. An efficient substitute that reduces evaporation inhibits weed development and enhances safety when utilizing reclaimed WW is subsurface irrigation. Salinity problems can be further addressed by desalinating WW. Additionally, fertigation—injecting fertilizers into the irrigation system—enhances nutrient efficiency, promoting maximum crop yields and improved soil health.

1.5.6 Pollution

Water pollution threatens both ecological balance and human well-being. The presence of inorganic and organic contaminants, alongside microbial agents, has raised significant concerns regarding their detrimental impact on ecosystems and living organisms. As a result, allocating resources to control and prevent water pollution is essential (Singh & Goyal, 2025). Pharmaceuticals, personal care items, plasticizers, surfactants, fire retardants, nanomaterials, and pesticides are examples of emerging contaminants (ECs) that have drawn more attention recently because of their possible risks (Li et al., 2024). Various chemical, physical, and biological treatment methods are being investigated to effectively eliminate these pollutants and mitigate their negative effects (Ahmed et al., 2021). Extensive research has been conducted to evaluate these treatment technologies and their efficiency in reducing water pollution (Priyadarshini et al., 2022; Sengupta et al., 2021).

1.5.7 Advancing Water Treatment Technologies

The development of cost-effective, efficient, and resilient technologies for water and WW treatment is essential for ensuring sustainable water management (Kumar et al., 2022). However, technological advancements alone are not sufficient. Economic and societal factors must also be considered, as they play a crucial role in the successful adoption and long-term viability of new technologies (Söderholm, 2020). Several barriers hinder innovation in the water sector. It is challenging to rapidly integrate new solutions due to the high expenses and lengthy lifespan of the current infrastructure. Furthermore, there is little money available for further investments due to the comparatively low cost of water. Existing restrictions frequently discourage innovation, and public health concerns further limit the willingness to take chances with new technologies. Even when regulatory frameworks support advancements, water utilities frequently deprioritize research and development (Ahmed et al., 2023).

Many novel technologies show promise at the laboratory scale but face challenges when moving towards large-scale implementation. These include low process efficiency, high energy demands, limited engineering expertise for upscaling, minimal economic incentives, and inadequate infrastructure. Furthermore, the specificity of these innovations often limits their adaptability to diverse needs (Shamshad & Ur Rehman, 2025). To address these challenges, fostering an organizational culture that embraces innovation is key. Encouraging policies should prioritize adaptability and technological advancement to tackle water-related challenges while driving economic growth. Key factors that promote innovation in the water sector include supportive regulatory frameworks, sufficient financial investment in research and development, a culture that values innovation, and strong public engagement (Zhuo Chen, 2016).

1.5.8 Strengthening Transboundary Water Cooperation

Managing transboundary water resources is crucial due to the complex interactions between economies, climates, ecosystems, political systems, and cultures within shared watersheds (Petersen-Perlman et al., 2017). Successful cooperation requires coordination and joint efforts between nations. Progress is frequently hampered by obstacles like concerns about national sovereignty, misunderstandings of the advantages and disadvantages of cooperation, and a lack of political will or capacity (Konecna, 2025). Effective transboundary water management brings numerous benefits, including economic growth in agriculture, mining, and energy sectors, as well as reduced financial losses from floods and droughts. Collaborative initiatives can lead to shared investments, improved resource preservation, greater ecological integrity, and enhanced political stability.

A notable example of such cooperation is the Jordan–Israel energy and water agreement. Under this arrangement, Jordan will build a large solar farm to supply electricity to Israel, while Israel will construct desalination plants along its Mediterranean coast to provide water to both nations. Further collaboration includes a new desalination facility in the southern region, ensuring a sustainable water supply for both countries. Adequate funding, data exchange, and regulatory enforcement are essential components of successful transboundary water governance (Mohammed et al., 2022; UN World Development Report, 2022). Table 1.1 presents the benefits of such collaboration, categorized into societal, economic, and environmental advantages.

1.5.9 Desalination

Membrane-based technologies play a crucial role in water and WW treatment, desalination, biofuel production, value-added material extraction, and the food industry. The primary membrane filtration methods include microfiltration (MF), ultrafiltration (UF), nanofiltration (NF), and reverse osmosis (RO) (Shehata et al., 2023) To overcome individual limitations, a combination of these methods is often employed. However, public concerns regarding environmental impact, environmental impact, and cost continue to pose challenges to their widespread adoption (Yusuf et al., 2020).

Desalination has become a dependable freshwater source in many regions, primarily for treating brackish water (BW), seawater (SW), and WW. Despite its increasing application, reducing costs remains a key challenge for broader implementation in drinking water supply and agricultural irrigation (Ayaz et al., 2022). Many people employed early desalination techniques such as thermal vapour compression, membrane distillation, multi-effect distillation (MED), and multi-stage flash (MSF) desalination. Over time, newer techniques, including solar still distillation, humidification–dehumidification desalination,

TABLE 1.1 Categories of potential benefits of transboundary water cooperation (UNICE, 2015)

ASPECT	*DETAILS*
Importance of transboundary water cooperation	Essential for integrated and sustainable management of shared waters
Potential benefits	• Accelerated economic growth • Improved human well-being • Enhanced environmental sustainability • Increased political stability
Challenges	• Incomplete or biased perception of benefits • Political and institutional barriers • Limited funding and resources
UN Economic Commission for Europe (UNECE) Convention	Encourages nations to work together to manage transboundary water resources in accordance with the International Lakes Convention and the Protection and Use of Transboundary Watercourses
Role of benefit assessment	• Helps countries understand and realize cooperation benefits • Identifies previously overlooked opportunities • Strengthens political support and funding for transboundary water cooperation
Objective of the publication	• Supports governments and stakeholders in harnessing cooperation benefits • Introduces various advantages of collaboration • Provides a step-by-step guide for conducting a benefit assessment
Key steps in benefit assessment	• **Identification**: Recognizing possible benefits of cooperation • **Assessment**: Evaluating the significance and impact of benefits • **Communication**: Presenting findings to stakeholders and decision-makers
Integration into policy	• Aligning benefit assessments with national and regional policies • Strengthening legal and institutional frameworks for cooperation

freezing desalination, and capacitive deionization, have emerged (Liponi et al., 2020). Membrane-based desalination, particularly RO, has gained prominence, accounting for nearly 80% of global desalination capacity (Tayeh, 2024).

As of 2020, the global desalination capacity was approximately 115 million m^3/day (42 billion m^3/year) (Dhakal et al., 2022). Initially, thermal desalination technologies dominated, particularly in the Middle East. But since the 1980s, thermal techniques have gradually been supplanted by RO technology. By 2000, 93% of all desalinated water was produced by thermal desalination (mostly MSF) and RO, which had approximately identical production capacities of 11.6 and 11.4 million m^3/day, respectively. The quantity and capability of RO plants have greatly increased, although thermal technologies have only seen slight advancements.

Saline water contains a variety of dissolved organic and inorganic substances, silt, colloidal suspensions, nanoparticles, viruses, bacteria, and other impurities (Dugan, 2024). Salts are eliminated by desalination, but impurities like silica and boron frequently need further treatment. At quantities below 1 mg/L, boron is necessary for plant growth, but at greater levels, it can be hazardous. The World Health Organization (WHO) recommends a boron concentration limit of 2.4 mg/L in desalinated water, but certain countries impose stricter standards (WHO, 2017). To improve boron removal, desalination plants employ techniques such as a second RO pass with pH adjustment, ion exchange, and electrodialysis (ED) (Landsman et al., 2020). Silica scaling is another common challenge in desalination processes. High silica levels can cause membrane fouling, reducing efficiency. Silica and hardness ions are typically removed through techniques such as lime softening, electro-coagulation, coagulation, adsorption, ion exchange, and seeded precipitation. Ion exchange softeners combined with pH adjustments are among the most commonly used pre-treatment methods to mitigate silica scaling (Warsinger et al., 2015).

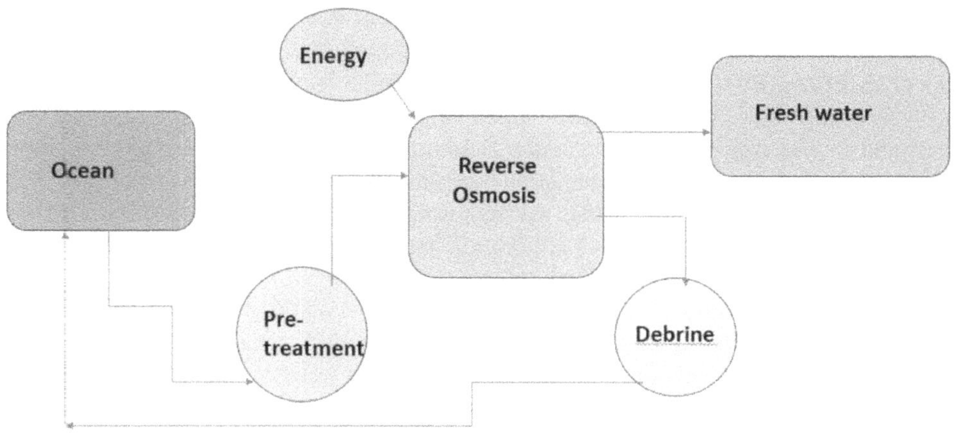

FIGURE 1.2 Desalination process for obtaining freshwater.

Desalination offers a viable solution to water scarcity, particularly for coastal countries with access to SW (Figure 1.2). Desalination can also be used to recover and purify low-quality water for sustainable usage in isolated and inland locations. This technique can assist in meeting the increasing demand for freshwater worldwide by increasing desalination efficiency and reducing costs.

1.5.10 Water Reclamation

The process of preparing industrial and municipal WW for reuse in a variety of uses is known as water reclamation. By offering substitute water sources, this approach improves resilience, sustainability, and water security. Agricultural practices, landscape irrigation, potable water supply, industrial operations, groundwater recharge, artificial lakes, and ecosystem restoration all make extensive use of reclaimed water (Shahid et al., 2022). In areas with limited resources, treated WW discharge into streams can assist in maintaining aquatic ecosystems and restoring river flow, but it can also have a detrimental effect on biological communities and water quality. For example, using recycled water in agriculture can save about 1.7 tons of CO_2-equivalent per hectare, which is a significant reduction in carbon emissions (Lahlou et al., 2022).

Currently, 63% of the world's WW is collected, and 52% is treated, meaning 48% is discharged untreated. This represents a decline from previous estimates of 80%, yet untreated municipal WW remains one of the most significant threats to water ecosystems due to its high organic and nutrient content (Jones et al., 2021). Only about 20% of treated WW is reused worldwide, with North Africa and the Middle East leading at 15%, followed by Western Europe at 16%. Up to 50% of Israel's agricultural water needs are met by the centrally treated 93% of WW, of which 86% is used for agriculture, especially during droughts (EU Policy on the Environment, 2018).

Whereas desalination plants are found along beaches, wastewater treatment plants (WWTPs) are usually found on the outskirts of cities. Strategic plant locations are crucial since the cost of delivering desalinated water more than 200 km inland is equal to the cost of producing desalinated WW. Proper infrastructure development enables the efficient distribution of reclaimed water to meet urban and agricultural demands. Additionally, reclaimed water derived from desalinated sources has lower salinity and total dissolved solids (TDS), reducing salinity-related problems in irrigation (Shemer & Semiat, 2017).

Despite public concerns, advanced treatment technologies enable the production of potable water from WW through processes such as RO, NF, and environmental buffering with rivers, lakes, or groundwater (Warsinger et al., 2018). In hot climates, frequent utilization of storage lakes is necessary to prevent excessive salt accumulation from evaporation. Covering storage lakes with flexible surfaces equipped with solar panels can mitigate evaporation while generating renewable energy (Zhao et al., 2025).

Properly treated reclaimed water poses no higher risk of produce-related illnesses than conventional irrigation sources. More than 30% of the global reclaimed water market is dedicated to agriculture (Helmecke et al., 2020). However, regulations on permissible levels of salts, pollutants, and metals vary across countries. Reclaimed water originates from multiple sources, including municipal WW, industrial process and cooling water, stormwater, and agricultural runoff (Capodaglio, 2021). Membrane-based technologies play a vital role in water reclamation by eliminating solids and salts from WW. MF and UF remove suspended solids, while RO ensures salt removal. Gravity-driven membrane (GDM) and membrane bioreactors (MBRs) offer alternatives to conventional activated sludge processes, providing superior pollutant retention with reduced fouling (Shehata et al., 2023).

Recent research continues to explore innovations in membrane-based water reclamation, focusing on improving efficiency, environmental sustainability, and cost-effectiveness. As global water scarcity intensifies, these technologies will play a crucial role in securing water resources for future generations.

1.6 FUTURE PROSPECTIVE

- The growing threat of water scarcity due to population growth, climate change, and unsustainable consumption necessitates both immediate and long-term strategic interventions. Future efforts must integrate technological advancements, policy frameworks, and community participation to ensure sustainable water resource management. As demand continues to rise, innovative solutions such as water reclamation, desalination, and advanced agricultural water-use efficiency techniques will be crucial in addressing shortages.
- Advancements in desalination technologies, particularly RO and energy-efficient membrane systems, will lead to cost reductions and environmental improvements. Future research should focus on developing high-performance membranes with improved permeability, fouling resistance, and extended lifespans. Integration of renewable energy sources like solar and wind power into desalination plants will help mitigate energy consumption concerns, while enhanced brine discharge management strategies, including dilution systems and eco-friendly disposal methods, will minimize environmental impacts.
- Water reclamation and reuse will become increasingly vital for water security. The adoption of advanced membrane technologies such as UF and MBRs will enable the production of high-quality reclaimed water for industrial, agricultural, and even potable uses. Greater regulatory support and public awareness campaigns will be necessary to overcome societal concerns and enhance the acceptance of reclaimed water. Decentralized WW treatment systems in urban and rural areas will further optimize water reuse and reduce dependence on freshwater sources.
- Sustainable agricultural practices will play a crucial role in mitigating water scarcity, especially in water-stressed regions like India, where agriculture accounts for 80% of total water consumption. Precision agriculture, drip irrigation, fertigation, and drought-resistant crop varieties will improve water-use efficiency. Integration of soil moisture sensors, artificial intelligence, and remote sensing technologies will enable real-time monitoring and optimization of irrigation, reducing water wastage. Conservation techniques such as mulching, crop rotation, and CT will enhance soil water retention and decrease reliance on irrigation.
- Policy frameworks and transboundary water governance will be essential for equitable water distribution and sustainable resource management. Future policies must emphasize integrated water resource management (IWRM) and promote cross-border collaboration for managing shared water bodies. Strengthening water pricing mechanisms and market-based incentives will encourage conservation while ensuring economic feasibility. Governments and international organizations should invest in capacity-building initiatives to foster community participation in water conservation efforts and infrastructure development.

- The synergy between technological advancements, sustainable practices, and policy interventions will determine the success of global water management. AI-driven smart water grids, IoT-enabled water monitoring systems, and automated leak detection in urban water distribution networks will revolutionize water efficiency. Research into novel materials, such as superabsorbent polymer hydrogels and biochar, will enhance soil water retention and agricultural sustainability. A holistic approach combining engineering solutions, policy reforms, and behavioural changes will ensure a resilient and sustainable water supply for future generations.

1.7 CONCLUSION

Water scarcity is a growing global challenge driven by climate change, population growth, and unsustainable water usage. Addressing this crisis requires an integrated approach encompassing technological advancements, policy frameworks, and community engagement. The agricultural sector, particularly in water-stressed countries like India, plays a critical role in mitigating water shortages, given its significant share of total water consumption. Sustainable agricultural practices, improved irrigation systems, and innovative water management strategies are essential to reducing the strain on freshwater resources.

The United Nations' SDG 6 provides a comprehensive framework for ensuring the availability and sustainable management of water and sanitation. Achieving these targets requires the adoption of water-efficient technologies, improved governance, and investments in sustainable water infrastructure. Measures such as water pricing, conservation education, and centralized water governance can enhance resource management and resilience against extreme climate events.

Desalination and water reclamation have emerged as viable solutions for augmenting water supply. Despite concerns about energy consumption and environmental impact, advancements in membrane technology and renewable energy integration have significantly improved the efficiency and cost-effectiveness of desalination processes. Properly managed brine discharge can minimize ecological harm and, in some cases, enhance marine biodiversity. Similarly, WW reclamation using UF and RO has proven to be a sustainable means of producing high-quality water for reuse in agriculture, industry, and portable applications.

A multifaceted approach is necessary to ensure long-term water sustainability. This includes fostering international cooperation in transboundary water management, investing in infrastructure for water catchment and distribution, and implementing stringent pollution control measures. Governments, researchers, and policymakers must work collaboratively to bridge knowledge gaps, develop equitable water policies, and implement sustainable solutions. With a combination of technological innovation, policy-driven water management, and active public participation, water security can be safeguarded for future generations. The transition towards sustainable water use will not only enhance global resilience to water crisis but also support economic growth, food security, and environmental preservation.

REFERENCES

Acharjee, T. K., van Halsema, G., Ludwig, F., Hellegers, P., & Supit, I. (2019). Shifting planting date of Boro rice as a climate change adaptation strategy to reduce water use. *Agricultural Systems*, *168*, 131–143. https://doi.org/10.1016/j.agsy.2018.11.006

Ahmed, F., Johnson, D., Hashaikeh, R., & Hilal, N. (2023). Barriers to innovation in water treatment. *Water*, *15*(4), 773. https://doi.org/10.3390/w15040773

Ahmed, S. F., Mofijur, M., Nuzhat, S., Chowdhury, A. T., Rafa, N., Uddin, Md. A., Inayat, A., Mahlia, T. M. I., Ong, H. C., Chia, W. Y., & Show, P. L. (2021). Recent developments in physical, biological, chemical, and hybrid treatment techniques for removing emerging contaminants from wastewater. *Journal of Hazardous Materials, 416*, 125912. https://doi.org/10.1016/j.jhazmat.2021.125912

Alim, M. A., Rahman, A., Tao, Z., Samali, B., Khan, M. M., & Shirin, S. (2020). Suitability of roof harvested rainwater for potential potable water production: A scoping review. *Journal of Cleaner Production, 248*, 119226. https://doi.org/10.1016/j.jclepro.2019.119226

Ayaz, M., Namazi, M. A., ud Din, M. A., Ershath, M. I. M., Mansour, A., & Aggoune, el -H. M. (2022). Sustainable seawater desalination: Current status, environmental implications and future expectations. *Desalination, 540*, 116022. https://doi.org/10.1016/j.desal.2022.116022

Baba, A. (2018). Developments in water dams and water harvesting systems throughout history in different civilizations. *International Journal of Hydrology, 2*(2). https://doi.org/10.15406/ijh.2018.02.00064

Basel, B., Hernández Quiroz, N., Velasco Herrera, R., Santiago Alonso, C., & Hoogesteger, J. (2021). Bee mietii rak rkabni nis (the people know how to seed water): A Zapotec experience in adapting to water scarcity and drought. *Climate and Development, 13*(9), 792–806. https://doi.org/10.1080/17565529.2020.1855100

Basel, B., Hoogesteger, J., & Hellegers, P. (2022). Promise and paradox: A critical sociohydrological perspective on small-scale managed aquifer recharge. *Frontiers in Water, 4*. https://doi.org/10.3389/frwa.2022.1002721

Bauer, S., & Wagner, M. (2022). Possibilities and challenges of wastewater reuse: Planning aspects and realized examples. *Water, 14*(10), 1619. https://doi.org/10.3390/w14101619

Bertule, M., Appelquist, L. R., Spensley, J., Trærup, S. L. M., & Naswa, P. (2018). *Climate Change Adaptation Technologies for Water: A Practitioner's Guide to Adaptation Technologies for Increased Water Sector Resilience.* UNEP DTU Partnership.

Biswas, A., Sarkar, S., Das, S., Dutta, S., Roy Choudhury, M., Giri, A., Bera, B., Bag, K., Mukherjee, B., Banerjee, K., Gupta, D., & Paul, D. (2025). Water scarcity: A global hindrance to sustainable development and agricultural production – A critical review of the impacts and adaptation strategies. *Cambridge Prisms: Water, 3*, e4. https://doi.org/10.1017/wat.2024.16

Boretti, A., & Rosa, L. (2019). Reassessing the projections of the World Water Development Report. *NPJ Clean Water, 2*(1), 15. https://doi.org/10.1038/s41545-019-0039-9

Bose, D., Bhattacharya, R., Kaur, T., Banerjee, R., Bhatia, T., Ray, A., Batra, B., Mondal, A., Ghosh, P., & Mondal, S. (2024). Overcoming water, sanitation, and hygiene challenges in critical regions of the global community. *Water-Energy Nexus, 7*, 277–296. https://doi.org/10.1016/j.wen.2024.11.003

Bwambale, E., Abagale, F. K., & Anornu, G. K. (2022). Smart irrigation monitoring and control strategies for improving water use efficiency in precision agriculture: A review. *Agricultural Water Management, 260*, 107324. https://doi.org/10.1016/j.agwat.2021.107324

Calero Preciado, C., Husband, S., Boxall, J., del Olmo, G., Soria-Carrasco, V., Maeng, S. K., & Douterelo, I. (2021). Intermittent water supply impacts on distribution system biofilms and water quality. *Water Research, 201*, 117372. https://doi.org/10.1016/j.watres.2021.117372

Capodaglio, A. G. (2021). Fit-for-purpose urban wastewater reuse: Analysis of issues and available technologies for sustainable multiple barrier approaches. *Critical Reviews in Environmental Science and Technology, 51*(15), 1619–1666. https://doi.org/10.1080/10643389.2020.1763231

Chen, Z., Ngo, H. H., Guo, W. S., & Eslamian, S. (2016). Water shortages. In *Urban Water Reuse Handbook* (Eslamian, S., Ed., pp. 1–12). CRC Press, Boca Raton.

Cusser, S., Jha, S., Lonsdorf, E., & Ricketts, T. (2023). Public and private economic benefits of adopting conservation tillage for cotton pollination. *Agriculture, Ecosystems & Environment, 342*, 108251. https://doi.org/10.1016/j.agee.2022.108251

Damkjaer, S., & Taylor, R. (2017). The measurement of water scarcity: Defining a meaningful indicator. *Ambio, 46*(5), 513–531. https://doi.org/10.1007/s13280-017-0912-z

Das, S., Christopher, J., Roy Choudhury, M., Apan, A., Chapman, S., Menzies, N. W., & Dang, Y. P. (2022). Evaluation of drought tolerance of wheat genotypes in rain-fed sodic soil environments using high-resolution UAV remote sensing techniques. *Biosystems Engineering, 217*, 68–82. https://doi.org/10.1016/j.biosystemseng.2022.03.004

Das, S., Singh, C. K., Sodhi, K. K., & Singh, V. K. (2023). Circular economy approaches for water reuse and emerging contaminant mitigation: Innovations in water treatment. *Environment, Development and Sustainability*. https://doi.org/10.1007/s10668-023-04183-z

Dhakal, N., Salinas-Rodriguez, S. G., Hamdani, J., Abushaban, A., Sawalha, H., Schippers, J. C., & Kennedy, M. D. (2022). Is desalination a solution to freshwater scarcity in developing countries? *Membranes, 12*(4), 381. https://doi.org/10.3390/membranes12040381

Di Vaio, A., Trujillo, L., D'Amore, G., & Palladino, R. (2021). Water governance models for meeting sustainable development goals: A structured literature review. *Utilities Policy, 72*, 101255. https://doi.org/10.1016/j.jup.2021.101255

du Plessis, A. (2022). Persistent degradation: Global water quality challenges and required actions. *One Earth*, *5*(2), 129–131. https://doi.org/10.1016/j.oneear.2022.01.005

Dugan, H. A. (2024). Salinity and ionic composition of inland waters. In *Wetzel's Limnology* (pp. 275–299). Elsevier. https://doi.org/10.1016/B978-0-12-822701-5.00012-4

Erickson, J. J., Smith, C. D., Goodridge, A., & Nelson, K. L. (2017). Water quality effects of intermittent water supply in Arraiján, Panama. *Water Research*, *114*, 338–350. https://doi.org/10.1016/j.watres.2017.02.009

EU Policy on the Environment. (2018). *Water Reuse*. https://www.water-reuse-europe.org/about-us/wre-activities/water-reuse-europe-review-2018/

FAO. (2020). *The State of Food and Agriculture 2020, Overcoming Water Challenges in Agriculture*. FAO.

Farmani, R., Dalton, J., Charalambous, B., Lawson, E., Bunney, S., & Cotterill, S. (2021). Intermittent water supply systems and their resilience to COVID-19: IWA IWS SG survey. *Journal of Water Supply: Research and Technology-Aqua*, *70*(4), 507–520. https://doi.org/10.2166/aqua.2021.009

Garg, M. C., Kumari, S., & Malik, N. (2024). Role of nanomaterials in advanced membrane technologies for groundwater purification. *Environmental Science: Water Research & Technology*, *10*(11), 2628–2645. https://doi.org/10.1039/D4EW00353E

Gora, S. L., Trueman, B. F., Anaviapik-Soucie, T., Gavin, M. K., Ontiveros, C. C., Campbell, J., L'Hérault, V., Stoddart, A. K., & Gagnon, G. A. (2020). Source water characteristics and building-specific factors influence corrosion and point of use water quality in a decentralized arctic drinking water system. *Environmental Science & Technology*, *54*(4), 2192–2201. https://doi.org/10.1021/acs.est.9b04691

Halim, A. A. (2021). Potential risk assessment of drinking water source exposed to contamination using microbial indicators and multiple antibiotic resistance index. *Iranian Journal of Energy and Environment*, *12*(1). https://doi.org/10.5829/IJEE.2021.12.01.10

Helmecke, M., Fries, E., & Schulte, C. (2020). Regulating water reuse for agricultural irrigation: Risks related to organic micro-contaminants. *Environmental Sciences Europe*, *32*(1), 4. https://doi.org/10.1186/s12302-019-0283-0

Jasechko, S., Seybold, H., Perrone, D., Fan, Y., Shamsudduha, M., Taylor, R. G., Fallatah, O., & Kirchner, J. W. (2024). Rapid groundwater decline and some cases of recovery in aquifers globally. *Nature*, *625*(7996), 715–721. https://doi.org/10.1038/s41586-023-06879-8

Jones, E. R., van Vliet, M. T. H., Qadir, M., & Bierkens, M. F. P. (2021). Country-level and gridded estimates of wastewater production, collection, treatment and reuse. *Earth System Science Data*, *13*(2), 237–254. https://doi.org/10.5194/essd-13-237-2021

Kader, M. A., Senge, M., Mojid, M. A., & Ito, K. (2017). Recent advances in mulching materials and methods for modifying soil environment. *Soil and Tillage Research*, *168*, 155–166. https://doi.org/10.1016/j.still.2017.01.001

Kapuria, P., & Banerjee, S. (2022). "Crop shifting for improved water use and nutritional productivity in the lower Indo-gangetic plains of West Bengal." *ORF Occasional Paper No. 358*, June 2022, Observer Research Foundation (p. 358).

Konecna, L. (2025). Unpacking water conflicts in Africa: Exploring key drivers. *Small Wars & Insurgencies*, *36*(1), 1–28. https://doi.org/10.1080/09592318.2024.2429050

Kumar, A., Behera, U. K., Dhar, S., Shukla, L., Singh, R., Babu, S., Sharma, V. K., Upadhyay, P. K., Bairwa, R. K., Gupta, G., Kumar, A., Singh, S., Bhandari, N., Qureshi, A. A., Gudade, B. A., Verma, G., Kumar, R., & Kumar, A. (2023). Conservation tillage and microbially mediated integrated phosphorus management enhance productivity, profitability and energy use efficiency of maize. *The Journal of Agricultural Science*, *161*(3), 373–386. https://doi.org/10.1017/S0021859623000266

Kumar, V., Bilal, M., & Ferreira, L. F. R. (2022). Editorial: Recent trends in integrated wastewater treatment for sustainable development. *Frontiers in Microbiology*, *13*. https://doi.org/10.3389/fmicb.2022.846503

Kumari, S., Kumar, M., Agrawal, S., Kaushik, A., Hashem, A., Abd_Allah, E. F., Kumar, A., & Garg, M. C. (2025). Research hotspot and trend analysis of heavy metals decontamination of wastewater in past two decades: A bibliometric analysis. *Journal of Hazardous Materials Advances*, *17*, 100532. https://doi.org/10.1016/j.hazadv.2024.100532

Lahlou, F.-Z., Mackey, H. R., & Al-Ansari, T. (2022). Role of wastewater in achieving carbon and water neutral agricultural production. *Journal of Cleaner Production*, *339*, 130706. https://doi.org/10.1016/j.jclepro.2022.130706

Landsman, M. R., Lawler, D. F., & Katz, L. E. (2020). Application of electrodialysis pretreatment to enhance boron removal and reduce fouling during desalination by nanofiltration/reverse osmosis. *Desalination*, *491*, 114563. https://doi.org/10.1016/j.desal.2020.114563

Lazoglou, G., Papadopoulos-Zachos, A., Georgiades, P., Zittis, G., Velikou, K., Manios, E. M., & Anagnostopoulou, C. (2024). Identification of climate change hotspots in the Mediterranean. *Scientific Reports*, *14*(1), 29817. https://doi.org/10.1038/s41598-024-80139-1

Li, H., Cohen, A., Li, Z., Lv, S., He, Z., Wang, L., & Zhang, X. (2020). Intermittent water supply management, household adaptation, and drinking water quality: A comparative study in two Chinese Provinces. *Water, 12*(5), 1361. https://doi.org/10.3390/w12051361

Li, R. A., McDonald, J. A., Sathasivan, A., & Khan, S. J. (2019). Disinfectant residual stability leading to disinfectant decay and by-product formation in drinking water distribution systems: A systematic review. *Water Research, 153*, 335–348. https://doi.org/10.1016/j.watres.2019.01.020

Li, X., Shen, X., Jiang, W., Xi, Y., & Li, S. (2024). Comprehensive review of emerging contaminants: Detection technologies, environmental impact, and management strategies. *Ecotoxicology and Environmental Safety, 278*, 116420. https://doi.org/10.1016/j.ecoenv.2024.116420

Liang, X., Rehman, S. U., Zhiqi, W., Raza, M. A., Haider, I., Khalid, M. H. B., Saeed, A., Iqbal, Z., Fatima, S., Siddiqa, A., Ansar, M., Ijaz, S. S., & Zhongming, M. (2024). Impacts of conservation tillage on agricultural land development: A review. *Journal of Soil Science and Plant Nutrition*. https://doi.org/10.1007/s42729-024-02142-9

Liponi, A., Wieland, C., & Baccioli, A. (2020). Multi-effect distillation plants for small-scale seawater desalination: Thermodynamic and economic improvement. *Energy Conversion and Management, 205*, 112337. https://doi.org/10.1016/j.enconman.2019.112337

Liu, G., Zhang, Y., Knibbe, W.-J., Feng, C., Liu, W., Medema, G., & van der Meer, W. (2017). Potential impacts of changing supply-water quality on drinking water distribution: A review. *Water Research, 116*, 135–148. https://doi.org/10.1016/j.watres.2017.03.031

Liu, W., He, C., Han, S., Lin, B., Liu, W., Dang, Y. P., Zhao, X., & Zhang, H. (2025). Enhancing soil ecosystem multifunctionality through combined conservation tillage and legume-based crop rotation in the North China Plain. *Agriculture, Ecosystems & Environment, 379*, 109355. https://doi.org/10.1016/j.agee.2024.109355

Liu, Z., Ying, J., He, C., Guan, D., Pan, X., Dai, Y., Gong, B., He, K., Lv, C., Wang, X., Lin, J., Liu, Y., & Bryan, B. A. (2024). Scarcity and quality risks for future global urban water supply. *Landscape Ecology, 39*(2), 10. https://doi.org/10.1007/s10980-024-01832-0

Loubser, C., Chimbanga, B. M., & Jacobs, H. (2021). Intermittent water supply: A South African perspective. *Water SA, 47*(1 January). https://doi.org/10.17159/wsa/2021.v47.i1.9440

Luo, L., Mei, H., Yu, X., Xia, H., Chen, L., Liu, H., Zhang, A., Xu, K., Wei, H., Liu, G., Wang, F., Liu, Y., Ma, X., Lou, Q., Feng, F., Zhou, L., Chen, S., Yan, M., Liu, Z., … Li, M. (2019). Water-saving and drought-resistance rice: From the concept to practice and theory. *Molecular Breeding, 39*(10–11), 145. https://doi.org/10.1007/s11032-019-1057-5

Mancosu, N., Snyder, R., Kyriakakis, G., & Spano, D. (2015). Water scarcity and future challenges for food production. *Water, 7*(3), 975–992. https://doi.org/10.3390/w7030975

Manga, M., Ngobi, T. G., Okeny, L., Acheng, P., Namakula, H., Kyaterekera, E., Nansubuga, I., & Kibwami, N. (2021). The effect of household storage tanks/vessels and user practices on the quality of water: A systematic review of literature. *Environmental Systems Research, 10*(1), 18. https://doi.org/10.1186/s40068-021-00221-9

Manungufala, T. (2021). *Water Scarcity: Classification, Measurement and Management* (pp. 1–14). https://doi.org/10.1007/978-3-319-70061-8_7-1

McGuinness, S. L., O'Toole, J., Barker, S. F., Forbes, A. B., Boving, T. B., Giriyan, A., Patil, K., D'Souza, F., Vhaval, R., Cheng, A. C., & Leder, K. (2020). Household water storage management, hygiene practices, and associated drinking water quality in Rural India. *Environmental Science & Technology, 54*(8), 4963–4973. https://doi.org/10.1021/acs.est.9b04818

Mekonnen, M. M., & Hoekstra, A. Y. (2016). Four billion people facing severe water scarcity. *Science Advances, 2*(2). https://doi.org/10.1126/sciadv.1500323

Mishra, B., Kumar, P., Saraswat, C., Chakraborty, S., & Gautam, A. (2021). Water security in a changing environment: Concept, challenges and solutions. *Water, 13*(4), 490. https://doi.org/10.3390/w13040490

Miyagi, K., Sano, K., & Hirai, I. (2017). Sanitary evaluation of domestic water supply facilities with storage tanks and detection of Aeromonas, enteric and related bacteria in domestic water facilities in Okinawa Prefecture of Japan. *Water Research, 119*, 171–177. https://doi.org/10.1016/j.watres.2017.04.002

Mohammadi, A., Jalili-Ghazizadeh, M., Moslehi, I., & Yousefi-Khoshqalb, E. (2020). Survival analysis of water distribution network under intermittent water supply conditions. *Water Supply, 20*(8), 3531–3541. https://doi.org/10.2166/ws.2020.228

Mohammed, I. N., Bolten, J. D., Souter, N. J., Shaad, K., & Vollmer, D. (2022). Diagnosing challenges and setting priorities for sustainable water resource management under climate change. *Scientific Reports, 12*(1), 796. https://doi.org/10.1038/s41598-022-04766-2

Mukherjee, B., Naskar, M. K., Basak, J. S., Ghosh, M., Nath, R., & Kumari, V. V. (2023). Impact of sowing time, tillage and variety management on growth indices, nodulation and yield parameters of chickpea (Cicer arietinum) grown after monsoon rice in Lower Gangetic Plains of West Bengal. *Indian Journal of Dryland Agricultural Research and Development, 38*(1), 15–20. https://doi.org/10.5958/2231-6701.2023.00003.9

Musie, W., & Gonfa, G. (2023). Fresh water resource, scarcity, water salinity challenges and possible remedies: A review. *Heliyon*, *9*(8), e18685. https://doi.org/10.1016/j.heliyon.2023.e18685

Nath, D., Peramaiyan, P., Kumari, V., Laik, R., & Dotaniya, M. L. (2024). Water resources of rural India: Challenges and management strategies for sustainable development. In *Water Resources Management for Rural Development* (pp. 191–200). Elsevier. https://doi.org/10.1016/B978-0-443-18778-0.00015-5

Niva, V., Cai, J., Taka, M., Kummu, M., & Varis, O. (2020). China's sustainable water-energy-food nexus by 2030: Impacts of urbanization on sectoral water demand. *Journal of Cleaner Production*, *251*, 119755. https://doi.org/10.1016/j.jclepro.2019.119755

Ochungo, E. A., Ouma, G. O., Obiero, J. P. O., & Odero, N. A. (2019). The implication of unreliable urban water supply service: The case of vendor water cost in Langata Sub County, Nairobi City, Kenya. *Journal of Water Resource and Protection*, *11*(07), 896–935. https://doi.org/10.4236/jwarp.2019.117055

Oladimeji, T. E., Oyedemi, M., Emetere, M. E., Agboola, O., Adeoye, J. B., & Odunlami, O. A. (2024). Review on the impact of heavy metals from industrial wastewater effluent and removal technologies. *Heliyon*, *10*(23), e40370. https://doi.org/10.1016/j.heliyon.2024.e40370

Orimoloye, I. R., Belle, J. A., Olusola, A. O., Busayo, E. T., & Ololade, O. O. (2021). Spatial assessment of drought disasters, vulnerability, severity and water shortages: A potential drought disaster mitigation strategy. *Natural Hazards*, *105*(3), 2735–2754. https://doi.org/10.1007/s11069-020-04421-x

Pal, D., Mondal, S., Banerjee, S., Saha, A., & Pramanick, M. (2022). Evaluating the influence of conservation agricultural practices and fertilizer on PAR use efficiency and biophysical parameters of maize grown under new alluvial zone of West Bengal. *Journal of Agrometeorology*, *24*(2). https://doi.org/10.54386/jam.v24i2.1548

Peter, A., & Routledge, E. (2018). Present-day monitoring underestimates the risk of exposure to pathogenic bacteria from cold water storage tanks. *PLoS ONE*, *13*(4), e0195635. https://doi.org/10.1371/journal.pone.0195635

Petersen-Perlman, J. D., Veilleux, J. C., & Wolf, A. T. (2017). International water conflict and cooperation: Challenges and opportunities. *Water International*, *42*(2), 105–120. https://doi.org/10.1080/02508060.2017.1276041

Pizzorni, M., Innocenti, A., & Tollin, N. (2024). Droughts and floods in a changing climate and implications for multi-hazard urban planning: A review. *City and Environment Interactions*, *24*, 100169. https://doi.org/10.1016/j.cacint.2024.100169

Priyadarshini, M., Das, I., Ghangrekar, M. M., & Blaney, L. (2022). Advanced oxidation processes: Performance, advantages, and scale-up of emerging technologies. *Journal of Environmental Management*, *316*, 115295. https://doi.org/10.1016/j.jenvman.2022.115295

Pugel, K., Javernick-Will, A., Peabody, S., Nyaga, C., Mussa, M., Mekonta, L., Dimtse, D., Watsisi, M., Buhungiro, E., Mulatu, T., Annis, J., Jordan, E., Sandifer, E., & Linden, K. (2022). Pathways for collaboratively strengthening water and sanitation systems. *Science of the Total Environment*, *802*, 149854. https://doi.org/10.1016/j.scitotenv.2021.149854

Qin, S., Li, S., Kang, S., Du, T., Tong, L., & Ding, R. (2016). Can the drip irrigation under film mulch reduce crop evapotranspiration and save water under the sufficient irrigation condition? *Agricultural Water Management*, *177*, 128–137. https://doi.org/10.1016/j.agwat.2016.06.022

Rijsberman, F R. (2006). Water scarcity: Fact or fiction? *Agricultural Water Management*, *80*(1–3), 5–22. https://doi.org/10.1016/j.agwat.2005.07.001

Salehi, M. (2022). Global water shortage and potable water safety; Today's concern and tomorrow's crisis. *Environment International*, *158*, 106936. https://doi.org/10.1016/j.envint.2021.106936

Saquib, S., Gupta, A., & Joshi, A. (2022). *Emerging Water Crisis: Impact of Urbanization on Water Resources and Constructed Wetlands as a Nature-Based Solution (NbS)* (pp. 447–468). https://doi.org/10.1016/B978-0-323-91838-1.00021-X

Schafer, C. A., & Mihelcic, J. R. (2012). Effect of storage tank material and maintenance on household water quality. *Journal AWWA*, *104*(9). https://doi.org/10.5942/jawwa.2012.104.0125

Seddon, N., Chausson, A., Berry, P., Girardin, C. A. J., Smith, A., & Turner, B. (2020). Understanding the value and limits of nature-based solutions to climate change and other global challenges. *Philosophical Transactions of the Royal Society B: Biological Sciences*, *375*(1794), 20190120. https://doi.org/10.1098/rstb.2019.0120

Sengupta, A., Jebur, M., Kamaz, M., & Wickramasinghe, S. R. (2021). Removal of emerging contaminants from wastewater streams using membrane bioreactors: A review. *Membranes*, *12*(1), 60. https://doi.org/10.3390/membranes12010060

Shahid, M. K., Kashif, A., Pathak, P., Choi, Y., & Rout, P. R. (2022). Water reclamation, recycle, and reuse. In *Clean Energy and Resource Recovery* (pp. 39–50). Elsevier. https://doi.org/10.1016/B978-0-323-90178-9.00028-7

Shamshad, J., & Ur Rehman, R. (2025). Innovative approaches to sustainable wastewater treatment: A comprehensive exploration of conventional and emerging technologies. *Environmental Science: Advances*, *4*(2), 189–222. https://doi.org/10.1039/D4VA00136B

Sharma, M. D., Mishra, P., Ali, A., Kumar, P., Kapil, P., Grover, R., Verma, R., Saini, A., & Kulshrestha, S. (2023). Microbial waterborne diseases in India: Status, interventions, and future perspectives. *Current Microbiology, 80*(12), 400. https://doi.org/10.1007/s00284-023-03462-2

Shehata, N., Egirani, D., Olabi, A. G., Inayat, A., Abdelkareem, M. A., Chae, K.-J., & Sayed, E. T. (2023). Membrane-based water and wastewater treatment technologies: Issues, current trends, challenges, and role in achieving sustainable development goals, and circular economy. *Chemosphere, 320*, 137993. https://doi.org/10.1016/j.chemosphere.2023.137993

Shemer, H., & Semiat, R. (2017). Sustainable RO desalination – Energy demand and environmental impact. *Desalination, 424*, 10–16. https://doi.org/10.1016/j.desal.2017.09.021

Shemer, H., Wald, S., & Semiat, R. (2023). Challenges and solutions for global water scarcity. *Membranes, 13*(6), 612. https://doi.org/10.3390/membranes13060612

Singh, S., & Goyal, M. K. (2025). A review of India's water policy and implementation toward a sustainable future. *Journal of Water and Climate Change*. https://doi.org/10.2166/wcc.2025.560

Slavik, I., Oliveira, K. R., Cheung, P. B., & Uhl, W. (2020). Water quality aspects related to domestic drinking water storage tanks and consideration in current standards and guidelines throughout the world: A review. *Journal of Water and Health, 18*(4), 439–463. https://doi.org/10.2166/wh.2020.052

Söderholm, P. (2020). The green economy transition: The challenges of technological change for sustainability. *Sustainable Earth, 3*(1), 6. https://doi.org/10.1186/s42055-020-00029-y

Srisawat, T., Sakprom, S., Kunsawat, P., Praksong, K., Suchat, S., & Muangprathub, J. (2025). IoT-enabled agricultural environmental monitoring: Enhancing growth and yield using natural-rubber straw and mulching experiment. *Industrial Crops and Products, 225*, 120524. https://doi.org/10.1016/j.indcrop.2025.120524

Srivastava, R. K., Purohit, S., Alam, E., & Islam, M. K. (2024). Advancements in soil management: Optimizing crop production through interdisciplinary approaches. *Journal of Agriculture and Food Research, 18*, 101528. https://doi.org/10.1016/j.jafr.2024.101528

Sukanya, S., & Joseph, S. (2023). Climate change impacts on water resources: An overview. In *Visualization Techniques for Climate Change with Machine Learning and Artificial Intelligence* (pp. 55–76). Elsevier. https://doi.org/10.1016/B978-0-323-99714-0.00008-X

Tabari, H. (2020). Climate change impact on flood and extreme precipitation increases with water availability. *Scientific Reports, 10*(1), 13768. https://doi.org/10.1038/s41598-020-70816-2

Tapiador, F. J., Moreno, R., Navarro, A., Sánchez, J. L., & García-Ortega, E. (2019). Climate classifications from regional and global climate models: Performances for present climate estimates and expected changes in the future at high spatial resolution. *Atmospheric Research, 228*, 107–121. https://doi.org/10.1016/j.atmosres.2019.05.022

Tayeh, Y. A. (2024). A comprehensive review of reverse osmosis desalination: Technology, water sources, membrane processes, fouling, and cleaning. *Desalination and Water Treatment, 320*, 100882. https://doi.org/10.1016/j.dwt.2024.100882

Tripathi, D. K., Singh, S., Singh, V. P., Prasad, S. M., Dubey, N. K., & Chauhan, D. K. (2017). Silicon nanoparticles more effectively alleviated UV-B stress than silicon in wheat (Triticum aestivum) seedlings. *Plant Physiology and Biochemistry, 110*, 70–81. https://doi.org/10.1016/j.plaphy.2016.06.026

UN World Development Report. (2022). *The United Nations World Water Development Report 2022: Groundwater: Making the Invisible Visible*. UNESCO World Water Assessment Programme.

UNICEF and WHO. (2019). *Progress on Household Drinking Water, Sanitation and Hygiene 2000-2017. Special Focus on Inequalities*. United Nations Children's Fund (UNICEF) and World Health Organization, New York.

United Nations (UN) (2015). *Transforming Our World: The 2030 Agenda for Sustainable Development*. United nation, New York City.

van den Berg, H., Quaye, M. N., Nguluve, E., Schijven, J., & Ferrero, G. (2021). Effect of operational strategies on microbial water quality in small scale intermittent water supply systems: The case of Moamba, Mozambique. *International Journal of Hygiene and Environmental Health, 236*, 113794. https://doi.org/10.1016/j.ijheh.2021.113794

van Vliet, M. T. H., Jones, E. R., Flörke, M., Franssen, W. H. P., Hanasaki, N., Wada, Y., & Yearsley, J. R. (2021). Global water scarcity including surface water quality and expansions of clean water technologies. *Environmental Research Letters, 16*(2), 024020. https://doi.org/10.1088/1748-9326/abbfc3

Vanham, D., Hoekstra, A. Y., Wada, Y., Bouraoui, F., de Roo, A., Mekonnen, M. M., van de Bund, W. J., Batelaan, O., Pavelic, P., Bastiaanssen, W. G. M., Kummu, M., Rockström, J., Liu, J., Bisselink, B., Ronco, P., Pistocchi, A., & Bidoglio, G. (2018). Physical water scarcity metrics for monitoring progress towards SDG target 6.4: An evaluation of indicator 6.4.2 "Level of water stress." *Science of the Total Environment, 613–614*, 218–232. https://doi.org/10.1016/j.scitotenv.2017.09.056

Vollmer, D., & Harrison, I. J. (2021). $H_2O \neq CO_2$: Framing and responding to the global water crisis. *Environmental Research Letters, 16*(1), 011005. https://doi.org/10.1088/1748-9326/abd6aa

Warsinger, D. M., Chakraborty, S., Tow, E. W., Plumlee, M. H., Bellona, C., Loutatidou, S., Karimi, L., Mikelonis, A. M., Achilli, A., Ghassemi, A., Padhye, L. P., Snyder, S. A., Curcio, S., Vecitis, C. D., Arafat, H. A., & Lienhard, J. H. (2018). A review of polymeric membranes and processes for potable water reuse. *Progress in Polymer Science*, *81*, 209–237. https://doi.org/10.1016/j.progpolymsci.2018.01.004

Warsinger, D. M., Swaminathan, J., Guillen-Burrieza, E., Arafat, H. A., & Lienhard, V. J. H. (2015). Scaling and fouling in membrane distillation for desalination applications: A review. *Desalination*, *356*, 294–313. https://doi.org/10.1016/j.desal.2014.06.031

WHO. (2017). *Guidelines for Drinking-Water Quality: First Addendum*. 4th ed. WHO, Geneva.

Yildirim, G., Alim, M. A., & Rahman, A. (2022). Review of rainwater harvesting research by a bibliometric analysis. *Water*, *14*(20), 3200. https://doi.org/10.3390/w14203200

Yusuf, A., Sodiq, A., Giwa, A., Eke, J., Pikuda, O., De Luca, G., Di Salvo, J. L., & Chakraborty, S. (2020). A review of emerging trends in membrane science and technology for sustainable water treatment. *Journal of Cleaner Production*, *266*, 121867. https://doi.org/10.1016/j.jclepro.2020.121867

Zeng, L., Yan, C., Yang, F., Zhen, Z., Yang, J., Chen, J., Huang, Y., Xiao, Y., & Zhang, W. (2023). The effects and mechanisms of pH and dissolved oxygen conditions on the release of arsenic at the sediment–water interface in Taihu Lake. *Toxics*, *11*(11), 890. https://doi.org/10.3390/toxics11110890

Zhang, C., & Lu, J. (2021). Optimizing disinfectant residual dosage in engineered water systems to minimize the overall health risks of opportunistic pathogens and disinfection by-products. *Science of the Total Environment*, *770*, 145356. https://doi.org/10.1016/j.scitotenv.2021.145356

Zhang, H., Wang, K., Zhou, X., Zhu, W., & Wang, W. (2018). Water quality characteristics and corrosion potential in blending zones in X city drinking water distribution system. *Environmental Monitoring and Assessment*, *190*(9), 524. https://doi.org/10.1007/s10661-018-6901-2

Zhao, G., Gao, H., Li, Y., Tang, Q., Woolway, R. I., Merder, J., Rosa, L., & Michalak, A. M. (2025). Decoupling of surface water storage from precipitation in global drylands due to anthropogenic activity. *Nature Water*, *3*(1), 80–88. https://doi.org/10.1038/s44221-024-00367-7

Zhu, Y., Li, Z., Zhao, D., Zhang, B., Zhu, B., Yao, Z., Kiese, R., Butterbach-Bahl, K., & Zhou, M. (2025). Effects of conservation agriculture on soil N_2O emissions and crop yield in global cereal cropping systems. *Global Change Biology*, *31*(1). https://doi.org/10.1111/gcb.70048

Zribi, W., Aragüés, R., Medina, E., & Faci, J. M. (2015). Efficiency of inorganic and organic mulching materials for soil evaporation control. *Soil and Tillage Research*, *148*, 40–45. https://doi.org/10.1016/j.still.2014.12.003

Role of AI in Mitigating the Water Crisis

2

Charu Saxena

2.1 INTRODUCTION

Human existence, economic growth, and environmental sustainability all depend on water. But the global water issue is becoming worse, with billions of people suffering from water scarcity and without access to clean drinking water [1]. This dilemma is mostly caused by pollution, population increase, climate change, and ineffective water management. Innovative approaches are needed to address these issues, and artificial intelligence (AI) has become a game-changing tool for reducing water-related issues. Water distribution, conservation, quality monitoring, and catastrophe forecasting are all being transformed by AI-driven solutions. AI optimizes water use, finds leaks, forecasts floods and droughts, and improves wastewater treatment efficiency by utilizing machine learning (ML), predictive analytics, and Internet of Things (IoT)-enabled smart technologies. In order to alleviate water shortages, increase efficiency, and create climate-resilient water systems, this book examines how AI-powered solutions might benefit communities, businesses, and governments [2].

One of the most important resources for environmental sustainability, economic growth, and human life is water. But the global water issue is becoming worse; 2.2 billion people do not have access to clean drinking water, and more than 4 billion people face acute water scarcity at least once a year. The problem is getting worse due to factors including pollution, climate change, rapid population increase, and ineffective water management. Innovative solutions are needed to address this catastrophe, and AI has become a game-changing instrument in addressing water-related problems [3].

Water distribution, conservation, quality monitoring, and catastrophe forecasting are all being transformed by AI-driven technology. AI helps manage water use, identify distribution network leaks, forecast droughts and floods, and improve wastewater treatment efficiency by utilizing ML, predictive analytics, and IoT-enabled smart devices. By guaranteeing fair access to clean water and reducing their negative effects on the environment, these developments not only enhance water security but also promote sustainable development [4].

The numerous uses of AI in water management are examined in this chapter, emphasizing the ways in which AI-driven solutions may assist communities, businesses, and governments in addressing water scarcity, increasing productivity, and creating climate-resilient water systems. The use of AI in water resource management has enormous potential to address one of the most important global issues of our day as it develops further [5].

 DOI: 10.1201/9781003600718-3

2.2 UNDERSTANDING AI IN WATER MANAGEMENT

By increasing productivity, reducing waste, and guaranteeing sustainable use, AI has revolutionizing the management of water resources. Smarter decisions in water distribution, quality monitoring, and conservation initiatives are being made possible by AI-powered technologies such as ML, predictive analytics, the IoT, and big data [6].

2.2.1 AI in Water Management: What Is It?

In water management, AI refers to the monitoring, analysis, and optimization of water-related activities via the use of intelligent algorithms and automated systems. Leak detection, water demand forecasting, quality evaluation, and even prediction of natural disasters like droughts and floods are all capabilities of these systems. Governments, businesses, and communities may improve water sustainability and efficiency using AI. By increasing productivity, cutting expenses, and guaranteeing environmental sustainability, AI is revolutionizing wastewater treatment. One important use case is real-time monitoring and predictive analytics, where ML models forecast system failures and contamination hazards before they happen, while AI-powered IoT sensors monitor water quality indicators like pH, turbidity, and chemical oxygen demand (COD). AI also improves treatment procedures by instantly modifying sludge removal, aeration rates, and chemical doses. Digital twins increase overall efficiency by enabling simulation-based decision-making [7].

2.3 IMPORTANT AI TOOLS FOR WATER MANAGEMENT

2.3.1 Predictive Analytics and Machine Learning

In order to forecast future water demand, spot possible shortages, and find irregularities in water distribution, ML algorithms analyze massive information from water utilities, climate models, and remote sensors. For instance, AI models can forecast drought situations by analyzing historical rainfall patterns, assisting officials in putting conservation measures into place beforehand [8,9].

2.3.2 Intelligent Sensors and the Internet of Things (IoT)

Real-time data on water flow, pressure, and quality is continually gathered by IoT devices like smart water meters and sensors. AI then analyzes these data to identify inefficiencies and improve water use. As an illustration, smart irrigation systems in agriculture minimize waste by adjusting water distribution based on soil moisture and weather [10].

2.3.3 Image Recognition Driven by AI and Computer Vision

Drones with AI capabilities and satellite photos can monitor reservoir conditions, measure water levels, and identify contamination. For instance, AI can detect unlawful dumping in lakes and rivers, allowing for quicker pollution prevention measures [11].

2.4 AI APPLICATIONS IN WATER RESOURCE MANAGEMENT

AI is revolutionizing water resource management by changing the way we distribute, preserve, and maximize water use. Governments, businesses, and communities may better plan for disasters, monitor water quality, identify leaks, and estimate water demand with the use of AI technologies like ML, IoT, big data analytics, and computer vision. In a time of growing water shortage and climate change, these applications are essential to maintaining sustainable water management [12].

2.4.1 Improving Water Efficiency and Conservation

Incompetency is one of the major problems in water management or water consumption. Water distribution systems may be monitored in real time thanks to technologies like smart meters, sensors, and IoT devices. These technologies aid in leak detection, water loss reduction, and regional water flow optimization. To assist utilities in avoiding waste and guaranteeing dependable water delivery to customers, smart water networks, for example, may employ AI and big data to track pressure levels, identify pipe bursts, and examine water use trends. As an illustration, a pilot project in Singapore that included smart sensors in its water distribution systems greatly increased efficiency and decreased water losses by 10%–15% [13].

2.4.1.1 Intelligent water management systems

These systems improve water distribution, keep an eye out for leaks, and guarantee effective use of water resources via the use of sophisticated IoT-enabled sensors and AI-driven analytics. As an illustration, AI is used by smart irrigation systems to assess soil moisture content and modify water consumption, minimizing agricultural waste. Smart water distribution systems increase the sustainability and efficiency of water delivery networks by utilizing cutting-edge technology like IoT sensors, AI, and data analytics. Real-time optimization is made possible by these systems, which continually monitor water flow, pressure, quality, and consumption trends across a distribution network. AI algorithms examine the data to forecast changes in demand, identify any issues, and modify the water supply appropriately, guaranteeing effective resource allocation and reducing waste. Smart systems may save energy use, improve storage and pumping schedules, and automatically balance water pressure. Smart water distribution systems save water loss, improve service dependability, and boost overall operational efficiency by better managing and monitoring water networks [14].

2.4.1.2 Finding and stopping leaks

Predictive maintenance driven by AI finds pipeline breaches before they become serious, minimizing water loss. As an illustration, water utilities employ ML algorithms to detect leaks in real time by analyzing pressure differences in water delivery networks. Real-time data from sensors installed in water pipes is used by AI-powered leak detection and prevention systems to identify abnormalities like pressure decreases or odd flow patterns that point to breaches. In order to prevent water loss and enable timely maintenance, ML algorithms examine the data to find minor leaks or possible pipeline breakdowns before they become serious problems. Based on previous data and environmental parameters like soil conditions or pipeline age, AI can also forecast when and where leaks are likely to occur. Utilities may proactively fix water distribution system flaws with these predictive capabilities, lowering repair costs, averting interruptions, and saving water. In order to reduce waste and guarantee the long-term viability of water delivery networks, this technology is crucial [15].

2.4.1.3 Anticipating analytics for water utilization and supply

Using AI and ML, predictive analytics for water supply and demand forecasts future water requirements based on a variety of variables, including population growth, weather patterns, seasonal variations, and past consumption data. These insights assist communities and water utilities in preparing for times of high demand, such as hot summers or drought conditions, and guarantee that sufficient infrastructure and water supply are available. Additionally, supply shortfalls may be predicted using predictive models, allowing for preemptive actions like infrastructure improvements, policy changes, and water conservation programs. Predictive analytics helps manage resources effectively, save waste, and guarantee that the water supply is dependable and sustainable despite shifting environmental and demographic conditions by optimizing water allocation and predicting demand changes [16].

2.4.2 Improving Water Quality and Pollution Control

To enhance the quality and purity of water, pollution control is an important parameter. Because contaminants from cities, industries, and agriculture are entering water sources, water pollution is a serious worldwide issue. Drones that monitor water bodies and real-time water quality monitors are examples of technological solutions that can assist in the detection and analysis of pollutants. Rapid reactions are made possible by the early identification of dangerous substances, infections, or heavy metals, averting extensive contamination. AI may also be used to analyze trends in water quality, spot patterns, and guide policy changes. As an illustration, IoT sensors have been used in the Ganges River to monitor water quality giving authorities information on pollution levels and facilitating focused clean-up operations [17].

2.4.2.1 AI-based water quality monitoring

While AI-driven systems evaluate real-time data from sensors to identify pollution before it becomes dangerous, traditional water quality testing is reactive and sluggish. For instance, AI models use rainfall data and patterns of industrial waste discharge to forecast the levels of pollution in rivers and lakes. By offering precise, real-time data on a range of water parameters, including pH, turbidity, temperature, dissolved oxygen, and the presence of contaminants, AI-based sensors are completely changing the way that water quality is monitored. These sensors, which are driven by ML algorithms, keep an eye on bodies of water and are able to identify minute variations in water quality that might be signs of contamination or pollution. By evaluating the data, finding trends, and forecasting possible quality problems, AI improves sensor capabilities. AI, for instance, may identify possible pollution incidents before they become widespread, sending early alerts to local authorities so they can act quickly. By enabling more effective water quality management, this real-time analysis reduces the possibility of contaminated drinking water or environmental harm [18].

2.4.2.2 Progressive filtration and purification technologies

Reverse osmosis systems and filters based on nanotechnology eliminate dangerous pollutants, including microplastics, heavy metals, and microorganisms. As an illustration, graphene-based membranes in desalination plants save energy consumption and boost efficiency. By analyzing vast amounts of environmental, chemical, and biological data to pinpoint contamination hazards, AI plays a crucial role in anticipating and avoiding water pollution. To predict locations at the highest risk of pollution, ML algorithms examine weather patterns, industrial discharges, historical contamination data, and other pertinent elements. AI can, for example, forecast dangerous algal blooms in lakes or reservoirs based on weather, nutrient levels, and water temperature. This enables authorities to take precautionary measures like water treatment or restricting access to regions that are at risk. Furthermore, AI-based prediction models assist in locating the sources of pollution, such as sewage, agricultural runoff, or industrial waste, allowing for early action to stop extensive contamination [19].

2.4.2.3 AI in water reclamation

AI has a revolutionary impact on wastewater treatment operations' sustainability and efficiency. AI can enhance many phases of the treatment process, from the initial screening and filtration to the biological treatment and nutrient removal, by incorporating ML algorithms into treatment plant operations. Real-time wastewater composition and flow data may be analyzed by AI models, which can then automatically modify chemical dosage, aeration levels, and filtering methods to enhance water quality and save energy. AI can also anticipate maintenance requirements or system breakdowns, enabling preventative measures and cutting downtime. Wastewater treatment facilities can treat more water with fewer resources because of AI's capacity to optimize procedures and lower operating costs, which improves treatment efficiency and environmental friendliness. AI also promotes a circular water economy by enhancing the quality of treated wastewater, making it fit for industrial, agricultural, or even drinking purposes. In conclusion, the way we forecast pollution, treat wastewater, and monitor and manage water quality is being completely transformed by AI-based sensors, predictive models, and optimization approaches. By enabling proactive management of water resources, these technologies guarantee water systems that are safer, cleaner, and more sustainable [20].

2.4.3 Intelligent Irrigation and Sustainable Agriculture

Seventy percent of the water used worldwide is used for agriculture, and a large portion of that is squandered because of antiquated irrigation methods. Farmers can maximize water use thanks to technological advancements like smart agricultural systems and precise irrigation. Drip irrigation systems, for instance, save waste by delivering water straight to the roots using sensors that detect soil moisture. Furthermore, weather patterns may be predicted by satellite imagery and AI-powered technologies, guaranteeing that water is only utilized when required. As an illustration, precision irrigation technology has helped farmers in Israel increase crop yields while using up to 50% less water [21].

2.4.3.1 AI-powered precision irrigation

ML algorithms assess crop water requirements, soil health, and weather patterns to make sure the right quantity of water is used. As an illustration, farmers evaluate crop moisture levels using AI-powered drones and modify irrigation as necessary. Smart irrigation systems maximize agricultural water distribution by utilizing cutting-edge technologies such as sensors, meteorological information, and AI algorithms. To calculate the exact quantity of water required, these devices keep an eye on variables including temperature, humidity, precipitation, and soil moisture. Smart irrigation systems make sure that crops get the proper quantity of water at the right time by modifying watering schedules based on weather forecasts and real-time data collected from IoT sensors. By preventing overwatering or underwatering, this enhances agricultural yields while conserving energy and water. Additionally, AI-based systems have the ability to evaluate past data and gradually improve irrigation patterns, which leads to more productive agriculture and more economical use of water [22].

2.4.3.2 Cutting down on agricultural water waste

IoT-controlled valves used in drip irrigation make sure water gets to plant roots effectively, minimizing runoff and evaporation. As an illustration, Israel's smart irrigation technologies have increased crop output while reducing agricultural water consumption by 50%. By improving water consumption and avoiding inefficiencies in farming methods, AI significantly contributes to the reduction in agricultural water waste. AI algorithms can estimate and control the precise quantity of water required for various crops, guaranteeing that water is utilized only when required, by integrating data from smart irrigation systems, soil moisture sensors, and weather predictions. AI also aids in the analysis of crop species, soil types, and development phases in order to tailor irrigation plans that minimize water waste. AI-based solutions may also identify irrigation system leaks or inadequate water distribution, enabling fast repairs. By using these technologies, farmers may conserve precious water resources and reduce the environmental effect of agriculture while increasing crop yields while using less water [22].

2.4.3.3 *AI in soil moisture monitoring*

By analyzing data from soil sensors to determine water content and forecast crop irrigation requirements, AI is improving soil moisture monitoring. In order to identify the areas of the soil that require the most water, ML algorithms examine the moisture content of various depths and locations. AI models can also integrate weather data to predict future moisture changes and adjust irrigation schedules accordingly, ensuring optimal soil hydration. AI helps farmers prevent water waste, enhance crop health, and boost production by giving them accurate moisture data. Additionally, AI can detect early signs of soil dehydration or overwatering, allowing for timely interventions to prevent crop stress and maximize water efficiency [23].

2.4.4 Disaster Mitigation and Climate Change Adaptation

Technologies like weather forecasting, climate modeling, and drought monitoring systems assist communities get ready for extreme weather occurrences as climate change exacerbates water shortages. Governments and businesses may modify their water usage policies in response to these instruments' precise forecasts of droughts, floods, and shifts in rainfall patterns. Water managers may make plans for future water supply with the use of technologies that monitor rainfall, groundwater levels, and snowmelt. As an illustration, sophisticated drought predicting models in Australia assist farmers in controlling water consumption and reducing crop losses during dry times [24].

2.4.4.1 *AI-driven drought and flood prediction*

By combining satellite data with AI models, governments are able to forecast droughts and floods months ahead of time, improving preparation. As an illustration, AI-based climate models predict extreme weather, assisting at-risk areas in creating adaptation plans. By evaluating enormous volumes of environmental data, AI plays a vital role in forecasting and controlling catastrophic weather occurrences like droughts and floods. To find trends and forecast possible drought or flood hazards, ML models employ weather patterns, historical climate data, soil moisture levels, and data from water bodies. AI is able to predict long-term hazards like drought caused by prolonged dry conditions as well as short-term ones like floods caused by unexpected heavy rains. Early detection of these trends enables AI to assist decision-makers in planning by putting in place suitable flood prevention measures, water management plans, and relief operations, thereby reducing the negative effects of natural catastrophes on the economy and environment [25].

2.4.4.2 *Flood early warning systems*

To give vulnerable areas early warnings, ML algorithms examine previous flood data, river levels, and rainfall patterns. By offering real-time forecasts and alerts on upcoming dangers, like floods, droughts, and other climate-related hazards, AI-based early warning systems improve readiness for disasters. In order to identify early warning indicators of disasters, including increasing river levels or extreme weather patterns, these systems combine data from sensors, satellites, weather stations, and other sources. This data is processed by AI algorithms that provide precise, localized alerts by predicting when and where a

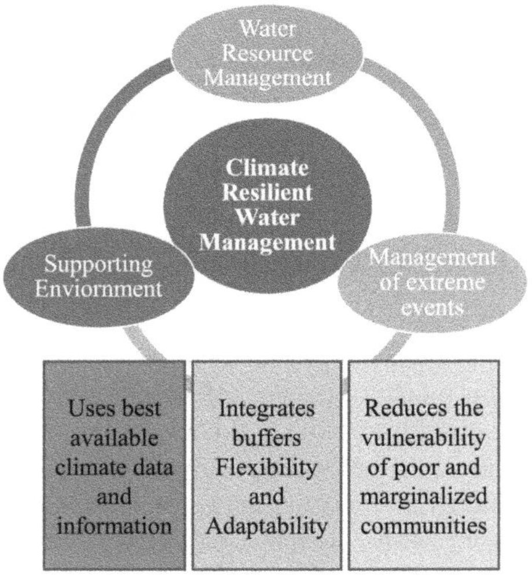

FIGURE 2.1 Climate-resilient water management.

disaster could occur. AI, for example, may forecast flooding in certain areas and issue early warnings for evacuation or suggest preventative actions, such as dam releases. These early warnings minimize damage, fatalities, and economic disruptions by providing communities, governments, and companies with crucial time to take action [26].

2.4.4.3 Climate-resilient water policies

By offering predictive insights that guide long-term water resource management plans, AI aids in the creation of climate-resilient water policy. AI can estimate the effects of different climatic scenarios on water availability and demand by examining a variety of elements, such as population expansion, changing weather patterns, and evolving agricultural demands. This makes it possible for policymakers to create climate change-responsive adaptation measures, such improving water storage and distribution systems, investing in infrastructure to withstand extreme events, and allocating water as efficiently as possible (Figure 2.1). AI also aids in assessing the efficacy of current regulations, directing changes in response to fresh difficulties brought on by climate change. In the end, AI aids in the development of regulations that guarantee fair and sustainable water distribution despite erratic and shifting environmental circumstances [27,28].

2.4.5 Sustainable Water Supply Solutions

To guarantee long-term water availability, sustainable water supply solutions combine conservation, effective resource management, and cutting-edge technologies. Water-efficient farming methods like drip irrigation and drought-tolerant crops, water-saving urban technologies like smart meters and grey water recycling, and industrial closed-loop waste reduction systems are important tactics. Groundwater recharging, desalination, and rainwater collection are examples of alternative water sources that lessen reliance on conventional sources. Water efficiency is improved via smart technology, such as IoT monitoring and AI-based analytics, and advanced wastewater treatment. Maintaining a resilient water system requires strong policy, community involvement, and climate adaptation techniques like drought and flood control. We can minimize our influence on the environment and ensure future generations have access to clean, sustainable water by combining these strategies [28].

2.4.5.1 Recycling of wastewater and desalination

Particularly in areas with limited water resources, desalination and wastewater recycling are essential technologies for a sustainable water supply. Reverse osmosis and solar-powered distillation are two techniques used in desalination, which turn saltwater into drinkable water and offer a dependable substitute for freshwater supplies. To reduce the impact on the environment, however, a lot of energy and cautious brine disposal are needed. In contrast, wastewater recycling uses membrane bioreactors, sophisticated filtration, and artificial wetlands to purify and repurpose spent water, making it fit for industrial operations, irrigation, and in certain situations, drinking water. Communities may address climate change issues while reducing their reliance on freshwater, improving water security, and fostering a circular water economy by combining the two technologies [28].

2.4.5.2 Planning for water resources using AI

AI revolutionizes the planning of water resources by increasing sustainability, efficiency, and water distribution. AI-driven algorithms evaluate water usage, forecast shortages, and find leaks to reduce waste using real-time data from IoT devices. In order to assist authorities make proactive judgments, ML algorithms are used in weather forecasting, flood risk assessment, and groundwater management. Smart irrigation systems driven by AI improve agricultural efficiency by modifying water use according to soil conditions and climatic trends. AI also improves wastewater treatment by automating water quality monitoring and filtering procedures. Governments and businesses may guarantee more intelligent, robust, and sustainable planning of water resources by using AI into water management. A good example is California's AI Water Management. To forecast drought conditions, maximize reservoir performance, and oversee groundwater sustainability, the California Department of Water Resources uses AI-powered simulations. During shortages, these models assist policymakers in more effectively allocating water. Another example is the use of AI in the Netherlands for flood prevention. To evaluate the hazards of extreme weather and sea level rise, the Netherlands government uses AI-based flood simulations. In order to reduce the effects of flooding, these models serve as a reference for the construction of drainage systems, flood barriers, and urban planning [18].

2.4.6 Bring Down the Water Disparity

In order to provide safe and reasonably priced water for everyone, reducing water inequality and increasing access need a mix of legislative changes, infrastructural development, and cutting-edge technology. Investments in decentralized water systems, such solar-powered desalination and community-based rainwater collection, facilitate the availability of clean water in isolated and underserved places. In order to minimize misuse and maintain affordability, governments and non-governmental organizations are essential in providing water service subsidies to low-income groups and enacting equitable water pricing. Smart water management systems powered by AI and IoT increase distribution efficiency, minimize losses, and maximize supply. Programs for education and community involvement also enable the local populace to efficiently manage and conserve water resources. We can close the gap in water access and advance global water fairness by combining these measures [29].

2.4.6.1 AI-enabled dissemination of water

AI-enabled water distribution optimizes supply networks in real time, increasing efficiency, reducing waste, and guaranteeing fair access. AI tracks water flow, finds leaks, and forecasts demand variations using IoT sensors and ML algorithms, allowing utilities to more efficiently distribute water. In order to modify pressure levels, lower energy consumption, and avoid shortages in high-demand locations, smart grid systems examine consumption trends. Predictive maintenance powered by AI reduces water loss by identifying pipeline problems before they become serious interruptions. AI also makes it possible for automated resource allocation and dynamic water pricing, which guarantees equitable distribution, particularly in areas with limited water supplies. Cities and rural communities may attain more egalitarian, resilient, and sustainable water management by incorporating AI into their water distribution systems.

AI systems effectively distribute resources and forecast water demand in both urban and rural locations. As an illustration, smart water systems in India are decreasing shortages in low-income regions and guaranteeing an equal distribution of drinking water [30].

2.4.6.2 Economic water desalination for deprived communities

Using reasonably priced and environmentally friendly technology, low-cost water purification systems are crucial to supplying underprivileged populations with clean drinking water. While bio-sand filters and ceramic filters eliminate bacteria and pollutants through natural filtering, simple techniques like sun disinfection (SODIS) employ UV rays to destroy hazardous germs. In an emergency, chlorine pills and powdered disinfectants provide a low-cost method of purifying water. Community-scale UV purification systems and solar-powered portable membrane filtration machines are more sophisticated but still reasonably priced options. In order to ensure long-term water security, governments and non-governmental organizations are essential in putting these solutions into practice through educational initiatives and subsidies. Millions of people in isolated and underdeveloped regions can now have access to safe and clean drinking water by utilizing low-cost solutions.

To supply safe drinking water in isolated locations, portable solar-powered water purifiers employ AI-driven monitoring systems. For instance, more than a million people in Africa now have access to clean water thanks to AI-powered portable water filtration devices [31].

2.4.7 Public Awareness and Involvement

Water conservation awareness may be increased by technologies like Internet platforms and smartphone applications. These systems allow people and companies to monitor local water availability, manage their water usage, and get advice on how to cut back. By encouraging users to lessen their water impact, gamification features may promote a sustainable culture.

As an illustration, the Jal Jeevan Mission in India has created a smartphone app that assists rural communities in monitoring water supply and use, enhancing water management at the local level [32].

2.4.8 Information and Making Decisions

In order to improve water management, data collection and analysis are essential. Technologies that offer important insights into water availability, consumption trends, and infrastructure include satellite imaging, remote sensing, and Geographic Information Systems (GIS). This information may be used by decision-makers to optimize infrastructure investment, prioritize water allocation, and develop regulations. As an illustration, satellite data has been used in Kenya to map water resources in dry regions, enhancing rural residents' access to water [33].

2.4.9 Extended Sustainability

Through encouraging conservation, cutting waste, and boosting efficiency, technological solutions help ensure the long-term sustainability of water supplies. In areas where freshwater resources are severely limited, emerging technologies like water collection systems, artificial aquifers, and desalination facilities driven by renewable energy provide creative solutions to augment them. As an illustration, rainfall harvesting technologies have been used in California to lessen dependency on conventional water sources by collecting and storing water for use during dry seasons [33].

2.4.10 Adaptation and Resilience to Climate Change

2.4.10.1 AI for predicting droughts and floods

By evaluating massive datasets from several sources, including satellite imaging, weather stations, and historical climate data, AI plays a crucial role in forecasting and controlling catastrophic weather events like droughts and floods. By identifying trends in temperature, rainfall, and soil moisture, ML algorithms can forecast drought conditions and possible flooding incidents. AI models are capable of producing precise short- and long-term forecasts by examining historical weather trends and anomalies. This makes it possible for communities and governments to take preventative actions, such bolstering flood defenses in areas that are susceptible to flooding or putting water conservation plans into place during droughts. AI algorithms, for instance, can forecast which regions are most likely to experience water scarcity in the upcoming months, enabling improved management of water resources [34].

2.4.10.2 AI-powered early warning systems

AI-based early warning systems forecast and notify populations about possible natural catastrophes, such as floods, droughts, and other climate-related occurrences, using real-time data from sensors, weather stations, and satellite feeds. Large volumes of data are processed by ML algorithms, which also identify danger indicators and produce very accurate projections, allowing authorities to react with preparatory measures. By giving timely warning to evacuate at-risk regions, modify agricultural methods, or reroute water resources, these technologies are essential for reducing damage and fatalities. AI systems, for instance, can forecast flood threats in real time, enabling prompt infrastructure modifications or evacuations to safeguard agricultural and populations [35].

2.4.10.3 Climate-resilient water policies using AI

By modeling different scenarios based on shifting climatic circumstances, AI can assist in the development of climate-resilient water policy. AI-driven models can help guide policy decisions to maintain sustainable water supply and consumption by evaluating past climate data, present water resource utilization, and future estimates. Better flood control plans, drought readiness techniques, and efficient water distribution methods are a few examples of these policies. Additionally, AI assists in forecasting the effects of various policies, such as putting water conservation measures into place or funding green infrastructure like permeable surfaces. Policymakers can develop adaptive measures to address the uncertainties offered by climate change thanks to these forecasting capacities. In areas that are most likely to experience flooding or water shortages, this allows for more focused policy changes and resource allocations to protect water supplies for future generations [36].

2.5 CHALLENGES AND LIMITATIONS OF AI IN WATER MANAGEMENT

High upfront infrastructure and technological expenditures, problems with data availability and quality, and a lack of system standardization are some of the obstacles to the use of AI in water management. Furthermore, in underdeveloped regions, where data are frequently obsolete or missing, AI's dependence

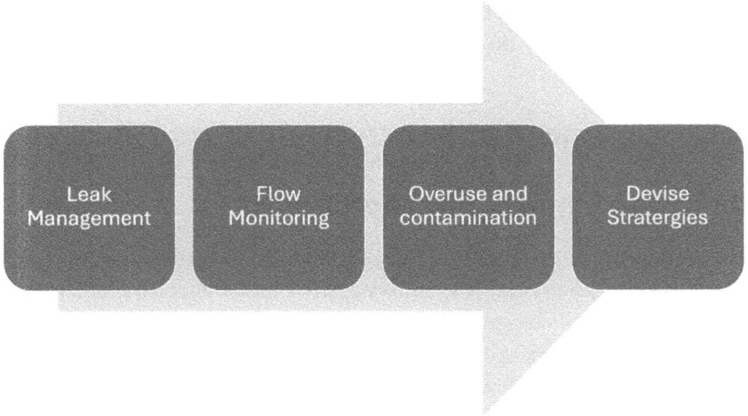

FIGURE 2.2 Intelligent system for water management.

on vast volumes of correct data may be hampered. Adoption of AI is made more difficult by ethical worries about privacy, the possibility of bias in decision-making, and the high energy consumption of AI models. Furthermore, obstacles to wider adoption are created by the difficulty of integrating AI into current infrastructure, the requirement for specialized knowledge, and the absence of clear laws and rules (Figure 2.2). To fully utilize AI's promise to improve water management, these constraints must be overcome. While there are many benefits to using AI in water management, there are also many drawbacks and restrictions. Among the main concerns are [37,38]:

2.5.1 Quality and Availability of Data

Large volumes of data are necessary for AI models to work well, yet data in water management systems is frequently inconsistent, out of date, or incomplete. The creation of trustworthy AI models may be hampered in rural or underprivileged areas by a lack of access to precise, real-time data. Furthermore, inaccurate forecasts and decisions may result from low-quality data from sensors or monitoring systems. Accuracy and data availability are major obstacles to using AI effectively for environmental applications such as water management. To create predictions and streamline procedures, AI systems use enormous volumes of data from sensors, weather stations, satellite photos, and other sources. However, inconsistent, out-of-date, or missing data might result in erroneous models and judgments. Building reliable AI models may be more difficult in some places due to a lack of data, especially in rural or poor areas. Errors can also occasionally result from malfunctioning sensors or data gathering techniques. The success of AI-driven systems depends on ensuring high-quality, accurate, and complete data, which calls for constant investment in data infrastructure and routine validation to preserve the dependability of operations and forecasts [39].

2.5.2 High Costs of Implementation

It costs a lot of money up front to adopt AI technologies, which include installing sensors, data infrastructure, and sophisticated AI algorithms. It may be too expensive for towns to create or upgrade infrastructure, particularly in rural or low-income areas. Despite AI's long-term promise for cost savings and increased efficiency, these high implementation costs may prevent its broad use in water management. High upfront expenditures are frequently associated with the implementation of AI technologies, such as predictive analytics, smart water systems, and AI-powered monitoring. These costs cover the installation of sensors, infrastructure for gathering data, and the creation or acquisition of AI systems and software. There can also be expenses for cyber security, hardware maintenance, and employee training. These expenses may be expensive for smaller utilities or in areas with limited resources, making the adoption of AI solutions difficult. The initial outlay needed to construct the essential infrastructure can be a major obstacle, even if AI systems may ultimately save money through increased efficiency and less waste. These high prices are still a major factor, but they may be reduced with the use of public-private partnerships, government subsidies, and scalable solutions [40].

2.5.3 Insufficient Standardization

Water management AI systems are frequently developed using proprietary technology and processes, which may result in a lack of uniformity among utilities or areas. Compatibility problems resulting from this lack of standardization might make it challenging to expand solutions across multiple locations or incorporate AI systems into current infrastructure. For smooth interoperability and wider acceptance of AI technologies, a more standardized approach is required.

2.5.4 Privacy and Ethical Issues

Data at the individual and community levels, including water consumption trends, are gathered and analyzed as part of the application of AI in water management. This brings up issues with data security, privacy, and the possibility of sensitive information being misused. Furthermore, if the data used to train AI systems are biased or inadequate, they may occasionally reinforce biases or uneven results. It is crucial to set up moral standards and legal frameworks to protect privacy and guarantee equitable application of AI. It brings up a number of privacy and ethical issues. Personal and community-level data, including geographic location, water consumption trends, and even behavioral data, are frequently used by AI systems. There is a chance that these data will be utilized in ways that discriminate against particular groups or breach privacy. Third parties, like marketers or insurance firms, may abuse water use data if they are connected to specific people or families. Concerns have also been raised regarding the opaqueness of AI decision-making processes, which may result in prejudices or unjust consequences like an uneven allocation of water supplies. Clear ethical standards and legal frameworks are essential for ensuring data security, transparency, and the equitable application of AI in water management while striking a balance between privacy and equity issues and innovation.

2.5.5 Excessive Energy Use

AI models demand a lot of processing power and energy, especially those employed for predictive analytics and real-time monitoring. Particularly in energy-intensive applications like data storage and ML, this can have a significant negative impact on the environment. If not properly controlled, this energy usage in water management systems may partially counteract the environmental advantages of AI.

2.5.6 Intricacy and Insufficient Experience

Data science, ML, and water management are among the specific knowledge and abilities needed to develop and implement AI solutions. Many utilities might not have the technical know-how to install and operate cutting-edge AI systems, especially in rural or underdeveloped regions. AI adoption may encounter opposition, be implemented slowly, or fall short of its full potential in the absence of adequate training and experience.

2.5.7 Scalability and System Integration

It can be difficult to incorporate AI technologies into current water management systems, particularly when working with legacy infrastructure that was not created with AI in mind. It takes careful planning, customization, and maybe equipment modification to ensure that new AI systems can integrate seamlessly with existing systems. Another issue is scalability, as AI solutions that perform well in pilot or small-scale projects cannot translate well to bigger, more complicated systems.

2.5.8 Issues with Trust and Reliability

AI systems need to be extremely dependable, especially when it comes to making decisions in crucial domains like flood forecasting and water distribution. There may be serious repercussions, such as water shortages or greater damage from natural catastrophes, if AI systems are unable to foresee flood threats or water supply requirements. Ensuring transparency, accountability, and dependability in AI models is crucial to gaining the public's and governments' trust in AI-driven systems.

2.5.9 Policy and Regulatory Difficulties

The legal environment around AI in water management is continually developing. To control the application of AI technology in water management, governments and regulatory agencies must create thorough regulations and guidelines. The deployment of AI technologies may be hampered by unclear rules that raise questions about responsibility, compliance, and ethical norms. In conclusion, AI has a lot of promise to enhance water management, but it also has drawbacks, including high prices, problems with data quality, ethical dilemmas, and integration hurdles. Governments, IT companies, and communities must work together to overcome these constraints, and further investment in infrastructure, research, and capacity building will be necessary.

2.6 FUTURE PROSPECTS AND INNOVATIONS

AI in water management has enormous promise for innovation and game-changing solutions in the future, which will increase water resource management's resilience, sustainability, and efficiency. Among the most important opportunities and innovations are the following [14]:

2.6.1 Advanced Integration of AI and IoT

Water management is about to undergo a revolution thanks to the combination of AI with IoT sensors. AI will process and analyze the continuous, real-time data on water quality, flow rates, and consumption that IoT-enabled sensors will give in order to improve distribution, identify problems, and forecast demand trends. Water systems will become more intelligent and responsive as a result of this integration, which may save expenses, enhance service quality, and limit waste.

2.6.2 AI-Powered Climate Adaptability

AI will become more crucial in assisting systems in adapting as climate change continues to affect water resources. AI can forecast how climate change would affect local water supply and demand, which can help build policies and infrastructure that are climate resilient. More precise forecasting of catastrophic weather occurrences, such droughts and floods, will be made possible by AI-based simulations, improving preparedness and enabling more focused resource allocation.

2.6.3 Systems for Independent Water Management

Autonomous systems driven by AI are poised to revolutionize water management. Without the need for human interaction, these devices will employ ML algorithms to identify leaks, modify irrigation schedules, and monitor water quality and distribution. Particularly in urban water systems and agriculture, such automation will increase water usage accuracy, lower labor costs, and improve operational efficiency.

2.6.4 Smart and Decentralized Water Systems

Decentralized water systems, which use AI to improve optimization and administration of modest, community-based solutions like solar-powered desalination or water purification technologies, will also become more prevalent in the future of water management. Granular, localized data from smart water meters and sensors will enable more effective and fair water distribution, especially in places with inadequate infrastructure or water shortage.

2.6.5 Cutting-Edge Water Reuse and Recycling Technologies

By streamlining wastewater treatment procedures, AI will support advancements in water recycling and reuse. Algorithms that apply ML can increase treatment plant efficiency, forecast when maintenance is necessary, and improve treated water quality so that it can be used for a variety of purposes. AI developments in circular water economies—where wastewater is treated, cleaned, and repurposed for industrial, agricultural, or even drinking purposes—will be fueled by the rising need for sustainable water solutions.

2.6.6 AI in Smart Agriculture

Precision agriculture, where farmers can use AI-driven systems to track crop health, soil moisture, and water consumption in real time, will continue to progress thanks to AI. In order to increase agricultural productivity and water efficiency, ML algorithms will suggest the best irrigation and water-saving techniques. Farmers will be able to better manage water resources with the use of drones and AI to monitor vast agricultural areas, particularly in areas experiencing drought or water stress. By increasing production, cutting down on resource waste, and guaranteeing sustainable farming methods, AI is transforming the agricultural industry. Precision farming is one significant use case, where AI-driven drones and IoT sensors track crop conditions, moisture content, and soil health in real time. Farmers may take prompt action by using predictive analytics to foresee weather trends, insect infestations, and disease outbreaks. By evaluating environmental factors, AI-driven automated irrigation systems maximize water use while preserving crop health.

2.6.7 Potential of AI in Attaining Sustainable Development

The Sustainable Development Goals (SDGs) of the UN are greatly advanced by AI, especially SDG 6, which is concerned with guaranteeing access to clean water and sanitation. AI contributes to water conservation and improved water management techniques by enhancing water efficiency through predictive analytics, smart irrigation, and optimized water distribution systems. By forecasting droughts, floods, and other climate-related water effects, it helps communities adapt and become more resilient, which advances climate action (SDG 13). AI helps manage urban water supply systems, optimize water consumption, and identify leaks or pollution to increase access to clean drinking water in the context of sustainable cities and communities (SDG 11). By bringing together governments, non-governmental organizations, and the commercial sector in joint endeavors to create and deploy smart water technology, AI also promotes innovative collaborations (SDG 17). Additionally, by guaranteeing sustainable water usage and reducing waste, AI in wastewater treatment and water recycling supports the development of a circular economy and is consistent with SDG 12 (Responsible Consumption and Production).

2.7 CONCLUSION

In summary, AI is quickly revolutionizing water management by providing creative answers to problems like pollution, shortage, and wasteful resource use. AI has the potential to optimize water consumption, improve sustainability, and increase resistance to climate change through a variety of applications, including real-time water quality monitoring, smart distribution systems, predictive analytics, and enhanced wastewater treatment. AI greatly aids in accomplishing the SDGs of the UN, especially in guaranteeing access to clean water and sanitation, by facilitating more precise forecasting, automation, and data-driven decision-making. However, there are obstacles to the broad use of AI in water management, such as problems with data quality and availability, expensive implementation costs, and moral dilemmas with prejudice and privacy. Governments, the corporate sector, and communities must work together to overcome these challenges, as well as continue to invest in infrastructure, standards, and regulatory frameworks. AI's contribution to the development of effective, just, and sustainable water management systems will only increase as it develops further, opening up new avenues for tackling the most important water-related issues facing the globe. By transforming monitoring, predictive analysis, and sustainable water management, AI is reinventing the fight against the water crisis. AI is improving water distribution and conservation through its use in policymaking, wastewater treatment, and agriculture. To improve AI's capabilities, incorporate it with cutting-edge technology, and guarantee worldwide accessibility, further research is necessary. We can get closer to a future where sustainability is attained for future generations and water security is reinforced by developing AI-driven solutions. The various key findings are real-time monitoring, predictive analysis, smart irrigation, wastewater management, and desalination and purification. The importance of AI in water management is still developing despite major achievements, requiring more study creating AI models that combine water management techniques with estimates of climate change.

REFERENCES

1. Sun, Z., Sandoval, L., Crystal-Ornelas, R., Mousavi, S. M., Wang, J., Lin, C., … & John, A. (2022). A review of earth artificial intelligence. *Computers & Geosciences*, *159*, 105034.
2. Sun, Y., Nan, Z., Yang, W., & Li, L. (2023). Projecting China's future water footprints and water scarcity under socioeconomic and climate change pathways using an integrated simulation approach. *Climate Services*, *30*, 100385.
3. Talaviya, T., Shah, D., Patel, N., Yagnik, H., & Shah, M. (2020). Implementation of artificial intelligence in agriculture for optimisation of irrigation and application of pesticides and herbicides. *Artificial Intelligence in Agriculture*, *4*, 58–73.

4. Blake, R. W., Mathew, R., George, A., & Papakostas, N. (2021). Impact of artificial intelligence on engineering: Past, present and future. *Procedia CIRP, 104*, 1728–1733.
5. Jayakumar, D., Bouhoula, A., & Al-Zubari, W. K. (2024). Unlocking the potential of artificial intelligence for sustainable water management focusing operational applications. *Water, 16*(22), 3328.
6. Krysanova, V., Donnelly, C., Gelfan, A., Gerten, D., Arheimer, B., Hattermann, F., & Kundzewicz, Z. W. (2018). How the performance of hydrological models relates to credibility of projections under climate change. *Hydrological Sciences Journal, 63*(5), 696–720.
7. Imrie, C. E., Durucan, S., & Korre, A. (2000). River flow prediction using artificial neural networks: Generalisation beyond the calibration range. *Journal of Hydrology, 233*(1–4), 138–153.
8. Krishnan, S. R., Nallakaruppan, M. K., Chengoden, R., Koppu, S., Iyapparaja, M., Sadhasivam, J., & Sethuraman, S. (2022). Smart water resource management using artificial intelligence: A review. *Sustainability, 14*(20), 13384.
9. Rizal, N. N. M., Hayder, G., & Yusof, K. A. (2022). Water quality predictive analytics using an artificial neural network with a graphical user interface. *Water, 14*(8), 1221.
10. Podder, A. K., Al Bukhari, A., Islam, S., Mia, S., Mohammed, M. A., Kumar, N. M., … & Abdulkareem, K. H. (2021). IoT based smart agrotech system for verification of Urban farming parameters. *Microprocessors and Microsystems, 82*, 104025.
11. Iqbal, U., Riaz, M. Z. B., Barthelemy, J., Perez, P., & Idrees, M. B. (2023). The last two decades of computer vision technologies in water resource management: A bibliometric analysis. *Water and Environment Journal, 37*(3), 373–389.
12. Yereseme, A. K., Surendra, H. J., & Kuntoji, G. (2022). Sustainable integrated urban flood management strategies for planning of smart cities: A review. *Sustainable Water Resources Management, 8*(3), 85.
13. Bajaj, A., Singh, S. P., & Nayak, D. (2022). Impact of water markets on equity and efficiency in irrigation water use: A systematic review and meta-analysis. *Agricultural Water Management, 259*, 107182.
14. Chen, J., Chen, S., Fu, R., Li, D., Jiang, H., Wang, C., … & Hicks, B. J. (2022). Remote sensing big data for water environment monitoring: Current status, challenges, and future prospects. *Earth's Future, 10*(2), e2021EF002289.
15. Park, Y., Cho, K. H., Park, J., Cha, S. M., & Kim, J. H. (2015). Development of early-warning protocol for predicting chlorophyll-a concentration using machine learning models in freshwater and estuarine reservoirs, Korea. *Science of the Total Environment, 502*, 31–41.
16. Chen, K., Chen, H., Zhou, C., Huang, Y., Qi, X., Shen, R., … & Ren, H. (2020). Comparative analysis of surface water quality prediction performance and identification of key water parameters using different machine learning models based on big data. *Water Research, 171*, 115454.
17. Kalfarisi, R., Chew, A., Cai, J., Xue, M., Pok, J., & Wu, Z. Y. (2022). Predictive modeling framework accelerated by GPU computing for smart water grid data-driven analysis in near real-time. *Advances in Engineering Software, 173*, 103287.
18. Alam, G., Ihsanullah, I., Naushad, M., & Sillanpää, M. (2022). Applications of artificial intelligence in water treatment for optimization and automation of adsorption processes: Recent advances and prospects. *Chemical Engineering Journal, 427*, 130011.
19. Albergamo, V., Blankert, B., Cornelissen, E. R., Hofs, B., Knibbe, W. J., van der Meer, W., & de Voogt, P. (2019). Removal of polar organic micropollutants by pilot-scale reverse osmosis drinking water treatment. *Water Research, 148*, 535–545.
20. Bhagat, S. K., Tung, T. M., & Yaseen, Z. M. (2020). Development of artificial intelligence for modeling wastewater heavy metal removal: State of the art, application assessment and possible future research. *Journal of Cleaner Production, 250*, 119473.
21. Elbeltagi, A., Kushwaha, N. L., Srivastava, A., & Zoof, A. T. (2022). Artificial intelligent-based water and soil management. In Poonia, R. C., Singh, V., & Nayak, S. R. (Eds.) *Deep Learning for Sustainable Agriculture* (pp. 129–142). Academic Press.
22. Berbel, J., & Expósito, A. (2022). A decision model for stochastic optimization of seasonal irrigation-water allocation. *Agricultural Water Management, 262*, 107419.
23. Adarsh, V., Parida, B., Balo, S., & Choudhury, R. K. (2024). Role of Artificial Intelligence (AI) and machine learning in soil moisture determination. *Vigyan Varta, 5*(5), 179–185.
24. Olabi, A. G., & Abdelkareem, M. A. (2022). Renewable energy and climate change. *Renewable and Sustainable Energy Reviews, 158*, 112111.
25. Xie, K., Liu, P., Zhang, J., Han, D., Wang, G., & Shen, C. (2021). Physics-guided deep learning for rainfall-runoff modeling by considering extreme events and monotonic relationships. *Journal of Hydrology, 603*, 127043.
26. Golding, B. W. (2009). Long lead time flood warnings: Reality or fantasy? *Meteorological Applications: A Journal of Forecasting, Practical Applications, Training Techniques and Modelling, 16*(1), 3–12.

27. Srivastav, A.L., Dhyani, R., Ranjan, M., Madhav, S., & Sillanpää, M. (2021). Climate-resilient strategies for sustainable management of water resources and agriculture. *Environmental Science and Pollution* Research, *28*, 41576–41595. https://doi.org/10.1007/s11356-021-14332-4
28. Wan Rosely, W.I.H., & Voulvoulis, N. (2024). System thinking for sustainable water management: The use of system tools in sustainability transitions. *Water Resour Manage*, *38*, 1315–1337. https://doi.org/10.1007/s11269-023-03723-6
29. Meneses, Y. E., Stratton, J., & Flores, R. A. (2017). Water reconditioning and reuse in the food processing industry: Current situation and challenges. *Trends in Food Science & Technology*, *61*, 72–79.
30. Tran, H. N., Nguyen, N. B., Ly, N. H., Joo, S. W., & Vasseghian, Y. (2023). Core-shell Au@ ZIF-67-based pollutant monitoring of thiram and carbendazim pesticides. *Environmental Pollution*, *317*, 120775.
31. Medjili, C., Lakhdari, N., Lakhdari, D., Berchi, A., Osmani, N., Laourari, I., ... & Berkani, M. (2023). Synthesis of novel PANI/PVA-NiCu composite material for efficient removal of organic dyes. *Chemosphere*, *313*, 137427.
32. Parisot, A., Bento, L. M., & Machado, R. C. (2021, June). Testing and selecting lightweight pseudo-random number generators for IoT devices. In *2021 IEEE International Workshop on Metrology for Industry 4.0 & IoT (MetroInd4. 0&IoT)* (pp. 715–720). IEEE.
33. Rawat, S., Tateh, S., & Dubey, S. (2024). Geographical information system (GIS) Applications for Water Resources Management. *Agriculture & Food: E-Newsletter*, 7(1), 613–618.
34. Adikari, K. E., Shrestha, S., Ratnayake, D. T., Budhathoki, A., Mohanasundaram, S., & Dailey, M. N. (2021). Evaluation of artificial intelligence models for flood and drought forecasting in arid and tropical regions. *Environmental Modelling & Software*, *144*, 105136.
35. Beiranvand, B., & Rajaee, T. (2022). Application of artificial intelligence-based single and hybrid models in predicting seepage and pore water pressure of dams: A state-of-the-art review. *Advances in Engineering Software*, *173*, 103268.
36. Zhang, S., Jin, Y., Chen, W., Wang, J., Wang, Y., & Ren, H. (2023). Artificial intelligence in wastewater treatment: A data-driven analysis of status and trends. *Chemosphere*, *336*, 139163.
37. Ingrao, C., Strippoli, R., Lagioia, G., & Huisingh, D. (2023). Water scarcity in agriculture: An overview of causes, impacts and approaches for reducing the risks. *Heliyon*, *9*(8), e18507.
38. Song, M., Tao, W., Shang, Y., & Zhao, X. (2022). Spatiotemporal characteristics and influencing factors of China's urban water resource utilization efficiency from the perspective of sustainable development. *Journal of Cleaner Production*, *338*, 130649.
39. Kamyab, H., Khademi, T., Chelliapan, S., SaberiKamarposhti, M., Rezania, S., Yusuf, M., ... & Ahn, Y. (2023). The latest innovative avenues for the utilization of artificial Intelligence and big data analytics in water resource management. *Results in Engineering*, *20*, 101566.
40. Essamlali, I., Nhaila, H., & El Khaili, M. (2024). Advances in machine learning and IoT for water quality monitoring: A comprehensive review. *Heliyon*, *10*(6), e27920.

PART II

AI Applications in Water Quality Monitoring

Synergy of AI and IoT in Water Monitoring

3

Preeti Karki, Priyanjali Chowdhury, Joel Benny, Alakto Choudhury, Abhishek Sengupta, and Priyanka Narad

3.1 INTRODUCTION

The availability of clean and safe water has become an increasingly critical issue due to the impacts of industrialization, urbanization, and global warming, which contribute to contamination and water shortage. Traditional methods for water quality monitoring (WQM) depend on manual sampling and laboratory tests, which usually consume time, are reactive, rather than proactive, and have no real-time data information. Consequently, there is a growing need for technological advances to improve water quality management. Artificial intelligence (AI) and Internet of Things (IoT) transformed water monitoring, allowing predictive collection, analysis, and modeling. These technologies facilitate continuous tracking of the main parameters of water quality, with AI-oriented models processing data to predict contamination levels and recommend appropriate treatment measures. As a result, water management systems have become more efficient, accurate, and proactive, minimizing the need for human intervention. This chapter explores the integration of AI and IoT in water monitoring, emphasizing their applications, advances, and the ways in which these technologies increase the response capacity and effectiveness of water management systems in addressing emerging water quality challenges.

3.2 COMPONENTS OF IOT-BASED WATER MONITORING SYSTEMS

Traditional water monitoring techniques are very much time-consuming and not suitable to give real-time data (Lakshmikantha et al., 2021). With the emergence of IoT technology, WQM has been revolutionized, allowing for continuous, automated, and remote evaluation of several water parameters. IoT-based systems can combine several technical elements in order to provide effective, real-time, and predictive analysis, tackling issues including population expansion, scarce water supplies and therefore the need for safe water distribution.

DOI: 10.1201/9781003600718-5

Generally, these systems are known to include several smart sensor interfaces that are able to detect several key water parameters such as testing pH, turbidity, temperature, and dissolved oxygen. Real-time monitoring and administration are made possible by the transmission of the data gathered by these sensors to web platforms (Mahbubur Rahman et al., n.d.) The technologies, making use of cloud computing and the IoT, not only improve the effectiveness of monitoring water quality but also offer predictive insights essential for decisions meant to protect water quality, making it possible to pinpoint locations needing urgent care and place preventative measures in place to stop contamination (Essamlali, Nhaila, & El Khaili, 2024).

An IoT-based water monitoring system depends on the integration of various technological components like sensors and gadgets essential for recording data in real time. To guarantee a smooth data flow, communication methods are essential. 5G and LPWAN have enabled energy-efficient transmission, and this is further improved by cloud computing via quick data processing and storage and guarantees prompt water management responses.

To get a clearer picture and to understand how these elements work and interact with each other, let us explore the sections discussed below in order to find out and further learn about the architectural framework of these IoT-based water monitoring systems with an in depth analysis on several sensors and devices being used along with the communication protocols, data collection strategies, and the role of AI in data analysis. Together, these two have enabled easy automation, precision, and made modern water quality management seamless and scalable.

3.2.1 IoT Architecture for Water Monitoring

IoT-enabled water monitoring system is designed to monitor the water level and their connected tanks. The system includes two major monitoring units: lower tank monitoring unit and upper tank monitoring unit. Collectively, they communicate with each other in a wireless manner using a 2.4 GHz RF communication module, to allow real-time monitoring and automation (Singh et al., 2021).

3.2.1.1 Upper tank monitoring unit

Real-time water quality and level monitoring is the responsibility of this unit. It includes a power supply unit, RF module, Wi-Fi module, pH sensor, water level sensor, flow sensor, and controller unit. The controller unit processes the sensor data, and the water level of the tank detects sensor variations. If the water level falls below the range, the lower tank monitoring unit is activated. It is possible to transfer automatic water by installing communication with the lower tank through the 2.4 GHz RF module. ESP8266 Wi-Fi module transmits real-time data to the cloud, allowing remote accessibility.

3.2.1.2 Lower tank monitoring unit

Primarily responsible for ensuring a steady water supply to the upper tank, this unit comprises essential components such as a controller unit, RF module, water level sensor, water pump, relay, and power supply unit. Incoming data and device operations are controlled by the controller unit. The RF module establishes a communication with the upper tank, getting signals to activate the water pump whenever needed. The pump's functionality is regulated by the relay system, preventing overflows ensuring an automated water transfer process while also maintaining system efficiency (Singh et al., 2021) (Figures 3.1 and 3.2).

3.2.1.3 pH sensor

A solution's acidity or alkalinity is determined by its pH level. The pH sensor, preamplifier, and analyzer (transmitter) are the parts of the pH measurement loop. A temperature sensor, reference electrode, and measuring electrode are all part of the pH sensor. The hydrogen ion causes the electrode to react. The voltage generated by the measuring electrode is directly proportional to the concentration

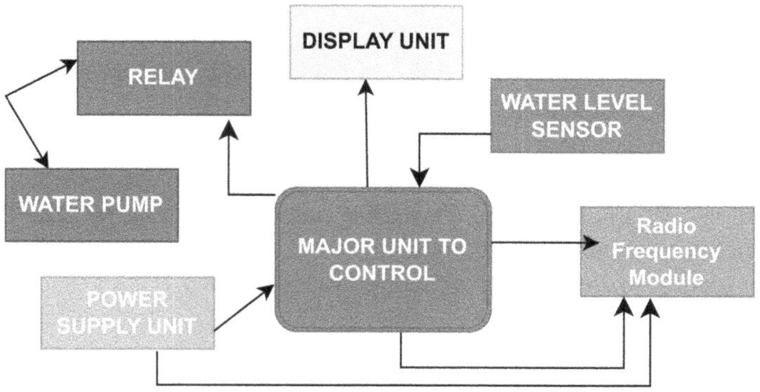

F GURE 3.1 An upper tank monitoring unit.

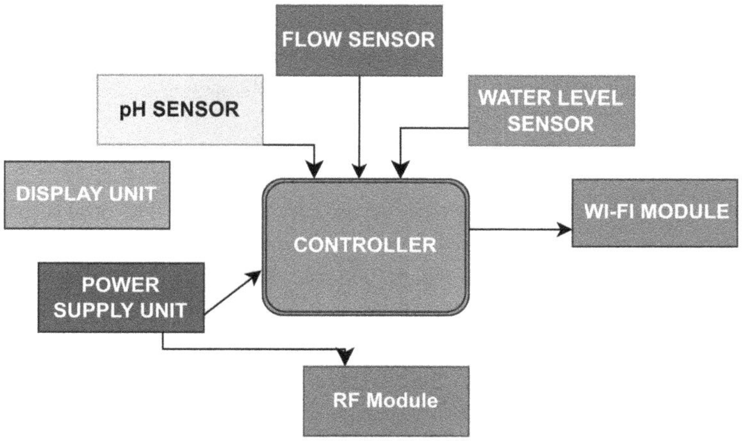

F GURE 3.2 A lower tank monitoring unit.

of hydrogen ions in the solution. In contrast to the electrode measurement, the reference electrode provides a significant potential. A pH electrode combination is the name of the single device that makes up the pH sensor components.

3.2.1.4 Flow sensor

The flow sensor is positioned skillfully along the water supply pipe and is intended to measure water flow. It uses a pinwheel mechanism, wherein a tiny magnet is fastened to the pinwheel as it rotates. Accurate flow monitoring is made possible by a Hall effect magnetic sensor that is positioned on the other side of the plastic tube and counts the number of rotations the pinwheel makes through the plastic wall. The sensor has three wires: yellow (Hall effect pulse output), black, and red. The pulses from the sensor output are measured to track the fluid's velocity.

3.2.1.5 Turbidity sensor

Turbidity is the degree of cloudiness or haziness in water; hence, it indicates the presence of suspended particles. This serves as a key parameter to check water quality as high turbidity levels may obstruct light penetration, which negatively impacts aquatic life. Moreover, suspended particles at the surface absorb more heat from sun, thereby increasing water temperature beyond normal levels. Turbidity sensors are necessary in monitoring water clarity and ensuring compliance with environmental and safety standards (Daigavane & Gaikwad, 2017).

3.2.1.6 Temperature sensor

Temperature influences several chemical reactions, biological processes, as well as aquatic life sustainability, thereby becoming an essential parameter. The DS18B20 temperature sensor is used for water monitoring for its accuracy and an operating range, from –55°C to +125°C. It utilizes a one-wire communication protocol, allowing multiple sensors to be connected on a single data line, thus ensuring efficiency in large-scale water monitoring applications. Variations in temperature in the water bodies can seriously affect oxygen levels, pollutant solubility, and microbial activity; thus, real-time monitoring becomes essential to maintain water quality in environmental and industrial settings (Daigavane & Gaikwad, 2017).

3.2.1.7 ESP 8266 Wi-Fi modem

Being a self-contained system-on-chip (SoC), this modem allows for wireless connectivity for microcontrollers. It comes together with a TCP/IP protocol stack, allowing a smooth Wi-Fi communication. The module is loaded with an AT command set firmware, which allows easy interfacing with various devices. It handles data transmission, allowing robust processing and storage capabilities. Connected to sensors via GPIO pins, an external logic level converter is required for voltage compatibility. It features a total of eight pins, allowing multiple configurations for diverse IoT applications.

3.2.1.8 RF module

This module can send and receive data at any time because it operates in half-duplex mode. It immediately transitions to receiver mode after a broadcast is finished. The module's present operation is visually shown by the TX (transmit) and RX (receive) LEDs, which indicate whether data are being sent or received. If any data are entered into the transmitter during transmission, they are stored in the buffer for the subsequent transmission cycle. In order to prevent data loss while the module is actively transmitting data, the incoming data are kept in the incoming buffer, and they will be supplied at the time of submission (Singh et al., 2021).

3.2.1.9 Communication protocols

LPWAN is widely used as it supports long-range, low-power devices, hence making it suitable for smart homes. 5G is also becoming more popular for high-speed, high-capacity needs like industrial automation. These two technologies work in tandem; LPWAN offers energy-efficient and wide coverage for devices that must function over longer distances with less data, while 5G oversees quick and dependable connections. Combining the two provides a complete answer for a range of IoT requirements. Unlike traditional methods, continuous monitoring, real-time anomaly detection, cost-effectiveness, and thorough data collecting are all made possible by IoT-based WQM, improving management. Water Quality (WQ) monitoring has been greatly improved by LPWAN and cellular networks, as well as IoT wireless technologies like Wi-Fi, Zigbee, Radio Frequency Identification (RFID), and Bluetooth.

The establishment of cellular networks has been essential for IoT connection because they serve voice communication and power billions of smart devices. Mobile communication networks became possible through early 1G and 2G systems, though the development of IoT began with 3G, as it enabled mobile Internet and data transmission capabilities. The evolution of IoT technology received significant advancement when 4G rolled out since it brought high-speed data transmission, which real-time applications needed. Modern healthcare facilities and smart cities together with industrial automation, benefit from seamless IoT integration because 5G delivers ultra-low latency coupled with huge connections and elevated bandwidth (Table 3.1).

3.2.1.10 Data storage and cloud computing

Effective management of water resources is necessary to ensure sustainability, and optimizing its use saves water for the next generations. Sensor integration provides real-time monitoring of hydraulic data, automated control, and notifications through cloud-based platforms in the event of leaks, overuse, or other irregularities.

TABLE 3.1 Overview on IoT technologies used for successful water quality management

REFERENCES	*YEAR*	*COMMUNICATION*	*MONITORED PARAMETERS*
Hartono et al. (2022)	2022	LoRaWAN, 5G	Temperature, humidity, and pH
Akhter, Siddiquei, Alahi, Jayasundera, and Mukhopadhyay (2022)	2022	LoRaWAN	Temperature, nitrate, phosphate, and pH
Saravanan, Anusuya, Kumar, and Son (2018)	2018	4G/5G/GSM	Temperature, color, and TUR
González, Gonzales, González, and Cartuche (2024)	2024	LoRaWAN	Temperature, DO, ORP, pH
Budiarti, Sukaridhoto, Hariadi, and Purnomo (2019)	2019	4G	IP

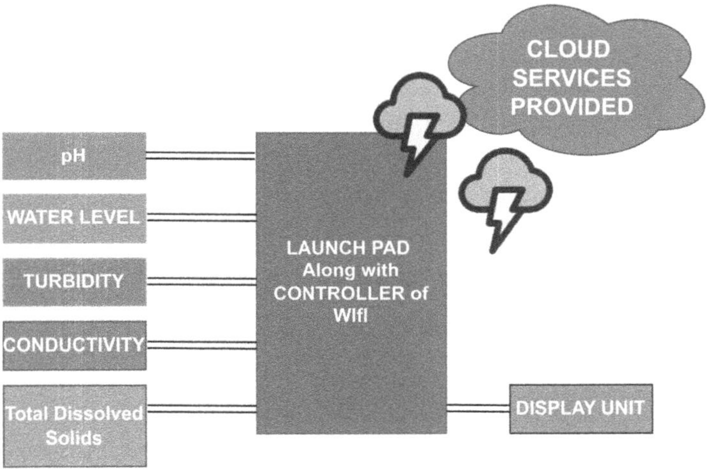

FIGURE 3.3 The proposed system architecture.

Sensor data is processed and analyzed in a cloud environment; for example, Ubidots is capable of distributed computing for optimized data processing. Cloud computing provides seamless integration with display devices and mobile phones to enable real-time monitoring and remote access. Ubidots is a Platform as a Service (PaaS) cloud-based system that enables effortless communication between networked sensors and devices. It offers web services to collect, visualize, and analyze data in real time. With its cloud-based system, scalability is increased to manage large networks of sensors and provide secure and reliable data transfer over the internet (Jha et al., 2020) (Figure 3.3).

3.2.2 Data Collection and Transmission

Any machine learning (ML) model's development starts with data collecting, and the model's effectiveness is directly impacted by the quality of the data. To monitor the quality of the water, the QoW-Pro system uses a variety of sensors to measure temperature, pH, conductivity, and turbidity. Cloud platforms like ThingSpeak and Firebase receive real-time data from these sensors for processing and storing. Real-time water quality status reporting is made possible by the unbroken data flow, which lets users monitor conditions in real time. Moreover, IoT sensors, together with remote sensing technologies, allow the gathering of data from hard-to-reach areas and provide valuable information for analyzing water quality (Guezouli, Guezouli, Djeghaba, & Bentahrour, 2024).

3.2.3 Role of AI in IoT Systems

AI and IoT together have the ability to analyze big data which were generated through several environmental sensors. ML algorithms have been capable of identifying certain patterns, thereby allowing the forecasting of future trends from complex data available. AI-driven models have been proven handy for predicting pollution levels and forecasting climate anomalies, allowing policymakers to respond efficiently to day-to-day environmental challenges (Li et al., 2025).

In AI-based WQM, several ML algorithms are applied, each selection being made for data types, computational cost, and interpretability. Random forests and support vector machines (SVMs) are typical in supervised learning operations, that is, discriminating water quality from known degrees of contamination previously established. K-means clustering has enabled the detection of anomalies from the data available, thereby assisting in identifying pollution patterns in the environment without any data available. Convolutional neural networks (CNNs) have been utilized for satellite image processing, speeding up the large-scale monitoring of water quality. Long short-term memory (LSTM) networks are well suited to process time-series data, e.g., forecasting water quality trends over time by identifying long-term patterns in sensor readings. Decision trees are employed where there is limited computing capability on edge devices to facilitate real-time monitoring of water quality. Finally, logistic regression is typically applied in policymaking and regulation because it is interpretable and therefore easier to justify decision-making. All these algorithms combined constitute a strong system for efficient, effective, and scalable WQM.

3.3 AI-DRIVEN DATA ANALYSIS FOR WATER QUALITY MONITORING

Integrating AI and ML with WQM assures a significant shift in environmental data treatment and analysis. While conventional systems take sample points intermittently and resort to laboratory trials, which sometimes can be assumed as cost- and time-intensive and could also make errors based on the technician, the AI-based technology can make it possible to perform water quality testing very quickly and determine its state more accurately and distinctly. As far as the general quality of ML algorithms is concerned in water quality, the results have been quite impressive regarding the complexities of handling data due to their learning from previous data in order to anticipate later state conditions, categorize water into different classes with regard to its quality, and identify anomalies or trends that may indicate pollution incidence. AI application in WQM in real-time, predictive analysis, scalability, and cost-effectiveness are some of the immediate advantages of implementing AI in sustainable projects.

Delving into the peculiarities of water quality data by learning from current data to try and predict what situations will arise when categorizing water into the different classes of quality and the corresponding detections of anomalies.

3.3.1 Machine Learning Algorithms for Water Quality Monitoring

Classification of water quality samples based on pH, turbidity, and contamination using the ML technique is increasingly utilized, and particularly, ML learning algorithms have been evident. Decision trees and SVM are two classification models that most engineers' resort to because they can deal with large or messy datasets that have confusing or nonlinear interrelationships. These models, on the one hand, use the energy or value of the feature, and thus, a decision tree is known to be a simple basic model that predicts water quality in the form of classification by forming some kind of tree structure from the data. Because such models are able to build easy ensembles of headings, they are useful in application since one model can be divided by neurons in the network and other model can be divided into inputs and rules to be processed separately.

The opposite, however, SVMs are other ML topics that are well researched especially when they deal with complex, extensive data with numerous classes, but where the borders are not linear. Each of the two classes is important. Experts in both classes are retained in quality control. Either can be applied if they are able to select the appropriate model for their particular piece of data (Figures 3.4 and 3.5).

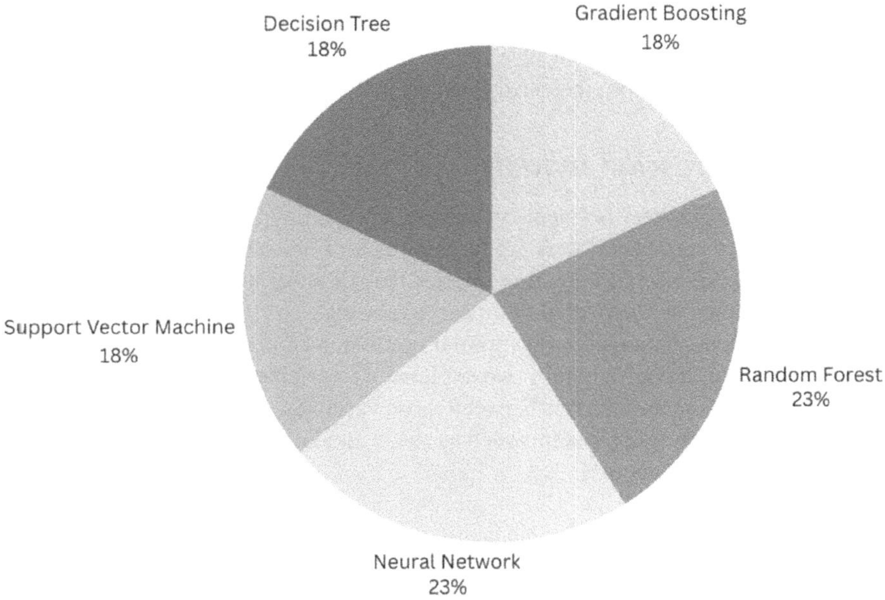

FIGURE 3.4 Supervised learning models for water quality evaluations.

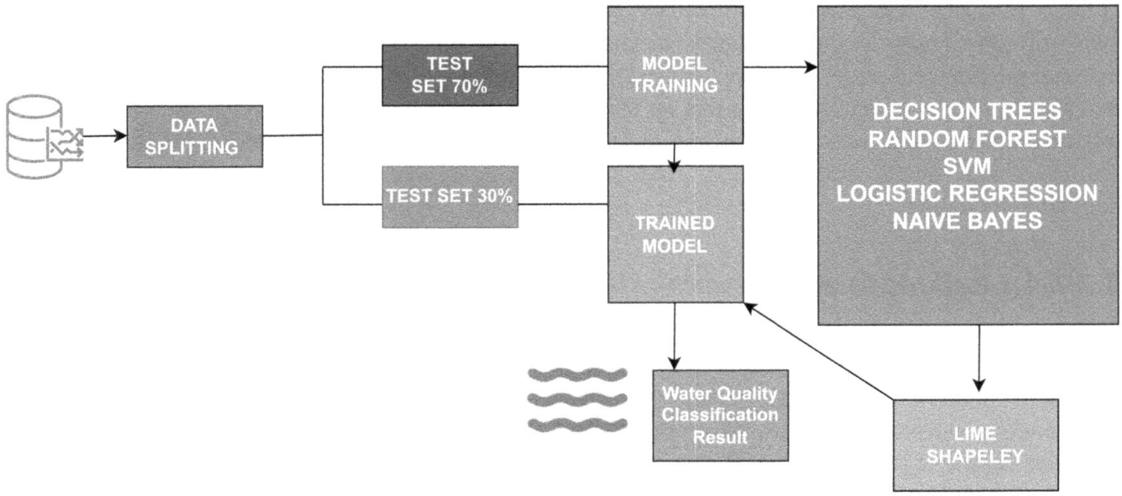

FIGURE 3.5 Linking the ML algorithms with Explainable Artificial Intelligence (XAI).

3.3.1.1 Multiple linear regression

Interpretations of water quality show considerable relationships with one another, clustering multiple water quality indices into describing factors. Multiple linear regression (MLR) stands as a retrogression technique to study further interrelationships. The MLR model relationship among one or more predictor variables (such as turbidity, pH, discharge) and a target variable (e.g., substrate concentration) and its linear coefficients. MLR estimates chemical parameters through continuous measures and discrete and manually taken water samples to develops predictive models to get constituent concentration and loading estimates.

The first step toward success starts with data examination to determine its underlying structures in relation to each potential explanatory variable and the dependent variable. For example, it is observed that suspended sediment concentration (SSC) increases linearly with turbidity (Rasmussen et al., 2009). By refining models, the real-time assessment of water quality can boost the environmental assessment for making better decisions that are more inclusive (Iowa Real-Time Water Quality, n.d.).

3.3.1.2 Types of regression models

The most basic form of regression is linear regression, and this type is often used in the presence of explanatory and dependent variables if they are largely linear. For example, when turbidity is predicated upon the streamflow or suspended solids, this implies that time series can be utilized. However, some relationships are too complex to be covered by linear regression.

In the case of nonlinear relationship, polynomial regression can be used. And in the more advanced settings, multiple regression models contain two explanatory variables or more that seek to make better predictions and incorporate the interactions between those variables. The model is executed in the estimation of water quality in moving water systems such as rivers and lakes, which have several factors controlling the quality of water simultaneously.

3.3.1.3 Predicting pH and turbidity

To predict pH, some of the variables, concentrations of organic matter, temperature, dissolved oxygen, etc., are regressed against any measure of acidity or alkalinity in water. To predict what could happen in the water in response to pollution or changes in the environment, the relationships should be known. For example, a rise in organic matter usually results from agricultural runoff in some regression model or other. It was hypothesized that an increase in organic matter would affect water pH in each stream or lake. Turbidity, based on the suspended particulate material, can be associated with regression models that will provide calibration functions to quantify the parameter of water clarity. Rainfall storm intensity into a stream and concentration of sediments are forms of data that usually inform the other kinds of factors input into these models. For example, rainfall storm event disturbance introduces higher suspended sediments with runoff, resulting in low water quality during an increase in turbidity quantified through regression modeling.

3.3.1.4 Challenges and considerations

Regrettably, previously mentioned regression models are helpful in making predictions about water quality. However, they still face some limits. Although beneficial, the regression model needs to rely on good data conditions. The prediction error rests on how good the data fed in the model are and, therefore, how representative they are. Since the quality of water spreads from one region to another and from one season to the next large extent, it is necessary to adjust the model such asymmetry. External influences, say, from human activities or climate change, can also bring considerable noise in data; thus, the model may face even more challenges. Multicollinearity is another problem that can plague regression, where explanatory variables can be heavily correlated. Such a situation will not allow the model to explain the model contributing independently to the factors. In this situation, one should handle model tuning very carefully and regularization techniques such as ridge regression or lasso region to avoid such errors.

3.3.2 Deep Learning in Water Monitoring

3.3.2.1 Unsupervised learning in water quality monitoring

Unsupervised learning is solid in WQM in that it can detect concealed-regularity patterns, relationships, and outliers in a huge dataset without the use of any labeled data. Unlike supervised learning, which depends on prelabeled data, unsupervised methods postulate unexploited data space for the discovery of evolved insights into complex water dynamics, point sources of contamination, and seasonal/spatial trends. These types of methods can include clustering and anomaly detection, some of the methods that aid in understanding water quality.

Deep learning (DL) and generative models have elevated unsupervised learning for WQM. The methods facilitate the tasks of dimension reduction, cluster analysis, and anomaly detection necessary for the identification of emerging contaminants, spatial trends, and pollution sources. The most solid contribution of DL to WQM, which helps in early warning systems and better utilization of resources, is for prototypical water quality parameters for assessing anomalies or outliers.

3.3.2.2 Neural networks for anomaly detection

Water quality management may also employ the use of neural networks, especially in detecting anomalies, wherein the networks are able to detect abnormal patterns in the water quality data. It is invaluable in locating pollution sources or sudden changes in water quality that cannot be easily noticed. It is possible to teach neural networks to detect minor changes in parameters such as pH, turbidity, and dissolved oxygen, which are all indicative of increased contamination or environmental changes made by the individuals. The autoencoder, for example, is a type of network that is used primarily for data compression and easier encoding and that employs high reconstruction errors for spotting anomalies.

3.3.2.3 Clustering techniques and PCA

Two of the widely used methods of unsupervised learning in WQM are principal component analysis (PCA) and k-means clustering. PCA is used for reducing dimensions to reveal hidden factors and intercorrelation among several water quality parameters. Its name is e-means clustering because it groups data together according to patterns to make potential detection of outliers or hotspots easier. These two methods have been used in water quality studies throughout the world, including application in river basins and groundwater studies that identify sources of contamination and then finally facilitate tracing linear and temporal patterns. When embedding supervised learning algorithms into IoT-based WQM systems, it enables them to detect and act to manage water resources before time, forming a comprehensive solution to water quality and sustainability (Table 3.2 and Figure 3.6).

3.3.3 AI for Predictive Maintenance

PdM uses the port environment with real time data in addition to historical patterns. This prevents equipment failure before it actually happens' leading to unplanned downtimes at interventions. In contrast to statically preventive maintenance cycles, PdM manages with dynamic trends; PdM tries to increase its maintenance level with the real condition of an asset, especially in mission-critical infrastructures, such as those suffering from failure of equipment due to security concerns related to service delivery.

TABLE 3.2 Machine learning techniques for WQM: a table overview

REFERENCES	*YEAR*	*TECHNIQUES*	*PURPOSE*
Jamshidzadeh et al. (2024)	2024	SVM, BiLSTM	TDS, EC
Khullar and Singh (2022)	2024	SVM, RF, linear regression, naive Bayes, and DT	Predicting water needs in agriculture
Uddin, Nash, Rahman, and Olbert (2023)	2023	GPR	Predicting WQI
Hu et al. (2023)	2023	PCR, resilient backpropagation (RPROP), generalized regression neural network (GRNN), bidirectional recurrent neural network (BRNN), RF, SVR, GPR, MLR, least absolute shrinkage, and selection operator (lasso)	Predicting when disinfection byproducts (DBPs) will occur.
Omeka (2024)	2023	MLP-ANNs, MLR	Predicting WQI
Narita, Matsui, Matsushita, and Shirasaki (2023)	2023	RF, XGBoost, LightGBM	Predicting the detectability of pesticides in surface water
Chen et al. (2023)	2023	RF, GBRT, XGBoost, deep neural network (DNN)	Predicting chlorophyll (Chla), TUR, ammonia nitrogen
Khullar and Singh (2022)	2022	Bidirectional LSTM(Bi-LSTM)	Predicting BOD and COD.

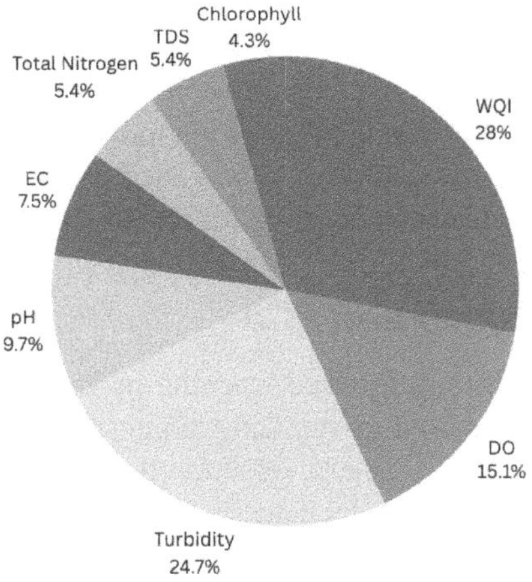

FIGURE 3.6 Percentage distribution in WQM (predicted pollutants).

3.4 CURRENT APPLICATIONS OF AI-BASED PREDICTIVE MAINTENANCE IN CRITICAL INFRASTRUCTURE

3.4.1 Energy Sector

AI predictive maintenance is currently the norm within the energy sector, particularly on renewable assets such as solar panels and wind turbines. These types of assets generate high volumes of operational data that can be processed using AI models like recurrent neural networks and DL to project component failure beforehand, for example. AI-based data analytics applications even enable monitoring for various turbine components like blades and bearings to seek out potential defects beforehand, along with reducing maintenance costs and optimizing operational efficiency.

3.4.2 Technical Feasibility Assessment

In the first step for enabling AI-based Predictive Maintenance (PdM) systems, it is relevant to understand if the technical infrastructure can support these systems. One of the most important factors is data, as the operational data taken from sensors is required to train the AI models, as well as to make sure that the model is effective in taking judgments. Data quality should also be as good, as only real and high-quality data can result in reliable and accurate predictions by AI, whereas poor data would lead to the inaccuracy of the final output and jeopardize the reliability of the system. In this line, it is necessary for an overall view of the computation resources to determine if sufficient computing power exists for real-time processing and analysis of Big Data: a prerequisite for a successful AI PdM. Without these fundamental elements, any incorporation of AI-driven PdM could never achieve the intended result.

3.4.3 Economic Analysis

It is of paramount importance in the justification of any investment for AI-based PdM that an economic analysis is performed. Thus, the upfront expenses, which involve the cost of hardware, software, and the training of personnel, have to be taken into account to mount the framework. Meanwhile, the projected future benefits that PdM can offer should be weighed against potential operational savings from PdM, such as reducing maintenance costs and downtime to internally long-term financial benefits. Also, it is inarguably imperative to conduct comprehensive Return on Investment (ROI) calculations, considering that the same has to dangle a total cost-benefit analysis over the life of the asset to ascertain ROI. Estimation of the ROI enabled by such an evaluation tool offers organizations the ability to make informed decisions regarding the implementation of AI-based PdM.

3.4.4 Regulatory and Safety Considerations

Safety standards must be conformed to for the right working of AI systems. Then, AI-based systems should comply with statutory requirements of data safety and operational safety. This ensures that the systems work within norms and secure information that is sensitive. Apart from this, it is also needed that AI systems are transparent since it allows the operators to know how the prediction is arrived at and then they can trust the outcome as predicted by the model. Lack of transparency leads to apprehension in making the operators depend upon AI-driven insights, which can ultimately lessen the effect of the system. It is not sufficient to fulfil these minimum standards because in that case AI systems can be implemented effectively and safely. This not only ensures users acceptability but also achieves compliance in totality.

3.4.5 Pilot Testing and Scalability

It is by conviction that pilot testing of the ML application should be performed to assess the feasibility of the large-scale deployment of AI-based PdM strategies before moving on to the next stage. The test should preferably begin with a limited number of machines or some geographical location in order to assess whether the system produces desired results, if any issues arise, or if the model needs modification. Furthermore, this systematic strategy makes it possible to correct the issues with minimum risk. Adding to the previous point, the system should possess sufficient scalability so that it can be spread across the whole infrastructure without compromising performance, as observed in most IT systems.

3.4.6 Water Quality Monitoring

AI has been deployed in the process of evaluating and forecasting water quality key indicators: this includes several measures such as pH, DO, and different types of pollutants. These models rely on remote sensing and sensor network inputs and provide instantaneous determinations, which are extremely important for water resource management and hygiene purposes. The AI-based monitoring of environment protection is not limited to traditional spheres like bird and animal migration (through the usage of GPS regarding man-made projects) but it is also extended to a more promising area of searching for particular chemical compounds. Health protection is a major consideration driving the advancement of IoT technology in that it complements AI in health checks of contamination, which is not just restricted to management of water quality only but generally in all facets of public health (Alotaibi & Nassif, 2024).

3.5 USE CASES OF AI AND IOT IN WATER MONITORING

3.5.1 Water Quality Monitoring in Agriculture: Use Cases of AI and IoT

Water management is necessary to achieve an optimal crop yield. Farmers utilize advanced systems which irrigate automatically through IoT tech. These systems can in real time track soil moisture, crop needs, and weather; these data are harvested using sensors and then communicated using various tech like Wi-Fi so that farmers remotely manage irrigation systems.

AI helps these systems by data analysis, making precise recommendations for water management. IoT sensors monitor temperature, moisture, and environmental factors, while AI models use this data to predict when irrigation is needed, how much water is required, and when to stop. AI can provide more accurate predictions and insights, making agriculture more sustainable and efficient.

3.5.2 AI-Based Water Quality Monitoring

In agriculture, water quality is typically outdone by excellent irrigation management. Major parameters such as pH, dissolved oxygen, and pollutants are tracked in real time with the help of AI-based models in conjunction with IoT systems. Assembling sensor into irrigation systems or water source provides continuous actual data, which helps AI to detect early issues and before they damage crops by harmful water and learn about the clean crop.

By observing environmental parameters, AI could identify potential risks of contamination and would be able to signal the farmer to take anticipatory steps. This reduces the scale of physical inspections and, in turn, enhances water management. AI-powered monitoring systems monitoring technology is developing day by day and will be more affordable and far-reaching so as to help farmers manage water resources efficiently while improving crop health.

3.5.3 AI-Based Water Conservation Strategies

AI-based irrigation systems focus on the exploitation of favorable technology in water usage. Thus, in terms of analyzing and optimizing the previously set irrigation water schedules with ML algorithms, data-driven irrigation systems are getting adjusted with time for enhanced efficiencies. Also, forecasting models are employed to give predictive support to proactive decision-making based on monitoring variables such as the need for water for crops, rainfall, and drought risk perception. They give prediction results for automatic control systems adjusting irrigation levels through real-time sensing, thus ensuring the delivery of water in the right amount according to the individual requirements of crops. When these are coupled, it assists in making the entire water management system sustainable by reducing the overuse of water, thus improving crop productivity (Figure 3.7).

3.5.4 Smart Water Management in Agriculture

Water conservation most significantly underpins the use of efficient land, and IoT has begun to revolutionize crop selection, irrigation management, and disease detection. Such practices are combined smart farming alongside AI-powered analysis and live monitoring, which can immediately save a lot of water for farmers, thus cutting waste. Some crops consume a lot of water, while others do not. Rice and sugarcane drink a lot, but millet and beets can make do with less. The possibly best irrigation instruction is now incorporated in very intelligent crop advice systems checking soil moisture, weather trends, and water sources for the right crops that should be grown in a certain area. Good advice saves water and immediately increases the quality, if not quantity, of the harvest. On top of that, matching irrigation methods for each crop can make water use more efficient.

3.5.5 Leaf Disease Detection and Water Conservation

IoT-based disease detection provides multiple advantages for the agricultural sector and significantly enhances efficiency and sustainability. Diseases are detected at the earlier stages, which minimizes the loss of crops and decreases the application of excessive pesticides. Through real-time monitoring of water quality and soil moisture levels, the need for water is optimized, as irrigation is improved. Automation

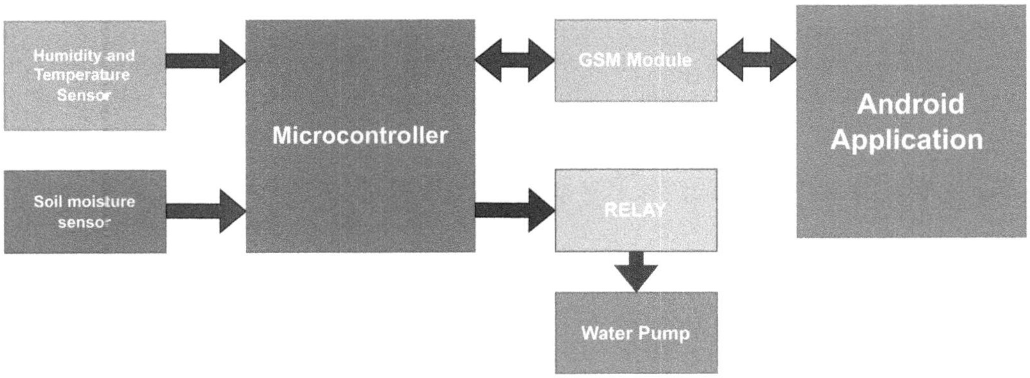

FIGURE 3.7 Architecture of water management system.

of data collection and analysis reduces the manual effort farmers must put into the traditional method of performing day-to-day tasks, thus unfastening them to prepare for other admittedly more critical activities. Additional advantages are achieved when merging disease detection with water management, among which are lesser water use, less operational cost, and increased productivity.

3.5.6 Web-Based and Mobile Applications

AI-powered web and mobile applications can provide farmers with real-time data about the humidity of soil, when they should water their crops, and leaf content. They help in comprehensively screening out non-legitimate users by promoting strategic water and other resource use practices. Thus, sustainable agricultural methods can be now actively embraced since, through AI, there is scope for irrigating the plot without waste of water and this helps in water-saving agriculture.

3.5.7 Crop Water Demand Prediction

ML algorithms analyze historical climate data, crop growth patterns, and soil conditions to predict future water needs. This accurate water demand forecasting enables farmers to plan irrigation cycles more efficiently, minimizing water wastage and reducing water stress on crops. By leveraging predictive insights, farmers can optimize water usage, improve crop health, and enhance overall agricultural sustainability.

3.5.8 Crop Health Monitoring

Drones, cameras, and IoT sensors are used to capture high-resolution photos of crops grown at different stages. AI-based image analysis allows early response to be taken, given that agronomic pests, nutrition deficiency, and sicknesses can be easily detected. This preventive strategy of addressing problems before they escalate is a guarantee that the crop produce is maintained, and resource utilization is optimized aside from the control of water and other wastages.

3.5.9 Decision Support Systems

AI-based decision support systems help farmers in agricultural water management through IoT data and predictive analytics. Such systems give advice on irrigation timings, crop types, and fertilizers, which results in enhanced efficiency and output (Geetha Rani et al., 2024).

3.5.10 Urban and Smart City Water Management

The rapid pace in which urban areas are being highlighted that the cities are constantly dealing with the problem of allocating and using water efficiently. Other physical technologies such as smart water grid (SWG) typically include the most recent IoT, AI and advanced data analysis technologies, which offer wise management and adaptation to water resources to cities. Smart water grids take it one step more to use real-time monitoring, forecasting, and automatic control strategies for efficient use of water, and there is also a better decrease in water waste to achieve climate change policies. Some examples are presented below.

3.5.10.1 Korea Water Resources Corporation & IWRA Collaboration

With the support of the International Water Resources Association (IWRA), the Korea Water Resources Corporation has established a cutting-edge Smart Water Management (SWM) system using information and communication technology (ICT) to track and analyze water-related data in real time. The cutting-edge technology uses satellite data, GIS monitoring engines, and IoT sensors to offer automated, AI-driven feedback for resolving major water management issues. Through AI-based decision-making, the SWM system tackles issues such as leakage detection, WQM, and flood handling effectively and sustainably, leading to better efficient and sustainable management of water resources.

3.5.10.2 *Verizon grid wide intelligent water solutions (USA)*

Verizon has brought a cloud-based smart water-metering solution, which can suit small towns all over the Southeastern USA. What's more, this intelligent system is based on a meter that can provide entire water consumption in real time. It can also detect leaks and other anomalies, using advanced algorithms of ML. The solution enables resource optimization by efficient multipoint secure communications in the water supply network, thereby aiding in the distribution of water and minimizing wastage needed in every single residence. This, in addition to greater water management, also encourages sustainability and salvages costs and benefits for the society.

3.5.10.3 *Global smart water management initiatives*

In Mexico, a project called PUMMAGU aims to address water supply and quality issues using a water supply smart infrastructure. It is said that South Korea has developed a concept of hydrointelligent toolkit, designed to monitor hydrological data, rainfall, flood risk maps, and groundwater level monitoring for effective city water supply. These cases, in turn, demonstrate how inventive technologies can contribute to the use of water in communities, given the improvement in risk and environmental pressures.

3.5.10.4 *Future outlook: sustainable and resilient water management*

SWG plays an important role in making smart, flexible, and skilled, yet durable, supporting cities, energy-saving and climate flexibility through efficient adaptation of water distribution. SWG utilities, coordinated water consumers, and infrastructure through smart analysis of data enhance costs significantly, adapt businesses, and offer increased service distribution. As climate change, urbanization increases a sharp rate, and water speed is lacking, the future will depend on much more SWM system, in which SWG is led. They are essential for creating water-safe smart cities that are flexible, permanent water resources to increase the population among environmental challenges.

3.5.11 Industrial Water Monitoring

3.5.11.1 *AI models for detecting industrial pollutants*

Applications of AI include several advanced applications, including the development of the future model of algae blooming, development of neural network to identify specific pollutants such as toxic metals and infections, and automatic decision-making systems (Rajitha et al, 2024). The application of AI systems is increasing in water quality management, particularly for both predicting and reacting to events through monitoring. The water resources management system provides advance monitoring for water resources, and it is easy to detect potential risks in an early stage, thus reducing risks for the public and environment for environment.

3.5.11.2 *IoT-based automation in industrial water systems*

The V-NET software covers the V-Box device, and the V-Box device provides an online setup interface of web-based. Data parameters such as real-time data, alarms, and history data are set on V-Box devices. According to a IoT cloud server of System IoT, the V-BOX is self-spanning and controlled by the server.

The combination of AI technology with the IoT system produces a more efficient water treatment process. A well-structured ML algorithm can check real-time sensor reading and developing complex strategies for detection and deficiency in industrial water systems (Forhad et al., 2024).

3.6 CHALLENGES AND LIMITATIONS

IoT-based water monitoring systems collect and send data using network devices such as sensors, actuators, and communication modules. These systems are unsafe for cyberattacks, even if they provide real-time monitoring and data-powered insights. The water management system can be seriously in danger from IoT network weaknesses that endanger the availability, privacy, and integrity of important data.

3.6.1 Vulnerabilities in Water Monitoring Systems

Integration of sensor technologies in the IoT includes various challenges in the water monitoring system, mainly interoperability, and data security and privacy. Scalability is also an important issue, as in addition to many devices, network surcharge, delay problems, and ineffective data management can be created. In addition, it is important to protect against data security and privacy violations, as unauthorized access, cyberattacks, and data theft attacks can endanger the reliability and privacy of water monitoring systems. All these challenges can be solved by implementing standard communication protocols, safe encryption algorithms, and sophisticated cybersecurity designs (Rajitha et al., 2024). Security weaknesses in IoT-based water monitoring systems are usually caused by weaknesses in hardware and software elements. Manufacturers transfer safety risks to consumers by launching devices with hardware barriers, unpacked software, and old operating systems, which may be a possible entry point for all cyberattacks. Manufacturers can avoid such weaknesses by ensuring safe software development practices, establishing strong firmware updates, and using good authentication and encryption practices (Jan, Min-Allah, & Düştegör, 2021).

3.6.2 Encryption and Secure Communication Protocols

The IoT-based water monitoring system requires wireless connection technology to enable simple data interchange. They facilitate data flow between the controller and cloud-based data collection as well as the sensor and the controller. Zigbee, which provides low-power and short-distance connections; universal asynchronous receiver-transmitter (Uart), a famous serial communication protocol; global system for mobile communication (GSM); general packet radio service (GPRS), supporting data transmission on cellular network; and Wi-Fi modules supporting IoT offer high-speed internet connectivity to monitor the real-time monitoring of data and remote accessibility. The option of an efficient communication technology depends on the power efficiency, data range for transmission, bandwidth requirements, and the ideas of scalability of the system (Geetha & Gauthami, 2016). IoT effectively addresses safety challenges in based water monitoring systems, researchers recommend implementing sophisticated encryption techniques, well-established security protocols, and scalable system design. Encryption technology ensures that the data transmitted between sensors, controller, and cloud storage is protected from confidential and unauthorized access (Rajitha et al., 2024).

3.6.3 Data Scarcity and Model Training Issues

AI algorithm performance is often forced by the limited availability of training data in several studies found in literature. Many research papers have used only a small number of samples, which can negatively affect the normalization capacity of the AI model. Due to insufficient data, ML algorithms struggle to catch any complex mentor, leading to overfitting, low accuracy, and unreliable predictions when applied. A more diverse dataset is required to improve the strength of the model, increase future-stating accuracy, and ensure reliable real-world applications in different fields. To address these limitations requires a collection of comprehensive datasets, with the use of data enlargement techniques-the implementation of the implementation of advanced ML approaches such as transfer learning and ensemble methods to compensate for data shortage (Tiyasha, Tung, & Yaseen, 2020).

3.6.4 Complexity of AI Model Integration with IoT Devices

The integration of AI models with the IoT platforms is difficult because they have huge computational requirements, demand for high processing, and limited adaptability regarding the change in the environment in real time. The AI model, especially based on deep education, requires heavy computing machinery to process abundant data, and therefore embedding these models in computationally resource-limited IoT devices and embedding of these models in efficient and effective etiquette is challenging. In addition, most of the AI techniques used for assessing water quality are computationally intensive and require high-end hardware and cloud computing, which may limit the delay in real-time monitoring applications. The second main challenge is the overfitting trend of these models so that they normalize on training sets but struggle to normalize the new. Overfitting will result in wrong predictions and inconvenient decision-making, especially in dynamic conditions where the condition of water quality often changes (Pérez-Beltrán, Robles, Rodriguez, Ortega-Gavilán, & Jiménez-Carvelo, 2024).

3.6.5 Infrastructure and Cost Barriers

Challenges in implementing IoT in rural or underdeveloped areas: Implementation in rural and poor areas is challenging due to insufficient infrastructure, poor Internet use, and expensive cost of IoT adoption. Another major obstacle is the cost of establishing and maintaining the IoT network, including sensors and cloud infrastructure.

3.6.6 High Costs of AI-Driven Monitoring Systems

AI-powered WQM systems require high investment in cloud storage, computing hardware and professionals. They require strong data management to support real-time sensor data processing and AI-operated analysis. Heavy deployment expenses, sensors, network equipment, and maintenance create economic barriers, especially for broad adoption (Rajitha et al., 2024). Although IoT-based WQM is a lot of profit but will always have problems with effectiveness and reliability. Regulation compliance also increases complexity, in which health standards should be followed along with the environment. These problems can be handled to enable energy-efficient sensors, flexible networks, flexible security, and uninterrupted integration using a competent regulatory environment.

3.7 FUTURE TRENDS AND OPPORTUNITIES

AI, ML, and IoT are completely a major component of future development under water quality supervision (Rajitha et al., 2024). Using AI model is being used more and more to analyze large data, find pollutants, and improve water treatment processes, which reduces the need for human intervention by increasing accuracy (Rajitha et al., 2024). Integration of remote sensing and geographic information system (GIS) increases the spatial analysis of sources of pollution and benefits regulatory bodies in policy development (Pérez-Beltrán, Robles, Rodriguez, Ortega-Gavilán, & Jiménez-Carvelo, 2024). The combination of analytics is going to undergo a revolution, which protects both public health and environment. The combined capacity of the IoT has been exposed by the integration of sensor technologies to the ability to bring revolution in data collection, analysis, and applications in industries. With the emphasis on the development of significant technical and infrastructure, it examines the infection in the integrated IoT network from the standalone sensor. Using a multiguardian approach supported by ML algorithms provides a broader perspective on water quality (Rajitha et al., 2024). Measures to monitor water quality include turbidity, portraying the optimal characteristics of water sources (Zulkifli et al., 2022).

3.7.1 Potential of Edge Computing for Faster Data Processing

The combination of edge computing with IoT has reduced delay and enhanced performance and moreover also reduced the reaction time and dependence on cloud services. Studies have shown how sensors and computing allowed the power to rapid data collection and analysis (Rajitha et al., 2024). The technique of processing data locally before sending it over is suitable for remote and real-time monitoring applications as this system improves reliability and reduces energy consumption (Rana et al., 2023). In addition, progress in the edge computing models has allowed automated responses and timely intervention in environmental management systems. The AI is expected to revolutionize the real-time analytics dependent industries, making the processes more scalable, safe, and energy-efficient.

3.7.2 Benefits of Decentralized AI Models in IoT Ecosystems

The decentralized AI model has helped distribute real time and make decisions by going with centralized cloud infrastructure. This has increased the flexibility of the system and has reduced the congestion of any network that improves security. According to research, decentralized AI can increase productivity by dividing processing demands between edge equipment, which will shorten the processing time (Nallakaruppan et al., 2024). The decentralized AI model enabled the localized analysis of data, along with reducing dependence on distance servers, allowing continuous monitoring of significant water parameters. Through better scalability, the method has enabled the system to adapt to external conditions (Zulkifli et al., 2022).

3.7.3 Securing Water Data and Transactions Using Blockchain

Blockchain technology increases water data security and transparency in transactions using a tampering-proof, decentralized account book system. The use of blockchain in WQM systems ensures that all the data collected are proofs, which enable pollution sources and real-time tracking of compliance. IoT-A research study on based water monitoring systems emphasizes the need to use blockchain in the protection of environmental data such as water resource management can remain transparent and verified

(Zulikifali et al., 2022). Without middlemen, blockchain applications in water rule have enabled stakeholders to monitor and audit water use, contamination levels and treatment processes (Lin et al., 2020).

3.7.4 Potential Applications in Water Resource Trading and Rights

Blockchain enabled the possibility of a decentralized and safely trading water assets, eliminating the requirement of central regulatory bodies, and facilitating colleague-to-colleague transactions. The blockchain enables effective tracking and enforcement of allocation in real time by digitizing contracts and water rights. Blockchain makes transparent water trade platforms easier for high water scarcity areas, where municipal, industrial, and agricultural users can buy or sell water rights based on credential use information. The irreversibility of the blockchain does not guarantee no false claims, and smart contracts have facilitated automatic settlements based on pre-ended terms (Lin et al., 2020). With increasing water scarcity, blockchain-based trading systems provide a permanent solution for resource distribution, providing equal and responsible access to water resources.

3.8 CONCLUSION

Under water quality supervision, AI and IoT, have changed the traditional method for real-time data acquisition, forecasting analytics and self-sufficient decisions. The IoT sensor has enabled the wireless network on cloud platforms (LPWAN, Wi-Fi, Zigbee, Cellular) to feed the most parameters of pH, turbidity, temperature, and dissolved oxygen in real time to feed back on the wireless network (LPWAN, Wi-Fi, Zigbee, Cellular). AI-operated algorithms analyze this information to identify discrepancies, forecast incidents of contamination, and optimize water management. Applications include accurate irrigation in agriculture to smart city water network and industrial contaminated monitoring, efficiency, reducing water loss and completing environmental control. Edge computing, DL (CNNs, LSTM), and predictive maintenance also enhance the performance of the system by enabling decentralized, rapid data processing, and more accurate forecast of water quality patterns. AI-IT-based water monitoring is still significant disadvantages, including cybersecurity hazards, deployment costs, and challenges in integrating the AI model with IoT devices. Data privacy, interoperability, and resource limitations in rural areas move forward. However, upcoming technologies such as blockchain for secure exchange of water data, AI models that increase scalability, and are bound to run the AI-powered decision-making systems with even more progress. As climate change, industrialization, and water scarcity becomes at the forefront of global issues, A-IIT integration will play a fundamental role in facilitating permanent water resources management, maintaining public health, and increasing environmental flexibility.

REFERENCES

Akhter, F., Siddiquei, H. R., Alahi, M. E. E., Jayasundera, K. P., & Mukhopadhyay, S. C. (2022). An IoT-enabled portable water quality monitoring system with MWCNT/PDMS multifunctional sensor for agricultural applications. *IEEE Internet of Things Journal*, *9*(16), 14307–14316. https://doi.org/10.1109/JIOT.2021.3069894

Singh, R., Baz, M., Gehlot, A., Rashid, M., Khurana, M., Akram, S. V., … & Alghamdi, A. S. (2021). Water quality monitoring and management of building water tank using industrial internet of things. *Sustainability, 13*(15), 8452. https://doi.org/10.3390/SU13158452

Alotaibi, E., & Nassif, N. (2024). Artificial intelligence in environmental monitoring: In-depth analysis. *Discover Artificial Intelligence*, *4*(1), 1–26. https://doi.org/10.1007/s44163-024-00198-1

Budiarti, R. P. N., Sukaridhoto, S., Hariadi, M., & Purnomo, M. H. (2019). Big Data Technologies using SVM (case study: Surface water classification on regional water utility company in Surabaya). *Proceedings-2019 International Conference on Computer Science, Information Technology, and Electrical Engineering, ICOMITEE 2019*, 94–101. https://doi.org/10.1109/ICOMITEE.2019.8920823

Chen, P., Wang, B., Wu, Y., Wang, Q., Huang, Z., & Wang, C. (2023). Urban river water quality monitoring based on self-optimizing machine learning method using multi-source remote sensing data. *Ecological Indicators*, *146*, 109750. https://doi.org/10.1016/J.ECOLIND.2022.109750

Daigavane, V. V, & Gaikwad, M. A. (2017). *Water Quality Monitoring System Based on IOT* (Vol. 10). Retrieved from https://www.ripublication.com

Essamlali, I., Nhaila, H., & El Khaili, M. (2024). Advances in machine learning and IoT for water quality monitoring: A comprehensive review. *Heliyon*, *10*(6), e27920. https://doi.org/10.1016/j.heliyon.2024.e27920

Forhad, H. M., Uddin, M. R., Chakrovorty, R. S., Ruhul, A. M., Faruk, H. M., Kamruzzaman, S., Sharmin, N., Jamal, A. S. I. M., Haque, M. M., & Morshed, A. M. (2024). IoT based real-time water quality monitoring system in water treatment plants. *Heliyon*, *10*(23), e40746. https://doi.org/10.1016/j.heliyon.2024.e40746

Geetha Rani, E., Tukkoji, C., Anusha, D., Dhanalakshmi, M., & Bharath, B. (2024). Water management for IoT-based smart agriculture using machine learning algorithms. *Lecture Notes in Electrical Engineering, 1220 LNEE*, 351–363. https://doi.org/10.1007/978-981-97-4650-7_26

Geetha, S., & Gouthami, S. (2016). Internet of things enabled real time water quality monitoring system. *Smart Water*, *2*, 1. https://doi.org/10.1186/s40713-017-0005-y

González, L., Gonzales, A., González, S., & Cartuche, A. (2024). A low-cost IoT architecture based on LPWAN and MQTT for monitoring water resources in Andean Wetlands. *SN Computer Science*, *5*(1), 1–13. http://dx.doi.org/10.1007/s42979-023-02515-4

Guezouli, L., Guezouli, L., Djeghaba, M. B. E., & Bentahrour, A. (2024). IoT and AI for real-time water monitoring and leak detection. *Journal of Renewable Energies*, *27*(2). https://doi.org/10.54966/JREEN.V27I2.1210

Hartono, R., Safi'ie, M. A., Yoeseph, N. M., Bawono, S. A. T., Hartatik, & Aziz, A. (2022). Improved data transmission of smart water quality sensor devices in Bengawan Solo River with LoRa. *APICS 2022–2022 1st International Conference on Smart Technology, Applied Informatics, and Engineering, Proceedings*, 171–174. https://doi.org/10.1109/APICS56469.2022.9918709

Hu, G., Mian, H. R., Mohammadiun, S., Rodriguez, M. J., Hewage, K., & Sadiq, R. (2023). Appraisal of machine learning techniques for predicting emerging disinfection byproducts in small water distribution networks. *Journal of Hazardous Materials*, *446*, 130633. https://doi.org/10.1016/J.JHAZMAT.2022.130633

Jha, B. K., Sivasanikari, G. G., & Venugopal, K. R. (2020). Cloud-Based Smart Water Quality Monitoring System using IoT Sensors and Machine Learning. *International Journal of Advanced Trends in Computer Science and Engineering*, *9*(3), 1419–1432. https://doi.org/10.30534/ijatcse/2020/141932020

Iowa Real-Time Water Quality. (n.d.). Retrieved February 21, 2025, from https://nrtwq.usgs.gov/ia/methods/

Jamshidzadeh, Z., Latif, S. D., Ehteram, M., Sheikh Khozani, Z., Ahmed, A. N., Sherif, M., & El-Shafie, A. (2024). An advanced hybrid deep learning model for predicting total dissolved solids and electrical conductivity (EC) in coastal aquifers. *Environmental Sciences Europe*, *36*(1), 1–19. https://doi.org/10.1186/s12302-024-00850-8

Jan, F., Min-Allah, N., & Düştegör, D. (2021). IoT based smart water quality monitoring: Recent techniques, trends and challenges for domestic applications. *Water*, *13*(13), 1729. https://doi.org/10.3390/w13131729

Khullar, S., & Singh, N. (2022). Water quality assessment of a river using deep learning Bi-LSTM methodology: Forecasting and validation. *Environmental Science and Pollution Research*, *29*(9), 12875–12889. https://link.springer.com/article/10.1007/s11356-021-13875-w

Lakshmikantha, V., Hiriyannagowda, A., Manjunath, A., Patted, A., Basavaiah, J., & Anthony, A. A. (2021). IoT based smart water quality monitoring system. *Global Transitions Proceedings*, *2*(2), 181–186. https://doi.org/10.1016/j.gltp.2021.08.062

Li, W., Zhang, H., Zhan, T., Zhang, C., Miller, T., Durlik, I., … & Nowy, A. (2025). Integrating Artificial Intelligence agents with the Internet of Things for enhanced environmental monitoring: Applications in water quality and climate data. *Electronics 2025, 14*(4), 696. https://doi.org/10.3390/ELECTRONICS14040696

Lin, Y. P., Mukhtar, H., Huang, K. T., Petway, J. R., Lin, C. M., Chou, C. F., & Liao, S. W. (2020). Real-time identification of irrigation water pollution sources and pathways with a wireless sensor network and blockchain framework. *Sensors 2020, 20*(13), 3634. https://doi.org/10.3390/S20133634

Mahbubur Rahman, M., Bapery, C., Jamal Hossain, M., Hassan, Z., Jamil Hossain, G., & Muzahidul Islam, M. (n.d.). Internet of Things (IoT) based water quality monitoring system. *International Journal of Multidisciplinary and Current Educational Research (IJMCER)*, 2. Retrieved from www.ijmcer.com

Nallakaruppan, M. K., Gangadevi, E., Shri, M. L., Balusamy, B., Bhattacharya, S., & Selvarajan, S. (2024). Reliable water quality prediction and parametric analysis using explainable AI models. *Scientific Reports*, *14*(1), 1–24. https://doi.org/10.1038/s41598-024-56775-y

Narita, K., Matsui, Y., Matsushita, T., & Shirasaki, N. (2023). Screening priority pesticides for drinking water quality regulation and monitoring by machine learning: Analysis of factors affecting detectability. *Journal of Environmental Management*, *326*, 116738. https://doi.org/10.1016/J.JENVMAN.2022.116738

Omeka, M. E. (2024). Evaluation and prediction of irrigation water quality of an agricultural district, SE Nigeria: An integrated heuristic GIS-based and machine learning approach. *Environmental Science and Pollution Research*, *31*(41), 54178–54203. https://doi.org/10.1007/s11356-022-25119-6

Pérez-Beltrán, C. H., Robles, A. D., Rodriguez, N. A., Ortega-Gavilán, F., & Jiménez-Carvelo, A. M. (2024, March 1). Artificial intelligence and water quality: From drinking water to wastewater. *TrAC-Trends in Analytical Chemistry*, *172*. https://doi.org/10.1016/j.trac.2024.117597

Rasmussen, P.P., Gray, J.R., Glysson, G.D., and Ziegler, A.C. (2009). *Guide-lines and Procedures for Computing Time-series Suspended-sediment Concentrations and Loads from In-stream Turbidity-sensor and Streamflow Data*. U.S. Geological Survey Techniques and Methods, book 3, chap. C4, 52 p. (PDF).

Ravitha, A., Aravinda, K., Nagpal, A., Kalra, R., Maan, P., Kumar, A., & Abdul-Zahra, D. S. (2024). Machine learning and AI-driven water quality monitoring and treatment. *E3S Web of Conferences*, *505*. EDP Sciences. https://doi.org/10.1051/e3sconf/202450503012

Rana, R., Kalia, A., Boora, A., Alfaisal, F. M., Alharbi, R. S., Berwal, P., … & Qamar, O. (2023). Artificial Intelligence for surface water quality evaluation, monitoring and assessment. *Water*, *15*(22). https://doi.org/10.3390/w15223919

Saravanan, K., Anusuya, E., Kumar, R., & Son, L. H. (2018). Real-time water quality monitoring using Internet of Things in SCADA. *Environmental Monitoring and Assessment*, *190*(9), 1–16. https://doi.org/10.1007/s10661-018-6914-x

Singh, R., Baz, M., Gehlot, A., Rashid, M., Khurana, M., Akram, S. V., … & Alghamdi, A. S. (2021). Water quality monitoring and management of building water tank using industrial internet of things. *Sustainability*, *13*(15). https://doi.org/10.3390/SU13158452

Tiyasha, N., Tung, T. M., & Yaseen, Z. M. (2020). A survey on river water quality modelling using artificial intelligence models: 2000–2020. *Journal of Hydrology*, *585*, 124670. https://doi.org/10.1016/j.jhydrol.2020.124670

Uddin, M. G., Nash, S., Rahman, A., & Olbert, A. I. (2023). A novel approach for estimating and predicting uncertainty in water quality index model using machine learning approaches. *Water Research*, *229*, 119422. https://doi.org/10.1016/j.watres.2022.119422

Zulkifli, C. Z., Garfan, S., Talal, M., Alamoodi, A. H., Alamleh, A., Ahmaro, I. Y. Y., … & Chiang, H. H. (2022, November 1). IoT-based water monitoring systems: A systematic review. *Water*, *14*. https://doi.org/10.3390/w14223621

Detecting Chemical Contaminants in Water Using AI

4

Dibyarupa Pal, Partha Sarathi Mandal, Ritam Paul, Avishek Kumar Kaser, and Pranabi Maji

4.1 INTRODUCTION

Water is the basic prerequisite for the existence of all lifeforms. But with the progress in industrialization, urbanization, and population expansion, exposure to fresh water has become one of the major challenges facing the world (Tozer, 2023). Pollutants emerge from untreated industrial effluent, sewage outlets, mining activities, farming discharges, and other anthropogenic activities (Lu et al., 2024). Common pollutants of water bodies are commercial pesticides, herbicides, inorganic pollutants, heavy metals, quaternary ammonium compounds (QACs), microplastics, and others. These are charged particles obtained from diverse uses of humans, but with depredation activities. The aquatic ecosystem harbors heavy metals of diverse range, which leads to the death of aquatic life (Karaouzas et al., 2021). Eutrophication, which results from pesticide contamination, may harm fish and other creatures by causing the growth of algae and lowering the amount of dissolved oxygen in the water (Ly et al., 2023). Thus, the identification of contaminants in lakes and rivers may be used for early warning and damage prevention. Chromatography, mass spectrometry (MS), atomic absorption, and spectrophotometry are examples of conventional techniques for assessing the water quality (Albignac et al., 2023; Pouyanfar et al., 2022). Researchers have used the features of pollutants and biomarkers in wastewater sludge to gain insight into the daily activities and general health of the local people. Technological challenges exist along with financial constraints and prolonged testing procedures, which are difficult to execute and require highly qualified trainers to operate them. It becomes highly crucial when the water sample is compounded with complex and advanced micropollutants generated through human activities. These pollutants endanger the lives of humans, aquatic microbiota, and macrobiota of the water system. Thus, it becomes imperative to find a fast, reliable, and efficient technology that can solve the problem. Recently, great interest has been given to microfluidics technology and artificial intelligence (AI), separately as well as synergistically to monitor water quality parameters (Pouyanfar et al., 2022; Cardoso et al., 2023; Aryal et al., 2024). In this advanced microfluidic technology, samples in the range of micron or nanometer scale which are in liquid state are evaluated. When compounded with AI technology, the detection level can be brought under one single umbrella which is scalable, high sensitivity as well as programmable microplatform.

DOI: 10.1201/9781003600718-6

In microfluidics, various tasks can be achieved like chemical testing of the water sample which includes pesticides, hormones, and pharmaceutical drugs by monitoring the nanoquantity and regulating flow rate. The water samples to be tested are initially split into many individual responses in real time, and the data are analyzed concurrently. Thus, microfluidic channels and microvalves are involved in this process dealing with very small quantities of the water sample (Ma et al., 2023; Chen et al., 2023; Mao and Zhang, 2022 Zhai et al., 2024).

A new field of technology called AI aims to broaden our understanding of human intelligence's theories, practices, and application. It was created so that computer systems could gain knowledge, logic, perceive, and comprehend language—tasks that call for human intelligence. AI has advanced quickly in the last few years because of the creation of several neural networks and algorithms. It has already achieved significant progress in the fields related to ecological monitoring, material processing, biomedical science, and agriculture (Manickam et al., 2022; Himeur et al., 2022; Banerjee et al., 2024; Borah et al., 2025a). Without the need for specific programming, computer systems can gather information from data and enhance its efficiency through machine learning (ML), a significant area of AI. Supervised, unsupervised, semi-supervised, and reinforcement learning are among its algorithms. With increasing amounts of data and computing power, deep learning (DL) has also gained popularity as an ML approach. DL models the ordered structure of the human nervous system employing neural network topologies, enabling effective analysis of complicated data. Two crucial areas of DL are convolutional neural networks (CNNs) and recurrent neural networks (RNNs), which are used in image processing and sequence data modeling, respectively (Himeur et al., 2022; Poole, 2022).

The ability to discriminate between significant changes in the measured value of variables and the usual regular and/or periodic variations in such values is crucial when trying to identify contamination situations. Consequently, the usage of some algorithms—most of which are created using probabilistic or randomized techniques—is imperative. In the last few years, artificial neural networks (ANNs), a data-driven model type that emphasizes modeling and prediction, have been effectively used for water quality evaluation in an effort to enhance the water distribution system (WDS). This is because ANNs are based on information and prioritize modeling and prediction. Following the creation of a model using the application of multivariate ANNs, the events were categorized using Bayesian theory (Ahmad et al., 2017). Bayesian belief networks (BBNs) were employed to detect *E. coli* events in combination with conductivity, viscosity, and pH. The binary signals produced by the individual detection of incidents carried out by each sensor were combined with a source recognition model based on a backtracking algorithm to create a method that can be used throughout the entire system. The chance that an incident will occur was calculated by the computing system by analyzing the variations in sensor warnings. Adaptive updating dynamic thresholds can be used to enhance the caliber of the research on the topic. The dimension of the sliding window affects these thresholds, which are established as a function of the sliding window's error standard deviation.

It is important to remember that, despite its many benefits, AI has certain drawbacks. For instance, algorithms must be taught using vast volumes of data, and their effectiveness may suffer from insufficient data collection. This is often an extremely difficult and challenging task. Microfluidic chip technology, which offers the benefits of automated processes, huge scale, and affordability, may surpass the constraints of AI. Combining the two makes it possible to collect data more effectively and provides a strong tool for examining water pollutants.

This chapter's goal is to develop and demonstrate the viability of an intelligent system for tracking water quality indicators that can detect biocontamination early on. The notion of dynamic thresholds, the display of abnormalities found on ANNs, and the evaluation of the interrelated changes of several water quality measures (such as the concentration of free chlorine as well, pH, alkalinity, and total organic carbon (TOC) produced by the contaminant) are further objectives (Lai et al., 2011). All of them are instances of goals that have to do with the contaminant. Additionally, this study's goal is to investigate the connections among the various changes in water quality that have been noted. Over the past 5 years, a number of detection methods have been investigated for the quality of the environment monitoring systems based on microfluidics; the most popular ones are colorimetric, electrochemical, and fluorescence detection. Additional techniques include absorbance-based, chemiluminescence (CL), photoelectrochemical, surface plasmon resonance (SPR), and quartz crystal microbalance.

Conductometry, potentiometry, voltammetry, polarography, amperometry, and coulometry are some of the methods used in electrochemical analysis; each is reliant on certain electrical properties. A working electrode, a counter electrode, and a reference electrode make up the typical three-electrode arrangement for an electrochemical system. Numerous analytes may be identified using their corresponding uniform redox potentials on blank electrodes, which enables the creation of distinct signals. Low detection limits are provided via electrochemical analysis, which can reach the picomole level. Compared to commonly used technologies like colorimetric or fluorescence, which have not yet reached comparable low limits of detection, this is a substantial benefit. Many times, employing transportable and/or reusable electrochemical devices to identify heavy metals provides additional benefits over electrochemical sensing. Many electrodes have been investigated, such as carbon paste electrodes, screen-printed electrodes, disposable glass electrodes, and disposable paper electrodes (Chen et al., 2024). Portable setups, such as smartphone-based systems, portable potentiostats, and handmade electronics, have also been investigated.

Effective monitoring of environmental analytes (mainly in water) is made possible by fluorescence-based environmental monitoring approaches, which involve the development and production of fluorophores that integrate recognition components such as ligands, enzymes, aptamers, etc. These methods identify a particular analyte that is relevant by combining fluorescence detector with a variety of appropriate probes, including rhodamine, pyrene, anthracene, naphthylamide, aminoquioline, bithiophene, etc. A typical method for integrating fluorescent detection into microfluidics is to combine a sample with a specific fluorescent probe designed for the target analytic compound, utilize it using the apparatus, illuminate it to cause fluorescence, and then use a photodetector to measure the intensity of the light that is released. This measurement is directly related to the quantity of the analyte. However, some techniques prepare probes by putting them within the microfluidic channel, where fluorescence is seen as the required analyte flows into the microchannels.

Furthermore, flow-based CL experiments using microfluidic systems have been the focus of current research (Zhang et al., 2022). CL, which is characterized by electromagnetic radiation with a certain wavelength that primarily falls between the visible and near-infrared (vis–NIR) range, is released when a molecule experiences exothermic excitation to achieve the singlet excited state. The degree of brightness of the emitted light can be linked to the quantity of the analyte. In contrast to conventional CL tests, flow-based CL uses programmed, small, and extensively integrated flow detection devices, significantly minimizes sample and reagent use, and reduces analytical time. The introduction of colorimetric systems for ecological analyte detection has exploded in the last 5 years. The majority of studies have concentrated on creating paper-based microfluidic colorimetric detectors as viable methods for low-cost and accessible detection. The sample is directed to a paper pad prepared with a colorimetric reagent for analysis via a flow channel made by adding hydrophobic coatings to both faces of the paper. The sample reacts with the deposited colorimetric reagent when it is driven through the paper pad by capillary action, producing a colorimetric signal. Wax printing, plasma treatment, and photolithography are some of the methods that may be used to include hydrophobic coatings onto paper substrates in order to regulate flow (Chen et al., 2024). Figure 4.1 shows the role of AI and ML to detect and control the pollutants in water and its algorithms.

4.2 MICROFLUIDIC PLATFORM DETECTION TECHNOLOGY

Rapid monitoring and automated technology of many chemical processes are made possible by microfluidics, which has become an achievable solution. These platforms use a range of substrates, including silicon, ceramics, glass, paper, thermoplastics, polydimethylsiloxane (PDMS), and hydrogels,

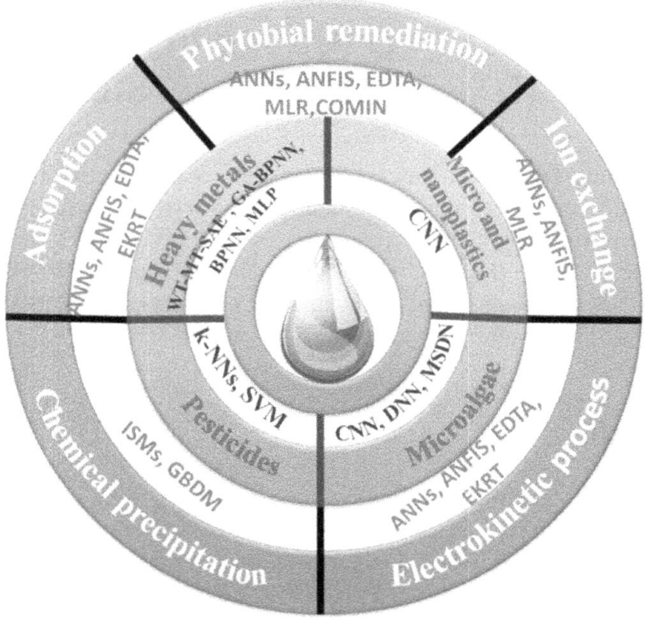

FIGURE 4.1 Role of AI and ML to detect and control the pollutants in water and its algorithms.

to accurately manage microliters of liquid in small flow pathways. Here, samples are pretreated and concentrated at the same time which adds a different magnitude to the microfluidic devices. Furthermore, different combinations of materials increases the adaptability and versatility of the devices in processing the samples. The easy-to-work approach and small volume of sample materials have added to the advantages in the detection and monitoring of water pollutants. Water, being an universal solvent, has inherent fluidic properties like low surface tension and viscosity, which encourages rapid mixing with chemicals and easy flow inside the microchannels. These advantages allow for point-of-use water monitoring in resource-limited areas which are far off from laboratories and also are cost effective in action. Conventional methods include costly approaches like nuclear magnetic resonance (NMR), MS, and gas/liquid chromatography (Albignac et al., 2023; Chen et al., 2023). Thus, utilizing microfluidic systems offers a workable approach to fill the lacunae in conventional procedures, saving money and time.

In case studies relevant to growing agricultural crops such as vegetables (lettuce, coriander, tomato, and other leafy vegetables) and cereal crops, it is significant to monitor the incorporation of micropollutants in foods. These micropollutants impact the nutritional properties of these vegetables and crops which are consumed by humans through the food chain and accumulate in their system (Kamat et al., 2023; Borah et al., 2025b). Moreover, accumulation of chemical contaminants in soil affects the agro-system by changing soil properties like soil health and soil microbiota. It is crucial to monitor these chemical pollutants in the ecosystem for saving different forms of lives. Traditional techniques require heavy instruments and labor- and time-consuming sample collection, which are only functional in laboratory setups. Again application of optimum fertilizer or pesticides is necessary, which can be only possible through real-time and on-site analysis. Microfluidics methods with portable instruments like portable potentiostats, colorimetric devices with smartphone-based color measurement systems, or distance-based detectors, or fluorescence-based microfluidic devices with lightweight fluorometers, etc., are highly demanding at this juncture. Based on this method, several research works have been done to detect analytical substances such as heavy metals, pesticides, phenolic chemicals, and microbes present in these samples.

4.3 ADVANCED WATER POLLUTION CONTROL TECHNOLOGIES

Physical separation, chemical breakdown processes, and biological decomposition are some of the methods used in the literature to manage water contamination. Adsorption, phytobiological remediation, ion exchanges, electrocoagulation, electrokinetic processes, phytoremediation, nano-phytoremediation, and chemical precipitation are a few of the most recent treatment techniques in use. Additionally, new developments in nanotechnology are thought to be affordable that can regulate the many issues that have arisen due to existing therapeutic approaches.

4.3.1 Adsorption

The most well-known water treatment technology is adsorption, which uses activated carbon as a universal adsorbent for eliminating various contaminants from wastewater, the ground, and surface by absorbing dangerous metal ions. The main purpose of adsorption is the removal or reduction of different contaminant levels in water or wastewater that contains dissolved pollutants. One common adsorbent used in the adsorption process to eliminate a variety of contaminants is activated carbon (Modak et al., 2023). According to Alka et al. (2021), adsorbents have been examined and employed as a form of treatment and control method in order to provide an eco-friendly substitute that can eliminate the effectiveness of contaminants. But again the adsorption process, though achievable, cannot be self-monitored. For instance, adsorption of arsenic from wastewater by iron oxide-coated rice husk biochar was successfully reported by Nath et al. (2019) and Lakshmi et al. (2021). The experiment was optimized on the parameters of pH, contact time, adsorbent, and adsorbate dosage using modeling techniques. Though utilizing rice husk biochar was a successful operation, time required for experimental validation could be overcome by the intervention of automated technology like inclusion of AI in water treatment technologies.

4.3.1.1 Applications of AI in adsorption technique

AI plays a key role in developing techniques for restoring contaminated water resources resulting from unauthorized mining operations since it can optimize and analyze the relationships among various factors. Due to the impact of unauthorized mining industry over lands, undergrounds, and waterbodies, the extraction of minerals using mercury leads to ecological harm (Nti et al., 2023). Adsorption is one of the safest techniques for removing metallic compounds and heavy metal contaminants from groundwater.

The purpose of the adsorption procedure is to eliminate a large number of contaminants in the water using absorbent material. The adsorption efficiency is guided by several operating parameters such as temperature, exposure time, pH of the reacting solution, surface area of the adsorbent, dose and concentration of adsorbent, and particle size (Nti et al., 2023). In this scenario, AI model tools like ANNs, multiple linear regression (MLR), and fuzzy inference systems (FIS) are efficacious in monitoring the adsorption processes from heavily contaminated water bodies. For instance, Rambutan peel biochar was developed by Lakshmi et al. (2021) and Wong et al. (2020) to extract Cu(II) from contaminated water sources in the presence of optimized operational parameters like experimental contact duration, temperature, dose, and initial adsorbate concentration. Additionally, Li et al. (2019) employed support vector machine (SVM) technique to design an AI model for Pb (II) removal using EDTA-modified biochar. Since the AI model assists them in establishing appropriate adsorption processes, they all attain an excellent efficiency in removal rate.

4.3.1.2 *Adsorption technology gaps and difficulties*

Adsorption is limited by its low capacity for high contaminant levels, minimal surface area that is not self-monitored (Alka et al., 2021), adsorbent loss, and the high cost of adsorbent renewal. This effectiveness of the adsorption method might be increased by combining it with other deletion technologies.

4.3.2 Ion Exchange

The physicochemical method of ion exchange, which employs solid-phase ions with identical ion numbers as those from contaminated water as a component of its process, is used to eliminate the majority of arsenic contaminants. In other words, undesirable ions in water will just be swapped out for new ones. Electrostatically trapped ions are utilized in ion exchange processes to lower the hardness of water and eliminate impurities from waste or polluted water. This technology's drawbacks include its high cost and the frequent need for regeneration procedures to get rid of all contaminants. Although its benefits are even larger since it leads to both the elimination and reuse of metal objects, for instance, it has an effective removal rate of 97.9% in the pH range from 3.5 to 7 (Nti et al., 2023). The use of ion exchange for removing and restoring metal objects has the ability to contribute to the achievement of both SDGs 3 and 6 by ensuring the availability of clean drinking water and sanitation and improving the well-being of people. Numerous medical issues, including cancer, neurological diseases, and hormonal difficulties, can be brought on by prolonged contact with heavy metals. Ion exchange, on the other hand, can assist increase accessibility to safe water and enhance people's health by eliminating heavy metals from water supplies.

4.3.2.1 *Applications of artificial intelligence (AI) in ion exchange*

Ion exchange is one of the most effective methods for eliminating arsenic contaminants from the exchange process between a solution and an insoluble solid. This method may be used for ion recovery as well as removal. Ion exchange is utilized to mainly replace undesirable ions from the water source with suitable ions in order to remediate water that has been contaminated by illegal mining (galamsey). Top AI technology can be used for this. This technique provides a practical means of isolating, purifying, and eliminating pollutants that include ionic compounds. Ion exchange substances are used in water filtration to eliminate harmful heavy metals, including lead (Pb) and mercury, since illegal mining processes sometimes pollute water. For example, the ion concentration in water may be changed using AI models like ANNs and Adaptive-Network-based Fuzzy Inference System (ANFIS). This would enable water to pass through the ion-exchange substance, which swaps ions with identical charges. This could also result in the more efficient removal of calcium and magnesium ions.

4.3.2.2 *Ion exchange technology gaps and difficulties*

The excessive operating expenses of the ion exchange system are another main drawback. Additionally, it is not self-regenerating since it needs constant regeneration.

Pretreatments such as filtering can be used to remove organic compounds before the waste product is passed through substances since the substances used in this method are costly.

4.3.3 Electrokinetic Processes or Methods

Electrokinetic technology describes the flow of ions and other comparable particles by an electric field, as well as the transfer of water and other fluids with an electric field. In particular, by using electrical adsorptions to remove and isolate contaminants and electrically charged particles from soil and water constituents, this approach aids in the management of contamination of water. Nevertheless, due to the danger of heavy metals and the challenging soluble phases that need to be discussed, the electrokinetic technique is insufficient for the removal of arsenic contaminants (Nti et al., 2023).

4.3.3.1 Applications of artificial intelligence (AI) in electrokinetic procedures or techniques

The electrokinetic method is essential for cleaning up contaminated waterways. The contaminants due to such illegal mines (galamsey) operations include heavy metals like both mercury and lead as well as other ionic chemicals. Comparing electrokinetic methods to other methods, Vocciante et al. (Vocciante et al., 2021) claim that they are very effective and eco-friendly. They maintained that the primary benefit of the method is that electrons are the only reagent required for it to work. However, these electrolytes are required to achieve a quantifiable impact in a reasonable duration along with renewable power usage. The technique, for example, may be used in initial and temporal states and offers improved effectiveness, since the applied electric fields can readily reach contaminants in the water surface or below.

For instance, an electric potential gradient was applied to a section of polluted soil to create low electricity using the electrokinetic remedial technology (EKRT) technique. Thus, even though the polluted soil would not readily infiltrate, an electric potential gradient was used to create a small amount of electricity across a section of it using the EKRT methods. During induction, several chemical, physical, and electrochemical reactions depended heavily on the system's properties and the strength of the generated electricity. Eventually, they could be removed by adsorption, electrodeposition, filtration of the contaminated electrolyte solution, and/or precipitation. This would allow important species to move through the water and into the wells (electrode housing and electrolyte solution).

4.3.3.2 Electrokinetic processes gaps and difficulties

The possibility of having a high pH during functioning may reduce the capacity to eliminate contaminants. The method breaks down heavy elements, including arsenic, into its fluid phase, which makes treatment challenging. During the process, heavy metal elimination also rises.

Combining electrokinetic technology with additional innovations, such permeable-reactive barriers, will increase its effectiveness because it is a relatively recent invention.

4.3.4 Chemical Precipitation

This process of turning elements that dissolve in water into solid components is known as solidification (Nti et al., 2023). Chemical precipitation is the most widely used method for handling contaminants like wastewater or arsenic. Therefore, this procedure helps to transform soluble arsenic into less soluble compounds and remove ions such fluoride, cyanide, phosphate, and phenols from the precipitate (Nti et al., 2023). Alka et al. (2021) have observed that chemical precipitation has been used to treat gold-mining waste using a two-stage nanofiltration method to remove calcium and arsenic.

4.3.4.1 Applications of artificial intelligence (AI) in chemical precipitation

AI may be used in chemical precipitation to help remove heavy metals like lead and mercury, as well as arsenic pollution, from water bodies that result from illegal mining (galamsey) operations. As a water remediation method, chemical precipitation techniques are used to transform soluble contaminants in water bodies into insoluble forms. Based on their observations, AI-based methods may be utilized to identify the contaminants in the accumulated particles for prospective precipitation estimates. This is accomplished by atmospheric reanalyzing and the combination of precipitation data collected from many satellites. For instance, improved precipitation projections across the Upper Blue Nile Basin were produced using AI algorithms (Khan and Bhuiyan, 2021). For upcoming adsorption or ion exchange operations, AI will therefore assist in forecasting the heavy metal elements in the resultant solid particles.

4.3.4.2 Chemical precipitation gaps and difficulties

Large amounts of less thick waste can be produced when hydroxide is used as a precipitant. Water extraction and eventually disposal procedures are hampered by this. Another difficulty is that complex chemicals can prevent metal hydroxides from precipitating if they are present in the solution being used. Colloidal precipitates are created when metal sulfides are present, which leads to filtering issues following precipitation.

The procedure can be enhanced by combining chemical precipitation with ion exchange and nanofiltration methods. The difficulty of achieving the process's environmental emission restrictions can also be eliminated with the introduction of chelating compounds (Alka et al., 2021; Nti et al., 2023).

4.3.5 Phytobial Remediation

Phytobial remediation encompasses the utilization of both phyto- and bioremediation processes of both plants and microbes to mitigate environmental pollution (Alka et al., 2021). The synergistic action of microbes and rhizosphere benefits plants by augmenting their growth, productivity, improved stress tolerance, reduced pathogen attack, and nutrient availability. The process relies on plant-microbial interactions through which several small molecules like sugar, hormones, and amino acids are secreted which are absorbed by the plants to aid in bioremediation. For example, removal of Arsenic from terrestrial and aqueous environment is achieved by phytobial remediation process. Since it benefits the environment while producing long-lasting, efficient processes, the procedure is both inexpensive and highly environmentally friendly. This process does not harm the environment, but has its own challenges and constraints. Phytobial cleanup takes a long period because plants need time to grow before they can start uptaking heavy metals from polluted sources. The root systems are extremely vulnerable to metal concentrations before they acclimatize, which can lead to stunted growth and eventually metal insufficiency.

4.3.5.1 Applications of artificial intelligence (AI) in phytobial remediation

Phytobial remediation can be accelerated with the help of ML and AI systems which are robust, efficient, and process massive amounts of data within a short time. They learn and predict from the data they are fed, finding patterns and making predictions based on those patterns. These systems are extremely intricate, with many layers of neural networks and algorithms (Yaqoob et al., 2023). For example, in the case of arsenic removal from terrestrial and aqueous environments by Phytobial remediation, it is aided by plant growth promoting rhizobacteria (PGPR) growing attached to the roots. PGPR aids the plant synergistically to behave as hyperaccumulators of heavy metals, but they are subjected to seasonal variation and nutrition availability. Similarly, algae and fungi by biosorption and bioaccumulation processes are involved in bioremediation. AI methods assist to detect, track movement, measure bioavailability, examine, and assess plant reactions to heavy metals. This technology can monitor plant development in different polluted environments using AI-based cutting-edge technology like remote sensing tools, sensors, and cameras to track plant development (Kumar et al., 2024).

Random forest model works on several decision trees (DTs)entailing both regression and classification to make a collective decision which increases prediction accuracy (Speiser et al., 2019). It is very flexible, scalable, interpretable, and can manage big multi feature datasets with ease. Usually, the algorithm processes a considerable number of DTs and combines their predictions. In this approach, random selection of a subset of trees during training is done, which helps to reduce overfitting and reduce performance efficiency during the generalization process. After building each DT, the random forest algorithm selects the most popular class in a classification task and averages out all the predictions in a regression task. Investigators can optimize performance by adjusting flexibility and the model's accuracy by changing parameters like the amount of data used to train each tree or the total number of trees in the forest. In this way, the model can then predict the most effective plant species for absorption of heavy metals based on the characteristics of the plants and the chemical properties of the contaminants.

Neural networks, a subset of ML models, consists of interconnected data processing and communication nodes. During training of the model, it improvises on the inter-neurons weights to make better predictions. The first step in training an ANN for phytoremediation is collecting data on which plant species are most successful at eliminating pollutants from water or soil. With this information, the neural network may be trained to find correlations between different species and chemical contaminants removal from polluted sites. Soil or water samples taken from a polluted area can be analyzed using ANN to determine which chemicals are present. This allows for the selection of the most suitable plant species for phytoremediation. The best circumstances for phytoremediation and the possible environmental impact of the pollutants can be predicted by the neural network by examining data on the efficacy of various plant species under different circumstances.

The SVM is a support vector machine which excels in tackling regression and classification issues using a binary classifier. The algorithm divides the data into two groups using a hyperplane, and the model seeks to maximize the distance between the hyperplane and the nearest data points. Since phytoremediation applications frequently require massive volumes of data, the SVM model's ability to manage high-dimensional data is one of its key features that can manage overlapping and noisy data. For example, the contaminants are the input features, while the output labels show how well the phytoremediation worked. Applying the SVM model, the entered data are categorized based on where they fall on the hyperplane. A data point located on the positive hyperplane indicates success; whereas data on the negative hyperplane reflects failure.

DTs are another class of ML algorithms that finds useful applications in classification and regression analysis. One common application in phytoremediation is to determine which plant species are expected to be effective. The foundation of decision-tree models is a model that illustrates actions and their possible outcomes (Li et al., 2022). Each node in the tree stands for a test of a different data property or aspect. From one node in the tree, branches lead to the next node, and each branch represents a potential test result. Because of their visual nature and ease of interpretation, DT models are great for determining which features are crucial for making accurate predictions. Before training a DT model for phytoremediation, data is collected for effective plant species which are hyperaccumulators of chemical contaminants. To find correlations between plant traits and phytoremediation efficacy, this data is split during training, into new nodes in the tree. The process then keeps going until a stopping condition—like an optimum number of samples per leaf or an appropriate depth is satisfied.

Nanotechnology and genetically modified microbes can enhance the plant/root system, allowing it to overcome the aforesaid challenges.

4.4 APPLICATION OF AI-BASED MICROFLUIDIC PLATFORMS IN ENVIRONMENTAL ANALYSIS

4.4.1 Heavy Metal Detection

HMs can be found in environmental samples such as soil, plants, water, etc. Commonly employed methods were Atomic Absorption Spectroscopy (AAS) and inductively coupled plasma atomic emission spectrometry (ICP-AES). The data analysis had inaccuracies because of neighboring elements' signal and background interference. AI techniques including ML, FIS, back-propagation neural network (BPNN), successive projections algorithm (SPA), and generic algorithms are utilized for high efficacy during validation of the input data (Zhou et al., 2023; Liu et al., 2021; Zhang et al., 2021). These algorithms incorporate spectral data from the input feed and break it into multiple protected layers with interruption data, and the output results reflect the optimum heavy metals (HM) concentration. For example, to identify the HM mercury from soil sample, an Ava field spectrometer was utilized. In another set of investigations to detect copper (Cu) and cadmium (Cd) from rice leaves, BPNN was used along with AAS. This integration highlighted a fantastic level of efficacy in comparison to the conventional statistical model (Liu et al., 2021). In conjunction, different approaches of AI like multilayer perceptron (MLP), radial basis function network (RBFN), and FIS were used to remove Cu, Ni, and Pb from water sample (Choe et al., 2008). Several other studies utilized the BPNN and Generalized Regression Neural Network (GRNN) algorithms to identify the presence of Cu, Fe, Mn, and Zn from drainage water, based on AAS data (Rooki et al., 2011). ANN and Many Linear Regression (MLR) were employed to identify the presence of Hg, Fe, Zn, and Pb from soil samples using Vis-IR spectrometry data (Kemper and Sommer, 2002). In a separate investigation of mining soil, Pb was identified utilizing conventional spectrometry with MLR and regression analysis and better results were observed (Goodarzi et al., 2015). In another case, Pb from mining soil was identified through AAS, coupled with AI-based hyperspectral imaging, followed by analysis through ANN and Feedforward Neural Network (FNN). In this context, AI demonstrated greater accuracy compared to the traditional detecting system of HM (Zhang et al., 2021).

4.4.2 Pesticide Detection

Insects and other pests are difficult to control in farming and lead to massive plant and food loss for which the farmers have to rely on pesticides. Contrarily traces of pesticide remnants in food or raw vegetables can pose serious health risks to humans like respiratory and neurological diseases. Other symptoms like tension, rage, headaches, and depression are quite common. Thus, pesticide detection is necessary for improving human health. Traditional methods for detection of pesticide residue on vegetable and fruit relied on simple techniques like scrubbing, washing, applying baking soda, and vinegar which removed 70% of it. Laboratory techniques involved Liquid chromatography and Gas chromatography after the pesticide has been extracted using various solvents like ethyl acetate, acetonitrile, and hexane. But these are costly, inaccessible to common man and time-consuming. The sole option for improving pesticide detection methods are AI and ML algorithms, which use modern methods like sensors and automation for quick assessment of data. For example, Fluorescent spectroscopy is a reliable tool for identifying pesticides by analyzing their light reflections. The sample's molecules release light when excited by the light beam and can detect one pesticide at a time. Using BPNNs coupled with fluorescence enhances the system's dependability and decreases the inaccuracy. The AI technique employs hidden layers with weights to detect multiple pesticide residues. Modifying the weight values appropriately can analyze multiple data to handle the pest management problem, including the detection, monitoring, and control of pests to reduce their negative effects on crop yields (Kaur et al., 2024). Early pest infestations in an agricultural field can be assessed through satellite images, sensors present in the field, and unmanned aerial vehicles (UAVs). By analyzing obtained images from the crops, minute disruption and change in leaf morphology, these models can identify the early sign of pest attract. These AI-based models like ML can detect early signs and symptoms of potential pest or pathogen attackers, even before any symptoms are visible. For instance, DTs, SVMs, and random forest; a part of ML and DL techniques are used frequently for agricultural crop protection and detection of plant disease (Li et al., 2024). For example, the use of hyperspectral and fluorescence spectra allows for the detection of pesticide residue in lettuce. An algorithm for consecutive projections and wavelets are used to measure the output data. In another instance, Screening cards were developed which were able to detect pesticide residue by change in color intensity in a quick and cheap way. But the process is not so accurate because the difference in shade has to be examined by the naked eye. To accurately detect low and high amounts of pesticide residue based on color change, the color card-based technique is integrated with ML and yields reliable findings (Ly et al., 2023; Watthaisong et al., 2022). In another application, The Tamil Nadu Agricultural University has just released an Android app for use with integrated pest management (IPM) that can detect the fall of armyworms infesting maize crops. With a validation accuracy of 93.47% and a training accuracy of 98.57%, the app successfully detects infected areas using transfer learning and deep CNNs. The AI algorithms of the app can identify armyworm infestations during the fall season and make the farmers vigilant of the future problem so that they can take necessary intervention processes. For example, optimum pesticide application during pest attack following its life cycle pattern, a part of integrated pest control management. This digital solution equips farmers with the necessary tools to protect their maize crops and themselves against pests, an issue that may be solved by innovative technology.

4.4.2.1 Pesticide detection and AI

UAVs can execute diverse agricultural activities like mapping fields, collection of crop related data, and identification of pathogen infested crops (Hafeez et al., 2023). Farmers can analyze crop health by tracking the net sunshine reaching different parts of the crop plant, locate pests, and plants facing abiotic stress. Here, in conjunction with this data farmers can use AI-based pesticide allocation, pest oriented spraying, and focus on interventions targeting specific pests. Thus with help of AI pesticide consumption can be reduced benefiting safety of that environment. Conventional techniques like pheromone traps can kill grape moths present in vineyards and protect the grapes and production of the wine. With help of AI these traditional techniques are incorporated with smart insect monitoring systems, which include highly sensitive cameras and sensors. In a reported investigation by Teixeira et al. (2022), DL algorithm was used to detect grape moths by data segmentation, a pre-processing step in initial detection. They verified with ResNet18 and InceptionV3 backbones utilizing YOLO V5s and YOLO V8s algorithms where they obtained significant output data. Thus AI networks based smart traps increased crop production and lessened the use of pesticide.

Data integration is another feature to create prediction models for forecasting pest attack. By integrating data, AI uses several facts, counting weather patterns, agricultural phenology, and insect life cycles, to build prediction models for pest forecasting. These models incorporate both past data and real-time inputs to forecast when insect outbreaks may occur. Insect counts, their stages, movement and behaviors recorded by Internet of Things (IoT) based cameras and sensors allow for quick tracking and accurate pest and disease detection. This leads to proactive management strategies like preventing pesticides overuse, timely use, environmental predictions, and pest dynamics which all interact in complex ways. In the context, scientific investigators have developed an IPM platform that makes use of AI, data analytics and various data sources to provide farmers with better and more long-term solutions to safeguard their crops, all while reducing harm to the environment and the rise of chemical resistance. For both testing convenience and accuracy, modern technology for detecting pesticide residues would be a great boon. Due to the rapid and accurate outcomes that may be predicted with the use of AI, ML, and automation, their potential applications in the future are vast. Even if laboratory procedures are improving, testing and analysis still take the same amount of time as they do with machines. Nevertheless, issues with current advanced techniques can be addressed in the future, leading to improvements in these areas and others, such as detection time minimization, accuracy, and repeatability.

4.4.2.2 Precision monitoring

Drones equipped with AI capabilities, high-definition cameras, and detection systems can effectively survey agricultural fields. To identify areas of concern, assess crop health, and detect pests, computer vision algorithms process the collected data. This analysis facilitates targeted interventions and optimal resource distribution. The AI tool analyzes data related to crop health, pest management, and environmental factors to reduce pesticide usage, thereby lessening environmental impact. In grape cultivation, the presence of moths poses a significant threat to wine production, necessitating the implementation of pheromone traps for monitoring and pest control. There has been a significant increase in the adoption of smart pest management systems that incorporate cameras, sensors, and AI software due to their effectiveness. Teixeira et al. investigate the use of segmentation as a preliminary processing step to improve the identification of grape moths in trap images through DL techniques. The research utilizes U-Net with Res Net18 and InceptionV3 backbones to assess the performance of YOLO V5s and YOLO V8s sensors. The findings indicate that this differentiation enhances identification rates by a factor of 3 for YOLO V5 and 1.2 for YOLO V8, highlighting its commitment to enhancing crop management and preserving grape yields. The cotton bollworm poses a significant challenge to sustainable cotton production. To minimize pesticide use, we can once again depend on AI-powered pest surveillance networks that utilize smart traps and advanced algorithms for support. Still caution is required in AI's implementation in agree works as it can possibly harm the human workforce by reducing the manual activities. AI combines a variety of data, similar as meteorological patterns, agrarian phenology, and life cycles of the pests etc., for the production of prediction models for pest forecasting. For making these models efficient both literal data and current inputs are used. This allows for visionary operation ways and instant responses. Data regarding the quantity of insects, their movements, and behavioral patterns are collected through IoT cameras and detectors, facilitating accurate and rapid identification and monitoring of pests and their associated impacts. Farmers face the complex interactions of pest dynamics, environmental factors, and excessive fungicide application. For resolving the issue, experimenters have cooked an IPM platform employing AI, data analytics, and different data sources. The objective of this groundbreaking outcome is to enhance pest management techniques in agriculture, providing farmers with more efficient and sustainable methods for crop protection, while also mitigating environmental harm and the emergence of fungicide resistance.

4.4.2.3 Automation

Automated monitoring and identification of pests, with the help of AI has significantly reduced the need of manual activities enhancing crop productivity. Detector-equipped independent robots can move around fields and identify pest hotspots, taking real- time decisions instantly (Chandra, 2023; Borah et al., 2025b). AI systems take data from various sources and analyze them to support both integrated and sustainable pest management. Rustia and colleagues (2021) created an automated system utilizing images from sticky paper traps for the identification of pests. Their methodology incorporates a CNN as an object sensor, along with image classifiers, to assist in recognizing various insect species. The algorithm achieved high delicacy whiles tested in various greenhouse environments with average F1-scores of 0.92 and 0.90, along with mean counting rigor values of 0.90 and 0.91 across different datasets. This automated system provides timely information regarding pest situations, thereby facilitating the development of comprehensive IPM strategies in agriculture. The Google Net used with CNN model identifies 93.78% plant pests directly, this helps the farmers to use the Smartphone operation for proper identification of the threats and take required action to protect their crop fields (Yulita et al., 2023).

4.4.2.4 Decision making activities

Drones or UAV, improved performance by incorporating AI-powered computer vision algorithms, which played a crucial role in agricultural activities specifically in terms of pest monitoring and control by aerial mapping of fields (Borah et al., 2025b). They perform spraying activity on plants based on gaseous, kinetic, centrifugal, or hydraulic methods. With AI software tool, these devices can assist farmers to observe the areas receiving sunlight and identify the areas where pesticide spraying is needed. Chen et al. (2021) employed AI to develop an advanced pest recognition system. For instance, a detection drone captures images of plant pests like *Tessaratoma papillosa* in real-time by utilizing the Tiny-YOLO V3 neural network model. This system is capable of optimizing pesticide spraying routes based on the locations of the pests and transmits pest data to the cloud platform for long-term analysis and monitoring purposes. This results in preventive measures that decrease pesticide application and lessen environmental impact and increases agricultural production. Farmers are provided suggestions from the experts via an Internet based on output data derived from the interaction with Decision tools (Rossi et al., 2023). For example,

Trimble's Farmer Core software is a good example of AI oriented software that enables farmers with actionable awareness for improving yields of the crop and efficiency. Taranis is a precision agriculture platform powered by AI that leverages weather information, satellite imagery, and ML algorithms to identify pest infestations, crop diseases, and nutrient deficiencies in horticultural crops. This technology empowers farmers to make informed, data-driven decisions. The AI-driven platform developed by Descartes Labs examines satellite images and various spatial data to evaluate forest health, identify deforestation, and measure biodiversity. This technology aids forest managers and conservationists in making well-informed decisions. City Zenith's Smart World Pro platform combines AI and IoTs technologies to oversee urban settings for pest issues, including rodents and mosquitoes. It supplies city officials with real-time information to implement focused pest control strategies. Additionally, Verily has initiated the Debug Project, which employs AI algorithms to evaluate data regarding mosquito populations and environmental conditions, aiming to forecast and avert disease outbreaks like Zika and dengue fever. Drone Flight Sensor from ground based detectors monitor the pest activity 24 hours cycle. The system can classify insects precisely as all the entities work in an AI-based cloud network.

4.4.3 Micro- and Nanoplastic Detection

Plastic pollution is a compound scenario where diverse fragments of plastic are visualized as macro (>5 mm), micro (<5 mm) and nanoplastics (within size range of 1 nm to 1 mm). Nanoplastics are considered more potent and dangerous as they can enter human systems permeating through the skin. Anthropogenic causes are the main drivers of plastic pollution whereas the secondary sources originate from breaking down of macroplastics by physical and chemical agents (Mariano et al., 2025). Again on the basis of chemical composition, poly(ethylene), poly(vinyl chloride), poly(methyl methacrylate), polypropylene, polystyrene are the major ones (Campanale et al., 2022).

Specialized microscopic techniques like stereomicroscope, fluorescence microscopy, Atomic spectroscopy, help in detection of microplastics but with limitations like high cost and availability. For example, in fluorescence microscopy, samples are excited at specific wavelengths from where the fluorescence rays are captured back. Another constraint is the presence of chemical additives inherent in the plastic material which can interfere with the fluorescence analysis. Modern detection methods rely on digital holography (DH), capable of marking out microplastics from other phytoplanktons (Bianco et al., 2024). To make the microplastic detection easier, AI is coupled with holographic microscopy which makes the identification precise and offers scalability in this area.

AI-driven methods such as DL and ML enable self-detection, classifications, and quantification of microplastics from complex and varied datasets. These advanced AI techniques analyze vast amounts of spectral data revealing intricate patterns, connections and correlations which are overlooked by traditional methods. Since an extensive huge number of data sets are needed to be handled, advanced algorithms are necessary to scrutinize and evaluate, exposing hidden relationships among the input data. For example, AI-driven detection of chemical contaminants like microplastics helps in detection of different polymers involved in the manufacturing and contributes to deeper ecological interactions. In this context ML enhances efficiency and accuracy of input data by analyzing different data based on the chemical composition of the materials. For instance, Guo et al. (2022) utilized ANN to correlate experimental data of HM adsorption on nanomaterials with environmental monitoring parameters. Challenges related to the sample size used for adsorption and variability in the chemical composition of non-microplastic particles makes the AI-based prediction complicated. Ng et al. (2021) utilized this CNNs algorithm to assess these contamination level in soil samples, employing vis–NIR reflectance spectra.

Traditional techniques like Raman spectroscopy were extensively utilized for the identification of nanoplastic particles (NMPs) across various ranges, soil (150–200 μm) and aquatic environments (10 μm). For example, Xie et al. (2023) demonstrated the application of Raman spectroscopy along with a ML-based automatic detection for NMPs. Their study of the data set including random forest model revealed significant accuracy of 98.8%. However, challenges like weak Raman signals and the complexity

TABLE 4.1 Application of AI and ML for detecting pollutants in water and its algorithms

SL. NO.	*SAMPLE*	*POLLUTANTS*	*AI–ML/ALGORITHM*	*REFERENCES*
1	Heavy metals	Pb	WT-MT-SAE (Wavelet transform Monte Carlo coupled to stacked auto-encoders)	Zhou et al. (2022)
2	Heavy metals	Hg	GA-BPNN (Genetic algorithm back-propagation neural network) BPNN (Back-propagation neural network)	Zhao et al. (2018)
3	Heavy metals	Cu, Ni, Pb	MLP (Multilayer perceptron) RBN (Radial basis function network) FIS (fuzzy inference system)	Ucun Ozel et al. (2020)
4	Micro- and nanoplastics	Polystyrene, polypropylene, polyethylene, polyamide, polyester, polyethylene terephthalate, polyvinyl chloride, polyurethane, polycarbonate, poly(methyl methacrylate), and cellulose acetate	Convolutional neural networks	Gong et al. (2023)
5	Microalgae	Pseudo-nitzschia	Convolutional neural network	Göröcs et al. (2018)
6	Microalgae	*Anabaena*, *Quadricauda*, *Closterium*	Deep neural network	Liao et al. (2021)
7	Microalgae	Cosmarium, Closterium, Micrasterias, Haematococcus pluvialis	MSDN (based on the YOLOv5 architecture)	Zheng et al. (2024)
8	Pesticides	Abamectin, myclobutanil, phosmet, thiram	k-Nearest neighbors (k-NNs)	Sahin et al. (2022)
9	Pesticides	Polyethylene (PE), polystyrene, polyethylene terephthalate, polypropylene, polyvinyl chloride (PVC)	Support vector machine (SVM)	Valentino et al. (2022)

of environmental samples are present. In an another approach to identify NMPs, Weber et al. (2014) established a platform by integrating ML with micro-Raman spectroscopy where they obtained significant output data. Table 4.1 shows the role of AI and ML for detecting pollutants in the water and its algorithms.

4.4.4 Microalgae Detection

Microalgae, as the name suggests are microscopic phytoplankton that float independently on water bodies. They photosynthesize and make their own food and are considered superior in terms of their versatility in different sectors like biofuel generation and food industry. But they propagate quickly and are an indicator of a highly enriched aquatic body endangering aquatic lives by depleting dissolved oxygen. Some strains of algae produce toxic algal blooms that further deteriorates the ecosystem and impose health hazard to human life (Göröcs et al., 2018). Apart from naked eye visibility, advanced techniques like flow cytometry and chlorophyll autofluorescence are employed to identify microalgae and evaluate their nature. The intracellular chlorophyll quantity can be measured by chlorophyll autofluorescence and flow cytometry measured the photosynthetic activity of microalgae. ML algorithms can help to detect and analyze microalgae in the laboratory quickly and robustly (Liao et al., 2021). But challenges lie in the in situ and real-time analysis where the samples are widespread and huge in number. Göröcs et al. (2018) created an image flow cytometer compounded with AI-based DL algorithm for in situ plankton detection. The whole setup operated on the principle of diffraction patterns from moving microelements along with holographic microscopy technology. CNN algorithm was used to process colored images of the data without any need for external validation. Additionally, the sample was concurrently illuminated with pulsed red, green, and blue light-emitting LEDs to remove background noise. Zheng et al. (2024) successfully investigated on an simpler and economical Automated Intelligent Microfluidic Platform (AIMP) that combines intelligent data evaluation and ML to recognize and classify *Cosmarium*, *Closterium*, *Micrasterias*, and *Haematococcus pluvialis* microalgae. Here on the basis of the algae segmentation and algae classification algorithm, the distinction between the species was achieved without relying on the traditional microscopic techniques.

4.5 CONCLUSIONS AND FUTURE PERSPECTIVE

AI-based chemical contaminant detection from polluted bodies offers an exceptional and easy way to combat ecosystem pollution. Conventional techniques are able to detect pollutants but robust data needs to be handled through AI algorithms for more accuracy and efficiency. Future investigation should focus on creating AI models in conjunction with available techniques like AI-based phytoremediation which can foretell the water quality parameters in advance and select plant species accordingly. Similarly, pesticide, microalgae, and heavy metal, which are major environmental pollutants, can be detected in advance by processing huge datasets base on image classification or analyzing chemical composition by the help of AI. This approach can improvise water quality assessment with speed, accuracy, and cost effectiveness, enabling pollutant identification with support of informed decisions for quick and preventive water management restoration.

REFERENCES

Ahmad, Z., Rahim, N. A., Bahadori, A., & Zhang, J. (2017). Improving water quality index prediction in Perak River basin Malaysia through a combination of multiple neural networks. *International Journal of River Basin Management*, 15(1), 79–87.

Albignac, M., de Oliveira, T., Landebrit, L., Miquel, S., Auguin, B., Leroy, E., … & ter Halle, A. (2023). Tandem mass spectrometry enhances the performances of pyrolysis-gas chromatography-mass spectrometry for microplastic quantification. *Journal of Analytical and Applied Pyrolysis*, 172, 105993.

Alka, S., Shahir, S., Ibrahim, N., Ndejiko, M. J., Vo, D. V. N., & Abd Manan, F. (2021). Arsenic removal technologies and future trends: A mini review. *Journal of Cleaner Production*, 278, 123805.

Aryal, P., Hefner, C., Martinez, B., & Henry, C. S. (2024). Microfluidics in environmental analysis: Advancements, challenges, and future prospects for rapid and efficient monitoring. *Lab on a Chip*, 24(5), 1175–1206.

Banerjee, D., Adhikary, S., Bhattacharya, S., Chakraborty, A., Dutta, S., Chatterjee, S., … & Rajak, P. (2024). Breaking boundaries: Artificial intelligence for pesticide detection and eco-friendly degradation. *Environmental Research*, 241, 117601.

Bianco, S., Bordiga, M., Langer, G., Ziveri, P., Cerino, F., Di Giulio, A. S., & Lupi, C. (2024). Low sensitivity of a heavily-calcified coccolithophore under increasing CO_2: The case study of Helicosphaera carteri. *EGUsphere*, 2024, 1–25.

Borah, S. K., Dutta, C., Pal, D., Sarkar, S., & Sethi, L. N. (2025a). Harnessing Artificial Intelligence in agriculture for a sustainable future. In Phukan, B. C., & Rongpipi, D. (Eds.), *Environmental Issues and Sustainable Development* (Vol. 1, pp. 120–159). Purbayon Publication.

Borah, S. K., Pal, D., Sarkar, S., & Sethi, L. N. (2025b). AI-powered drones for sustainable agriculture and precision farming. In Sadiq, S. M., Ahmad, M. M., Karunakarn, N., & Atapattu, A. J. (Eds.), *Advancing Global Food Security with Agriculture 4.0 and 5.0* (pp. 69–98). IGI Global.

Campanale, C., Savino, I., Massarelli, C., & Uricchio, V. F. (2022). Assessment of metals associated with virgin pre-production and freshwater microplastics collected by an Italian river. *NanoImpact*, 28, 100438.

Cardoso, B. D., Castanheira, E. M., Lanceros-Méndez, S., & Cardoso, V. F. (2023). Recent advances on cell culture platforms for in vitro drug screening and cell therapies: From conventional to microfluidic strategies. *Advanced Healthcare Materials*, 12(18), 2202936.

Chandra, V. S. (2023). Role of artificial intelligence in Indian agriculture: A review. *Agricultural Reviews*, 44(4), 558–562.

Chen, C. J., Huang, Y. Y., Li, Y. S., Chen, Y. C., Chang, C. Y., & Huang, Y. M. (2021). Identification of fruit tree pests with deep learning on embedded drone to achieve accurate pesticide spraying. *IEEE Access*, 9, 21986–21997.

Chen, X., Zhang, D., Wang, X., Liu, Z., Kang, H., Liu, C., & Chen, F. (2023). Preparation of porous GelMA microcarriers by microfluidic technology for stem-cell culture. *Chemical Engineering Journal*, 477, 146444.

Chen, W. H., Maheshwaran, S., Park, Y. K., & Ong, H. C. (2024). Iron-based electrode material composites for electrochemical sensor application in the environment: A review. *Science of the Total Environment*, 953, 176128.

Choe, E., van der Meer, F., van Ruitenbeek, F., van der Werff, H., de Smeth, B., & Kim, K. W. (2008). Mapping of heavy metal pollution in stream sediments using combined geochemistry, field spectroscopy, and hyperspectral remote sensing: A case study of the Rodalquilar mining area, SE Spain. *Remote Sensing of Environment*, 112(7), 3222–3233.

Gong, L., Martinez, O., Mesquita, P., Kurtz, K., Xu, Y., & Lin, Y. (2023). A microfluidic approach for label-free identification of small-sized microplastics in seawater. *Scientific Reports*, 13(1), 11011.

Goodarzi, M., Sharma, S., Ramon, H., & Saeys, W. (2015). Multivariate calibration of NIR spectroscopic sensors for continuous glucose monitoring. *TrAC Trends in Analytical Chemistry*, 67, 147–158.

Göröcs, Z., Tamamitsu, M., Bianco, V., Wolf, P., Roy, S., Shindo, K., … & Ozcan, A. (2018). A deep learning-enabled portable imaging flow cytometer for cost-effective, high-throughput, and label-free analysis of natural water samples. *Light: Science & Applications*, 7(1), 66.

Guo, W., Liu, J., Dong, F., Chen, R., Das, J., Ge, W., … & Hong, H. (2022). Deep learning models for predicting gas adsorption capacity of nanomaterials. *Nanomaterials*, 12(19), 3376.

Hafeez, A., Husain, M. A., Singh, S. P., Chauhan, A., Khan, M. T., Kumar, N., … & Soni, S. K. (2023). Implementation of drone technology for farm monitoring & pesticide spraying: A review. *Information Processing in Agriculture*, 10(2), 192–203.

Himeur, Y., Rimal, B., Tiwary, A., & Amira, A. (2022). Using artificial intelligence and data fusion for environmental monitoring: A review and future perspectives. *Information Fusion*, 86, 44–75.

Kamat, V., Burton, L., Venkadesh, V., Jayachandran, K., & Bhansali, S. (2023). Enabling smart agriculture through sensor-integrated microfluidic chip to monitor nutrient uptake in plants. *ECS Sensors Plus*, 2(4), 043201.

Karaouzas, I., Kapetanaki, N., Mentzafou, A., Kanellopoulos, T. D., & Skoulikidis, N. (2021). Heavy metal contamination status in Greek surface waters: A review with application and evaluation of pollution indices. *Chemosphere*, 263, 128192.

Kaur, R., Choudhary, D., Bali, S., Bandral, S. S., Singh, V., Ahmad, M. A., … & Chandrasekaran, B. (2024). Pesticides: An alarming detrimental to health and environment. *Science of the Total Environment*, 915, 170113.

Kemper, T., & Sommer, S. (2002). Estimate of heavy metal contamination in soils after a mining accident using reflectance spectroscopy. *Environmental Science & Technology*, 36(12), 2742–2747.

Khan, R. S., & Bhuiyan, M. A. E. (2021). Artificial intelligence-based techniques for rainfall estimation integrating multisource precipitation datasets. *Atmosphere*, 12(10), 1239.

Kumar, A., Dadhwal, M., Mukherjee, G., Srivastava, A., Gupta, S., & Ahuja, V. (2024). Phytoremediation: Sustainable approach for heavy metal pollution. *Scientifica*, 2024(1), 3909400.

Lai, Y. C., Yang, C. P., Hsieh, C. Y., Wu, C. Y., & Kao, C. M. (2011). Evaluation of non-point source pollution and river water quality using a multimedia two-model system. *Journal of Hydrology*, 409(3–4), 583–595.

Lakshmi, D., Akhil, D., Kartik, A., Gopinath, K. P., Arun, J., Bhatnagar, A., … & Muthusamy, G. (2021). Artificial intelligence (AI) applications in adsorption of heavy metals using modified biochar. *Science of the Total Environment*, 801, 149623.

Li, H., Lu, Q., Shi, J., Zhang, X., Sun, P., Yan, X., & Lu, G. (2024). Enzyme-engineered metal-organic frameworks for the construction of organophosphorus pesticide biosensor. *Advanced Functional Materials*, 34(6), 2309383.

Li, M., Wei, D., Liu, T., Liu, Y., Yan, L., Wei, Q., … & Xu, W. (2019). EDTA functionalized magnetic biochar for Pb(II) removal: Adsorption performance, mechanism and SVM model prediction. *Separation and Purification Technology*, 227, 115696.

Li, X., Yi, S., Cundy, A. B., & Chen, W. (2022). Sustainable decision-making for contaminated site risk management: A decision tree model using machine learning algorithms. *Journal of Cleaner Production*, 371, 133612.

Liao, Y., Yu, N., Tian, D., Wang, Y., Li, S., & Li, Z. (2020). Toward embedded sensing automation and miniaturization for portable smart cost-effective algae monitor. *IEEE Sensors Journal*, 21(4), 5230–5239.

Liu, W., Xu, Y., Fan, D., Li, Y., Shao, X. F., & Zheng, J. (2021). Alleviating corporate environmental pollution threats toward public health and safety: The role of smart city and artificial intelligence. *Safety Science*, 143, 105433.

Lu, X., Fan, Y., Hu, Y., Zhang, H., Wei, Y., & Yan, Z. (2024). Spatial distribution characteristics and source analysis of shallow groundwater pollution in typical areas of Yangtze River Delta. *Science of the Total Environment*, 906, 167369.

Ly, N. H., Nguyen, N. B., Tran, H. N., Hoang, T. T. H., Joo, S. W., Vasseghian, Y., ... & Klemeš, J. J. (2023). Metal-organic framework nanopesticide carrier for accurate pesticide delivery and decrement of groundwater pollution. *Journal of Cleaner Production*, 402, 136809.

Ma, L., Li, N., Wang, J., Ma, C., Hu, X., Li, M., & Wu, Z. (2023). Advances in application and innovation of microfluidic platforms for pharmaceutical analysis. *TrAC Trends in Analytical Chemistry*, 160, 116951.

Manickam, P., Mariappan, S. A., Murugesan, S. M., Hansda, S., Kaushik, A., Shinde, R., & Thipperudraswamy, S. P. (2022). Artificial intelligence (AI) and internet of medical things (IoMT) assisted biomedical systems for intelligent healthcare. *Biosensors*, 12(8), 562.

Mao, X., & Zhang, C. (2022). A microfluidic cloth-based photoelectrochemical analytical device for the detection of glucose in saliva. *Talanta*, 238, 123052.

Mariano, G., Magro, C., Urbanski, B. Q., & Nogueira, M. G. (2025). Microplastic contamination in the highly polluted Tietê River (São Paulo, Brazil): An unsustainable human-nature relationship. *Environmental Monitoring and Assessment*, 197(4), 1–15.

Modak, S., Kasula, M., & Esfahani, M. R. (2023). Nanoplastics removal from water using metal–organic framework: Investigation of adsorption mechanisms, kinetics, and effective environmental parameters. *ACS Applied Engineering Materials*, 1(2), 744–755.

Nath, B. K., Chaliha, C., & Kalita, E. (2019). Iron oxide permeated mesoporous rice-husk nanobiochar (IPMN) mediated removal of dissolved arsenic (As): Chemometric modelling and adsorption dynamics. *Journal of Environmental Management*, 246, 397–409.

Ng, K., Stenzl, A., Sharma, A., & Vasdev, N. (Eds.) (2021, January). Urinary biomarkers in bladder cancer: A review of the current landscape and future directions. In *Urologic Oncology: Seminars and Original Investigations* (Vol. 39, No. 1, pp. 41–51). Elsevier.

Nti, E. K., Cobbina, S. J., Attafuah, E. E., Senanu, L. D., Amenyeku, G., Gyan, M. A., ... & Safo, A. R. (2023). Water pollution control and revitalization using advanced technologies: Uncovering artificial intelligence options towards environmental health protection, sustainability and water security. *Heliyon*, 9(7), e18170.

Poole, W. (2022). In vitro convolutional neural networks. *Nature Machine Intelligence*, 4(7), 614–615.

Pouyanfar, N., Harofte, S. Z., Soltani, M., Siavashy, S., Asadian, E., Ghorbani-Bidkorbeh, F., ... & Hussain, C. M. (2022). Artificial intelligence-based microfluidic platforms for the sensitive detection of environmental pollutants: Recent advances and prospects. *Trends in Environmental Analytical Chemistry*, 34, e00160.

Rooki, R., Doulati Ardejani, F., Aryafar, A., & Bani Asadi, A. (2011). Prediction of heavy metals in acid mine drainage using artificial neural network from the Shur River of the Sarcheshmeh porphyry copper mine, Southeast Iran. *Environmental Earth Sciences*, 64, 1303–1316.

Rossi, V., Caffi, T., Salotti, I., & Fedele, G. (2023). Sharing decision-making tools for pest management may foster implementation of Integrated Pest Management. *Food Security*, 15, 1459–1474.

Rustia, D. J. A., Chao, J. J., Chiu, L. Y., Wu, Y. F., Chung, J. Y., Hsu, J. C., & Lin, T. T. (2021). Automatic greenhouse insect pest detection and recognition based on a cascaded deep learning classification method. *Journal of Applied Entomology*, 145(3), 206–222.

Sahin, F., Celik, N., Camdal, A., Sakir, M., Ceylan, A., Ruzi, M., & Onses, M. S. (2022). Machine learning-assisted pesticide detection on a flexible surface-enhanced Raman scattering substrate prepared by silver nanoparticles. *ACS Applied Nano Materials*, 5(9), 13112–13122.

Speiser, J. L., Miller, M. E., Tooze, J., & Ip, E. (2019). A comparison of random forest variable selection methods for classification prediction modeling. *Expert Systems with Applications*, 134, 93–101.

Teixeira, A. C., Ribeiro, J., Neto, A., Morais, R., Sousa, J. J., & Cunha, A. (2022, July). Using deep learning for detection and classification of insects on traps. In *IGARSS* 2022-2022 *IEEE International Geoscience and Remote Sensing Symposium*, Kuala Lumpur, Malaysia (pp. 5746–5749). IEEE.

Tozer, L. (2023). Water pollution 'timebomb' threatens global health. *Nature*, 18 Jul 2023.

Ucun Ozel, H., Gemici, B. T., Gemici, E., Ozel, H. B., Cetin, M., & Sevik, H. (2020). Application of artificial neural networks to predict the heavy metal contamination in the Bartin River. *Environmental Science and Pollution Research*, 27, 42495–42512.

Valentino, M., Pirone, D., Béhal, J., Itri, S., Miccio, L., Memmolo, P., ... & Ferraro, P. (2022, October). Intelligent Digital Holographic systems to counteract microplastic pollution in marine waters. In *2022 IEEE International Workshop on Metrology for the Sea; Learning to Measure Sea Health Parameters (MetroSea)*, Milazzo, Italy (pp. 344–348). IEEE.

Vocciante, M., Dovì, V. G., & Ferro, S. (2021). Sustainability in electrokinetic remediation processes: A critical analysis. *Sustainability*, 13(2), 770.

Watthaisong, P., Kamutira, P., Kesornpun, C., Pongsupasa, V., Phonbuppha, J., Tinikul, R., ... & Chaiyen, P. (2022). Luciferin synthesis and pesticide detection by luminescence enzymatic cascades. *Angewandte Chemie International Edition*, 61(16), e202116908.

Weber, I., Böttger, U., Pavlov, S. G., Jessberger, E. K., & Hübers, H. W. (2014). Mineralogical and Raman spectroscopy studies of natural olivines exposed to different planetary environments. *Planetary and Space Science*, 104, 163–172.

Wong, Y. J., Arumugasamy, S. K., Chung, C. H., Selvarajoo, A., & Sethu, V. (2020). Comparative study of artificial neural network (ANN), adaptive neuro-fuzzy inference system (ANFIS) and multiple linear regression (MLR) for modeling of Cu (II) adsorption from aqueous solution using biochar derived from rambutan (Nephelium lappaceum) peel. *Environmental Monitoring and Assessment*, 192(7), 439.

Xie, B., Liu, R., Han, Y., Xu, H., Jiang, C., Wu, H., & Wang, H. (2023). The polarized Raman spectra of N-methylpyrrolidone-an effective method for determination of intermolecular interaction and H-bond formation. *Spectrochimica Acta Part A: Molecular and Biomolecular Spectroscopy*, 284, 121808.

Yaqoob, A. A., Ibrahim, M. N. M., & Al-Zaqri, N. (2023). A pilot trial in the remediation of pollutants simultaneously with bioenergy generation through microbial fuel cell. *Journal of Environmental Chemical Engineering*, 11(5), 110643.

Yulita, I. N., Rambe, M. F. R., Sholahuddin, A., & Prabuwono, A. S. (2023). A convolutional neural network algorithm for pest detection using GoogleNet. *AgriEngineering*, 5(4), 2366–2380.

Zhai, J., Liu, Y., Ji, W., Huang, X., Wang, P., Li, Y., ... & Jia, Y. (2024). Drug screening on digital microfluidics for cancer precision medicine. *Nature Communications*, 15(1), 4363.

Zhang, D., Li, C., Ji, D., & Wang, Y. (2022). Based microfluidic sensors for onsite environmental detection: A critical review. *Critical Reviews in Analytical Chemistry*, 52(6), 1432–1449.

Zhang, S., Chen, Y., Yang, Z., & Gong, H. (2021). Computer vision based two-stage waste recognition-retrieval algorithm for waste classification. *Resources, Conservation and Recycling*, 169, 105543.

Zhao, L., Hu, Y. M., Zhou, W., Liu, Z. H., Pan, Y. C., Shi, Z., ... & Wang, G. X. (2018). Estimation methods for soil mercury content using hyperspectral remote sensing. *Sustainability*, 10(7), 2474.

Zheng, J., Cole, T., Zhang, Y., Yuan, D., & Tang, S. Y. (2024). An automated and intelligent microfluidic platform for microalgae detection and monitoring. *Lab on a Chip*, 24(2), 244–253.

Zhou, W., Zhao, L., Huang, H., Chen, Y., Xu, S., & Wang, C. (2023). Automatic waste detection with few annotated samples: Improving waste management efficiency. *Engineering Applications of Artificial Intelligence*, 120, 105865.

Zhou, X., Sun, J., Tian, Y., Yao, K., & Xu, M. (2022). Detection of heavy metal lead in lettuce leaves based on fluorescence hyperspectral technology combined with deep learning algorithm. *Spectrochimica Acta Part A: Molecular and Biomolecular Spectroscopy*, 266, 120460.

5 Artificial Intelligence Applications in Water Monitoring and Risk Assessment

Minh-Ky Nguyen

5.1 INTRODUCTION

Water functions as an essential natural resource, supporting life on Earth. Worldwide, approximately 7.5×10^8 people, primarily in developing countries, lack access to safe drinking water, 2.5×10^9 lack proper sanitation, and 8.4×10^5 die each year from diarrhea caused by polluted water (Tarhule, 2017). By 2050, the world must feed and supply energy to an additional $2–2.5\times10^9$ people while addressing the current needs of many billions of people today (Cosgrove & Loucks, 2015). Water quality is essential for environmental preservation, human well-being, and economic growth. Water is becoming a key policy priority on the global stage. We face numerous challenges, particularly with an unpredictable climate future and a rapidly escalating population driving heightened economic and social development, urbanization, or globalization (Cosgrove & Loucks, 2015). The water sector faces numerous challenges in providing sustainable, reliable, resilient, and affordable water and related services that align with future customer expectations (Brears, 2024). Water resource management involves planning, developing, distributing, and overseeing available water resources (Brears, 2024). Nevertheless, the global community encounters significant challenges in efficiently monitoring and managing water resources (Essamlali et al., 2024).

In recent years, artificial intelligence (AI) (including machine learning—ML, neural networks—NNs, deep learning—DL, and robotics), big data, information security, the internet, cloud computing, and forensic science have become prominent and innovative themes in information and communication technology (ICT) (Abiodun et al., 2018). AI, a field of machine intelligence established in the 1950s, has been effectively applied in numerous areas, both academia and industry (Hoang et al., 2022). Numerous studies at the forefront of AI and the internet of things (IoT) reveal their transformative impact on water resource management (Essamlali et al., 2024; Kumar et al., 2024). Wagle et al. (2020) explored using ML and remote sensing (RS) data to estimate water quality parameters. The combination of regression algorithms and artificial neural networks (ANNs) and the prospect of real-time AI-assisted water quality

DOI: 10.1201/9781003600718-7

monitoring systems are explored. Water quality monitoring is a vital task that demands an interdisciplinary method, encompassing insights into factors influencing water quality, the implementation of intelligent strategies to improve and supervise water resources, and the deployment of regulations and policies to control human impacts that could affect water quality (Essamlali et al., 2024). Zhu et al. (2022) investigated the use of ML algorithms in assessing water quality across various environmental matrices, e.g., groundwater, surface water, drinking water, seawater, and sewage. They focused on overcoming challenges related to data gathering, the applicability of algorithms, and the necessity for cross-disciplinary knowledge.

According to researchers, advancements in ML methods can effectively record and assess enormous amounts of data to address the multifaceted and large-scale requirements of water quality investigation. In ML, a subset of AI, algorithms are employed to analyze data and identify potential patterns, aiming to foresee forthcoming information. Wederni et al. (2024) highlight the effectiveness of ML models, especially the Classification and Regression Tree (CART), in accurately assessing groundwater quality. AI in environmental monitoring provides precise disaster predictions, contamination source identification, and thorough water and air quality assessment (Olawade et al., 2024). Several notable AI-based solutions for environmental tracking are emphasized, illustrating their role in strategic environmental governance. AI methodologies, including Logistic Regression (LOR), K-Nearest Neighbors (K-NN), Decision Trees (DT), Support Vector Machine (SVM) variants, Adaptive Boosting (AdaBoost), and Extreme Gradient Boosting (XGBoost), are utilized (Panigrahi et al., 2023). Prediction results show that models accurately classified water quality classes, achieving up to 92%–98%. It can assist decision-makers in deciding on the most suitable water source, e.g., for drinking water (Panigrahi et al., 2023). ML approaches have also been reported for risk classification over the past several years (Brito et al., 2023). ML algorithms have been applied to help risk analysis in areas with large volumes of data. It demonstrates that ML algorithms could recognize specific words and word groups, influencing experienced insights in risk examination (Brito et al., 2023). Focusing on cutting-edge processes integrated with tertiary technologies that are low-cost, highly efficient, and energy-saving is essential for removing these emerging pollutants to ensure sustainable development. The collected data can be analyzed using AI techniques to make precise predictions related to water quality and risk management/assessment. AI-aided technologies improve environmental monitoring by facilitating deeper insights, prediction, and reduction of ecological risks (Olawade et al., 2024). This chapter aims to extensively analyze cutting-edge practices in water quality monitoring and risk assessment, specifically focusing on employing AI-aided solutions and technologies.

5.2 WATER POLLUTION AND POTENTIAL IMPACTS

The growing population and urbanization have led to emerging issues in recent years. As urban areas expand, the demand for water increases, placing further stress on water resources and treatment infrastructure. As a result, environmental pollution is escalating into a major global issue and continues to pose a significant challenge for humanity in this century; it has also emerged as a critical concern for water resource management (Ky et al., 2023; Le et al., 2024b). Specifically, severe pollution results in degraded water quality, thereby impacting the health of both humans and ecosystems. For instance, water contamination has been linked with waterborne diseases and has shown a detrimental impact on physical, biological issues, and mental health (Han & Yan, 2014).

Water pollution could be categorized into physical, chemical, and biological types (Huynh et al., 2020; Ky et al., 2022; Landrigan et al., 2018). Pollution sources can include industrial, municipal, agricultural, and natural origins, as well as stormwater, landfills, underground waste storage tanks, human sewage, and animal waste (Tran et al., 2022). Inorganic contaminants originate from agricultural and industrial processes, such as phosphates (PO_4^{3-}), nitrates (NO_3^-), heavy metal(loid)s (e.g., Cd, Cr, Cu, Zn, As, Pb, Hg, Ni), etc. (Ali & Chidambaram, 2021; Kaur et al., 2021). Nitrate, for example, is a primary drinking water pollutant linked to colorectal cancer and presents a significant environmental health risk (Nasseri Maleki et al., 2022). In addition, environmental pollution from heavy metal(loid) has detrimental effects, and these substances can threaten living organisms and human health (Thien-Khanh et al., 2024). Recent investigations have identified adverse health influences, including cancer, resulting from levels of arsenic in drinking water (Hong et al., 2017). Exposure to arsenic-contaminated groundwater via drinking water could lead to increased concentration in urine. Cadmium is a toxic metal that poses a threat to both human health and ecosystems. While cadmium is naturally present at low levels, increased concentrations in the environment are primarily due to human activities, as it spreads into the water and air as micro-contaminants from industrial processes, waste incineration, pollution, and e-waste recycling (Peana et al., 2022). Exposure to cadmium compounds is primarily linked to an increased risk of kidney, prostate, lung, and pancreatic cancers. Further, cadmium has also been associated with cancers of the urinary system, breast, and bladder (Peana et al., 2022).

Organic pollutants include a variety of compounds, such as polybrominated compounds, herbicides, pesticides, antibiotics, plastics (including microplastics and nanoplastics), etc. (Lee et al., 2024; Li et al., 2024; Miao et al., 2025; Van-Giang et al., 2024). The most crucial water pollutants that influence the ecosystems and human body are phthalates, pharmaceuticals, pesticides, bisphenol A (BPA), and polychlorinated biphenyls (PCBs) (Bochynska et al., 2024; Nguyen et al., 2024a; Tuan Tran et al., 2022). Pharmaceutically active compound residues are recognized as an emerging microcontaminant that enters aquatic environments, leading to toxic ecotoxicity (Nguyen et al., 2023a). PCBs are commonly found in human tissues, where they remain a substantial health risk to populations that are highly exposed or particularly vulnerable (Pessah et al., 2019). Also, the accumulation of polycyclic aromatic hydrocarbons (PAHs) in ecosystems and food chains/webs, along with their uptake, is increasingly recognized as a critical concern for environmental and human health (Le et al., 2024b). Biological contaminants, including bacteria and viruses, primarily enter the water bodies through animal and human fecal waste due to insufficient wastewater treatment (Bochynska et al., 2024; Ky & Lam, 2016).

In the modern era, micro- and nanoplastics (MNPs) are prominent markers of the plastic era, widely dispersed across various environmental matrices (Hoang et al., 2024). MNPs have become worldwide stressors because of their inherent physicochemical properties and potential effects on ecosystems and human beings (Kumar et al., 2022; Le et al., 2024a; Nguyen et al., 2024b). MNPs have been revealed *in vitro* and *in vivo* and identified as health risks to respiratory, reproductive, digestive systems, immune systems, and so on (Kumar et al., 2022). Polyvinyl chloride (PVC) and polystyrene (PS) are widespread MNPs found in humans through inhalation, ingestion, and dermal exposure, potentially leading to carcinogenesis. It highlights the significance of MNPs as an emerging pollutant and is expected to spark increased vital research in this area.

In summary, various pollutants, including heavy metals, pesticides, phthalates, MNPs, BPA, and pharmaceuticals, can be toxic to living organisms, resulting in population declines, genetic mutations, endocrine disruption, impaired development, inflammation, chronic biological effects, and reduced reproductive success. These pollutants can reduce biodiversity, disrupt food webs, and also lead to health issues in humans. Polluted water could carry pathogens like viruses, bacteria, and parasites, resulting in waterborne diseases. These diseases pose significant risks to public health, especially in regions without access to clean water and sanitation. This raises concerns about their potential health risks, necessitating international interest and thorough investigation. These challenges highlight the urgent need for sustainable water management practices and integrated pollution control policies. Carrying out an in-depth assessment of the long-term impacts of these pollutants and implementing impactful management strategies is crucial. Additionally, considerable attention must be given to developing economic and effective approaches for detecting and managing risks from pollution sources, as well as promoting environmentally friendly technologies to protect both the environment and human health.

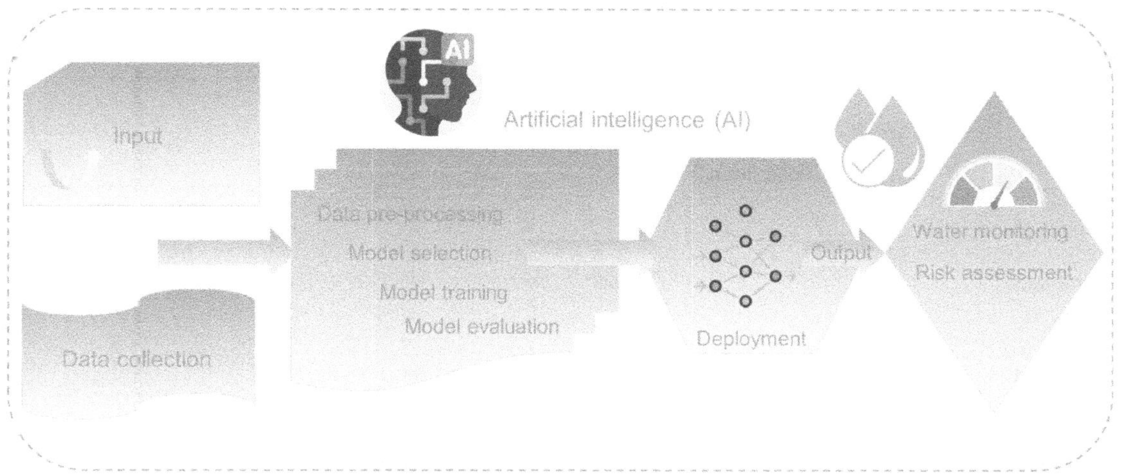

FIGURE 5.1 AI-driven application for water monitoring and risk assessment.

5.3 AI-DRIVEN SOLUTIONS FOR WATER MONITORING AND RISK ASSESSMENT

5.3.1 Solutions for Water Monitoring

Observing environmental pollutants is critical for understanding the harmful effects of potential toxins on humans, flora, and fauna (Artiola & Brusseau, 2019; Huang et al., 2023; Tran et al., 2023). The detection of contaminants is a key task for advanced technologies like AI and sensors in the context of pressing global environmental issues (Duc La et al., 2024; Zhang et al., 2021). AI has existed for many years, with the field covering a broad range of areas. AI has been illustrated and forecasted to play a key role in future pollution control, global industrial waste management, and achieving the "Sustainable Development Goals" (SDGs) remarked by the "United Nations Development Programme" (UNDP) (Hoang et al., 2022). AI could be considered a framework that includes ML and DL (Abiodun et al., 2018). ML is a subset of AI, while DL is a further subset of ML. DL refers to ANN with multiple complex layers (Albawi et al., 2017). Supervised learning ability can be applied to data classification and regression, with a range of algorithms developed, such as Linear Regression (LR), ANN, DT, SVM, KNN, Naive Bayes, and Random Forests (RF) (Zhu et al., 2022). In contrast, in unsupervised learning, the training data are classified into various categories according to their unique characteristics, primarily through dimensionality reduction and/or clustering methods. Algorithms such as K-Means and Principal Component Analysis (PCA) are widely used.

Table 5.1 demonstrates the use of AI-based models in monitoring and water management. AI-based technology is expected to substantially impact water quality management, as the increased deployment of devices for real-time, continuous monitoring of water quality indicators will contribute to enhanced water resource protection and management. For instance, Banna et al. (2014) examined novel technologies for online water quality monitoring, focusing on the importance of low-cost sensors. Recently, Durgun (2024) introduced an innovative approach for assessing water quality, using AI technologies and integrated with advanced sensor systems. Key to the work is the application of ML algorithms, e.g., SVM, RF, and NNs, for processing and classifying spectral data. Compared to conventional water quality monitoring methods, technology provides continuous remote observation for immediate detection of anomalies, offers

TABLE 5.1 Application of AI-based models in monitoring and water management

WATER SOURCES	*APPLICATIONS*	*ALGORITHMS*	*REMARKS*	*REFERENCES*
Groundwater	To evaluate the water quality	LOR, SVM, K-NN, DT, XGBoost, AdaBoost	Accurately classified: 92% (AdaBoost, XGBoost) and 98% (SVM)	Panigrahi et al. (2023)
Surface water	Dissolved oxygen (DO) prediction	PNN	Excellent interpolation performance (R^2=0.82)	Tomić et al. (2018)
Inland lakes	Monitoring water pollution	CNN, RF, SVM	Accuracy: CNN (97.12%) – SVM (96.89%) – RF (86.21%)	Pu et al. (2019)
Urban water systems	Real-time water quality monitoring	AI-enabled sensors (SVM, RF, NNs)	• A promising solution • Remarkable accuracy in real-time classification	Durgun (2024)
Drinking water	Predict the water quality	LSTM, DNN	• Good accuracy • Feasibility and effectiveness	Liu et al. (2019)
River water	Water quality monitoring and management	CNN- LSTM	CNN-LSTM model achieved 98.40% accuracy	Chellaiah et al. (2024)
Clean and polluted water	Water pollution monitoring	Attention NN	• Under IoT environment • Accuracy: 71.2% (clean water), 73.6% (polluted water)	Wu et al. (2020)
Irrigation water	Groundwater quality prediction	CART, SVM	• Using AI-driven predictions • CART model: R^2 (0.99), root mean square error (RMSE) (0.43) • SVM model: R^2 of 0.87	Wederni et al. (2024)
Water bodies	Examination of heavy metal pollution	PCA	Heavy metal-polluted areas were identified	Kumar et al. (2020)
Rivers and lakes	Predicting water quality levels	RF, DT, DCF, and other models	DT, RF, and DCF deliver superior predictions	Chen et al. (2020)
Groundwater	Prediction of the arsenic (As) contamination	MLR, PCR, ANN, PC-ANN	Efficiency coefficients of PC-ANN: 0.98 for the training step and 0.71 for the validation step	Cho et al. (2011)
River water	Water quality prediction	SVM, ANN	Machine learning: SVM>ANN	Haghiabi et al. (2018)
City harbor water	Biochemical oxygen demand (BOD) prediction	DMF, DNN	• DMF part achieved 11.5%–17.2% • DMF-DNN achieved 19.2%–25.2%	Ma et al. (2020)
Groundwater	Assessing water quality	RF, LOR, SVM, K-NN, DT, ANN	High accuracy: RF (R^2= 0.986), LOR (R^2=0.98), DT (R^2=0.979), K-NN (R^2=0.968), ANN (R^2=0.955), SVM (R^2=0.928)	Maleky et al. (2024)

AI, artificial intelligence; MLR, multiple linear regression; DCF, deep cascade forest; PCR, principal component regression; PNN, polynomial neural network; ANN, artificial neural network; PC-ANN, principal components artificial neural network; RF, random forest; LSTM, long short-term memory; LOR, logistic regression; IoT, internet of things; DMF, deep matrix factorization; CNN, convolutional neural networks; DNN, deep neural network; DT, decision tree; CART, classification and regression tree; PCA, principal component analysis; SVM, support vector machine; K-NN, K-nearest neighbor; NNs, neural networks; ANN, artificial neural network.

economical and extensive reach, and collects data to provide analytical support and guide management tactics. Wu et al. (2020) developed an attention NN as a crucial component of a water image-based system for contamination monitoring, helping users detect water quality degradation breaks in real time and can have emerging action to manage contaminants.

To evaluate water quality, Chellaiah et al. (2024) proposed a new hybrid approach that combines Long Short-Term Memory (LSTM) networks and Convolutional Neural Networks (CNNs). This combined model obtained a validation accuracy of up to 98.4%, outperforming other cutting-edge approaches. For instance, ML applications for water quality prediction revealed that K-NN reached an accuracy of 84%, whereas DT had an accuracy of 77% (Abbas et al., 2024). Ighalo et al. (2021) conducted a systematic review of AI models in surface water quality management. The results have highlighted typically used models, including the ANNs and Adaptive Neuro-Fuzzy Inference System (ANFIS). ANNs, a subset of ML, consist of traditional and deep neural networks (Abiodun et al., 2018; Hoang et al., 2022). ANN is an enhanced model based on the structure of a natural neural system, consisting of interconnected units with a specific topology that autonomously trains itself using various sets of training data. This points out the knowledge gaps in the field and recommends potential areas for future research. Additionally, other authors offered an in-depth review of DL techniques in water quality modeling and prediction (Ewuzie et al., 2022). Research on their application in modeling water quality and prediction is intensifying, particularly DL, which employs large NNs that scale and perform better with increased training data (Ewuzie et al., 2022). Historically, field measurements have served as the primary method for estimating and monitoring water quality parameters, where water samples are collected and analyzed in a laboratory. While this technique can provide reliable results, it is usually costly, slow, and unable to detect real-time or spatial changes in water quality. One of the key advantages of AI applications is that they simplify model usage and enhance accuracy when dealing with complex natural systems and large datasets. The quick training phase (just a few seconds) makes them suitable for dynamic monitoring systems that track water quality (e.g., drinking) and immediate safety management (Zhu et al., 2022).

The integration of ML and RS data provides an effective approach for routinely assessing spatial and temporal changes in water quality parameters (Wagle et al., 2020). Authors predominantly use regression algorithms for estimating water quality parameters, with SVM regression frequently applied. In regression algorithms, output values are predicted from input features in the given data. The methodology builds a model using data training features and uses it to predict values for new information/data. For instance, real-time AI-assisted models offer valuable recommendations and insights into water quality monitoring, empowering farmers to make informed decisions and take practical actions in aquaculture (Wagle et al., 2020). In a study by Pu et al. (2019), in the context of water quality monitoring for inland lakes, CNN demonstrated superior classification accuracy compared to traditional ML techniques like SVM and RF. CNN could be applied to monitor the water quality of lakes, resulting in improved accuracy and expanded coverage of the monitoring efforts (Pu et al., 2019).

Findings from using several ML models on big data indicate that DT, RF, and Deep Cascade Forest (DCF), when combined with selected essential water parameters, should be prioritized and focused on for future water quality management/monitoring and timely alerts (Chen et al., 2020). Haghiabi et al. (2018) analyzed the effectiveness of AI techniques, including ANN and SVM, in predicting water quality indicators of the Tireh River, Iran. SVM and ANN results revealed that both models strongly predict water quality components. Lee et al. (2019) investigated the spatial pattern of groundwater quality in Seoul, Korea, through a combination of a self-organizing fuzzy neural network (SOFNN) and Fuzzy C-Means (FCM) clustering. The results provide valuable insights for managing urban groundwater in metropolitan areas prone to anthropogenic contamination. Cho et al. (2011) applied an ANN to estimate the potential for arsenic (As) contamination in groundwater across Laos, Cambodia, and Thailand. The results could be used to predict As levels based on water quality data from field measurements, offering reliable predictive insights for public health management policies. Thus, AI-aided models, such as those combining fuzzy logic and ML algorithms, provide a new approach for more efficient and accurate groundwater quality prediction and management (Hajji et al., 2021).

Soft computing techniques are now recognized as a more effective solution for ensuring sustainability and water safety, supporting both ecological and human health. Leveraging the strong performance of LSTM-DNNs in time-series prediction, a water quality model, such as drinking water was developed using advanced DL theory to predict. For instance, Liu et al. (2019) employed these models to predict time-series data, developing models to provide relatively accurate forecasts for drinking water quality in the next six months. The findings highlighted the potential of integrated approaches for water quality investigation (Durgun, 2024).

5.3.2 Solutions for Risk Assessment

Environmental contamination is regarded as one of the most pressing issues, and ecological risk assessment is crucial for managing its impact (Nguyen et al., 2023c; Tran et al., 2024). For example, severe risks may arise from heavy metals and MNPs through food chains and drinking water supply (Nguyen et al., 2023b; Rakib et al., 2024). Polymers serve as human-caused stressors by interacting with hazardous micropollutants. Their co-occurrence and distribution in ecosystems could pose substantial threats to freshwater, marine ecology, and human health (Nguyen et al., 2023d). It is essential to provide robust quantitative insights into hazard levels and ecological impacts. There is also a need to develop innovative tools for the risk assessment of micropollutants in ecosystems.

Overall, ML can offer approaches for controlling water contamination, improving water quality, and managing watershed ecosystem safety (Zhu et al., 2022). Significant advancements in ML techniques, improvements in computer performance, and enhanced data collection processes have promoted the demand and capability to use ML approaches to risk measurement (Paltrinieri et al., 2019). Results illustrate a high suitability and degree of accuracy of a Deep Neural Network (DNN) for risk assessment. Furthermore, advanced AI techniques, i.e., ML, were applied to evaluate the model's effectiveness. Uddin et al. (2024) developed the "Root Mean Square Water Quality Index" (RMS-WQI) model to investigate pollution scenarios from potential toxic elements (PTEs) in groundwater, while the human health risk evaluation framework was used to estimate the toxicity risk related to these PTEs. Similarly, results from Huang and Zhang (2024), based on an ML method, provide key findings for forecasting eutrophication and performing risk evaluations for lake water blooms. The authors developed ML ensemble models using Multi-Layer Perceptron (MLP), Multiple Linear Regression (MLR), and Support Vector Regression (SVR). The results indicate that 8.8% of random samples showed a low/limited risk level, with a likelihood of algal bloom ranging from 21.1% to 36.5%, while one sample displayed a moderate risk level with a 45.5% probability. They also provide reliable scientific support for decision-making regarding eutrophication in lakes and reservoirs. The research findings provide significant findings for predictions

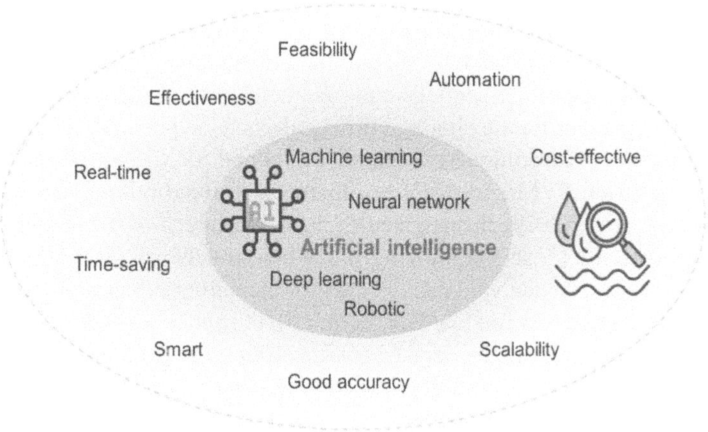

FIGURE 5.2 Advantages of AI applications in water monitoring and risk assessment.

and risk assessments related to water blooms. They suggest that these models have substantial theoretical relevance, practical utility, and broad applicability.

The accuracy of ML algorithms relies on consistent and reliable data. Therefore, Brito et al. (2023) introduced a framework for developing probabilistic risk assessment models using natural language and DL algorithms to simulate expert-generated risk assessments. This enables the risk analyst to reach a prior hazard examination even with limited textual and numerical data. When combined with universal sentence embedding and gradient-boosting regression trees trained on minimal structured data, they achieved the most effective outcomes (Brito et al., 2023). Significantly, by applying ML algorithms for probabilistic risk estimation, the methodology revealed in this study can be valuable for fields where narrative data enhance risk analysis (Brito et al., 2023). In a different context, combining multivariate methods and ML to examine microplastic contamination levels is crucial for evaluating ecological risks and human health (Jin et al., 2024). The findings reveal the potential of these applications to offer a comprehensive understanding of the current scenario in these field areas. Further studies on AI-based approaches in risk assessment and management are urgently needed to improve their effectiveness, focusing on potential methods to mitigate these critical environmental challenges. Designing an accurate and efficient prediction model for risk assessment highlighting health risks holds considerable scientific and practical value in preventing and managing pollution.

5.4 CONCLUSIONS AND OUTLOOKS

Water is an essential asset in our daily routines, emphasizing the necessity of consistently assessing its quality to maintain its suitability for use. Water scarcity is emerging as a key challenge in this new era. There is widespread concern about potential water shortages due to rising, predominantly population-driven water demands and their repercussions. We encounter numerous challenges in achieving this, particularly in the context of a changing and unpredictable future climate, coupled with a rapidly expanding population, accelerated socioeconomic progress, urbanization, globalization, and so on. Indeed, water pollution is progressively emerging as a global concern and continues to pose a significant challenge for humanity. Water is increasingly emerging as a priority policy issue on a worldwide scale, necessitating monitoring and risk assessment supported by advanced technologies. AI innovations serve as a robust method for boosting the efficiency and intelligence of our water systems. Predictive insights are essential in decision-making processes to protect water quality, enabling the identification of areas requiring immediate attention and the implementation of preventive measures to prevent contamination. These findings provide tangible benefits for sustainable management.

AI innovations have influenced decisions on water resource management, but they have not always contributed to greater security or sustainability. Interdisciplinary experts with knowledge across various fields must be trained to develop advanced ML techniques and integrate them into engineering practices. It has the potential to significantly contribute to ecological conservation by offering practical support for rational decision-making and prompt interventions. Yet, these obstacles to effective planetary management must be overcome. The future of AI in environmental monitoring appears promising, driven by advancements in AI algorithms and data collection techniques, and computing power is anticipated to enhance efficiency and accuracy in pollution management and monitoring (Olawade et al., 2024). Additional research is needed in areas such as algorithm optimization for adjusting network variables, which should be explored in greater detail. The challenges confronting water managers today and the depth of research necessary to guide those working toward a more sustainable and favorable future should be recognized. Furthermore, awareness-focused training on ANN models should be conducted regularly. This way, practical strategies and suitable approaches should be suggested to minimize risks and promote a sustainable future.

REFERENCES

Abbas, F., Cai, Z., Shoaib, M., Iqbal, J., Ismail, M., Alrefaei, A.F., Albeshr, M.F. 2024. Machine learning models for water quality prediction: a comprehensive analysis and uncertainty assessment in Mirpurkhas, Sindh, Pakistan. *Water*, **16**(7), 941.

Abiodun, O.I., Jantan, A., Omolara, A.E., Dada, K.V., Mohamed, N.A., Arshad, H. 2018. State-of-the-art in artificial neural network applications: a survey. *Heliyon*, **4**(11), e00938.

Albawi, S., Mohammed, T.A., Al-Zawi, S. 2017. Understanding of a convolutional neural network. *2017 International Conference on Engineering and Technology (ICET),* Antalya, Turkey.

Ali, A., Chidambaram, S. 2021. Assessment of trace inorganic contaminates in water and sediment to address its impact on common fish varieties along Kuwait Bay. *Environmental Geochemistry and Health*, **43**(2), 855–883.

Artiola, J.F., Brusseau, M.L. 2019. The role of environmental monitoring in pollution science. In: *Environmental and Pollution Science (Third Edition)*, (Eds.) M.L. Brusseau, I.L. Pepper, C.P. Gerba, Academic Press, London, pp. 149–162.

Banna, M.H., Imran, S., Francisque, A., Najjaran, H., Sadiq, R., Rodriguez, M., Hoorfar, M. 2014. Online drinking water quality monitoring: review on available and emerging technologies. *Critical Reviews in Environmental Science and Technology*, **44**(12), 1370–1421.

Bochynska, S., Duszewska, A., Maciejewska-Jeske, M., Wrona, M., Szeliga, A., Budzik, M., Szczesnowicz, A., Bala, G., Trzcinski, M., Meczekalski, B., Smolarczyk, R. 2024. The impact of water pollution on the health of older people. *Maturitas*, **185**, 107981.

Brears, R.C. 2024. *Water Resources Management: Innovative and Green Solutions*. Walter de Gruyter GmbH & Co KG, Berlin.

Brito, M.P., Stevenson, M., Bravo, C. 2023. Subjective machines: probabilistic risk assessment based on deep learning of soft information. *Risk Analysis*, **43**(3), 516–529.

Chellaiah, C., Anbalagan, S., Swaminathan, D., Chowdhury, S., Kadhila, T., Shopati, A.K., Shangdiar, S., Sharma, B., Amesho, K.T.T. 2024. Integrating deep learning techniques for effective river water quality monitoring and management. *Journal of Environmental Management*, **370**, 122477.

Chen, K., Chen, H., Zhou, C., Huang, Y., Qi, X., Shen, R., Liu, F., Zuo, M., Zou, X., Wang, J. 2020. Comparative analysis of surface water quality prediction performance and identification of key water parameters using different machine learning models based on big data. *Water Research*, **171**, 115454.

Cho, K.H., Sthiannopkao, S., Pachepsky, Y.A., Kim, K.-W., Kim, J.H. 2011. Prediction of contamination potential of groundwater arsenic in Cambodia, Laos, and Thailand using artificial neural network. *Water Research*, **45**(17), 5535–5544.

Cosgrove, W.J., Loucks, D.P. 2015. Water management: current and future challenges and research directions. *Water Resources Research*, **51**(6), 4823–4839.

Duc La, D., Khong, H.M., Nguyen, X.Q., Dang, T.-D., Bui, X.T., Nguyen, M.K., Ngo, H.H., Nguyen, D.D. 2024. A review on advances in graphene and porphyrin-based electrochemical sensors for pollutant detection. *Sustainable Chemistry One World*, **3**, 100017.

Durgun, Y. 2024. Real-time water quality monitoring using AI-enabled sensors: detection of contaminants and UV disinfection analysis in smart urban water systems. *Journal of King Saud University - Science*, **36**(9), 103409.

Essamlali, I., Nhaila, H., El Khaili, M. 2024. Advances in machine learning and IoT for water quality monitoring: a comprehensive review. *Heliyon*, **10**(6), e27920.

Ewuzie, U., Bolade, O.P., Egbedina, A.O. 2022. Application of deep learning and machine learning methods in water quality modeling and prediction: a review. In: *Current Trends and Advances in Computer-Aided Intelligent Environmental Data Engineering*, (Eds.) G. Marques, J.O. Ighalo, Academic Press, London, pp. 185–218.

Haghiabi, A.H., Nasrolahi, A.H., Parsaie, A. 2018. Water quality prediction using machine learning methods. *Water Quality Research Journal*, **53**(1), 3–13.

Hajji, S., Yahyaoui, N., Bousnina, S., Ben Brahim, F., Allouche, N., Faiedh, H., Bouri, S., Hachicha, W., Aljuaid, A.M. 2021. Using a mamdani fuzzy inference system model (MFISM) for ranking groundwater quality in an agri-environmental context: case of the hammamet-nabeul shallow aquifer (Tunisia). *Water*, **13**(18), 2507.

Han, Y., Yan, W. 2014. Bimetallic nickel–iron nanoparticles for groundwater decontamination: effect of groundwater constituents on surface deactivation. *Water Research*, **66**, 149–159.

Hoang, T.-D., Ky, N.M., Thuong, N.T.N., Nhan, H.Q., Ngan, N.V.C. 2022. Artificial intelligence in pollution control and management: status and future prospects. In: *Artificial Intelligence and Environmental Sustainability: Challenges and Solutions in the Era of Industry 4.0,* (Eds.) H.L. Ong, R.-A. Doong, R. Naguib, C.P. Lim, A.K. Nagar, Springer Nature Singapore, Singapore, pp. 23–43.

Hoang, V.-H., Nguyen, M.-K., Hoang, T.-D., Ha, M.C., Huyen, N.T.T., Bui, V.K.H., Pham, M.-T., Nguyen, C.-M., Chang, S.W., Nguyen, D.D. 2024. Sources, environmental fate, and impacts of microplastic contamination in agricultural soils: a comprehensive review. *Science of the Total Environment*, **950**, 175276.

Hong, Y.-S., Ye, B.-J., Kim, Y.-M., Kim, B.-G., Kang, G.-H., Kim, J.-J., Song, K.-H., Kim, Y.-H., Seo, J.-W. 2017. Investigation of health effects according to the exposure of low concentration arsenic contaminated ground water. *International Journal of Environmental Research and Public Health*, **14**(12), 1461.

Huang, C.-W., Lin, C., Nguyen, M.K., Hussain, A., Bui, X.-T., Ngo, H.H. 2023. A review of biosensor for environmental monitoring: principle, application, and corresponding achievement of sustainable development goals. *Bioengineered*, **14**(1), 58–80.

Huang, H., Zhang, J. 2024. Prediction of chlorophyll a and risk assessment of water blooms in Poyang Lake based on a machine learning method. *Environmental Pollution*, **347**, 123501.

Huynh, T.N., Hung, N.T.Q., Sac, T.C., Bang, N.H.K., Tri, T.Q., Hiep, N.T., Ky, N.M. 2020. Removal of nutrients and organic pollutants from domestic wastewater treatment by sponge-based moving bed biofilm reactor. *Environmental Engineering Research*, **25**(5), 652–658.

Igbalo, J.O., Adeniyi, A.G., Marques, G. 2021. Artificial intelligence for surface water quality monitoring and assessment: a systematic literature analysis. *Modeling Earth Systems and Environment*, **7**(2), 669–681.

Jin, H., Kong, F., Li, X., Shen, J. 2024. Artificial intelligence in microplastic detection and pollution control. *Environmental Research*, **262**, 119812.

Kaur, G., Kumar, R., Mittal, S., Sahoo, P.K., Vaid, U. 2021. Ground/drinking water contaminants and cancer incidence: a case study of rural areas of South West Punjab, India. *Human and Ecological Risk Assessment: An International Journal*, **27**(1), 205–226.

Kumar, A., Srivastav, A.L., Dubey, A.K., Dutt, V., Vyas, N. 2024. *Innovations in Machine Learning and IoT for Water Management*, IGI Global, Hershey, PA, pp. 1–312.

Kumar, R., Manna, C., Padha, S., Verma, A., Sharma, P., Dhar, A., Ghosh, A., Bhattacharya, P. 2022. Micro(nano) plastics pollution and human health: how plastics can induce carcinogenesis to humans? *Chemosphere*, **298**, 134267.

Kumar, V., Sharma, A., Kumar, R., Bhardwaj, R., Kumar Thukral, A., Rodrigo-Comino, J. 2020. Assessment of heavy-metal pollution in three different Indian water bodies by combination of multivariate analysis and water pollution indices. *Human and Ecological Risk Assessment: An International Journal*, **26**(1), 1–16.

Ky, N.M., Hung, N.T.Q., Nguyen, C.M., Lin, C., Nguyen, T.A., Nguyen, H.-L. 2023. Application of vetiver grass (Vetiveria Zizanioides L.) for organic matter removal from contaminated surface water. *Bioresource Technology Reports*, **22**, 101431.

Ky, N.M., Lam, N.H. 2016. Analysis of changing trend of fecal coliform levels at lakes in Hue Citadel, Vietnam. *Environment and Natural Resources Journal*, **14**(1), 1–7.

Ky, N.M., Nhut, H.T., Hiep, N.T., Lap, B.Q., Hung, N.T.Q., Lin, C., Tam, T.T.M., Ozaki, A. 2022. Investigation of nitrogen and phosphorus recovery from swine wastewater by struvite crystallization. *Journal of the Faculty of Agriculture Kyushu University*, **67**(1), 65–74.

Landrigan, P.J., Fuller, R., Acosta, N.J.R., Adeyi, O., Arnold, R., Baldé, A.B., Bertollini, R., Bose-O'Reilly, S., Boufford, J.I., Breysse, P.N. 2018. The Lancet Commission on pollution and health. *The Lancet*, **391**(10119), 462–512.

Le, V.-G., Nguyen, M.-K., Lin, C., Nguyen, H.-L., Nguyen, T.Q.H., Hue, N.K., Truong, Q.-M., Chang, S.W., Nguyen, X.H., Nguyen, D.D. 2024a. Review on personal protective equipment: emerging concerns in micro(nano)plastic pollution and strategies for addressing environmental challenges. *Environmental Research*, **257**, 119345.

Le, V.-G., Nguyen, M.-K., Nguyen, H.-L., Thai, V.-A., Le, V.-R., Vu, Q.M., Asaithambi, P., Chang, S.W., Nguyen, D.D. 2024b. Ecotoxicological response of algae to contaminants in aquatic environments: a review. *Environmental Chemistry Letters*, **22**(2), 919–939.

Lee, K.-J., Yun, S.-T., Yu, S., Kim, K.-H., Lee, J.-H., Lee, S.-H. 2019. The combined use of self-organizing map technique and fuzzy c-means clustering to evaluate urban groundwater quality in Seoul metropolitan city, South Korea. *Journal of Hydrology*, **569**, 685–697.

Lee, Y.-Y., Chang-Chien, G.-P., Li, I.C., Huang, B.-W., Nguyen, M.-K., Pham, M.-T., Wu, C.-H. 2024. Advanced ceramic membranes: novel strategies for sulfamethoxazole removal in wastewater treatment. *Journal of Environmental Chemical Engineering*, **12**(6), 114674.

Li, X., Shen, X., Jiang, W., Xi, Y., Li, S. 2024. Comprehensive review of emerging contaminants: detection technologies, environmental impact, and management strategies. *Ecotoxicology and Environmental Safety*, **278**, 116420.

Liu, P., Wang, J., Sangaiah, A.K., Xie, Y., Yin, X. 2019. Analysis and prediction of water quality using LSTM deep neural networks in IoT environment. *Sustainability*, **11**(7), 2058.

Ma, J., Ding, Y., Cheng, J.C.P., Jiang, F., Xu, Z. 2020. Soft detection of 5-day BOD with sparse matrix in city harbor water using deep learning techniques. *Water Research*, **170**, 115350.

Maleky, S., Faraji, M., Hashemi, M., Esfandyari, A. 2024. Investigation of groundwater quality indices and health risk assessment of water resources of Jiroft city, Iran, by machine learning algorithms. *Applied Water Science*, **15**(1), 3.

Miao, S., Zhang, Y., Yuan, X., Zuo, J. 2025. Antibiotic resistance evolution driven synergistically by antibiotics and typical organic pollutants in antibiotic production wastewater. *Journal of Hazardous Materials*, **483**, 136543.

Nasseri Maleki, G., Bayati Khatibi, M., Khamnian, Z., Jalali, Z., Dastgiri, S., Ghodrati Aroogh, H. 2022. Association between nitrate concentration in drinking water and rate of colorectal cancer: a case study in northwestern Iran. *International Journal of Environmental Health Research*, **32**(8), 1791–1800.

Nguyen, M.-K., Lin, C., Bui, X.-T., Rakib, M.R.J., Nguyen, H.-L., Truong, Q.-M., Hoang, H.-G., Tran, H.-T., Malafaia, G., Idris, A.M. 2024a. Occurrence and fate of pharmaceutical pollutants in wastewater: insights on ecotoxicity, health risk, and state–of–the-art removal. *Chemosphere*, **354**, 141678.

Nguyen, M.-K., Lin, C., Nguyen, H.-L., Hung, N.T.Q., La, D.D., Nguyen, X.H., Chang, S.W., Chung, W.J., Nguyen, D.D. 2023a. Occurrence, fate, and potential risk of pharmaceutical pollutants in agriculture: challenges and environmentally friendly solutions. *Science of the Total Environment*, **899**, 165323.

Nguyen, M.-K., Lin, C., Nguyen, H.-L., Le, V.-R., Kl, P., Singh, J., Chang, S.W., Um, M.-J., Nguyen, D.D. 2023b. Emergence of microplastics in the aquatic ecosystem and their potential effects on health risks: the insights into Vietnam. *Journal of Environmental Management*, **344**, 118499.

Nguyen, M.-K., Rakib, M.R.J., Lin, C., Hung, N.T.Q., Le, V.-G., Nguyen, H.-L., Malafaia, G., Idris, A.M. 2023c. A comprehensive review on ecological effects of microplastic pollution: an interaction with pollutants in the ecosystems and future perspectives. *TrAC Trends in Analytical Chemistry*, **168**, 117294.

Nguyen, M.-K., Rakib, M.R.J., Lin, C., Hwangbo, M., Kim, J. 2024b. Is micro(nano)plastic contamination in wet atmospheric deposition a prominent issue requiring heightened attention? *Journal of Hazardous Materials*, **476**, 135027.

Nguyen, M.K., Lin, C., Quang Hung, N.T., Hoang, H.-G., Vo, D.-V.N., Tran, H.-T. 2023d. Investigation of ecological risk of microplastics in peatland areas: a case study in Vietnam. *Environmental Research*, **220**, 115190.

Olawade, D.B., Wada, O.Z., Ige, A.O., Egbewole, B.I., Olojo, A., Oladapo, B.I. 2024. Artificial intelligence in environmental monitoring: advancements, challenges, and future directions. *Hygiene and Environmental Health Advances*, **12**, 100114.

Paltrinieri, N., Comfort, L., Reniers, G. 2019. Learning about risk: machine learning for risk assessment. *Safety Science*, **118**, 475–486.

Panigrahi, N., Patro, S.G.K., Kumar, R., Omar, M., Ngan, T.T., Giang, N.L., Thu, B.T., Thang, N.T. 2023. Groundwater quality analysis and drinkability prediction using artificial intelligence. *Earth Science Informatics*, **16**(2), 1701–1725.

Peana, M., Pelucelli, A., Chasapis, C.T., Perlepes, S.P., Bekiari, V., Medici, S., Zoroddu, M.A. 2022. Biological effects of human exposure to environmental cadmium. *Biomolecules*, **13**(1), 36.

Pessah, I.N., Lein, P.J., Seegal, R.F., Sagiv, S.K. 2019. Neurotoxicity of polychlorinated biphenyls and related organohalogens. *Acta Neuropathologica*, **138**(3), 363–387.

Pu, F., Ding, C., Chao, Z., Yu, Y., Xu, X. 2019. Water-quality classification of inland lakes using Landsat8 images by convolutional neural networks. *Remote Sensing*, **11**(14), 1674.

Rakib, M.R.J., Sarker, A., Nahida, Z.T., Islam, A.R.M.T., Mia, M.Y., Rahman, M.N., Ahsan, S.M., Idris, A.M., Nguyen, M.-K., Kumar, R., Malafaia, G. 2024. A critical review on heavy metal contamination in aquatic food webs by edible fish species: a special case concerning Bangladesh. *Environmental Monitoring and Assessment*, **196**(12), 1175.

Tarhule, A. 2017. Part 4 - The future of water: prospects and challenges for water management in the 21st century. In: *Competition for Water Resources*, (Eds.) J.R. Ziolkowska, J.M. Peterson, Elsevier, Amsterdam, pp. 442–454.

Thien-Khanh, T., Huynh, L., Nguyen, H.-L., Nguyen, M.-K., Lin, C., Hoang, T.-D., Hung, N.T.Q., Nguyen, X.H., Chang, S.W., Nguyen, D.D. 2024. Applications of engineered biochar in remediation of heavy metal(loid)s pollution from wastewater: current perspectives toward sustainable development goals. *Science of the Total Environment*, **926**, 171859.

Tomić, A.Š., Antanasijević, D., Ristić, M., Perić-Grujić, A., Pocajt, V. 2018. A linear and non-linear polynomial neural network modeling of dissolved oxygen content in surface water: inter-and extrapolation performance with inputs' significance analysis. *Science of the Total Environment*, **610**, 1038–1046.

Tran, H.-T., Dang, B.-T., Thuy, L.T.T., Hoang, H.-G., Bui, X.-T., Le, V.-G., Lin, C., Nguyen, M.-K., Nguyen, K.-Q., Nguyen, P.-T., Binh, Q.A., Bui, T.-P.T. 2022. Advanced treatment technologies for the removal of organic chemical sunscreens from wastewater: a review. *Current Pollution Reports*, **8**(3), 288–302.

Tran, H.-T., Hadi, M., Nguyen, T.T.H., Hoang, H.G., Nguyen, M.-K., Nguyen, K.N., Vo, D.-V.N. 2023. Machine learning approaches for predicting microplastic pollution in peatland areas. *Marine Pollution Bulletin*, **194**, 115417.

Tran, T.-K., Nguyen, M.-K., Lin, C., Hoang, T.-D., Nguyen, T.-C., Lone, A.M., Khedulkar, A.P., Gaballah, M.S., Singh, J., Chung, W.J., Nguyen, D.D. 2024. Review on fate, transport, toxicity and health risk of nanoparticles in natural ecosystems: emerging challenges in the modern age and solutions toward a sustainable environment. *Science of the Total Environment*, **912**, 169331.

Tran Tran, H., Lin, C., Bui, X.-T., Ky Nguyen, M., Dan Thanh Cao, N., Mukhtar, H., Giang Hoang, H., Varjani, S., Hao Ngo, H., Nghiem, L.D. 2022. Phthalates in the environment: characteristics, fate and transport, and advanced wastewater treatment technologies. *Bioresource Technology*, **344**, 126249.

Uddin, M.G., Imran, M.H., Sajib, A.M., Hasan, M.A., Diganta, M.T.M., Dabrowski, T., Olbert, A.I., Moniruzzaman, M. 2024. Assessment of human health risk from potentially toxic elements and predicting groundwater contamination using machine learning approaches. *Journal of Contaminant Hydrology*, **261**, 104307.

Van-Giang, L., Nguyen, M.-K., Ngo, H.H., Barceló, D., Nguyen, H.-L., Um, M.J., Nguyen, D.D. 2024. Microplastics in aquaculture environments: current occurrence, adverse effects, ecological risk, and nature-based mitigation solutions. *Marine Pollution Bulletin*, **209**, 117168.

Wagle, N., Acharya, T.D., Lee, D.H. 2020. Comprehensive review on application of machine learning algorithms for water quality parameter estimation using remote sensing data. *Sensors & Materials*, **32**, 3879–3892.

Wederni, K., Haddaji, B., Hamed, Y., Bouri, S., Colombani, N. 2024. Controversial insights into irrigation water quality in arid and semi-arid regions using AI driven predictions: case of southern Gabès. *Groundwater for Sustainable Development*, **27**, 101381.

Wu, Y., Zhang, X., Xiao, Y., Feng, J. 2020. Attention neural network for water image classification under IoT environment. *Applied Sciences*, **10**(3), 909.

Zhang, Y., Zhu, Y., Zeng, Z., Zeng, G., Xiao, R., Wang, Y., Hu, Y., Tang, L., Feng, C. 2021. Sensors for the environmental pollutant detection: are we already there? *Coordination Chemistry Reviews*, **431**, 213681.

Zhu, M., Wang, J., Yang, X., Zhang, Y., Zhang, L., Ren, H., Wu, B., Ye, L. 2022. A review of the application of machine learning in water quality evaluation. *Eco-Environment & Health*, **1**(2), 107–116.

Advanced Dynamic Water Quality Assessment Using Machine Learning Algorithms

6

Elif Esra Altuner, Hasan Ulus, Halil Burak Kaybal, and Fatih Sen

6.1 INTRODUCTION

6.1.1 Background: Global Water Quality Challenges

Water quality degradation presents a critical global challenge as anthropogenic pressures intensify and environmental resilience diminishes. Rapid industrialization, unregulated agricultural practices, and expanding urbanization significantly pollute freshwater systems. Industrial waste, agrarian runoff laden with pesticides and fertilizers, and untreated domestic sewage are primary pollutants contaminating rivers, lakes, and aquifers. These sources exacerbate water scarcity, undermine human health, and disrupt ecological balance. Over two billion people globally lack access to safe drinking water due to inadequate infrastructure and governance, particularly in low-income regions (Gleick, 1998; Makarigakis & Jimenez-Cisneros, 2019). For instance, industrial pollution has caused devastating impacts on significant water bodies, such as the Ganges in India and the Mississippi River in the USA. The Ganges, a river vital to millions, is severely polluted by untreated industrial waste, heavy metals, and sewage, leading to adverse health outcomes for dependent communities (Jahan & Singh, 2023). Similarly, agricultural runoff in the Mississippi River has created hypoxic zones in the Gulf of Mexico, endangering marine ecosystems and fisheries (Rabalais & Turner, 2019).

The economic consequences of such pollution are profound, affecting industries like fisheries, tourism, and agriculture. In the Ganges, cultural and spiritual activities are also jeopardized. Biodiversity loss and ecosystem degradation are common as aquatic species struggle to adapt to contaminated waters.

DOI: 10.1201/9781003600718-8

These pressing issues highlight the need for innovative solutions, including advanced water quality monitoring technologies. Integrating machine learning (ML) with environmental assessment can enhance the prediction of pollution patterns, enabling more effective mitigation strategies and safeguarding essential water resources.

The growing contamination of freshwater sources has led to a widening gap in global access to clean water. This disparity is especially severe in regions grappling with poverty and climatic variability, where waterborne diseases are prevalent. Additionally, the shared nature of many water resources, such as transboundary rivers and aquifers, creates geopolitical tensions and complicates equitable distribution. Climate change further amplifies these issues by intensifying droughts, flooding, and other extreme events, destabilizing water supply systems (MacAlister & Subramanyam, 2018; Yan, 2017).

Technological innovations are essential in overcoming these challenges. Advances in sensor technology, data analytics, and ML enable real-time tracking and evaluation of water quality metrics, providing actionable insights for mitigation. Remote sensing technologies, coupled with geographic information systems (GIS), allow for large-scale assessment of water resources, helping policymakers identify pollution hotspots and track changes over time. Moreover, the application of nanotechnology in water purification—using nanoparticles to remove contaminants like heavy metals and pathogens—holds promise for revolutionizing access to clean water (Abu-Dief & Hamdan, 2016; Bora & Dutta, 2014). By leveraging these technologies, global efforts to monitor, manage, and restore water quality can become more effective, contributing to improved water security and public health outcomes.

6.1.2 Nanotechnology in Water Quality Assessment

Nanotechnology revolutionizes water quality management through advanced detection, filtration, and purification processes. By manipulating materials at the nanoscale, it enables the development of highly efficient solutions to complex water treatment challenges (Darabi et al., 2023; Erduran, Altuner, & Şen, 2023; Kumar, 2023). Key nanomaterials, including carbon nanotubes (CNTs), graphene oxide (GO), halloysite nanotubes, boron nitride, and metal oxides like titanium dioxide and zinc oxide, play a pivotal role in detecting contaminants, adsorbing pollutants such as heavy metals and organic compounds and enhancing filtration processes (Ameen et al., 2023; Ansari, Shamim, Khan, Anwar, & Wani, 2023; Danyliuk, Tomaszewska, & Tatarchuk, 2020; Park et al., 2022; Salazar Sandoval et al., 2023; Wu, Altuner, et al., 2023; Wu, Tiri, et al., 2023). These nanostructures exhibit unique properties, including large surface areas (Mutlu, Yıldırım, Ulus, & Eskizeybek, 2023; Ulus, Kaybal, Cacık, Eskizeybek, & Avcı, 2022; Wu, Tiri, et al., 2023), high reactivity (Guerra, Attia, Whitehead, & Alexis, 2018; Park et al., 2022), and selective targeting of specific contaminants (Warner et al., 2010), making them ideal for improving water quality.

In filtration, nanomaterials enhance membrane performance by improving permeability, fouling resistance, and contaminant removal efficiency. For instance, nanoparticles like silver and copper impart antimicrobial properties, aiding in disinfection and pathogen inactivation. Additionally, catalytic nanomaterials, such as nano-zerovalent iron, further contribute by breaking harmful pollutants into benign substances, effectively addressing chemical contamination issues. A particularly transformative advancement in water treatment is the development of nanofiber-based membranes using electrospinning techniques, which allows precise control over fiber morphology and pore size distribution. This advancement has found numerous applications, including the production of nanofibers with tailored properties for various water treatment processes (Tınastepe et al., 2023; Ulus, Kaybal, Berber, et al., 2022). These membranes, characterized by their high surface-area-to-volume ratio and precise pore size control, excel in removing fine contaminants like nanoparticles, bacteria, and viruses. Furthermore, nanofiber structures offer superior permeability, reduced fouling, and enhanced contaminant removal efficiency compared to traditional membranes. Additionally, nanofiber membranes can be functionalized with titanium dioxide, silver nanoparticles, or halloysite nanotubes to provide antimicrobial properties or facilitate pollutant degradation (Spirescu, Chircov, Grumezescu, Vasile, & Andronescu, 2021).

Functionalized nanofibers improve performance and enable multifunctional applications, such as simultaneous filtration and chemical breakdown of contaminants. These characteristics make nanofiber membranes a promising technology for addressing current and emerging water quality challenges (Cao, Zhang, Ma, & Huang, 2023). For instance, Peter, Johns, Myung, and Cwiertny (2017) developed a polymer–metal oxide nanofiber filter via single-pot electrospinning for point-of-use water treatment targeting arsenate and chromate removal. The optimized composite (7 wt.% polyacrylonitrile, 3 wt.% ferrihydrite, 1 wt.% tetrabutylammonium bromide) combined dual mechanisms: iron oxide sites for arsenate sorption and quaternary ammonium salts for chromate ion exchange, achieving high efficiency even in complex water matrices. Their findings demonstrated the filter's potential to treat arsenic-contaminated water at levels significantly above regulatory limits, producing sufficient clean water for an individual's annual supply with just 130 g of the material. Zhou, Liu, and Yuan (2019) developed a functionalized poly(lactic acid) (PLA) nanofiber membrane with dual capabilities for oil/water separation and methylene blue adsorption. By coating the PLA nanofibers with titanium dioxide nanoparticles and a polysiloxane layer, the membrane achieved superhydrophobicity (water contact angle of 157.4°), high filtration flux (2,297.6 L/m²/h), and 98.4% filtration efficiency. This membrane also enabled rapid and reusable adsorption of toxic dyes, demonstrating its potential for versatile water purification applications. El-Aswar, Ramadan, Elkik, and Taha (2022) reviewed the potential of nanofiber membranes in wastewater treatment, highlighting their large surface area, high porosity, and flexibility for functionalization. While addressing challenges such as fouling and poor mechanical properties, they explored methods for enhancing membrane performance through functionalization with nanomaterials (e.g., nanoparticles, CNTs, and biomolecules) and microorganisms. The review also underscored recent advancements in pollutant removal, including heavy metals, dyes, and pharmaceuticals, as well as the antimicrobial capabilities of nanofiber membranes, offering insights to guide future research and industrial applications. He et al. (2014) developed a quaternary ammonium-functionalized cellulose nanofiber aerogel as an efficient adsorbent for heavy metal removal. The chemically crosslinked, highly porous aerogel demonstrated rapid and effective Cr(VI) adsorption, achieving over 99% removal from a 1 mg/L solution within 50 minutes using just 1 g of material. Additionally, the aerogel exhibited mechanical robustness, ease of separation, and excellent reusability, making it a promising solution for water purification. Kumar et al. (2019) fabricated bead-free electrospun nylon nanofiber membranes functionalized with thiol-modified SiO_2 to efficiently remove mercury ions from water. The membranes achieved an 88.77% removal efficiency within 15 minutes at pH 6, demonstrating their potential for rapid adsorption of mercury. Characterized using Fourier-transform infrared (FTIR) spectroscopy and scanning electron microscopy (SEM), this study highlights a promising approach for mercury removal and suggests future applications in treating mercury-contaminated seawater and freshwater systems. Yao, Li, Wang, Kong, and Wang (2015) developed composite membranes with high flux and mechanical robustness by functionalizing CNTs with an amphiphilic block copolymer and chemically crosslinking them onto porous substrates. Membrane thickness and permselectivity were tunable via CNT deposition, achieving complete removal of 10-nm particles while maintaining an impressive water flux exceeding 5,700 L/(m²·h·bar). This approach addresses CNT aggregation challenges, offering an efficient solution for purifying liquids contaminated with fine colloids or particulates. Thus, nanotechnology offers significant advantages in water quality management, providing cost-effective, scalable, and environmentally sustainable solutions. By tackling multiple contaminants simultaneously, nanotechnology outperforms many conventional methods in efficiency and adaptability. Despite these advantages, challenges such as potential toxicity to aquatic ecosystems and environmental impacts during production and disposal require careful management to ensure sustainable application.

6.1.3 Machine Learning Meets Nanotechnology

The convergence of ML and nanotechnology is revolutionizing nanomaterials' analysis, design, and real-time monitoring. With the exponential growth of data generated from nanomaterial synthesis, characterization, and applications, ML provides advanced tools for processing and interpreting complex datasets. In Figure 6.1, some characterizations and application techniques of ML are shown.

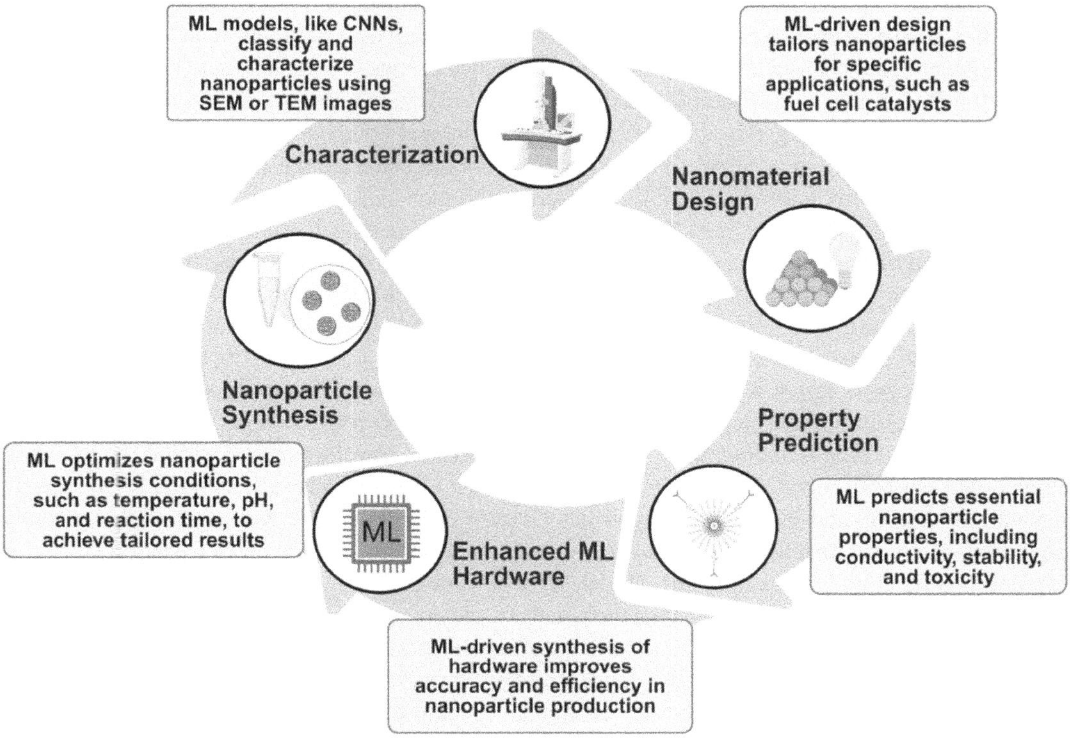

FIGURE 6.1 A cartoon depicting convergence and collaboration between ML and nanotechnology (Tripathy et al., 2024). (Reprinted (adapted) with permission from Tripathy et al. (2024). Copyright (2024) MDPI-International Journal of Molecular Sciences.)

By identifying patterns and building predictive models, ML enhances the efficiency and accuracy of nanotechnology research, unlocking new opportunities for innovation (Altuner et al., 2022; Konstantopoulos, Koumoulos, & Charitidis, 2022; Xi, 2022). One of the most impactful applications of ML in nanotechnology is its ability to analyze data related to nanomaterials' properties and performance (Al Janabi et al., 2024; Liu et al., 2022). For instance, ML algorithms like neural networks and support vector machines (SVMs) can predict mechanical, thermal, and electrical properties based on structural and compositional data (Dev, Rahman, Islam, Rahman, & Zhu, 2024). This capability reduces the reliance on exhaustive experimental testing, accelerating the design and optimization of nanomaterials. Studies have shown that ML models can optimize manufacturing parameters for electrospinning, ensuring precise control over fiber morphology and improving the quality of nanofiber-based membranes used in various applications. For instance, Sukpancharoen et al. (2024) developed a data-driven approach using ML to predict and optimize the diameter of electrospun polymer nanofibers. They compiled a dataset of around 430 data points from literature and evaluated six ML algorithms, finding that Random Forest and Extra Trees Regression provided the best predictive accuracy with R-squared values of 0.9468 and 0.9421, respectively. Their analysis identified polymer concentration, applied voltage, and feed rate as key factors influencing nanofiber diameter, and they created an interactive web server, Electrospun Nanofiber Diameter Predictor, to facilitate predictions based on user input. Golbabaei et al. introduced a data-driven approach combining the Simplified Molecular Input Line Entry System (SMILES) encoding method and ML algorithms to model the relationships between material properties, electrospinning parameters, and the physical properties of scaffolds. They successfully predicted fiber diameter and conductivity, achieving a balanced accuracy score of 0.9478 for the diameter classification task and an R-squared value of 0.723 for the regression task. Experimental validation using field-emission SEM images confirmed the model's efficacy in predicting nanofiber diameter and optimizing the electrospinning process for biopolymers (Golbabaei et al., 2024). Pervez et al. (2023) developed a methodology combining the design of experiments and ML models to predict the diameter of chitosan-based electrospun nanofiber membranes. They constructed a locally weighted kernel partial least squares regression (LW-KPLSR) model using Taguchi's statistical orthogonal design, which outperformed other regression models, including principal component regression and fuzzy modeling, in terms of prediction accuracy. The LW-KPLSR model achieved the highest R-squared value of 0.9996, with significantly lower root mean square error and mean absolute error compared to competing models. Trupp, Cibils, and Goyanes (2022) demonstrated that design of experiments, response surface methodology, and ML regressions can predict the bead size, size distribution, and bead density in electrospun poly(vinylidene fluoride) (PVDF) mats with fewer tests. They identified that PVDF concentration and solvent ratio are the primary factors influencing bead size and density, while the tip-to-collector distance helps reduce these properties, and voltage has minimal effect. The study also found that while response surface methodology is faster, ML regression achieves lower error and better generalization, providing valuable insights for improving electrospinning processes and model prediction performance.

Real-time monitoring is another area where ML demonstrates transformative potential. Nanomaterials integrated into sensors generate continuous data streams during applications such as water treatment, environmental monitoring, and biomedical diagnostics (Bassey & Sallis, 2020). ML algorithms can analyze these data in real time, enabling the rapid detection of contaminants, prediction of system behavior, and timely decision-making. For example, ML models have been employed in water treatment to interpret data from nanomaterial-based sensors, enhancing the efficiency of contamination detection and system responses (Lowe, Qin, & Mao, 2022). The integration of ML and nanotechnology also extends to predictive material design. By training algorithms on existing datasets, researchers can forecast the properties of novel nanomaterials, guiding the synthesis of tailored structures for specific needs (Lowe et al., 2022). This approach reduces resource consumption and bridges the gap between experimental and computational nanotechnology.

Expanding on these advancements, Kumar and Goyal (2025) highlighted the transformative role of artificial intelligence (AI) and ML in modernizing water and wastewater treatment processes by addressing challenges in energy efficiency, defect detection, infrastructure monitoring, and optimization. They provided case studies demonstrating practical applications and discussed future trends, emphasizing AI-driven solutions to meet sustainable water management needs. The study underscored that AI and ML offer groundbreaking potential to tackle 21st-century challenges in water and wastewater systems effectively. Sharmila, Kumar, and Tamilarasan (2024) reviewed the application of metal-organic frameworks (MOFs) as adsorbents for removing organic pollutants, such as dyes, toxins, and industrial chemicals, from wastewater. The study highlighted the unique features of MOFs, including their structural characteristics, water stability, and synthetic approaches for enhanced performance, along with the integration of ML to optimize pollutant degradation and removal. The authors discussed the environmental impact of organic pollutants, the challenges and opportunities in advancing MOF-based wastewater treatment, and the potential for ML to drive future industrial applications. Lin, Kim, Hua, Kang, and Park (2023) explored the application of ML models, including traditional and deep learning approaches, for predicting coagulant dosage and settled water turbidity in water treatment processes using 6 years of operational data. The study demonstrated that deep learning models significantly enhanced prediction accuracy by effectively handling temporal and sequential variations in raw water quality. The findings underscore the need for extensive data collection to improve model performance, offering valuable guidance on model selection, monitoring, and data management to optimize water treatment operations. These studies collectively underscore the pivotal role of ML and AI in revolutionizing environmental and industrial systems. By integrating advanced computational models with innovative materials, researchers are paving the way for more efficient, sustainable, and adaptive solutions.

6.2 FUNDAMENTALS OF NANOTECHNOLOGY AND MACHINE LEARNING IN WATER QUALITY

Nanotechnology has revolutionized water quality monitoring by developing highly sensitive and selective tools. Nanosensors, for instance, utilize materials like metal oxide nanoparticles, CNTs, and graphene-based sensors to detect pollutants at trace levels. These sensors excel in real-time monitoring, providing rapid and accurate data on water contaminants such as heavy metals and organic pollutants.

Nanocomposite membranes represent another significant advancement in water treatment technologies. These membranes, often integrated with nanomaterials such as titanium dioxide or silver nanoparticles, enhance filtration efficiency by offering improved antimicrobial properties and mechanical strength. They have been pivotal in removing contaminants from industrial and municipal wastewater and addressing global water scarcity and pollution (Ursino et al., 2018).

Nano-biosensing techniques further connect nanotechnology with biological systems, enabling the detection of pathogens and toxins. These techniques often employ nanomaterials functionalized with specific biomolecules, enabling precise identification of biological contaminants. For example, DNA-modified gold nanoparticles have been used in biosensors to monitor microbial water quality effectively (Koedrith et al., 2015; Kumar et al., 2020).

Integrating ML with nanotechnology revolutionizes water quality monitoring by enabling advanced data-driven insights. Nanosensors and other nanotechnology-based tools produce vast amounts of data, creating a need for robust computational approaches to extract meaningful information. In this context, supervised ML techniques, including decision trees and SVMs, are widely applied for classification tasks, like identifying pollutant types based on sensor data (Bhavsar & Panchal, 2012; Karimi et al., 2024; Zoubir, Elkhotfi, Tounsi, Assabbane, & Bakas, 2023). Similarly, unsupervised learning techniques, including clustering algorithms, are key in analyzing water quality trends, facilitating anomaly detection and long-term monitoring (Feldman et al., 2023; Mao & Lee, 2019). Additionally, real-time data analysis has emerged as a critical area where ML makes significant contributions. Predictive models allow water treatment facilities to anticipate contamination events and optimize filtration processes, enhancing operational efficiency. This proactive approach ensures higher water quality standards and reduces the environmental footprint of water management systems (Safder, Kim, Pak, Rhee, & You, 2022).

Despite its promise, integrating nanotechnology with ML poses several challenges. The diverse properties of nanomaterials, such as their size, shape, and surface chemistry, result in heterogeneous datasets that are difficult to standardize and interpret. Additionally, the dynamic nature of nanotechnology-based tools, which often involve interactions with complex water matrices, introduces variability that complicates data analysis (Von der Kammer et al., 2012). Another challenge lies in handling the sheer volume of data generated by nanosensors. Advanced computational resources are required to process and analyze these "big data" efficiently. Furthermore, developing ML models that generalize across various water quality scenarios remains an ongoing research area, as current models are often trained on localized datasets (Sagan et al., 2020). Finally, the ethical and regulatory aspects of deploying nanotechnology-based ML systems must be addressed. Data privacy, transparency in ML decision-making, and adherence to environmental safety standards are critical for the widespread adoption of these technologies (Holzinger, Weippl, Tjoa, & Kieseberg, 2021).

6.3 APPLICATIONS OF NANOTECHNOLOGY IN WATER QUALITY ASSESSMENT

6.3.1 Advanced Nanosensors for Real-Time Monitoring

The monitoring and analysis of water quality metrics are crucial for evaluating their effects and mitigating detrimental impacts. Traditional analytical techniques, such as gravimetric analysis, UV–Vis spectrometry, atomic absorption spectrometry, flow injection analysis, high-performance liquid chromatography, gas chromatography, electrochemical methods, and fluorescence spectroscopy, are well established and provide accurate, precise, and reproducible results. However, these methods face significant limitations, including their time-consuming nature, high energy and space requirements, and dependency on skilled analysts. These drawbacks present challenges, particularly for real-time or online water quality monitoring, rendering traditional methods less suitable for such applications (Zainurin et al., 2022).

To address these limitations, an ideal water quality monitoring system should perform all-in-one analyses, minimize reagent and sample volumes, reduce waste generation, and enable high-frequency, on-site assessments. Such systems should also be user-friendly, cost-effective, simple, sensitive, and resistant to false positives while delivering high-quality results. These requirements can be effectively met by leveraging nanomaterial-based and nanostructure-derived sensing strategies. The unique optical, electronic, thermal, magnetic, plasmonic, luminescent, and colorimetric properties of nanoparticles, combined with advanced handheld devices equipped with 3D printing technologies for flow injection analysis, sequential injection analysis, and batch analysis, offer transformative potential in this domain (Jiang et al., 2021; Thakur & Devi, 2024).

Nanoparticles, including dielectric, conductive, plasmonic, luminescent, and magnetic variants, can serve as core components of these advanced sensors. When functionalized with selective recognition elements or integrated into biomimetic, metallic, or conductive/metallic subsensors, such as chemosensors, screen-printed electrodes, and antenna-based nanosensors, they exhibit remarkable sensitivity, specificity, linear range, long-term stability, and low detection limits. For instance, surface-functionalized nanoparticles with coatings like dendrimers, chitosan, hydrogels, molecularly imprinted polymers, and inorganic oxide shells demonstrate enhanced performance in detecting a wide range of water contaminants (Howes, Chandrawati, & Stevens, 2014; Paimard, Ghasali, & Baeza, 2023).

Furthermore, nanostructures and thermosetting-based formulations can improve critical sensor properties, such as conductivity, porosity, selectivity, recyclability, and resistance to fouling, biofouling, swelling, or corrosion. The inherent large surface-area-to-volume ratio and surface rheology of nanoparticles contribute to enhanced sensor performance, enabling features such as boosted signal responses, self-regulating temperature control, and faster, more simultaneous, or highly sensitive detection capabilities. These functionalities are pivotal for developing advanced sensor configurations, including DNA-based, aptamer, and optical sensors. Particularly noteworthy are single-walled CNTs, whose properties can be tailored by adjusting their diameter to optimize electronic and optical functionalities. When combined with self-assembly techniques and DNA functionalization, these sensors gain the ability to detect hydrogen bonds, exhibit photoelectrochemical properties responsive to specific wavelengths, and leverage acid–base interactions for targeted sensing applications (Li et al., 2014; Wang & Lu, 2009). Figure 6.2 shows a figure of advanced sensor technology for real-time monitoring.

In summary, integrating nanomaterials and advanced fabrication techniques is revolutionizing water quality monitoring by enabling faster, more reliable, and cost-effective sensing technologies tailored to modern environmental and industrial needs.

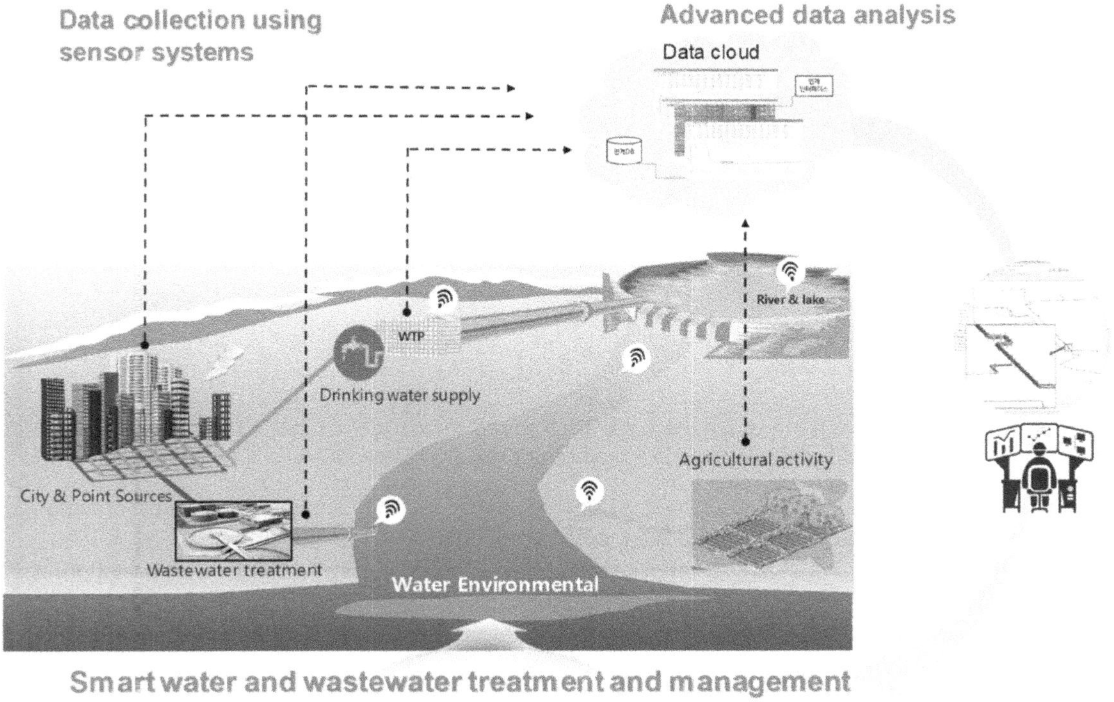

FIGURE 6.2 Schematic of the combination of real-time monitoring, transmission, and advanced data management system for smart water and wastewater treatment and management (Park et al., 2020). (Reprinted (adapted) with permission from Park et al. (2020). Copyright (2024) MDPI-Water.)

6.3.2 Nanocomposite Membranes for Water Filtration

The attainment of clean water for drinking and agricultural purposes has become a global challenge, encompassing both the provision of water to underserved populations and the preservation of vital ecosystems. Addressing this pressing issue requires the development of technologies that rely on sustainable and cost-effective materials. Recent advancements in nanotechnology have demonstrated that eco-friendly and widely available materials can significantly enhance water desalination and purification methods, catering to the increasing global demand for mobile water treatment systems in remote and rural areas (Ali & Ahmad, 2020).

In addition to portable solutions, some recently developed materials have been scaled up for implementation in centralized treatment plants, offering long-term, sustainable water treatment options. Nanocomposite membranes, in particular, have been at the forefront of these innovations, designed and fabricated for diverse water treatment applications (Ursino et al., 2018). Notable examples include:

- Nanocomposite catalytic membranes for enhanced contaminant removal (Wan et al., 2023).
- Zinc oxide–reinforced PVDF membranes, enabling efficient separation of mixtures (Wan et al., 2023).
- Amplified current transduction fibrous membranes, employing rare-earth-doped nanocomposites for ultrasensitive contaminant detection (Wang, 2005).
- Zinc phosphate-modified cellulose acetate membranes, demonstrating high separation efficiency (Asiri et al., 2021).

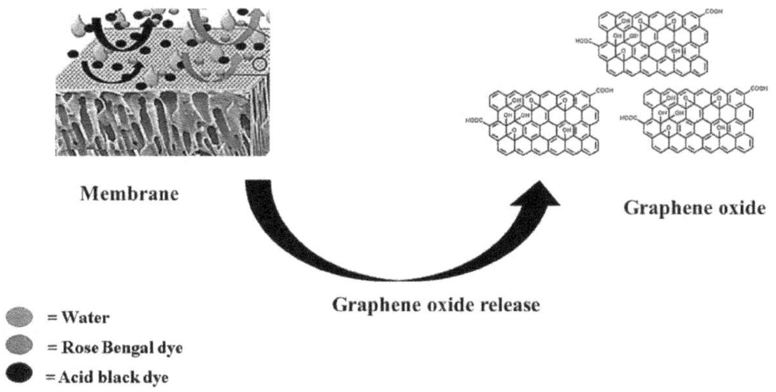

FIGURE 6.3 Dye removal using graphene oxide (GO) mixed matrix membranes (Davey, 2022). (Reprinted (adapted) with permission from Davey (2022). Copyright (2024) Azo Materials-An AzoNetwork Site.)

- Reduced GO-supported heterogeneous catalysts, used for effective removal of waterborne pollutants (Sreeprasad, Maliyekkal, Lisha, & Pradeep, 2011).
- Poly(ether-block-amide) nanocomposite membranes, incorporating functionalized mesoporous materials for pollutant capture (Song et al., 2020).
- Cellulose acetate/polyethylene glycol–polyvinylpyrrolidone–silver sulfadiazine membranes, featuring antibacterial properties and applications in wound healing (Aris, Fauzi, Tong, & Abdullah, 2019).
- Activated carbon–polyurethane composites, designed for the efficient removal of chemical substances from water (Gavabari, Goudarzi, Shahrousvand, & Asfaram, 2024).
- Hydrophilic hydrothermal carbon-embedded cellulose acetate microspheres/ultrafiltration ceramic hollow fiber membranes, optimized for bacterial removal (Yang et al., 2021).
- Nanofibrous membranes, particularly effective for oil–water separation (Hong et al., 2018).
- Copper–GO nanocomposites, serving as efficient catalysts for water treatment and electrode materials for environmental analysis (Zhao, Song, Song, & Yin, 2012).

The diagram showing that GO removes chemicals from water and performs filtration is presented in Figure 6.3.

These examples illustrate the versatility and efficiency of nanocomposites in water treatment applications, showcasing their potential to address water scarcity and pollution on both local and global scales. As such, nanotechnology continues to revolutionize water treatment, offering scalable and sustainable solutions that align with environmental and societal needs.

6.3.3 Machine Learning and Nanosensor Data Analysis for Environmental Monitoring

ML and nanosensor technologies have revolutionized environmental monitoring by providing efficient anomaly detection and pollution prediction solutions. Nanosensors, capable of real-time data collection, offer insights into ecological parameters such as heavy metal concentrations, pH levels, and other critical indicators. Integrating ML algorithms enhances the analytical capability of nanosensors, enabling more accurate detection and prediction outcomes.

Nanosensor data analysis is critical in identifying environmental anomalies like heavy metal leaks. These sensors collect real-time data from water and soil environments, enabling the identification of hazardous substances like lead (Pb), cadmium (Cd), and mercury (Hg) (Kim, Ren, Kim, & Yoon, 2012). ML algorithms, mainly unsupervised learning methods like clustering and Principal Component Analysis

(PCA), process the complex datasets generated by nanosensors (Arellano Vidal & Govan, 2024). These models identify patterns deviating from normal environmental baselines through anomaly detection techniques, signaling potential leaks. For instance, SVMs and Isolation Forest algorithms have been employed to detect abnormal spikes in metal concentrations (Susto, Beghi, & McLoone, 2017). Early detection of such leaks is crucial for mitigating environmental pollution, protecting water resources, supporting timely interventions, and ensuring sustainable management of ecological systems.

Predicting pollution concentration is another significant step in managing environmental health. Deep learning models such as Convolutional Neural Networks (CNNs) and Recurrent Neural Networks (RNNs) have shown remarkable performance in analyzing nanosensor data. These models predict the concentration levels of various pollutants by learning from historical and real-time datasets. Long Short-Term Memory (LSTM) networks, widely used for time-series analysis, enable accurate forecasting of pollutant levels over time. The ability of deep learning models to handle large and complex datasets ensures higher prediction accuracy than traditional methods (Figure 6.4). Predictive analytics powered by these models can inform proactive measures to curb pollution, aiding in developing environmental policies and improving public health outcomes (Kumar & Garg, 2018; Qin et al., 2019).

Combining nanosensors and ML has found practical applications in urban water systems and industrial wastewater management. Nanosensors embedded in pipelines and reservoirs in urban water systems provide continuous data on pH levels, turbidity, and metal ion concentrations (Skarbøvik & Roseth, 2015). ML models process these data to detect anomalies and predict water quality trends (Dogo, Nwulu, Twala, & Aigbavboa, 2019). For example, a study in smart cities employed nanosensors to monitor lead contamination in drinking water systems. Supervised learning models like Random Forest and Gradient Boosting identified contamination sources and provided actionable insights for remediation, improving water safety, and reducing operational costs.

Nanosensors integrated with ML models offer a comprehensive solution for monitoring and managing wastewater in industrial settings. In real time, these systems analyze parameters such as heavy metal concentrations, chemical oxygen demand, and total dissolved solids (Mitra, Kumar, Yadav, & Kaur, 2024). In a recent case study, deep learning models predicted heavy metal removal efficiency in treatment plants (Yuan et al., 2023). Integrating nanosensors and ML enables dynamic adjustments to treatment protocols, optimizing resource use and ensuring compliance with environmental regulations. This approach has proven particularly effective in industries such as mining, electronics, and textiles, where wastewater treatment is a critical challenge.

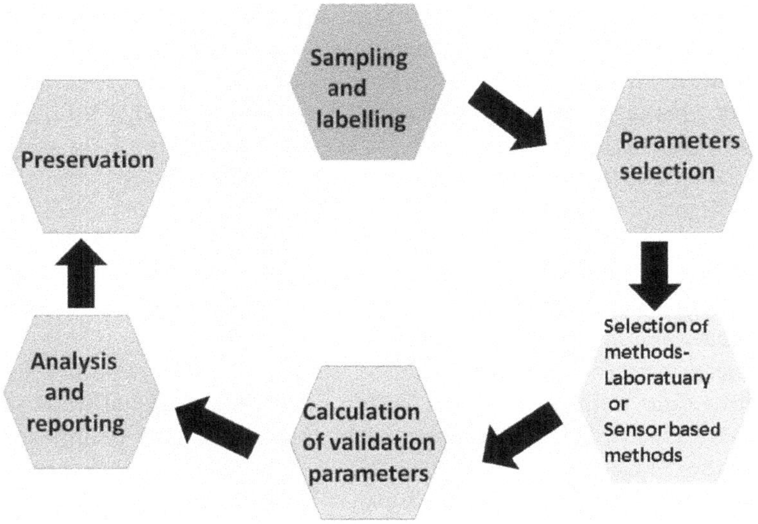

FIGURE 6.4 General flow of procedures for water quality analysis (Geetha et al., 2023). (Reprinted (adapted) with permission from Geetha et al. (2023). Copyright (2024) MDPI-Water.)

By leveraging the synergistic potential of ML and nanosensors, environmental monitoring systems can achieve unprecedented accuracy, efficiency, and reliability. These technologies improve our capacity to address environmental issues and contribute to building sustainable and resilient ecosystems.

6.4 MACHINE LEARNING ALGORITHMS FOR PREDICTING WATER QUALITY USING NANOTECHNOLOGY DATA

Nanotechnology has revolutionized environmental monitoring by providing highly sensitive and accurate nanosensors capable of detecting various water quality parameters. Researchers and practitioners can achieve predictive insights by integrating these sensors with ML algorithms, enabling proactive water management and pollution control. This chapter discusses key aspects of leveraging nanosensor data with ML, including data preprocessing, supervised and unsupervised learning models, and hybrid approaches that combine nano-enhanced sensors with ML to predict water quality effectively.

Nanosensor data often contain noise due to environmental factors, sensor limitations, and data acquisition errors. Effective data preprocessing begins with cleaning noisy data to ensure accurate ML predictions. Methods such as signal smoothing, outlier detection, and interpolation techniques are widely applied. For example, wavelet transforms and Savitzky–Golay filters have been used to denoise nanosensor data while preserving essential features of water quality signals (Bassey, Whalley, Sallis, & Prasad, 2020). Robust preprocessing pipelines include outlier removal using statistical methods like Z-scores or ML-based approaches such as Isolation Forests. Additionally, water quality monitoring often involves time-series data collected at regular intervals. Preparing these data for ML models requires handling missing values, ensuring uniform time intervals, and creating lagged features. Techniques like linear interpolation and forward filling commonly handle missing data, while resampling ensures consistent time intervals. Feature engineering, such as calculating moving averages or deriving frequency-domain features, enhances the predictive performance of ML models (Ali, 2023). Properly formatted time-series datasets allow ML algorithms to learn temporal patterns, improving predictions of water quality metrics like pH, turbidity, and dissolved oxygen concentration.

Regression models are crucial for forecasting continuous water quality metrics, such as pH, turbidity, and conductivity. Linear regression, support vector regression, and ensemble methods like Random Forest Regression have been successfully applied to nanosensor data. For instance, deep learning approaches, such as feedforward neural networks, have shown excellent performance in capturing nonlinear relationships between sensor inputs and water quality outputs (Chen, Song, Liu, Yang, & Li, 2020). These models leverage nanosensor data's high granularity, enabling precise forecasts for water resource management. On the other hand, classification models focus on identifying pollutants, such as heavy metals or harmful microbes, in water. Algorithms like Logistic Regression, Decision Trees, and Gradient Boosting classifiers are commonly used to analyze nanosensor data. Recent studies have shown the effectiveness of CNNs in classifying waterborne contaminants based on multispectral nanosensor readings (Chandanapalli, Reddy, & Lakshmi, 2019). These models' ability to process complex and high-dimensional data enables accurate pollutant detection, ensuring timely intervention.

Unsupervised learning techniques, particularly clustering algorithms, are invaluable for analyzing unlabeled nanosensor data. Methods such as K-Means, density-based spatial clustering of applications with noise (DBSCAN), and hierarchical clustering identify patterns and group similar data points, helping to detect anomalies (Seyyednourani, Akgun, Ulus, Yildiz, & Sas, 2023; Yildirim et al., 2023; Yildirim, Ulus, Beylergil, et al., 2024; Yildirim, Ulus, Sas, Topal, & Yildiz, 2024). For example, clustering has been used to identify unusual spikes in contaminant levels, such as lead or arsenic, in water samples (Karami & Guerrero-Zapata, 2015). By analyzing nanosensor data streams, these methods can flag potential pollution events for further investigation, efficiently monitoring water quality dynamically.

Hybrid approaches that combine nano-enhanced sensors and ML offer a comprehensive solution for water quality monitoring. These systems integrate multiple sensors measuring various parameters, such as temperature, pH, and heavy metal concentration, and use ML models to analyze the aggregated data. For example, hybrid networks employing nanosensors and Internet of Things (IoT) devices, combined with ML algorithms like LSTM networks, have been deployed to monitor urban water systems in real time (Liu, Wang, Sangaiah, Xie, & Yin, 2019). These systems' ability to predict pollution trends and detect anomalies ensures robust and adaptive water quality management. Ensemble learning methods, such as stacking or boosting, combine predictions from multiple ML models to improve accuracy. Ensemble models can integrate outputs from regression, classification, and clustering algorithms when applied to nanosensor data, providing a holistic view of water quality. Studies have demonstrated ensemble techniques' effectiveness in predicting continuous parameters and categorical pollution levels, outperforming individual models (Zheng, Cheng, & Li, 2020). This hybrid approach enhances the reliability and scalability of water quality prediction systems.

Combining ML algorithms with nanotechnology data has opened new avenues for water quality monitoring. From cleaning and preparing nanosensor data to leveraging advanced ML models and hybrid systems, this approach addresses key challenges in environmental monitoring. Future research should focus on developing more sophisticated algorithms and expanding the deployment of nano-enhanced ML systems to ensure sustainable water resource management worldwide.

6.5 CHALLENGES AND OPPORTUNITIES IN NANOTECHNOLOGY-DRIVEN WATER QUALITY MONITORING

Nanotechnology has emerged as a promising solution for water quality monitoring, yet several challenges hinder its seamless implementation in real-time scenarios (McGuinness et al., 2015). Among these, sensor sensitivity and stability are critical concerns. Due to their high surface area and unique physicochemical properties, nanomaterials are highly sensitive to environmental changes (O'Keefe & Haynes, 2024). However, their stability over prolonged usage or exposure to harsh conditions, such as fluctuating temperatures or varying pH levels, remains a significant limitation (Wang, Su, Brown, & Ohodnicki, 2015). For instance, prolonged exposure to high temperatures can lead to the degradation of specific nanostructures, reducing their efficacy in detecting contaminants.

Another pressing challenge involves the influence of environmental conditions on nanomaterials. Variations in parameters such as temperature and pH can drastically affect the performance and accuracy of nanosensors (Williams, Flynn, Xia, & Dunstan, 2016). For example, pH fluctuations may alter the surface charge of nanomaterials, leading to inconsistent detection results. Similarly, temperature variations can impact the kinetics of surface reactions on nanosensors, affecting their sensitivity.

Moreover, the complexities of data transmission and storage pose logistical barriers. Real-time water quality monitoring systems require continuous data collection and transmission, often from remote or inaccessible locations (Barabde & Danve, 2015). This necessitates the integration of reliable communication networks and robust data storage solutions, which can significantly increase operational costs. Furthermore, ensuring data integrity and security during transmission adds another layer of complexity.

Despite these challenges, significant opportunities exist to enhance nanotechnology-driven water quality monitoring systems. The development of smart nanosensing devices stands out as a promising avenue. These devices integrate advanced nanomaterials with intelligent systems, enabling real-time, accurate, and efficient monitoring. For instance, functionalized nanoparticles can be designed to bind to specific contaminants selectively, improving detection specificity and reducing false positives (Kuswandi, 2018).

ML-powered decision support systems offer another transformative potential. By leveraging ML algorithms, large volumes of sensor data can be analyzed in real time, enabling predictive analytics and automated decision-making. For instance, ML models can detect trends and irregularities in water quality data, delivering early alerts for contamination occurrences (Muharemi, Logofătu, & Leon, 2019). This enhances the responsiveness of monitoring systems and aids in resource optimization.

Furthermore, sustainability and cost optimization in nanotechnology present immense opportunities. Advances in green synthesis methods for nanomaterials can reduce environmental impacts and production costs. Additionally, the reuse and recycling of nanomaterials could be explored to enhance sustainability. For example, carbon-based nanomaterials such as graphene and CNTs have shown promise for reuse without significant loss of functionality (Guduru, Gupta, Pillai, & Dharaskar, 2021).

The integration of these advancements into water quality monitoring systems has the potential to revolutionize the field, making it more efficient, accessible, and environmentally friendly. By tackling the current challenges and leveraging these opportunities, nanotechnology can be instrumental in ensuring the global sustainability of water resources.

6.6 FUTURE PERSPECTIVES: NANOTECHNOLOGY AND AI SYNERGY IN WATER QUALITY MANAGEMENT

The combination of nanotechnology and AI represents a revolutionary approach to addressing water quality management challenges. This synergy enables innovative solutions that enhance detection, analysis, and remediation capabilities, paving the way for more sustainable and efficient water resource management. By leveraging the unique properties of nanoscale materials alongside AI's computational power, researchers and engineers are reimagining how water systems are monitored and maintained.

One of the most promising developments is the integration of nano-biosensors with AI. Nano-biosensors, which utilize nanomaterials such as graphene, CNTs, and gold nanoparticles, can detect contaminants at ultra-low concentrations with high specificity and sensitivity (Singh, 2011). These sensors gain enhanced functionality when paired with AI, such as real-time data processing and predictive analytics. AI algorithms can analyze the massive amounts of data generated by these sensors, identifying patterns and trends that might go unnoticed. For instance, ML models can be trained to distinguish between contaminants based on sensor data, enabling quicker and more accurate responses to water quality issues.

Autonomous water quality assessment systems are another significant trend. These systems combine AI-driven robotics with nanotechnology to offer self-operating solutions for monitoring water bodies. Equipped with nanosensors and advanced imaging technologies, autonomous devices can traverse large water systems, collecting and analyzing samples without human intervention (Diamond, Cleary, Maher, Kim, & Lau, 2011; Qadir et al., 2024). AI algorithms process these data to provide comprehensive assessments, flagging areas of concern and predicting potential future issues (Zsombók & Zsombók, 2023). Such systems are particularly valuable in remote or hazardous environments where manual monitoring is impractical or unsafe.

Developing next-generation nanomaterials is a critical area of exploration within this context. Intelligent polymers, for example, are gaining attention for their ability to respond to environmental stimuli such as pH, temperature, and chemical composition changes. These materials could be employed in filtration membranes or as part of reactive systems that neutralize specific pollutants (Dutta & De, 2017). Magnetic nanoparticles are another exciting avenue, as they can be engineered to target and remove contaminants like heavy metals or oil spills with high efficiency (Singh, Bhardwaj, Arya, & Khatri, 2020). These materials' behaviors can be optimized when integrated with AI systems, enabling tailored solutions for diverse water quality challenges.

Furthermore, the fusion of ML, big data analytics, and cloud computing with nanotechnology opens up new possibilities for water quality management. Cloud-based platforms can store and process the vast datasets generated by nanosensors, while ML algorithms extract actionable insights (Zhang, Liu, Zhou, Xu, & Lee, 2024). For instance, AI models can predict contamination events by analyzing historical and real-time data, enabling proactive interventions (Shang, Uber, Murray, & Janke, 2008). These systems can also optimize water treatment processes by continuously monitoring performance and suggesting adjustments based on nanosensor feedback. The scalability and adaptability of such integrations make them particularly appealing for urban and rural applications.

In conclusion, the convergence of nanotechnology and AI holds immense potential for revolutionizing water quality management. Emerging trends such as nano-biosensor integration and autonomous assessment systems showcase the power of this synergy, while research into advanced nanomaterials and ML-based solutions promises further advancements. As these technologies evolve, they will play a pivotal role in ensuring access to clean, safe water for communities worldwide.

6.7 CONCLUSION

Nanotechnology and ML have demonstrated transformative potential in water quality management, offering solutions that are efficient, scalable, and capable of addressing complex environmental challenges. The integration of these technologies has enabled innovative methods for monitoring, analyzing, and improving water systems, providing tools to safeguard water resources in the face of increasing global demand and environmental pressures. However, their full realization is still in progress, necessitating further research and a cohesive multidisciplinary approach.

The synergy between nanotechnology and ML has brought a paradigm shift in water quality assessment. Nanotechnology, with its ability to engineer materials at the atomic and molecular levels, offers unparalleled precision and sensitivity in detecting contaminants. Nanosensors and nanomaterials, for example, can detect pollutants at trace levels, allowing for real-time tracking of water quality parameters. When combined with the predictive and analytical power of ML, these technologies can transform raw data into actionable insights, facilitating timely interventions and proactive management. ML algorithms, trained on large datasets from water monitoring systems, can detect anomalies and predict trends, optimizing the decision-making process for water resource managers.

Despite these advancements, gaps in research and application persist. A significant limitation lies in the scalability of nanotechnology-based solutions. While laboratory-scale demonstrations have proven effective, the transition to field applications remains challenging. Factors such as cost, durability, and environmental impact of nanomaterials need to be addressed to ensure their widespread adoption. Similarly, developing ML models requires extensive, high-quality datasets, often unavailable in many regions, particularly developing countries. The lack of standardization in data collection and sharing further exacerbates this issue, creating a bottleneck in deploying AI-driven water quality management systems.

Another critical gap is the integration of these technologies into existing water management frameworks. Many current systems are not designed to incorporate advanced technologies, requiring substantial upgrades and policy adjustments. Moreover, the ethical and societal implications of using AI and nanotechnology, such as data privacy concerns and the potential environmental risks of nanomaterials, remain areas that demand careful consideration. Addressing these gaps will require collaborative efforts from researchers, policymakers, and industry stakeholders to develop both technologically advanced and socially and environmentally responsible holistic solutions.

The importance of a multidisciplinary approach in water quality management cannot be overstated. Global water systems' challenges are multifaceted, involving chemical, biological, physical, and socio-economic dimensions. Tackling these issues requires the combined expertise of scientists, engineers, environmentalists, and social scientists. For instance, while nanotechnologists may develop innovative materials for pollutant detection, environmental scientists can provide insights into ecosystem impacts, and data scientists can design robust ML algorithms for predictive analysis. Such cross-disciplinary collaborations ensure that solutions are comprehensive and contextually relevant, addressing the diverse needs of different regions and communities.

Furthermore, engaging stakeholders from various sectors, including government agencies, non-governmental organizations, and local communities, is crucial for successfully implementing advanced technologies. Public awareness and participation can significantly ensure that technological solutions are accepted and sustained over the long term. Education and capacity-building initiatives are also essential to equip future generations with the skills and knowledge needed to navigate the challenges of modern water quality management.

The continued advancement of nanotechnology and AI offers exciting possibilities for addressing water quality challenges. Emerging trends such as the development of nano-biosensors integrated with AI systems, autonomous water quality evaluation platforms, and next-generation nanomaterials like smart polymers and magnetic nanoparticles hold promise for revolutionizing the field. However, realizing this potential will require overcoming existing barriers through sustained investment in research, fostering global partnerships, and adopting inclusive policies that promote innovation while safeguarding ethical and environmental standards.

In conclusion, combining nanotechnology and ML represents a significant step forward in water quality management. While substantial progress has been made, considerable work remains to bridge the research, application, and integration gaps. By adopting a multidisciplinary approach and fostering collaboration among diverse stakeholders, we can harness the full potential of these transformative technologies to ensure the enduring management of water resources for future generations. As the global community grapples with the dual challenges of water scarcity and pollution, embracing advanced technologies with a holistic perspective offers a pathway to a resilient and water-secure future.

REFERENCES

Abu-Dief, A. M., & Hamdan, S. K. (2016). Functionalization of magnetic nano particles: synthesis, characterization, and their application in water purification. *Journal of Nanoscience and Nanotechnology*, *2*(3), 26–40.

Al Janabi, M. A. Y., El Houda Tiri, R. N., Cherif, A., Altuner, E. E., Lee, C.-J., Sen, F., … & Kalikeri, S. (2024). Hydrogen generation by methanolysis of $NaBH_4$ via efficient $CuFe_2O_4$ nanoparticle catalyst: a kinetic study and DNN model. *Topics in Catalysis*, *67*(9), 843–852.

Ali, A. A. (2023). The role of feature engineering in enhancing machine learning models.

Ali, Z., & Ahmad, R. (2020). Nanotechnology for water treatment. *Environmental Nanotechnology*, *3*, 143–163.

Altuner, E. E., Tiri, R. N. E. H., Aygun, A., Gulbagca, F., Sen, F., Iranbakhsh, A., … & Dragoi, E.-N. (2022). Hydrogen production and photocatalytic activities from $NaBH_4$ using trimetallic biogenic PdPtCo nanoparticles: development of machine learning model. *Chemical Engineering Research and Design*, *184*, 180–190.

Ameen, F., Altuner, E. E., Tiri, R. N. E., Gulbagca, F., Aygun, A., Sen, F., … & Dragoi, E. N. (2023). Highly active iron (II) oxide-zinc oxide nanocomposite synthesized Thymus vulgaris plant as bioreduction catalyst: characterization, hydrogen evolution and photocatalytic degradation. *International Journal of Hydrogen Energy*, *48*(55), 21139–21151.

Ansari, A. A., Shamim, M. A., Khan, A. M., Anwar, K., & Wani, A. A. (2023). Nanomaterials as a cutting edge in the removal of toxic contaminants from water. *Materials Chemistry and Physics*, *295*, 127092.

Arellano Vidal, C. L., & Govan, J. E. (2024). Machine learning techniques for improving nanosensors in agroenvironmental applications. *Agronomy*, *14*(2), 341.

Aris, F. A. F., Fauzi, F. N. A. M., Tong, W. Y., & Abdullah, S. S. S. (2019). Interaction of silver sulfadiazine with bacterial cellulose via ex-situ modification method as an alternative diabetic wound healing. *Biocatalysis and Agricultural Biotechnology*, *21*, 101332.

Asiri, A. M., Petrosino, F., Pugliese, V., Khan, S. B., Alamry, K. A., Alfifi, S. Y., … & Chakraborty, S. (2021). Synthesis and characterization of blended cellulose acetate membranes. *Polymers, 14*(1), 4.

Barabde, M., & Danve, S. (2015). Continuous water quality monitoring system for water resources at remote places. *International Journal of Innovative Research in Computer and Communication Engineering, 3*, 2320–9798.

Bassey, E., Whalley, J., Sallis, P., & Prasad, K. (2020). Wavelet transform smoothing filters for metal oxide gas sensor signal cleaning. *International Journal on Smart Sensing and Intelligent Systems*, *7*(5), 1–5.

Bassey, E. E., & Sallis, P. (2020). Integration of microfluidic sensors for interactive remote wireless data transmission. *Paper presented at the 7th International Conference on the Development of Biomedical Engineering in Vietnam (BME7)*. Springer, Singapore.

Bhavsar, H., & Panchal, M. H. (2012). A review on support vector machine for data classification. *International Journal of Advanced Research in Computer Engineering & Technology (IJARCET), 1*(10), 185–189.

Bora, T., & Dutta, J. (2014). Applications of nanotechnology in wastewater treatment: a review. *Journal of Nanoscience and Nanotechnology, 14*(1), 613–626.

Cao, W., Zhang, M., Ma, W., & Huang, C. (2023). Multifunctional electrospun nanofibrous membrane: an effective method for water purification. *Separation and Purification Technology*, *327*, 124952.

Chandanapalli, S. B., Reddy, E. S., & Lakshmi, D. R. (2019). Convolutional neural network for water quality prediction in WSN. *Journal of Networking and Communication Systems*, *2*(3), 40–47.

Chen, Y., Song, L., Liu, Y., Yang, L., & Li, D. (2020). A review of the artificial neural network models for water quality prediction. *Applied Sciences*, *10*(17), 5776.

Danyliuk, N., Tomaszewska, J., & Tatarchuk, T. (2020). Halloysite nanotubes and halloysite-based composites for environmental and biomedical applications. *Journal of Molecular Liquids*, *309*, 113077.

Darabi, R., Karimi-Maleh, H., Shahidi, S.-A., Gulbagca, F., Tiri, R. N. E., Sefidan, A. M., … & Altuner, E. E. (2023). Hexamethylenetetramine-based nanomaterial synthesis, characterization, investigation of photocatalytic activity against Rhodamine B dye and kinetic study for hydrogen fuel production. *Fuel*, *352*, 128841.

Davey, R. (2022). *Discussions About Water Filtration Using Adsorptive Nanocomposite Membranes*. Azo Materials. https://www.azom.com/news.aspx?newsID=59733

Dev, B., Rahman, M. A., Islam, M. J., Rahman, M. Z., & Zhu, D. (2024). Properties prediction of composites based on machine learning models: a focus on statistical index approaches. *Materials Today Communications*, *38*, 107659.

Diamond, D., Cleary, J., Maher, D., Kim, J., & Lau, K. T. (2011). Autonomous analyser platforms for remote monitoring of water quality. *Paper presented at the IECON 2011–37th Annual Conference of the IEEE Industrial Electronics Society*, Melbourne, VIC.

Dogo, E. M., Nwulu, N. I., Twala, B., & Aigbavboa, C. (2019). A survey of machine learning methods applied to anomaly detection on drinking-water quality data. *Urban Water Journal, 16*(3), 235–248. https://doi.org/10.1080/1573062X.2019.1637002

Dutta, K., & De, S. (2017). Smart responsive materials for water purification: an overview. *Journal of Materials Chemistry A*, *5*(42), 22095–22112.

El-Aswar, E. I., Ramadan, H., Elkik, H., & Taha, A. G. (2022). A comprehensive review on preparation, functionalization and recent applications of nanofiber membranes in wastewater treatment. *Journal of Environmental Management*, *301*, 113908.

Erduran, V., Altuner, E. E., & Şen, F. (2023). Stability and validation of bionanomaterials. In Öztürk, M. A., Roy, A., Bhat, R. A., Vardar-Sukan, F., & Tonelli, F. M. P. (Eds.) *Synthesis of Bionanomaterials for Biomedical Applications* (pp. 251–263): Elsevier, Amsterdam.

Feldman, A., Kendler, S., Marshall, J., Kushwaha, M., Sreekanth, V., Upadhya, A. R., … & Fishbain, B. (2023). Urban air-quality estimation using visual cues and a deep convolutional neural network in Bengaluru (Bangalore), India. *Environmental Science & Technology*, *58*(1), 480–487.

Gavabari, S. E., Goudarzi, A., Shahrousvand, M., & Asfaram, A. (2024). Preparation of novel polyurethane/activated carbon/cellulose nano-whisker nanocomposite film as an efficient adsorbent for the removal of methylene blue and basic violet 16 dyes from wastewater. *Separation and Purification Technology*, *330*, 125285.

Geetha, M., Bonthula, S., Al-Maadeed, S., Al-Lohedan, H., Rajabathar, J. R., Arokiyaraj, S., & Sadasivuni, K. K. (2023). Research trends in smart cost-effective water quality monitoring and modeling: special focus on artificial intelligence. *Water*, *15*(18), 3293. https://doi.org/10.3390/W15183293

Gleick, P. H. (1998). Water in crisis: paths to sustainable water use. *Ecological Applications*, *8*(3), 571–579.

Golbabaei, M. H., Varnoosfaderani, M. S., Hemmati, F., Barati, M. R., Pishbin, F., & Ebrahimi, S. A. S. (2024). Machine learning-guided morphological property prediction of 2D electrospun scaffolds: the effect of polymer chemical composition and processing parameters. *RSC Advances*, *14*(22), 15178–15199.

Guduru, R. K., Gupta, A. A., Pillai, P., & Dharaskar, S. (2021). Applications of carbon-based nanomaterials for wastewater treatment. In Shadpour Esteleky Mallakpour, S. E., & Hussain, C. M. (Eds.) *Environmental Applications of Carbon Nanomaterials-Based Devices* (pp. 87–133). Wiley-VCH, Weinheim.
Guerra, F. D., Attia, M. F., Whitehead, D. C., & Alexis, F. (2018). Nanotechnology for environmental remediation: materials and applications. *Molecules*, *23*(7), 1760.
He, X., Cheng, L., Wang, Y., Zhao, J., Zhang, W., & Lu, C. (2014). Aerogels from quaternary ammonium-functionalized cellulose nanofibers for rapid removal of Cr (VI) from water. *Carbohydrate Polymers*, *111*, 683–687.
Holzinger, A., Weippl, E., Tjoa, A. M., & Kieseberg, P. (2021). Digital transformation for sustainable development goals (SDGS): a security, safety and privacy perspective on AI. *Paper presented at the International Cross-Domain Conference for Machine Learning and Knowledge Extraction*, Vienna, Austria.
Hong, S. K., Bae, S., Jeon, H., Kim, M., Cho, S. J., & Lim, G. (2018). An underwater superoleophobic nanofibrous cellulosic membrane for oil/water separation with high separation flux and high chemical stability. *Nanoscale*, *10*(6), 3037–3045.
Howes, P. D., Chandrawati, R., & Stevens, M. M. (2014). Colloidal nanoparticles as advanced biological sensors. *Science*, *346*(6205), 1247390.
Jahan, S., & Singh, A. (2023). Causes and impact of industrial effluents on receiving water bodies: a review. *Malaysian Journal of Science and Advanced Technology*, *3*, 111–121.
Jiang, N., Tansukawat, N. D., Gonzalez-Macia, L., Ates, H. C., Dincer, C., Guder, F., ... & Yetisen, A. K. (2021). Low-cost optical assays for point-of-care diagnosis in resource-limited settings. *ACS Sensors*, *6*(6), 2108–2124.
Karami, A., & Guerrero-Zapata, M. (2015). A fuzzy anomaly detection system based on hybrid PSO-Kmeans algorithm in content-centric networks. *Neurocomputing*, *149*, 1253–1269.
Karimi, F., Altuner, E. E., Aygun, A., Bayat, R., Rajendran, S., & Sen, F. (2024). Synthesis of silver nanoparticles by biogenic methods: characterization and development of a sensor sensible to pharmaceutical medicine paracetamol. *Topics in Catalysis*, *67*(9), 585–593.
Kim, H. N., Ren, W. X., Kim, J. S., & Yoon, J. (2012). Fluorescent and colorimetric sensors for detection of lead, cadmium, and mercury ions. *Chemical Society Reviews*, *41*(8), 3210–3244.
Koedrith, P., Thasiphu, T., Weon, J.-I., Boonprasert, R., Tuitemwong, K., & Tuitemwong, P. (2015). Recent trends in rapid environmental monitoring of pathogens and toxicants: potential of nanoparticle-based biosensor and applications. *The Scientific World Journal*, *2015*(1), 510982.
Konstantopoulos, G., Koumoulos, E. P., & Charitidis, C. A. (2022). Digital innovation enabled nanomaterial manufacturing; machine learning strategies and green perspectives. *Nanomaterials*, *12*(15), 2646.
Kumar, H., Kuča, K., Bhatia, S. K., Saini, K., Kaushal, A., Verma, R., ... & Kumar, D. (2020). Applications of nanotechnology in sensor-based detection of foodborne pathogens. *Sensors*, *20*(7), 1966.
Kumar, P., Qureshi, U. A., Brohi, K. M., Ahmed, F., Khatri, Z., Kim, I. S., ... & Memon, A. H. (2019). Mercury (Hg) ions removal from aqueous solution using thiol functionalized electrospun nanofiber membranes. *AIP Conference Proceedings* (vol. 2119, p. 020020).
Kumar, R., & Goyal, M. K. (2025). Role of AI&ML in modernizing water and wastewater treatment processes. *Water, Air, & Soil Pollution*, *236*(1), 1–22.
Kumar, S. (2023). Smart and innovative nanotechnology applications for water purification. *Hybrid Advances*, *3*, 100044.
Kumar, V., & Garg, M. (2018). Deep learning as a frontier of machine learning: A. *International Journal of Computer Applications*, *975*, 8887.
Kuswandi, B. (2018). Nanobiosensors for detection of micropollutants. *Environmental Nanotechnology*, *1*, 125–158.
Li, N., Tittl, A., Yue, S., Giessen, H., Song, C., Ding, B., & Liu, N. (2014). DNA-assembled bimetallic plasmonic nanosensors. *Light: Science & Applications*, *3*(12), e226–e226.
Lin, S., Kim, J., Hua, C., Kang, S., & Park, M.-H. (2023). Comparing artificial and deep neural network models for prediction of coagulant amount and settled water turbidity: lessons learned from big data in water treatment operations. *Journal of Water Process Engineering*, *54*, 103949.
Liu, P., Wang, J., Sangaiah, A. K., Xie, Y., & Yin, X. (2019). Analysis and prediction of water quality using LSTM deep neural networks in IoT environment. *Sustainability*, *11*(7), 2058.
Liu, W., Wu, Y., Hong, Y., Zhang, Z., Yue, Y., & Zhang, J. (2022). Applications of machine learning in computational nanotechnology. *Nanotechnology*, *33*(16), 162501.
Lowe, M., Qin, R., & Mao, X. (2022). A review on machine learning, artificial intelligence, and smart technology in water treatment and monitoring. *Water*, *14*(9), 1384.
MacAlister, C., & Subramanyam, N. (2018). Climate change and adaptive water management: innovative solutions from the global South. *Water International*, 43, 133–144.
Makarigakis, A. K., & Jimenez-Cisneros, B. E. (2019). UNESCO's contribution to face global water challenges. *Water*, *11*(2), 388.

Mao, Y., & Lee, S. (2019). Deep convolutional neural network for air quality prediction. *Journal of Physics: Conference Series, 1302*, 032046.

McGuinness, N B., Garvey, M., Whelan, A., John, H., Zhao, C., Zhang, G., ... & Pillai, S. C. (2015). Nanotechnology solutions for global water challenges. In Ahuja, S., de Andrade, J. B., Dionysiou, D. D., Hristovski, K. D., & Loganathan, B. G. (Eds.) *Water Challenges and Solutions on a Global Scale* (pp. 375–411). ACS Publications, Washington, DC.

Mitra, M., Kumar, A., Yadav, L., & Kaur, M. (2024). Advancing industrial effluent monitoring systems for environmental protection and sustainable development. *Paper presented at the 2024 11th International Conference on Reliability, Infocom Technologies and Optimization (Trends and Future Directions)(ICRITO)*, Noida, India.

Muharemi, F., Logofătu, D., & Leon, F. (2019). Machine learning approaches for anomaly detection of water quality on a real-world data set. *Journal of Information and Telecommunication, 3*(3), 294–307.

Mutlu, G., Yıldırım, F., Ulus, H., & Eskizeybek, V. (2023). Coating graphene nanoplatelets onto carbon fabric with controlled thickness for improved mechanical performance and EMI shielding effectiveness of carbon/epoxy composites. *Engineering Fracture Mechanics, 284*, 109271.

O'Keefe, T. L., & Haynes, C. L. (2024). Review of oxidative dissolution and sulfidation of select nanoparticles in the environment: impact on applications. *ACS Applied Nano Materials, 7*(8), 8392–8406. https://doi.org/10.1021/acsanm.3c06227

Paimard, G., Ghasali, E., & Baeza, M. (2023). Screen-printed electrodes: fabrication, modification, and biosensing applications. *Chemosensors, 11*(2), 113.

Park, J., Kim, K. T., & Lee, W. H. (2020). Recent advances in information and communications technology (ICT) and sensor technology for monitoring water quality. *Water, 12*(2), 510. https://doi.org/10.3390/W12020510

Park, Y.-G., Nam, S.-N., Jang, M., Park, C. M., Her, N., Sohn, J., ... & Yoon, Y. (2022). Boron nitride-based nanomaterials as adsorbents in water: a review. *Separation and Purification Technology, 288*, 120637.

Pervez, M. N., Yeo, W. S., Mishu, M. R., Buonerba, A., Zhao, Y., Cai, Y., ... & Naddeo, V. (2023). Prediction of the diameter of biodegradable electrospun nanofiber membranes: an integrated framework of taguchi design and machine learning. *Journal of Polymers and the Environment, 31*(9), 4080–4096.

Peter, K. T., Johns, A. J., Myung, N. V., & Cwiertny, D. M. (2017). Functionalized polymer-iron oxide hybrid nanofibers: electrospun filtration devices for metal oxyanion removal. *Water Research, 117*, 207–217.

Qadir, M. I., Mumtaz, R., Manzoor, M., Saleem, M., Khan, M. A., & Charlesworth, S. (2024). Unmanned surface vehicle for intelligent water quality assessment to promote sustainable human health. *Water Supply, 24*, 2259–2270.

Qin, D., Yu, J., Zou, G., Yong, R., Zhao, Q., & Zhang, B. (2019). A novel combined prediction scheme based on CNN and LSTM for urban PM 2.5 concentration. *IEEE Access, 7*, 20050–20059.

Rabalais, N. N., & Turner, R. E. (2019). Gulf of Mexico hypoxia: past, present, and future. *Limnology and Oceanography Bulletin, 28*(4), 117–124.

Safder, U., Kim, J., Pak, G., Rhee, G., & You, K. (2022). Investigating machine learning applications for effective real-time water quality parameter monitoring in full-scale wastewater treatment plants. *Water, 14*(19), 3147.

Sagan, V., Peterson, K. T., Maimaitijiang, M., Sidike, P., Sloan, J., Greeling, B. A., ... & Adams, C. (2020). Monitoring inland water quality using remote sensing: potential and limitations of spectral indices, bio-optical simulations, machine learning, and cloud computing. *Earth-Science Reviews, 205*, 103187.

Salazar Sandoval, S., Bruna, T., Maldonado-Bravo, F., Jara, P., Caro, N., Rojas-Romo, C., ... & Urzúa, M. (2023). Nanomaterials for potential detection and remediation: a review of their analytical and environmental applications. *Coatings, 13*(12), 2085.

Seyednourani, M., Akgun, S., Ulus, H., Yildiz, M., & Sas, H. S. (2023). Experimental investigation on compression-after-impact (CAI) response of aerospace grade thermoset composites under low-temperature conditions assisted with acoustic emission monitoring. *Composite Structures, 321*, 117260.

Shang, F., Uber, J., Murray, R., & Janke, R. (2008). Model-based real-time detection of contamination events. In *Water Distribution Systems Analysis 2008* (pp. 1–5), South Africa.

Sharmila, V. G., Kumar, M. D., & Tamilarasan, K. (2024). Machine learning-driven advances in metal-organic framework nanomaterials for wastewater treatment: developments and challenges. *Separation & Purification Reviews*, 1, 1–21.

Singh, H., Bhardwaj, N., Arya, S. K., & Khatri, M. (2020). Environmental impacts of oil spills and their remediation by magnetic nanomaterials. *Environmental Nanotechnology, Monitoring & Management, 14*, 100305.

Singh, R. P. (2011). Prospects of nanobiomaterials for biosensing. *International Journal of Electrochemistry, 2011*(1), 125487.

Skarbøvik, E., & Roseth, R. (2015). Use of sensor data for turbidity, pH and conductivity as an alternative to conventional water quality monitoring in four Norwegian case studies. *Acta Agriculturae Scandinavica, Section B: Soil & Plant Science, 65*(1), 63–73. https://doi.org/10.1080/09064710.2014.966751

Song, X., Wang, Y., Jiao, C., Huang, M., Wang, G.-H., & Jiang, H. (2020). Microstructure regulation of polyamide nanocomposite membrane by functional mesoporous polymer for high-efficiency desalination. *Journal of Membrane Science*, *597*, 117783.

Spirescu, V. A., Chircov, C., Grumezescu, A. M., Vasile, B. Ş., & Andronescu, E. (2021). Inorganic nanoparticles and composite films for antimicrobial therapies. *International Journal of Molecular Sciences*, *22*(9), 4595.

Sreeprasad, T. S., Maliyekkal, S. M., Lisha, K. P., & Pradeep, T. (2011). Reduced graphene oxide–metal/metal oxide composites: facile synthesis and application in water purification. *Journal of Hazardous Materials*, *186*(1), 921–931.

Sukpancharoen, S., Wijakmatee, T., Katongtung, T., Ponhan, K., Rattanachoung, N., & Khojitmate, S. (2024). Data-driven prediction of electrospun nanofiber diameter using machine learning: a comprehensive study and web-based tool development. *Results in Engineering*, *24*, 102826.

Susto, G. A., Beghi, A., & McLoone, S. (2017). Anomaly detection through on-line isolation forest: an application to plasma etching. *Paper presented at the 2017 28th Annual SEMI Advanced Semiconductor Manufacturing Conference (ASMC)*, Saratoga Springs, NY.

Thakur, A., & Devi, P. (2024). A comprehensive review on water quality monitoring devices: materials advances, current status, and future perspective. *Critical Reviews in Analytical Chemistry*, *54*(2), 193–218.

Tınastepe, M. T., Kaybal, H. B., Ulus, H., Erdal, M. O., Çetin, M. E., & Avcı, A. (2023). Quasi-static tensile loading performance of bonded, bolted, and hybrid bonded-bolted carbon-to-carbon composite joints: effect of recycled polystyrene nanofiber interleaving. *Composite Structures*, *323*, 117445.

Tripathy, A., Patne, A. Y., Mohapatra, S., & Mohapatra, S. S. (2024). Convergence of nanotechnology and machine learning: the state of the art, challenges, and perspectives. *International Journal of Molecular Sciences*, 25(22), 12368. https://doi.org/10.3390/IJMS252212368

Trupp, F. J., Cibils, R., & Goyanes, S. (2022). Poly (vinylidene fluoride) electrospun nonwovens morphology: prediction and optimization of the size and number of beads on fibers through response surface methodology and machine learning regressions. *Journal of Industrial Textiles*, *51*(5_suppl), 9071S–9096S.

Ulus, H., Kaybal, H. B., Berber, N. E., Tatar, A. C., Ekrem, M., Ataberk, N., & Avci, A. (2022). An experimental evaluation on the dynamic response of water aged composite/aluminium adhesive joints: influence of electrospun nanofibers interleaving. *Composite Structures*, *280*, 114852.

Ulus, H., Kaybal, H. B., Cacık, F., Eskizeybek, V., & Avcı, A. (2022). Fracture and dynamic mechanical analysis of seawater aged aluminum-BFRP hybrid adhesive joints. *Engineering Fracture Mechanics*, *268*, 108507.

Ursino, C., Castro-Muñoz, R., Drioli, E., Gzara, L., Albeirutty, M. H., & Figoli, A. (2018). Progress of nanocomposite membranes for water treatment. *Membranes*, *8*(2), 18.

Von der Kammer, F., Ferguson, P. L., Holden, P. A., Masion, A., Rogers, K. R., Klaine, S. J., ... & Unrine, J. M. (2012). Analysis of engineered nanomaterials in complex matrices (environment and biota): general considerations and conceptual case studies. *Environmental Toxicology and Chemistry*, *31*(1), 32–49.

Wan, H., Xu, Z., Islam, M. S., Tarannum, T., Shi, K., Mills, R., ... & Ormsbee, L. (2023). Catalytic membrane for groundwater remediation of chlorinated organics: NF-assisted strategy and cost aspects. *SSRN Electronic Journal*. https://doi.org/10.2139/ssrn.4343259

Wang, C., Su, X., Brown, T. D., & Ohodnicki Jr, P. R. (2015). Novel sensing materials for harsh environment subsurface pH sensing applications. *Paper presented at the Fiber Optic Sensors and Applications XII,* Baltimore, MD.

Wang, J. (2005). Nanomaterial-based amplified transduction of biomolecular interactions. *Small*, *1*(11), 1036–1043.

Wang, Z., & Lu, Y. (2009). Functional DNA directed assembly of nanomaterials for biosensing. *Journal of Materials Chemistry*, *19*(13), 1788–1798.

Warner, C. L., Addleman, R. S., Cinson, A. D., Droubay, T. C., Engelhard, M. H., Nash, M. A., ... & Warner, M. G. (2010). High-performance, superparamagnetic, nanoparticle-based heavy metal sorbents for removal of contaminants from natural waters. *ChemSusChem*, *3*(6), 749–757.

Williams, A., Flynn, K. J., Xia, Z., & Dunstan, P. R. (2016). Multivariate spectral analysis of pH SERS probes for improved sensing capabilities. *Journal of Raman Spectroscopy*, *47*(7), 819–827.

Wu, Y., Altuner, E. E., Tiri, R. N. E. H., Bekmezci, M., Gulbagca, F., Aygun, A., ... & Karimi-Maleh, H. (2023). Hydrogen generation from methanolysis of sodium borohydride using waste coffee oil modified zinc oxide nanoparticles and their photocatalytic activities. *International Journal of Hydrogen Energy*, *48*(17), 6613–6623.

Wu, Y., Tiri, R. N. E., Bekmezci, M., Altuner, E. E., Aygun, A., Mei, C., ... & Sen, F. (2023). Synthesis of novel activated carbon-supported trimetallic Pt–Ru–Ni nanoparticles using wood chips as efficient catalysts for the hydrogen generation from $NaBH_4$ and enhanced photodegradation on methylene blue. *International Journal of Hydrogen Energy*, *48*(55), 21055–21065.

Xi, Z. (2022). Functional nanomaterials design in the workflow of building machine-learning models. *Paper presented at the Future of Information and Communication Conference,* Cornell University, Ithaca, NY.

Yan, W. (2017). Testing the waters: from moringa seeds to fruit peels, researchers are seeking out new ways to clean up the world? s drinking water. *IEEE Pulse*, *8*(6), 23–28.

Yang, S., Tang, R., Dai, Y., Wang, T., Zeng, Z., & Zhang, L. (2021). Fabrication of cellulose acetate membrane with advanced ultrafiltration performances and antibacterial properties by blending with HKUST-1@ LCNFs. *Separation and Purification Technology*, *279*, 119524.

Yao, X., Li, J., Wang, Z., Kong, L., & Wang, Y. (2015). Highly permeable and robust membranes assembled from block-copolymer-functionalized carbon nanotubes. *Journal of Membrane Science*, *493*, 224–231.

Yildirim, C., Ulus, H., Beylergil, B., Al-Nadhari, A., Topal, S., & Yildiz, M. (2023). Effect of atmospheric plasma treatment on Mode-I and Mode-II fracture toughness properties of adhesively bonded carbon fiber/PEKK composite joints. *Engineering Fracture Mechanics*, *289*, 109463.

Yildirim, C., Ulus, H., Beylergil, B., Al-Nadhari, A., Topal, S., & Yildiz, M. (2024). Tailoring adherend surfaces for enhanced bonding in CF/PEKK composites: comparative analysis of atmospheric plasma activation and conventional treatments. *Composites Part A: Applied Science and Manufacturing*, *180*, 108101.

Yildirim, C., Ulus, H., Sas, H. S., Topal, S., & Yildiz, M. (2024). Assessing the fracture and dynamic mechanical performance of CF/PEKK joints bonded with epoxy-based adhesive film for aerospace applications: impact of thermal and cycling hygrothermal conditions. *Composites Part A: Applied Science and Manufacturing*, *190*, 108659.

Yuan, X., Li, J., Lim, J. Y., Zolfaghari, A., Alessi, D. S., Wang, Y., ... & Ok, Y. S. (2023). Machine learning for heavy metal removal from water: recent advances and challenges. *ACS ES&T Water*, *4*(3), 820–836.

Zainurin, S. N., Wan Ismail, W. Z., Mahamud, S. N. I., Ismail, I., Jamaludin, J., Ariffin, K. N. Z., & Wan Ahmad Kamil, W. M. (2022). Advancements in monitoring water quality based on various sensing methods: a systematic review. *International Journal of Environmental Research and Public Health*, *19*(21), 14080.

Zhang, Z., Liu, X., Zhou, H., Xu, S., & Lee, C. (2024). Advances in machine-learning enhanced nanosensors: from cloud artificial intelligence toward future edge computing at chip level. *Small Structures*, *5*(4), 2300325.

Zhao, Y., Song, X., Song, Q., & Yin, Z. (2012). A facile route to the synthesis copper oxide/reduced graphene oxide nanocomposites and electrochemical detection of catechol organic pollutant. *CrystEngComm*, *14*(20), 6710–6719.

Zheng, H., Cheng, Y., & Li, H. (2020). Investigation of model ensemble for fine-grained air quality prediction. *China Communications*, *17*(7), 207–223.

Zhou, Z., Liu, L., & Yuan, W. (2019). A superhydrophobic poly (lactic acid) electrospun nanofibrous membrane surface-functionalized with TiO_2 nanoparticles and methyltrichlorosilane for oil/water separation and dye adsorption. *New Journal of Chemistry*, *43*(39), 15823–15831.

Zoubir, J., Elkhotfi, Y., Tounsi, A., Assabbane, A., & Bakas, I. (2023). A highly sensitive chloramphenicol detection sensor based on a carbon-graphite nanocomposite modified with silver nanoparticles application: human serum and urine. *Monatshefte für Chemie-Chemical Monthly*, *154*(8), 837–848.

Zsombók, A., & Zsombók, I. (2023). Revolutionizing predictive maintenance: how AI-driven solutions enhance efficiency and reduce costs across industries. *International Journal of Science and Research (IJSR)*, *12*(12), 1168–1171.

PART III

Predictive Modeling for Effective Water Resource Management

7 AI-Driven Predictive Maintenance and Downtime Optimization in Water Treatment Plants

Paramasivam Balakrishnan and Marimuthu Subramani

7.1 INTRODUCTION

7.1.1 Importance of Water Treatment Plants and Maintenance

Water treatment plants are fundamental to modern infrastructure, with their comprehensive operational framework depicted in Figure 7.1. The diagram illustrates the systematic integration of core water treatment operations with three primary impact areas, showing how these facilities serve multiple critical functions for society. The core water treatment operations encompass six essential processes: water intake, chemical treatment, filtration, disinfection, quality testing, and distribution. These operations form the foundation for ensuring public health, supporting economic activities, and maintaining environmental sustainability.

The figure also highlights how water treatment facilities contribute to public health through disease prevention and safe drinking water provision, while simultaneously supporting economic activities through industrial operations and agricultural support. The environmental conservation component, as depicted in the diagram, emphasizes ecosystem protection and pollution prevention measures. Particularly noteworthy is the AI integration systems layer, which represents the technological evolution in water treatment. This advancement manifests through three key components: predictive maintenance (PdM), process optimization, and early warning systems.

As demonstrated by Afan et al. (2024), the evaluation of four machine learning (ML) models, namely multilayer perceptron (MLP), Gaussian Fuzzy Forest Regression (GFFR), multilayer perceptron–principal component analysis (MLP-PCA), and Radial Basis Function (RBF), for predicting water parameters showed GFFR's superior performance. Their research revealed that the GFFR model achieved

DOI: 10.1201/9781003600718-10

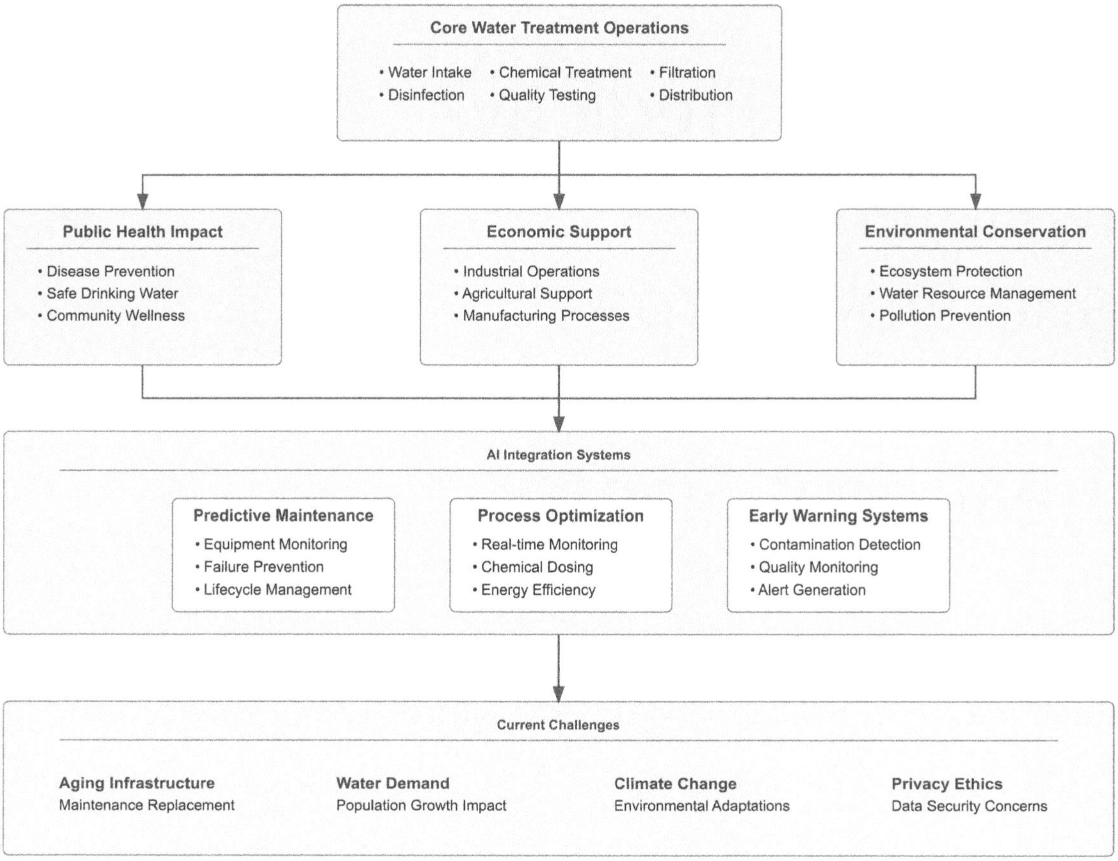

FIGURE 7.1 AI-driven framework for water treatment operations: benefits and challenges.

high accuracy in conductivity ($R=0.893$) with lower deviations (Normalized Root Mean Square Error (NRMSE)=0.091, Normalized Mean Absolute Error (NMAE)=0.071). The implementation of a multistep modeling approach significantly boosted prediction accuracy, averaging a 60% improvement. The exploration of two distinct modeling scenarios, with the second scenario incorporating intermediate process outputs, demonstrated enhanced predictive performance. These comprehensive findings underscore the critical importance of AI-driven optimization in wastewater management for sustainable water treatment.

The bottom section addresses current challenges facing water treatment facilities, including aging infrastructure, increasing water demand, climate change impacts, and privacy ethics concerns. This aligns with research by Gyparakis et al. (2024), which demonstrated how AI-driven PdM can help address these challenges through optimized chemical dosage estimation and improved operational efficiency. Their extensive study tested 304 artificial neural network (ANN) models, ultimately selecting an optimal configuration with 100 neural networks, 42 hidden nodes, 10 inputs, and 4 outputs. This sophisticated model demonstrated remarkable accuracy in predicting key operational parameters, thereby improving process reliability. The approach effectively utilized measurable *in situ* data from the Aposelemis Drinking Water Treatment Plant over 38 months, ensuring robust data-driven decision-making. The results conclusively suggest that ANNs can significantly reduce operational uncertainties, leading to better resource utilization. Future research directions may focus on refining these models with fewer parameters while maintaining high predictive accuracy.

Looking ahead, water treatment plants must strategically integrate these technological improvements while systematically addressing the challenges to ensure sustainable water management for future generations. The diagram effectively captures this complex interplay between traditional operations, technological advancement, and emerging challenges in the water treatment sector.

7.1.2 Evolution of Maintenance Strategies: Reactive vs. Preventive vs. Predictive

AI-driven PdM has fundamentally transformed water treatment operations by effectively reducing downtime and improving operational efficiency. In contrast to reactive and preventive approaches, PdM leverages sophisticated real-time sensor data and advanced ML algorithms to anticipate potential failures. According to comprehensive research by Mallioris et al. (2024), which reviewed 304 PdM applications, these systems can be systematically classified based on methodologies, algorithms, and industry-specific factors. Their findings convincingly highlight a 60% improvement in operational efficiency through AI-driven strategies. The research also effectively maps sector-specific challenges and readiness levels for PdM adoption, providing valuable insights for implementation. This data-driven approach significantly enhances decision-making processes, ensuring reliable and cost-effective water treatment management. Supporting these findings, Stengel et al. (2024) demonstrated that AI-driven PdM substantially enhances water treatment efficiency by proactively managing asset failures. Through the implementation of ML, specifically random forest (RF) classification, their study successfully identifies high-risk assets in wastewater distribution networks. The proposed predictive tool effectively minimizes environmental risks and financial losses, achieving up to a 60% reduction in asset failures. Their research emphasizes that traditional reactive strategies often lead to unexpected costs, whereas data-driven approaches optimize maintenance planning. The model, developed using Python, efficiently prioritizes critical maintenance actions, ensuring consistent operational reliability. This proactive framework demonstrates significant improvements in service delivery while reducing the likelihood of costly system failures.

7.1.3 Role of AI, ML, and Real-Time Analytics in Predictive Maintenance

AI-driven PdM has revolutionized water treatment processes by significantly enhancing decision-making capabilities and process control mechanisms. A comprehensive review of 91 peer-reviewed studies by Stengel et al. (2024) effectively highlights AI's crucial role in improving predictive capabilities across various aspects, including coagulant dosages, pollutant removal, and membrane filtration. Their research demonstrates that deep learning (DL) excels particularly in feature extraction, while hybrid AI–regression techniques consistently improve prediction accuracy. The implementation of image recognition-based DL frameworks further refines treatment efficiency through sophisticated analysis of floc shapes. These technological advancements consistently lead to more cost-effective operations and reduced downtime in water treatment plants. Future research directions aim to enhance AI-driven optimization for improved water quality management. Complementing these findings, Cechinel et al. (2024) demonstrated that AI-driven PdM significantly enhances wastewater treatment through optimized chemical oxygen demand (COD) predictions. Their detailed study evaluates various ML models, including MLP, Long Short-Term Memory (LSTM), Support Vector Machine (SVM), and RF, to accurately predict COD concentration in effluent. The research revealed that MLP performed optimally for daily datasets, while LSTM excelled with hourly data by effectively leveraging historical trends. Notably, SVM consistently outperformed traditional models when validated with real effluent measurements, highlighting influent total suspended solids (TSS) as a crucial factor. These advanced models demonstrably improve treatment efficiency while reducing operational downtime and associated costs. Future research initiatives aim to further refine AI-based solutions for enhanced water quality management.

7.2 FUNDAMENTALS OF PREDICTIVE MAINTENANCE (PDM) IN WATER TREATMENT

7.2.1 Definition and Core Principles of PdM

PdM is a proactive approach that leverages data-driven insights to anticipate equipment failures before they occur, reducing downtime and operational costs. Unlike traditional reactive maintenance, PdM uses advanced analytics, ML, and real-time monitoring to assess equipment health and optimize maintenance schedules. This enhances reliability, extends asset lifespan, and improves safety by minimizing unexpected failures. However, challenges such as data integration, system compatibility, and workforce training must be addressed for effective implementation. As industries embrace digital transformation, PdM plays a crucial role in achieving efficient, cost-effective, and sustainable operations.

This advanced methodology optimizes maintenance schedules through condition-based maintenance (CBM) of critical parameters such as temperature and vibration. The implementation of PdM significantly enhances asset reliability, extends equipment lifespan, and improves operational efficiency across various industries, particularly in water treatment and manufacturing sectors. Recent research by Yin et al. (2024) has introduced the physics-constrained dynamic mode decomposition (PCDMD) method, which enhances PdM in water treatment by seamlessly integrating physical constraints with data-driven modeling. Their findings demonstrate a notable reduction in reconstruction and prediction errors by 1%–10%, particularly improving accuracy in scenarios involving noisy or imperfect datasets. Through the refinement of predictions using a Kalman filter and physics-informed corrections, PCDMD ensures that results maintain both realism and reliability. J. Y. Li et al. (2024) further validate these findings, demonstrating that ML models analyzing historical and live data can improve failure prediction accuracy by up to 90%, while significantly reducing unexpected breakdowns and optimizing resource allocation.

7.2.2 Key Technologies: AI, ML, IoT, and Big Data

The integration of AI, ML, Internet of Things (IoT), and Big Data is transforming water treatment by enhancing efficiency, sustainability, and resource management. These technologies enable real-time monitoring, PdM, and early anomaly detection, ensuring continuous service delivery. IoT-driven smart metering improves water usage insights, while Big Data analytics optimize decision-making despite challenges in handling vast datasets. However, barriers like high implementation costs, data security concerns, and regulatory compliance slow widespread adoption. Overcoming these hurdles through innovation and collaboration will drive sustainable advancements in water management, improving reliability and public health outcomes. Figure 7.2 shows that the framework for PdM consists of four key layers: data collection layer, which gathers data from sensors, IoT devices, and monitoring systems; data integration layer, where enterprise resource planning (ERP) systems, Computerized Maintenance Management System (CMMS), and IoT platforms consolidate data; analytics layer, which utilizes ML, predictive models, and pattern recognition to derive insights; and implementation layer, where maintenance planning, resource allocation, and performance tracking drive actionable outcomes. This structured approach leads to cost reduction, extended equipment lifespan, improved safety, and enhanced reliability.

According to Xu et al. (2024), AI-driven PdM utilizes sophisticated ML models, including ANN, adaptive neuro-fuzzy inference system (ANFIS), and SVM, to optimize pollutant removal and water quality predictions. These AI techniques have proven particularly effective in predicting crucial parameters such as biochemical oxygen demand (BOD), COD, total nitrogen (TN), and total phosphorus (TP). Velásquez et al. (2024) expand on these findings, describing how a comprehensive visual analytics platform enables operators to build and implement ML models, resulting in improved energy efficiency and

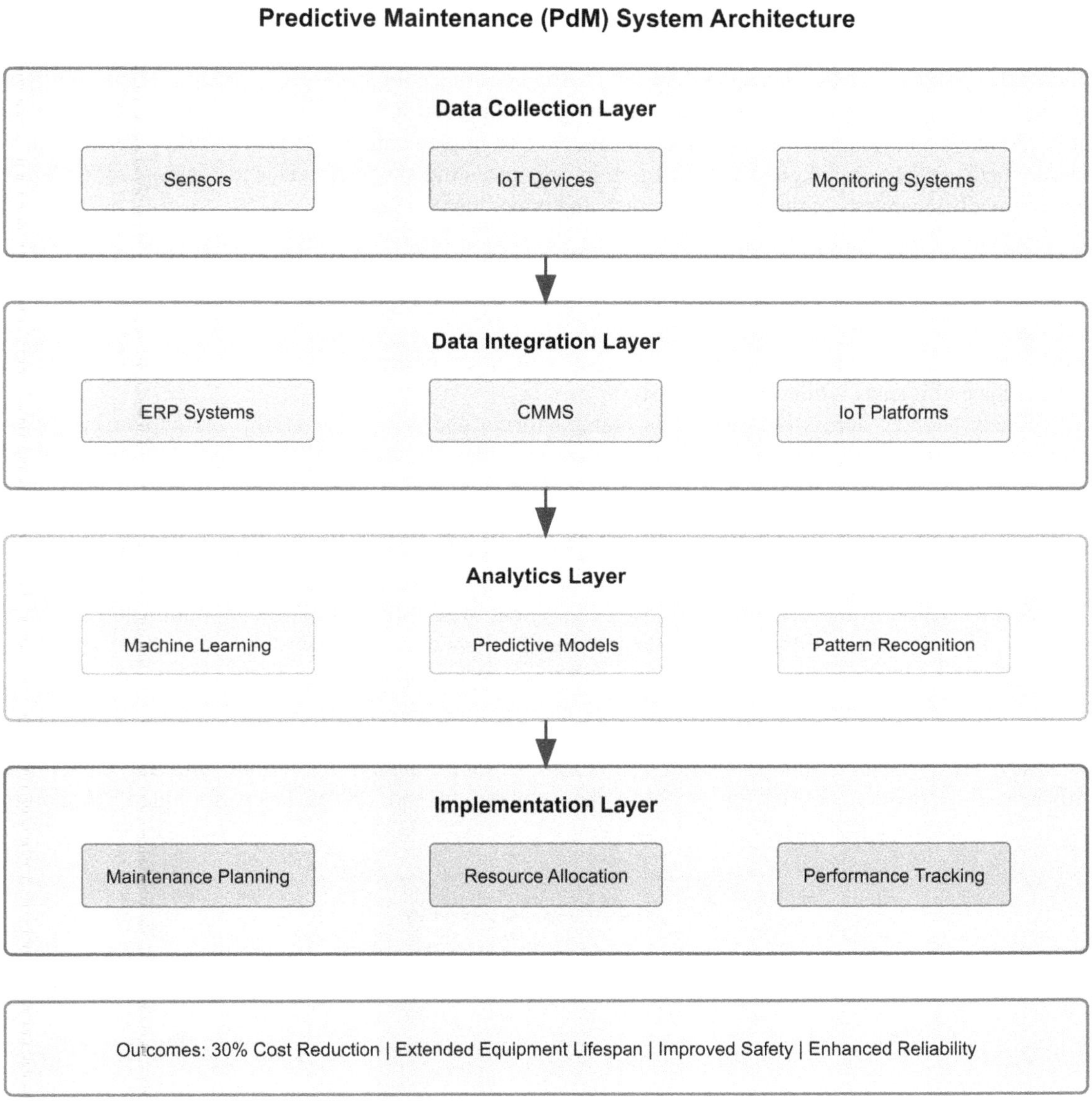

FIGURE 7.2 PdM framework: data processing and implementation layers.

optimized reagent consumption. Their research demonstrates that by analyzing vast datasets through integrated IoT and Big Data analytics, predictive insights significantly enhance sustainability while reducing operational downtime. The system's ability to identify key operational relationships ensures superior water quality management, leading to greater reliability and substantial cost savings.

7.2.3 Differences between Condition-Based, Preventive, and Predictive Maintenance

The evolution of maintenance strategies has led to distinct approaches, each with unique advantages. R. Li et al. (2024) demonstrate that CBM enhances system reliability by leveraging real-time monitoring and historical data for maintenance schedule optimization. Their research implements a Markov chain model that predicts state transitions, improving maintenance accuracy by up to 20%. This approach has proven particularly effective in complex multicomponent systems such as subsea production units. Building on these findings, Li and Wu (2024) explore condition-based opportunistic maintenance, which optimizes system reliability through scheduling based on actual component deterioration. Their improved genetic algorithm demonstrates superior performance compared to traditional methods, achieving up to 25% reduction in maintenance costs and downtime. The implementation of semi-Markov decision processes further refines maintenance scheduling by incorporating shock effects, establishing a foundation for future advancements in predictive accuracy and operational efficiency.

7.2.4 Benefits and Challenges of PdM Implementation

PdM leverages data analytics and ML to anticipate equipment failures, optimize maintenance schedules, and enhance efficiency while reducing costs.

Widely adopted across industries like manufacturing and utilities, it minimizes unplanned downtime, extends asset lifespan, and improves resource utilization. However, as illustrated in Figure 7.3, PdM implementation presents challenges such as data management issues, integration complexities, workforce training needs, organizational resistance, and high initial investment. Overcoming these barriers requires strategic planning, advanced analytics, IoT integration, AI/ML adoption, workforce development, and effective change management. Despite these challenges, PdM can yield substantial benefits, including a 25% cost reduction in the first year, improved equipment reliability, and optimized maintenance schedules. As industries embrace Industry 4.0, the increasing reliance on AI-driven solutions and IoT technologies underscores the necessity for continuous workforce upskilling and digital transformation.

While PdM offers significant advantages in detecting equipment failures preemptively, Hakami (2024) identifies key challenges, including data scarcity and temporal dependencies, that can affect prediction accuracy. Their research addresses these limitations through innovative solutions, utilizing Generative Adversarial Networks (GANs) for synthetic data generation and LSTM layers for temporal feature

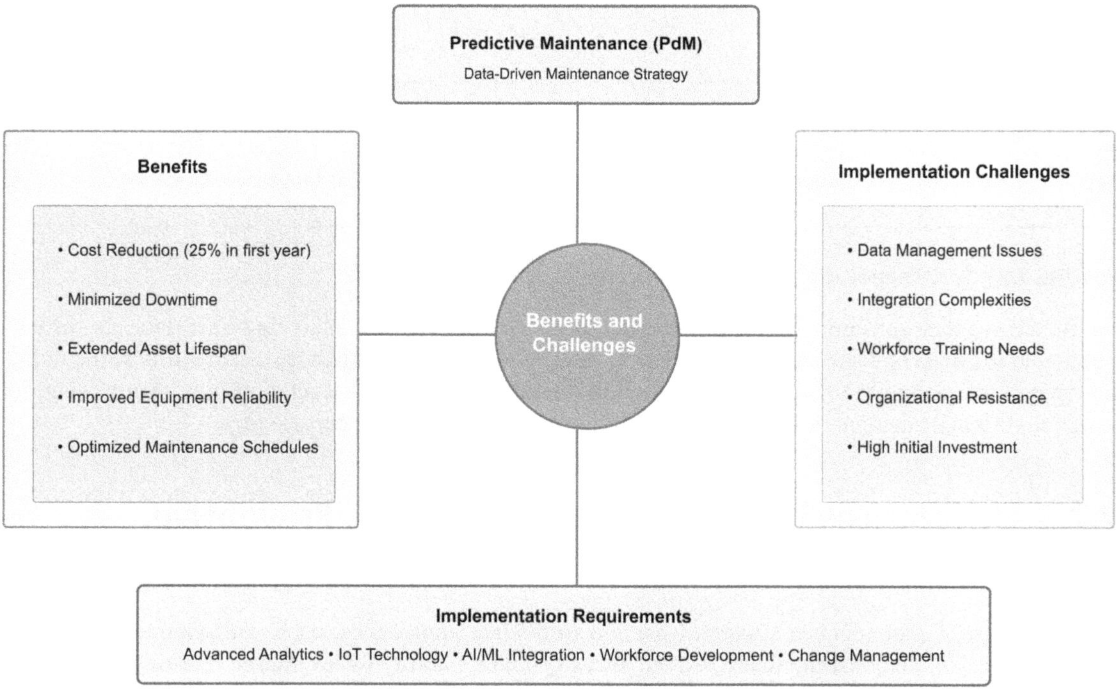

FIGURE 7.3 Benefits, challenges, and implementation requirements of PdM.

extraction. The results are promising, with ML models achieving impressive accuracy rates – ANN reaching 88.98% and RF attaining 74.15%. Skoumpopoulou et al. (2024) complement these technical findings by examining organizational challenges in PdM adoption. Through interviews with 15 industry experts, they identify critical barriers, including poor data integration and limited digital readiness. Their research emphasizes that successful implementation requires not only technical solutions but also fundamental shifts in management mindset and organizational culture. The findings underscore the importance of developing integrated systems that optimize both maintenance operations and spare parts inventory management, while highlighting the need for comprehensive approaches to digital transformation in maintenance practices.

7.3 AI AND ML IN PREDICTIVE MAINTENANCE

7.3.1 AI and ML Models Used in Anomaly Detection and Failure Prediction

Cutting-edge research in AI-driven PdM continues to advance through innovative approaches and methodologies. Chang et al. (2024) have developed the YC3Model, which enhances PdM in ship power systems by implementing Gaussian process regression for anomaly detection and predictive replacement. Their model employs a sophisticated dynamic triple sliding window mechanism to analyze the decay state coefficient, resulting in improved detection accuracy. Through the integration of local statistical measures and the Z-score method, the model achieves significant reductions in both false positives and negatives. Experimental results demonstrate the model's superior performance compared to traditional methods in detecting anomalies and monitoring system degradation. This AI-driven approach has proven effective in improving operational safety while reducing maintenance costs, with future research aimed at refining its adaptability for broader industrial applications. Complementing these findings, Mota et al. (2024) conducted a comprehensive comparison of four ML algorithms – ANN, Gradient Boosting, RF, and SVM – for PdM in detecting machine overstrain failures. Their research implemented a specialized training methodology for imbalanced data and automatic hyperparameter optimization, which significantly enhanced model performance. The results were particularly impressive, with ANN and SVM achieving remarkably high accuracies of 98.95% and 98.85%, respectively, demonstrating exceptional reliability in anomaly detection. This advanced approach has shown substantial benefits in enhancing machine stability and energy efficiency, particularly within manufacturing industries, while effectively preventing failures in advance and minimizing operational disruptions.

7.3.2 Role of Data Analytics and Sensor-Based Fault Detection

The integration of sophisticated data analytics and sensor-based fault detection has revolutionized PdM systems. Zhang et al. (2024) introduced an innovative hybrid data-driven model based on energy consumption correlation (ECC) for fault detection in **Heating, Ventilation, and Air Conditioning** (HVAC) system sensors. Their approach establishes a comprehensive mapping relationship between sensor data and energy usage, significantly enhancing PdM efficiency. The model demonstrates exceptional performance, achieving a fault diagnosis rate of 96.23% under single-operation conditions and 96.11% under dual-operation conditions. Notably, it maintains a remarkably low fault alarm rate ranging from 0% to 3.26%, showcasing superior reliability compared to traditional methods. This advanced approach has successfully improved fault detection interpretability while reducing dependency on sensor data quality. Building on these advancements, Wu et al. (2024) developed a novel sensor fault detection method that integrates graph neural networks (GNNs) and transformers to enhance PdM in networked systems. Their innovative approach leverages sensor interdependencies and system structure to improve information transmission and anomaly detection. The model achieves impressive performance metrics, including a high F1 score of 95.79% and an Area Under the Curve (AUC) of 98.83%, demonstrating exceptional fault detection accuracy. Through the strategic categorization of sensors into subgraphs and the implementation of an attention mechanism, the system effectively captures sensor features, with experimental results confirming its robustness across diverse real-world datasets.

7.3.3 Sparse Autoencoders and Their Effectiveness in Fault Identification

Recent developments in autoencoder technologies have significantly advanced fault identification capabilities in PdM systems. Lee et al. (2024) introduced a sophisticated fault diagnosis method that integrates a variational autoencoder (VAE) with a domain adaptation neural network (DANN) to enhance accuracy in challenging, noisy environments. Their innovative approach enables the VAE to identify faults based on reconstruction errors from normal-state acoustic data, while the DANN component adapts the model to accommodate varying noise levels. This integrated system maintains consistently high diagnostic accuracy, demonstrating improved classification of fault states even under fluctuating signal-to-noise ratios. Experimental results confirm that the DANN significantly enhances the VAE's performance in industrial settings, effectively addressing the complex challenges inherent in acoustic-based fault diagnosis. Further advancing the field, Eraliev et al. (2024) developed a stacked convolutional autoencoder (SCAE) model for fault diagnosis in hydraulic piston pumps, achieving remarkable accuracy rates exceeding 99.5% even when working with limited and noisy data. Their model enhances gradient information flow and extracts richer hierarchical features, consistently outperforming traditional DL models, including stacked sparse autoencoder (SSAE), convolutional neural network (CNN), and deep neural network (DNN). The SCAE approach demonstrates significant improvements in PdM efficiency compared to standard SSAEs, ensuring reliable fault detection despite various data acquisition challenges. The model's robust performance in handling compromised datasets makes it particularly well suited for real-world industrial applications, with future advancements aimed at further refining SCAE capabilities for broader fault diagnosis scenarios.

7.4 DIGITAL TWIN AND SCADA-BASED PREDICTIVE MAINTENANCE

Digital twin engineering represents an innovative approach to industrial asset management through the creation of dynamic data models. As described by Liu et al. (2021), these models utilize real-time sensory input to continuously adjust to operational and environmental changes, enabling accurate prediction of infrastructure performance. Yao et al. (2023) emphasize that digital twins leverage a sophisticated combination of data-driven analytics and physics-based models to anticipate remaining useful life and proactively identify potential issues in their physical counterparts, supported by an ecosystem comprising industrial IoT, simulation and modeling, ML, and advanced sensor technologies.

7.4.1 Concept and Importance of Digital Twin Technology

The development of digital twins follows two primary pathways, as outlined by Padovano et al. (2024): creating a digital twin for an existing physical asset or process, or simultaneously constructing both physical

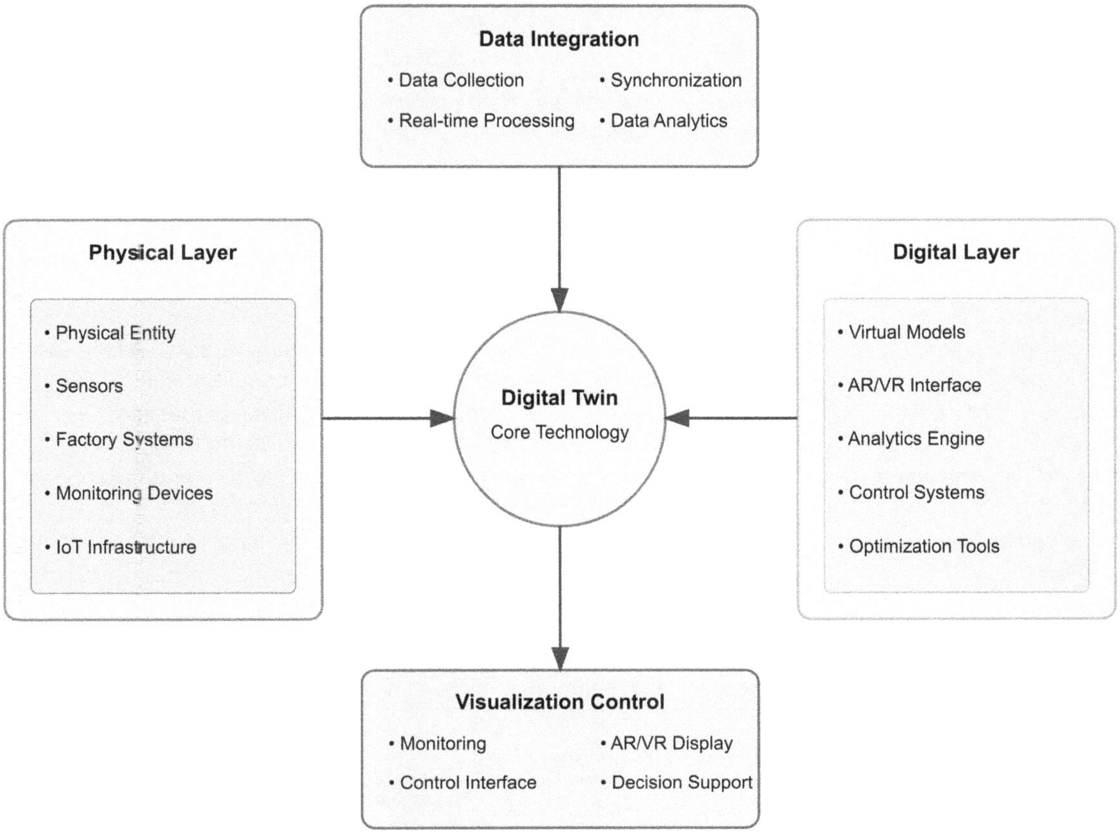

FIGURE 7.4 Digital twin technology framework.

and digital components on a digital thread. Melesse (2025) defines the digital thread as a data-driven architecture connecting information throughout the product lifecycle, facilitating effective lifecycle management, design, and evaluation.

Figure 7.4 illustrates the intricate relationship between the digital thread and digital twin technology within the circular economy's product lifecycle. The visualization demonstrates how all information collected along the digital thread can be utilized for new product design and optimization through digital simulation. This includes dynamic data-driven applications systems (DDDAS) paradigm for various sectors such as engineering (aerospace), medical (drug delivery), and science (weather forecasting).

While simulations operate offline during design stages to optimize parameters across various scenarios, digital twins establish real-time connections with their physical counterparts. When a real product is manufactured and linked with its virtual or digital counterpart, this virtual component becomes the "body" of a digital twin. Van Eyk and Heyns (2025) emphasize that the mindset, or "computational model," of a digital twin still requires development to fully encompass the product's lifecycle. A digital thread collects comprehensive information from raw materials to retirement, enabling new product design and optimization through digital simulation while creating a digital twin body. Akhator and Oboirien (2025) note that a live digital twin utilizes continuous data streaming to effectively mirror the product's lifecycle.

The importance of digital twin technology spans multiple dimensions, as highlighted by recent research. Kavousi-Fard et al. (2025) demonstrate its significance in real-time monitoring and data collection through IoT sensors, while Fu et al. (2025) emphasize its role in performance optimization across manufacturing, healthcare, and smart city applications. Cao et al. (2025) document its impact on cost reduction and risk mitigation, and Zhao et al. (2023) highlight its contribution to enhanced decision-making through accurate, real-time simulations. These are shown in Table 7.1.

TABLE 7.1 Key aspects and benefits of digital twin technology

KEY ASPECTS	*DESCRIPTION*
Real-time monitoring and data collection (Kavousi-Fard et al., 2025)	Continuously collects data from physical assets using IoT sensors. Helps detect anomalies and inefficiencies before they cause failures.
Predictive maintenance	Uses AI and ML to forecast potential failures. Reduces downtime and extends the lifespan of equipment.
Improved efficiency and performance optimization (Fu et al., 2025)	Simulates different scenarios to find the best operational settings. Enhances productivity in manufacturing, healthcare, and smart cities.
Cost reduction and risk mitigation (Cao et al., 2025)	Minimizes unexpected failures and reduces maintenance costs. Allows safe testing of changes before applying them to real-world systems.
Product and process innovation	Facilitates virtual prototyping and testing before actual production. Accelerates the development of smarter, more efficient products.
Sustainability and energy efficiency	Helps monitor energy consumption and optimize resource usage. Supports greener initiatives by reducing waste and carbon footprints.
Enhanced decision-making (Zhao et al., 2023)	Provides actionable insights for better strategic planning. Enables data-driven decisions with accurate, real-time simulations.
Applications across multiple industries (Arowoiya et al., 2024; Escribà-Gelonch et al., 2024)	Manufacturing: streamlines production processes and reduces defects. Healthcare: enables personalized treatment plans and patient monitoring. Smart cities: optimizes traffic, energy distribution, and infrastructure. Aerospace and automotive: improves safety, efficiency, and maintenance.

TABLE 7.2 Applications of SCADA in water treatment plants

KEY APPLICATIONS	*DESCRIPTION*
Water purification and filtration (Neelakandan et al., 2022)	Monitors intake, sedimentation, filtration, and disinfection processes to ensure water quality.
Chemical dosing control	Regulates chlorine and pH adjusters to maintain optimal chemical balance in water treatment.
Pumping station automation	Manages pump speed and water levels to ensure efficient water distribution.
Leak detection and pressure monitoring (Yousif and Abdalgader, 2022)	Identifies pipe leaks and monitors pressure fluctuations to prevent water loss and damage.
Energy optimization	Reduces power consumption through efficient control of pump operations.
Regulatory compliance reporting (Rohmingtluanga et al., 2023)	Automates the generation of reports required by environmental agencies for compliance.

7.4.2 SCADA (Supervisory Control and Data Acquisition) System Integration

Supervisory Control and Data Acquisition (SCADA) integration plays a pivotal role in modern industrial control systems. Kumar et al. (2023) describe its widespread application across manufacturing, energy, water treatment, and transportation sectors. Akhund et al. (2022) emphasize SCADA's crucial role in optimizing water utility operations, enhancing safety, reducing costs, and ensuring regulatory compliance.

Applications in water treatment plants span multiple areas, as documented by various researchers, as shown in Table 7.2. They enable precise control over water purification, chemical dosing, pump operations, leak detection, and energy consumption.

TABLE 7.3 Benefits of real-time monitoring and virtual simulation in predictive maintenance

KEY ASPECTS	*DESCRIPTION*
Reduced downtime	Identifies potential failures before they cause system disruptions, ensuring continuous operation.
Optimized maintenance scheduling (Matheri et al., 2022)	Ensures maintenance is performed only when needed, lowering operational costs.
Regulatory compliance	Helps wastewater plants maintain water quality standards by detecting issues early.
Extended asset lifespan (Wang et al., 2023)	Prevents excessive wear and tear by detecting early signs of degradation in pumps, valves, and aerators.
Virtual simulation in predictive maintenance (Matheri et al , 2022)	Enables engineers to model, analyze, and predict system behavior before real-world implementation.
Cost savings	Reduces reliance on physical testing and prevents unplanned repairs, minimizing expenses.
Improved decision-making (Aivaliotis et al., 2023)	Provides data-driven insights for better maintenance planning and operational efficiency.
Faster troubleshooting	Simulates various treatment conditions to determine optimal maintenance strategies.

7.4.3 Real-Time Monitoring and Virtual Simulation in Predictive Maintenance

The integration of real-time monitoring and virtual simulation has fundamentally transformed PdM practices. Mohanty et al. (2024) demonstrate how these technologies enable industries to detect potential failures preemptively, optimize maintenance schedules, and enhance overall operational efficiency. The evolution of condition-based maintenance, as described by Zhu et al. (2022) and Sharma et al. (2024), incorporates comprehensive monitoring of system components to forecast development trends and create proactive maintenance plans.

Real-time monitoring and virtual simulation enhance PdM by providing continuous system insights and enabling proactive decision-making. These technologies help reduce downtime, optimize maintenance schedules, and improve asset longevity while minimizing operational costs. Table 7.3 shows the detailed description of various key aspects.

7.4.4 Enhancing Decision-Making Using Digital Twins

The water and wastewater treatment sector is experiencing a significant digital transformation, as outlined by Wang et al. (2024), with new technologies driving improvements in productivity, cost reduction, and treatment efficacy. Chen and Kao (2025) emphasize the revolutionary impact of digital twins, defining them as virtual representations of physical systems that utilize real-time data flows for comprehensive analysis, optimization, and prediction.

Li et al. (2022) document how digital twins enhance operational efficiency through real-time insights and data analysis, leading to significant cost reductions. Sahu and Upadhyay (2024) demonstrate their effectiveness in process optimization, highlighting the technology's capability to simulate multiple scenarios and predict outcomes of various operational adjustments. Baena-Miret et al. (2024) further emphasize the importance of digital twins in enabling time-relevant and actionable decision-making, allowing operators to make better-informed choices based on real-time data and insights.

The comprehensive benefits of digital twin implementation in the water and wastewater industry extend beyond operational improvements. The technology enables PdM through continuous equipment monitoring, supports process optimization through scenario simulation, facilitates time-sensitive decision-making through real-time data analysis, and enhances safety through proactive risk assessment and mitigation strategies. These capabilities collectively contribute to more efficient, sustainable, and reliable water treatment operations, setting the foundation for continued advancement in the field.

This integration of digital twin technology, supported by sophisticated SCADA systems and real-time monitoring capabilities, represents a significant evolution in water treatment facility management, promising enhanced operational efficiency, improved maintenance practices, and more sustainable resource utilization for future generations.

7.5 METHODOLOGY

This methodology presents a comprehensive framework for implementing and evaluating AI-driven PdM systems in water treatment facilities. The approach integrates multiple components systematically to ensure robust performance assessment and practical implementation guidelines. Figure 7.5 illustrates the structured methodology and provides a concise overview of the key aspects discussed below.

7.5.1 Data Collection and Integration Framework

The data collection process begins with the strategic deployment of IoT sensors throughout the water treatment facility, focusing on critical equipment and processes. These sensors continuously monitor

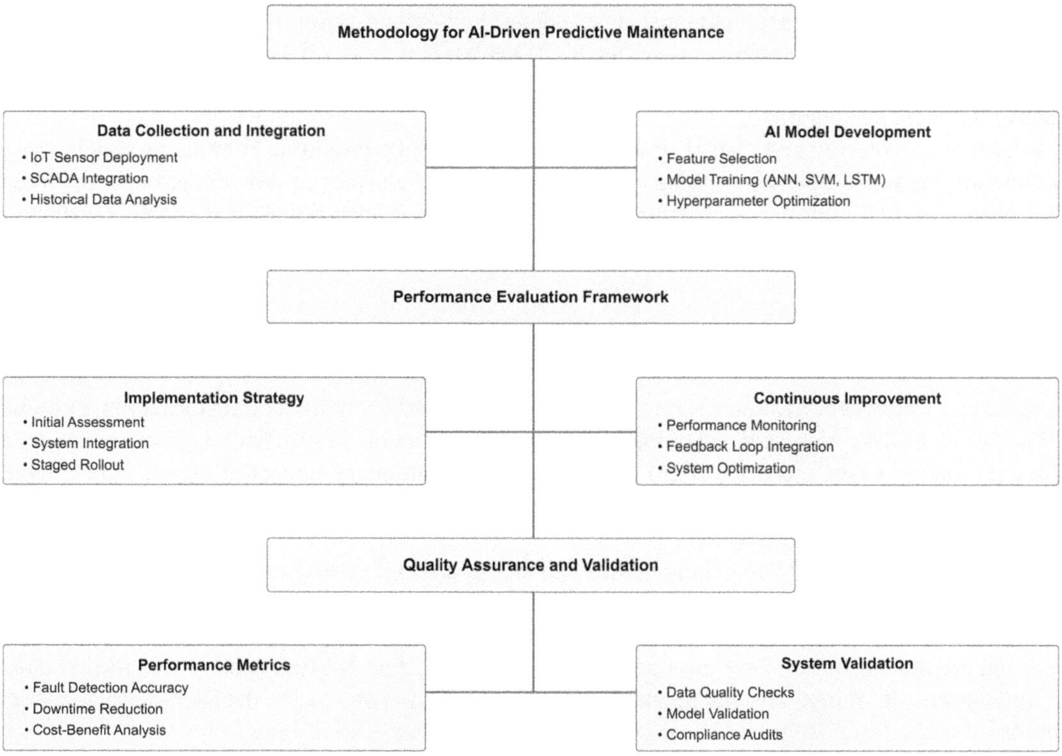

FIGURE 7.5 Structured methodology for PdM and analysis.

operational parameters, including temperature, pressure, vibration, and flow rates. The integration framework incorporates existing SCADA systems to ensure seamless data acquisition from both new and legacy infrastructure. Historical maintenance records are systematically analyzed to establish baseline performance metrics and identify recurring failure patterns. Environmental parameters and water quality measurements are collected through automated sampling systems and laboratory analysis, providing comprehensive contextual data for the AI models.

7.5.2 AI Model Development Architecture

The AI model development phase employs a structured approach to feature extraction and selection, utilizing advanced statistical techniques to identify relevant parameters from the collected data. Multiple ML models, including ANNs, SVMs, and LSTM networks, are developed and trained using historical operational data. The training process incorporates cross-validation techniques to ensure model robustness and generalization capability. Model optimization involves systematic hyperparameter tuning using grid search and Bayesian optimization methods. The integration of these models with existing maintenance management systems follows a modular architecture, allowing for flexible deployment and updates.

7.5.3 Performance Evaluation Framework

The evaluation framework encompasses multiple dimensions of system performance assessment. Fault detection accuracy is measured using standardized metrics such as precision, recall, and F1 score, with particular attention to false positive and negative rates. Downtime reduction is evaluated through comparative analysis of pre- and post-implementation periods, accounting for seasonal variations and operational load fluctuations. Cost-benefit analysis includes both direct maintenance costs and indirect operational impacts. Maintenance efficiency improvements are quantified through standardized key performance indicators (KPIs), including mean time between failures (MTBF) and mean time to repair (MTTR). System reliability testing incorporates stress testing and failure mode analysis to ensure robust performance under various operational conditions.

7.5.4 Implementation Strategy

The implementation follows a phased approach, beginning with a comprehensive initial assessment of facility requirements and existing infrastructure. The system integration phase focuses on establishing reliable data pipelines and ensuring seamless communication between different system components. Model deployment proceeds through staged rollouts, starting with non-critical systems to validate performance before expanding to critical infrastructure. Continuous performance monitoring enables ongoing system optimization and adaptation to changing operational conditions.

7.5.5 Continuous Improvement Mechanism

A feedback loop mechanism is integrated into the methodology to enable continuous system improvement. Regular performance reviews identify areas for enhancement, while automated monitoring systems detect deviations from expected behavior patterns. This iterative improvement process incorporates new operational data and failure patterns to refine model predictions and enhance system reliability. The methodology also includes provisions for incorporating emerging technologies and updated maintenance practices as they become available.

7.5.6 Quality Assurance and Validation

The methodology incorporates rigorous quality assurance measures throughout all phases. Data quality is ensured through automated validation checks and regular calibration of monitoring equipment. Model performance is continuously validated against actual operational outcomes, with systematic documentation of discrepancies and corrective actions. Regular audits of the entire system ensure compliance with industry standards and regulatory requirements, while also identifying opportunities for optimization and improvement.

7.6 RESULTS AND DISCUSSION

7.6.1 Effectiveness of AI and ML in Fault Detection and Prediction

AI and ML have fundamentally transformed fault detection and PdM practices in wastewater treatment plants (WWTPs), enabling sophisticated real-time monitoring, pattern recognition, and predictive analytics that collectively enhance operational efficiency, reduce downtime, and ensure regulatory compliance. Traditional WWTPs have historically faced significant operational challenges, including equipment failures, sensor malfunctions, and process inefficiencies, with maintenance strategies primarily relying on scheduled inspections or reactive approaches that often result in costly downtime and operational inefficiencies (Alprol et al., 2024).

The integration of AI in WWTPs has demonstrated remarkable potential in revolutionizing wastewater treatment, particularly in improving affordability, sustainability, and ecological integrity. This transformation is notably evident in the activated sludge process, a fundamental biological treatment method that utilizes microbes for organic material decomposition in wastewater. AI techniques have successfully optimized this process through the anticipation and control of critical process variables, including influent flow rate, sludge retention time (SRT), and dissolved oxygen (DO) content (Rahman et al., 2023). SVMs have proven particularly effective in predicting DO concentration within aeration tanks by integrating environmental parameters with historical process data (Hossain et al., 2022). Furthermore, neural networks have demonstrated significant capability in predicting essential effluent quality metrics, including TSS and BOD, thereby enhancing the overall effectiveness of the activated sludge process (Canals et al., 2023).

The implementation of AI and ML technologies has markedly improved fault detection and PdM capabilities in WWTPs, resulting in enhanced operational efficiency, reduced costs, and improved regulatory compliance through real-time anomaly detection and predictive analytics. While certain challenges persist, ongoing advancements in AI models, edge computing, and cybersecurity continue to strengthen the resilience and intelligence of wastewater treatment systems (Singh et al., 2023).

7.6.2 Impact of Predictive Maintenance on Downtime Reduction and Cost Optimization

The adoption of PdM has revolutionized industrial operations by significantly reducing unexpected failures and optimizing operational costs. Through the integration of real-time monitoring, AI-driven analytics, and ML models, organizations can now anticipate equipment failures, minimize downtime, and enhance operational efficiency with unprecedented accuracy (Alam et al., 2023). Traditional maintenance approaches, including reactive and preventive maintenance, often results in either excessive maintenance activities or unexpected breakdowns. However, PdM offers a sophisticated, data-driven alternative that leverages sensor data, advanced analytics, and ML to forecast failures before their occurrence, enabling proactive maintenance interventions (Ahuja and Gupta, 2024).

In terms of cost optimization, PdM has demonstrated significant advantages in optimizing operational expenses by preventing costly failures and reducing unnecessary maintenance efforts (Patil, 2024). Early fault detection capabilities have proven instrumental in preventing critical failures and reducing emergency repair expenses, while proper equipment maintenance has led to improved operational efficiency and reduced energy consumption (Udo et al., 2024). The implementation of AI-driven process optimization has further contributed to minimizing resource wastage in wastewater treatment facilities.

The role of PdM in industrial settings, particularly within WWTPs, has become increasingly crucial for minimizing downtime and optimizing operational costs. Through the strategic leveraging of AI, IoT, and data analytics, industries have achieved substantial improvements in efficiency, reduction in unexpected failures, and enhanced cost-effectiveness. While implementation challenges persist, continuous advancements in AI and automation technologies continue to strengthen the impact of PdM on industrial performance and sustainability (Narayanan et al., 2024; Yazdi, 2024).

7.6.3 ML Model Performance Comparison for Predictive Maintenance

The visualization presented in Figure 7.6 provides a comprehensive comparison of ML model accuracies in PdM applications for water treatment facilities. The analysis encompasses five distinct models, each evaluated through rigorous research studies conducted in 2024.

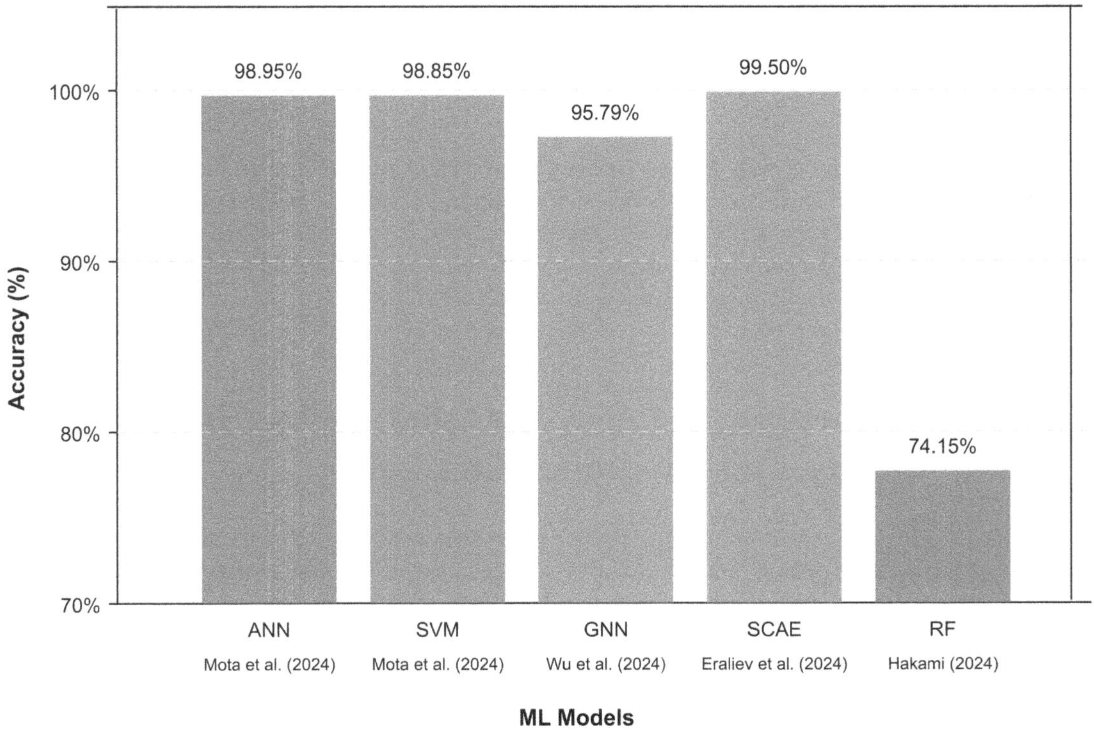

FIGURE 7.6 Performance comparison of ML models for PdM.

The SCAE model, developed by Eraliev et al., demonstrates superior performance with an accuracy rate of 99.50%, particularly noteworthy for its effectiveness with limited and noisy datasets. Following closely, the ANN and SVM implementations by Mota et al. achieve remarkable accuracy rates of 98.95% and 98.85%, respectively, showcasing the robustness of these traditional DL approaches in fault detection scenarios. The GNN methodology, introduced by Wu et al., maintains strong performance with a 95.79% accuracy rate, while the RF model studied by Hakami shows comparatively lower accuracy at 74.15%. This substantial performance differential between DL architectures and traditional ML approaches suggests that sophisticated neural network-based solutions may be more appropriate for complex PdM tasks in water treatment environments. The variation in performance metrics across these models provides valuable insights for practitioners in selecting appropriate methodologies for their specific maintenance monitoring requirements.

7.6.4 Challenges in Implementing PdM in Water Treatment Plants

The implementation of PdM in water treatment facilities presents both transformative opportunities and significant challenges in operational optimization. While PdM offers substantial potential for reducing downtime and improving efficiency, several critical obstacles must be addressed for successful implementation. Data integration complexities, substantial initial investments, workforce requirements, and cybersecurity considerations represent the primary challenges facing water treatment facilities in their PdM adoption journey (Hallaji et al., 2022).

A fundamental challenge lies in data integration and management, particularly regarding the compatibility of legacy systems with modern PdM technologies. Water treatment plants often operate with established infrastructure that requires extensive customization to integrate real-time sensor data with existing SCADA and control systems (Mohamed Almazrouei et al., 2023). This technical hurdle is compounded by the significant financial investment required for implementing PdM systems, including IoT sensors, cloud computing infrastructure, and AI-driven analytics platforms. Smaller and medium-sized facilities frequently encounter budget constraints that limit their ability to adopt comprehensive PdM solutions (Murtaza et al., 2024).

The human resource dimension presents another significant challenge, as effective PdM implementation demands specialized expertise in data science, AI, and maintenance engineering. Many facilities face a shortage of qualified personnel capable of interpreting PdM insights, necessitating continuous staff training to adapt to evolving technologies and workflows (Chen et al., 2023). Additionally, the integration of cloud-based PdM solutions with critical water infrastructure introduces cybersecurity vulnerabilities that require robust protection protocols and regulatory compliance measures (Almobarek, 2024).

7.6.5 Future Directions: Refining AI Models, Enhancing Data Quality, and Integration Strategies

The evolution of AI implementation in water treatment facilities continues to emphasize the importance of specialized model development and data quality management. Recent developments, such as Tiger Brokers' integration of DeepSeek's AI model into TigerGPT, demonstrate the growing trend toward industry-specific AI applications (Zha et al., 2025). However, ensuring sufficient domain expertise in AI models remains a significant challenge, as general models often lack the depth required for specialized applications (Gadde, 2022).

The quality of data processing has emerged as a critical factor in AI system effectiveness. Organizations are increasingly adopting ML and AI-driven automation to enhance data quality management, implementing solutions that ensure data reliability and analysis readiness throughout the lifecycle

(Barzamini et al., 2023). The emergence of data-centric AI (DCAI) represents a significant shift toward prioritizing data quality over model complexity (Mellal et al., 2024).

Success in AI implementation increasingly depends on seamless integration with existing processes (Nagpal et al., 2024). Direct integration of AI models into data pipelines and workflows has demonstrated improvements in automation and error reduction while maintaining real-time data quality. These advancements in AI applications have transformed data quality management, enabling organizations to deliver more efficient solutions while enhancing worker productivity (Santos et al., 2024).

7.7 CASE STUDIES AND INDUSTRIAL APPLICATIONS

7.7.1 Real-World Applications of PdM in Water Treatment Plants

Several innovative implementations of PdM in water treatment facilities have demonstrated significant operational improvements. Portuguese researchers developed an advanced predictive control policy for variable-frequency pumps in wastewater treatment facilities, combining deep reinforcement learning with statistical learning to forecast wastewater intake rates and optimize pump operations.

This implementation achieved remarkable results, reducing electrical energy consumption by 16.7% and decreasing high-level alarms by 97% during a 90-day evaluation period (Viveros et al., 2021). Figure 7.7 illustrates three performance improvements achieved through the implementation of PdM in water treatment facilities. The most significant impact is observed in the reduction in high-level alarms, which decreased by 97%, demonstrating substantial improvement in system reliability and operational stability. Energy consumption showed a notable reduction of 16.7%, indicating enhanced operational efficiency and sustainability in facility operations. Additionally, maintenance costs were reduced by 25%, highlighting the economic benefits of PdM implementation.

Australian facilities have pioneered the integration of digital twin technology in WWTP management. These sophisticated digital models incorporate real-time data, mathematical modeling, and advanced

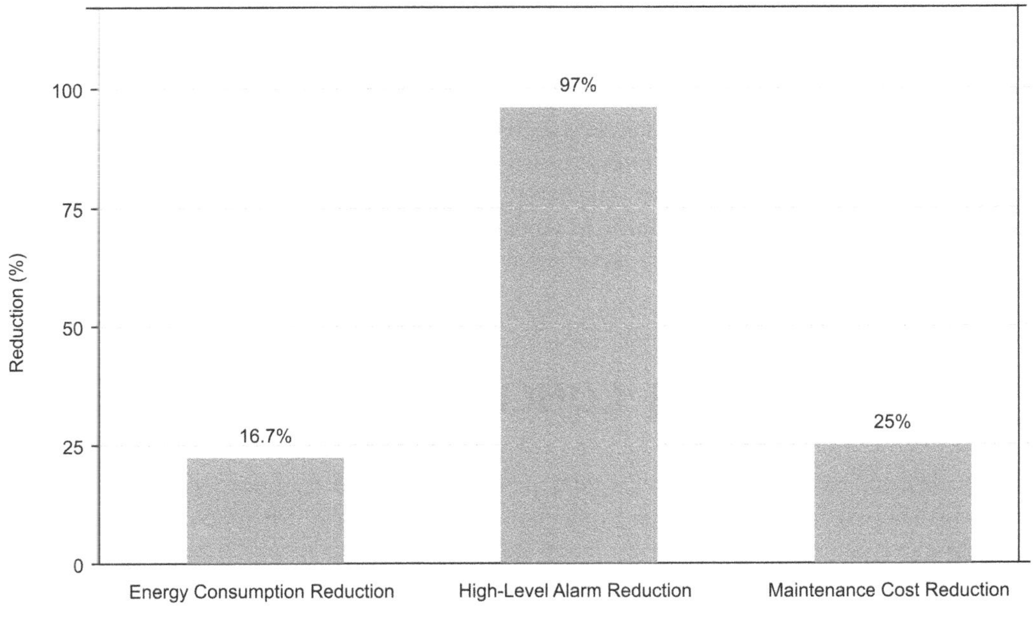

FIGURE 7.7 Performance improvements using PdM in water treatment facilities.

analytics to create virtual representations of physical assets, enabling optimized decision-making and predictive insights. The implementation of digital twins has significantly enhanced operational effectiveness through real-time monitoring and PdM capabilities (Wang et al., 2024).

A comprehensive Singaporean study examined the financial implications of implementing PdM for filtration units in water treatment facilities. The research utilized a Monte Carlo simulation combined with cost-benefit analysis and ML model training, demonstrating the economic viability of PdM programs while bridging the gap between technical implementation and business justification (Jeffrey et al., 2022).

7.7.2 Success Stories Demonstrating Downtime Reduction

Several notable success stories highlight the effectiveness of PdM implementations. The Sacramento Regional Wastewater Treatment Plant's $1.7 billion capital expansion incorporated reliability-centered design (RCD) and reliability-centered maintenance (RCM) principles, resulting in projected lifetime cost savings of $100 million while ensuring regulatory compliance and minimizing downtime (Escriva-Bou et al., 2019).

The Walla Walla Wastewater Treatment Plant successfully addressed struvite buildup challenges through targeted intervention strategies, achieving complete prevention of new buildup and progressive clearance of existing deposits within eight weeks of implementation (KingSlack, 2023). Similarly, a South Florida facility's transition to smart magnetic flow meters enhanced measurement accuracy and reduced operational disruptions (DeFosset and Cunningham, 2019), while a Colorado plant's implementation of modern trim blower technology significantly improved system reliability (Battaglin et al., 2023).

7.7.3 Key Takeaways from Implementations and Best Practices

The successful implementation of PdM systems has yielded valuable insights for future deployments. Advanced AI analysis of sensor data from critical components enables proactive failure prediction and maintenance scheduling, while DL-based simulators facilitate the development of sophisticated control strategies through reinforcement learning (Almobarek et al., 2022).

Continuous equipment performance monitoring through technologies such as power monitoring sensors has proven essential for maintaining optimal operating conditions. The Crestline Sanitation District's successful utilization of the Fluke 3540 FC Three-Phase Power Monitor exemplifies how advanced monitoring tools can inform equipment replacement and maintenance decisions (Mohanty et al., 2024).

The development of digital twins has revolutionized real-time monitoring and predictive analytics capabilities (Gaur et al., 2024). Furthermore, ensuring transparency in AI models has become increasingly important for building operator and stakeholder confidence (Rane et al., 2023). These implementations emphasize the importance of standardized guidelines for AI applications in wastewater treatment to ensure consistency and comparability across different facilities.

7.8 CONCLUSIONS

This comprehensive analysis of AI-driven PdM in water treatment facilities has revealed significant advancements in operational efficiency and system reliability through the integration of AI and ML technologies. Research demonstrates a clear evolution from traditional maintenance approaches to sophisticated predictive systems that leverage real-time data analytics and digital twin technology. Modern ML models, particularly SCAE, have achieved remarkable accuracy rates exceeding 99.50% in fault detection scenarios, while the implementation of digital twin technology with SCADA systems has enabled comprehensive real-time monitoring and virtual simulation capabilities.

The quantifiable benefits of these implementations are substantial, with documented improvements including a 16.7% reduction in energy consumption, a 97% decrease in high-level alarms, and a 25% reduction in maintenance costs. The methodological framework developed through this research provides a structured approach to implementing PdM systems, encompassing systematic data collection protocols, comprehensive AI model development architectures, and robust performance evaluation metrics. Case studies from various facilities demonstrate the practical applicability and success of this framework across different operational contexts.

Despite these achievements, several challenges persist in the widespread adoption of AI-driven PdM systems. Initial implementation costs remain a significant barrier, particularly for smaller facilities with limited resources. Technical challenges in integrating new systems with legacy infrastructure continue to pose difficulties for many organizations, while data quality and standardization issues across different facility types present ongoing challenges for model accuracy and reliability. Additionally, current AI models often require substantial customization to meet specific facility requirements, increasing implementation complexity and resource demands.

Looking toward the future, several key research directions emerge as priorities for advancing the field. These include the development of more cost-effective implementation strategies, enhancement of AI models' adaptability to varying operational conditions, and integration of emerging technologies such as edge computing and advanced IoT sensors. The standardization of data collection and analysis protocols across the industry, along with the investigation of hybrid approaches combining multiple AI techniques, shows potential for improving overall system accuracy and reliability.

In conclusion, while AI-driven PdM presents implementation challenges, its benefits in terms of operational efficiency, cost reduction, and system reliability make it an essential evolution in water treatment facility management. As implementation costs decrease and technology becomes more accessible, the adoption of these systems is expected to accelerate across the industry. The continued development of more sophisticated AI models, combined with standardized implementation frameworks and enhanced integration capabilities, will further strengthen the role of PdM in ensuring efficient and reliable water treatment operations for future generations.

REFERENCES

Afan, H. A., Melini Wan Mohtar, W. H., Khaleel, F., Kamel, A. H., Mansoor, S. S., Alsultani, R., Ahmed, A. N., Sherif, M., & El-Shafie, A. (2024). Data-driven water quality prediction for wastewater treatment plants. *Heliyon*, 10(18), e36940.

Ahuja, A., & Gupta, M. (2024). Optimizing predictive maintenance with machine learning and IoT: A business strategy for reducing downtime and operational costs. *Journal of Emerging Technologies and Innovative Research (JETIR)*, 11(10), b286–b292.

Aivaliotis, P., Arkouli, Z., Georgoulias, K., & Makris, S. (2023). Methodology for enabling dynamic digital twins and virtual model evolution in industrial robotics: A predictive maintenance application. *International Journal of Computer Integrated Manufacturing*, 36(7), 947–965.

Akhator, P., & Oboirien, B. (2025). Digitilising the energy sector: A comprehensive digital twin framework for biomass gasification power plant with CO_2 capture. *Cleaner Energy Systems*, 10, 100175.

Akhund, T. M. N. U., Newaz, N. T., Zaman, Z., Sultana, A., Barros, A., & Whaiduzzaman, M. (2022). IoT-based low-cost automated irrigation system for smart farming. In *Intelligent Sustainable Systems: Selected Papers of WorldS4 2021, Volume 1*, (pp. 83–91), London.

Alam, M., Islam, M. R., & Shil, S. K. (2023). AI-Based predictive maintenance for US manufacturing: Reducing downtime and increasing productivity. *International Journal of Advanced Engineering Technologies and Innovations*, 1(1), 541–567.

Almobarek, M. (2024). *Predictive Maintenance for Chilled Water System at Commercial Buildings: A Holistic Approach in Line with Industry 4.0.* Doctoral Thesis (Ph.D.), University of Strathclyde. University of Strathclyde Institutional Repository.

Almobarek, M., Mendibil, K., & Alrashdan, A. (2022). Predictive maintenance 4.0 for chilled water system at commercial buildings: A systematic literature review. *Buildings*, 12(8), 1229.

Alprol, A. E., Mansour, A. T., Ibrahim, M. E. E. D., & Ashour, M. (2024). Artificial intelligence technologies revolutionizing wastewater treatment: Current trends and future prospective. *Water*, 16(2), 314.

Arowoiya, V. A., Moehler, R. C., & Fang, Y. (2024). Digital twin technology for thermal comfort and energy efficiency in buildings: A state-of-the-art and future directions. *Energy and Built Environment*, 5(5), 641–656.

Baena-Miret, S., Puig, M. A., Rodes, R. B., Farran, L. B., Durán, S., Martí, M. G., ... & Valverde, A. C. (2024). Enhancing efficiency and quality control: The impact of digital twins in drinking water networks. *Water Environment Research*, 96(10), e11139.

Barzamini, F., Bagheri, A., Saeedi, R., Hejazy, T. H., & sadat Khaloo, S. (2023). Forecasting the effluent COD of refinery wastewater treatment plant using artificial intelligence. *Journal of Health in the Field*, 11(3), 19–30.

Battaglin, W., Bradley, P., Weissinger, R., Blackwell, B., Cavallin, J., Villeneuve, D., & Kinsey, J. (2023). Changes in chemical occurrence, concentration, and bioactivity in the Colorado River before and after replacement of the Moab, Utah wastewater treatment plant. *Science of the Total Environment*, 904, 166231.

Canals, J., Cabrera-Codony, A., Carbó, O., Torán, J., Martín, M., Baldi, M., & Monclús, H. (2023). High-rate activated sludge at very short SRT: Key factors for process stability and performance of COD fractions removal. *Water Research*, 231, 119610.

Cao, J., Spulbar, C., Eti, S., Horobet, A., Yüksel, S., & Dinçer, H. (2025). Innovative approaches to green digital twin technologies of sustainable smart cities using a novel hybrid decision-making system. *Journal of Innovation & Knowledge*, 10(1), 100651.

Cechinel, M. A. P., Neves, J., Fuck, J. V. R., de Andrade, R. C., Spogis, N., Riella, H. G., Padoin, N., & Soares, C. (2024). Enhancing wastewater treatment efficiency through machine learning-driven effluent quality prediction: A plant-level analysis. *Journal of Water Process Engineering*, 58, 104758.

Chang, X., Yan, X., Qiu, B., Wei, M., Liu, J., & Zhu, H. (2024). Anomaly detection and confidence interval-based replacement in decay state coefficient of ship power system. *IET Intelligent Transport Systems*, 18, 2409–2439.

Chen, K. J., & Kao, C. M. (2025). Optimization of municipal wastewater treatment plants management through digital twin modeling. *Journal of Environmental Engineering*, 151(4), 01825001.

Chen, Q., Cao, J., & Zhu, S. (2023). Data-driven monitoring and predictive maintenance for engineering structures: Technologies, implementation challenges, and future directions. *IEEE Internet of Things Journal*, 10(16), 14527–14551.

DeFosset, K. L., & Cunningham, K. J. (2019). *Hydrogeologic Characterization of Part of the Lower Floridan Aquifer at the South District Wastewater Treatment Plant, Miami-Dade County,* Florida. Report 2019-1034. US Geological Survey.

Eraliev, O., Lee, K. H., & Lee, C. H. (2024). SCAE—Stacked convolutional autoencoder for fault diagnosis of a hydraulic piston pump with limited data samples. *Sensors*, 24(14), 4661–4661.

Escribà-Gelonch, M., Liang, S., van Schalkwyk, P., Fisk, I., Long, N. V. D., & Hessel, V. (2024). Digital twins in agriculture: Orchestration and applications. *Journal of Agricultural and Food Chemistry*, 72(19), 10737–10752.

Escriva-Bou, A., Gray, B., Green, S., Harter, T., Jezdimirovic, J., Lund, J., ... & Seavy, N. (2019). *Water and the Future of the San Joaquin Valley. Technical Appendix A: Updated Assessment of the San Joaquin Valley's Water Balance* (p. 26).Public Policy Institute of California, San Francisco, CA.

Fu, X., Song, H., Li, S., & Lu, Y. (2025). Digital twin technology in modern machining: A comprehensive review of research on machining errors. *Journal of Manufacturing Systems*, 79, 134–161.

Gadde, H. (2022). Integrating AI into SQL query processing: Challenges and opportunities. *International Journal of Advanced Engineering Technologies and Innovations*, 1(3), 194–219.

Gaur, A., Pooja, Hooda, S., & Saya, L. (2024). Recommendations for implementation of AI: Steps toward industry-wide adoption. In Gulati, S. (Ed.) *Application of Artificial Intelligence in Wastewater Treatment* (pp. 281–298). Springer Nature.

Gyparakis, S., Trichakis, I., & Diamadopoulos, E. (2024). Using artificial neural networks to predict operational parameters of a drinking water treatment plant (DWTP). *Water*, 16(19), 2863–2863.

Hakami, A. (2024). Strategies for overcoming data scarcity, imbalance, and feature selection challenges in machine learning models for predictive maintenance. *Dental Science Reports*, 14(1), 9645.

Hallaji, S. M., Fang, Y., & Winfrey, B. K. (2022). Predictive maintenance of pumps in civil infrastructure: State-of-the-art, challenges and future directions. *Automation in Construction*, 134, 104049.

Hossain, S. Z., Sultana, N., Mohammed, M. E., Razzak, S. A., & Hossain, M. M. (2022). Hybrid support vector regression and crow search algorithm for modeling and multiobjective optimization of microalgae-based wastewater treatment. *Journal of Environmental Management*, 301, 113783.

Jeffrey, P., Yang, Z., & Judd, S. J. (2022). The status of potable water reuse implementation. *Water Research*, 214, 118198.

Kavousi-Fard, A., Dabbaghjamanesh, M., Sheikh, M., & Jin, T. (2025). A novel deep learning based digital twin model for mitigating wake effects in wind farms. *Renewable Energy Focus*, 53, 100686.

KingSlack, D. L. C. (2023). *Endangered Species Act* Section *7(a)(2) Biological Opinion, Concurrence Letter, and Magnuson–Stevens Fishery Conservation and Management Act Essential Fish Habitat Response for the Salmon City Sewer Pipeline Replacement; HUC 170602030403; Lemhi County, Idaho.* U.S. Army Corps of Engineers, Walla Walla District, National Marine Fisheries Service (NMFS), West Coast Region. https://repository.library.noaa.gov/view/noaa/49691

KumarA. R., Metwally, A. S. M., Ashraf, G. A., & Thamineni, B. L. (2023). Smart IoT-based water treatment with a supervisory control and data acquisition (SCADA) system process. *Water Reuse*, 13(3), 411–431.

Lee, J. G., Kim, K. S., & Lee, J. H. (2024). Sound-based unsupervised fault diagnosis of industrial equipment considering environmental noise. *Sensors*, 24(22), 7319–7319.

Li, J. Y., Mei, Y., Li, F., Shen, J., Zhou, F. J., & Sun, Y. H. (2024). An adaptive coupling of PD-CCM model involving a viscoelastic constitutive relation for quasi-brittle material dynamic failure. *Engineering Fracture Mechanics*, 298, 109904.

Li, M., & Wu, B. (2024). Optimal condition-based opportunistic maintenance policy for two-component systems considering common cause failure. *Reliability Engineering & System Safety*, 250, 110269.

Li, R., Cai, B., Zhao, Y., Liu, Y., Zhang, Y., Kong, X., & Liu, Y. (2024). Condition-based maintenance method for multi-component systems under discrete-state condition: Subsea production system as a case. *Ocean Engineering*, 306, 118166.

Li, X., Luo, J., Li, Y., Wang, W., Hong, W., Liu, M., & Lv, Z. (2022). Application of effective water-energy management based on digital twins technology in sustainable cities construction. *Sustainable Cities and Society*, 87, 104241.

Liu, M., Fang, S., Dong, H., & Xu, C. (2021). Review of digital twin about concepts, technologies, and industrial applications. *Journal of Manufacturing Systems*, 58, 346–361.

Mallioris, P., Aivazidou, E., & Bechtsis, D. (2024). Predictive maintenance in Industry 4.0: A systematic multi-sector mapping. *CIRP Journal of Manufacturing Science and Technology*, 50, 80–103.

Matheri, A. N., Mohamed, B., Ntuli, F., Nabadda, E., & Ngila, J. C. (2022). Sustainable circularity and intelligent data-driven operations and control of the wastewater treatment plant. *Physics and Chemistry of the Earth, Parts A/B/C*, 126, 103152.

Melesse, T. Y. (2025). Digital twin-based applications in crop monitoring. *Heliyon*, 11, e42137.

Mellal, N. E. H., Tahar, W., Boumaaza, M., Belaadi, A., & Bourchak, M. (2024). Prediction of purified water quality in industrial hydrocarbon wastewater treatment using an artificial neural network and response surface methodology. *Journal of Water Process Engineering*, 58, 104757.

MohamedAlmazrouei, S., Dweiri, F., Aydin, R., & Alnaqbi, A. (2023). A review on the advancements and challenges of artificial intelligence based models for predictive maintenance of water injection pumps in the oil and gas industry. *SN Applied Sciences*, 5(12), 391.

Mohanty, A., Mohanty, S. K., & Mohapatra, A. G. (2024). Real-time monitoring and fault detection in AI-enhanced wastewater treatment systems. In Garg, M. C. (Ed.) *The AI Cleanse: Transforming Wastewater Treatment Through Artificial Intelligence: Harnessing Data-Driven Solutions* (pp. 165–199). Springer Nature.

Mota, B., Faria, P., & Ramos, C. (2024). Machine overstrain prediction for early detection and effective maintenance: A machine learning algorithm comparison. *Logic Journal of the IGPL/Bulletin of the IGPL*. https://doi.org/10.1093/jigpal/jzae055

Murtaza, A. A., Saher, A., Zafar, M. H., Moosavi, S. K. R., Aftab, M. F., & Sanfilippo, F. (2024). Paradigm shift for predictive maintenance and condition monitoring from Industry 4.0 to Industry 5.0: A systematic review, challenges and case study. *Results in Engineering*, 24, 102935.

Nagpal, M., Siddique, M. A., Sharma, K., Sharma, N., & Mittal, A. (2024). Optimizing wastewater treatment through artificial intelligence: Recent advances and future prospects. *Water Science & Technology*, 90(3), 731–757.

Narayanan, D., Bhat, M., Samual, N. R., Khatri, N., & Saroliya, A. (2024). Artificial intelligence driven advances in wastewater treatment: Evaluating techniques for sustainability and efficacy in global facilities. *Desalination and Water Treatment*, 320, 100618.

Neelakandan, S., Prakash, M., Geetha, B. T., Nanda, A. K., Metwally, A. M., Santhamoorthy, M., & Gupta, M. S. (2022). Metaheuristics with deep transfer learning enabled detection and classification model for industrial waste management. *Chemosphere*, 308, 136046.

Padovano, A., Sammarco, C., Balakera, N., & Konstantinidis, F. (2024). Towards sustainable cognitive digital twins: A portfolio management tool for waste mitigation. *Computers & Industrial Engineering*, 198, 110715.

Patil, D. (2024). Artificial intelligence-driven predictive maintenance in manufacturing: Enhancing operational efficiency, minimizing downtime, and optimizing resource utilization. https://doi.org/10.2139/ssrn.5057406

Rahman, T. U., Roy, H., Islam, M. R., Tahmid, M., Fariha, A., Mazumder, A., & Islam, M. S. (2023). The advancement in membrane bioreactor (MBR) technology toward sustainable industrial wastewater management. *Membranes*, 13(2), 181.

Rane, N., Choudhary, S., & Rane, J. (2023). Leading-edge artificial intelligence (AI), machine learning (ML), blockchain, and Internet of Things (IoT) technologies for enhanced wastewater treatment systems. https://papers.ssrn.com/sol3/papers.cfm?abstract_id=4641557

Rohmingtluanga, C., Datta, S., Sinha, N., & Ustun, T. S. (2023). SCADA based intake monitoring for improving energy management plan: Case study. *Energy Reports*, 9, 402–410.

Sahu, Y., & Upadhyay, V. K. (2024). Environmental application of digital twins: A review. In Talukdar, S., Rahman, A., Bera, S., Ramana, G. V., and Prashar, A. (Eds.) *Environmental Risk and Resilience in the Changing World: Integrated Geospatial AI and Multidimensional Approach* (pp. 287–295). Springer Nature Switzerland.

Santos, A. V., Amaral, M. C. S., & Oliveira, S. M. A. C. (2024). Artificial neural networks and multivariate statistical process control to improve ammonia removal on membrane bioreactors treating refinery wastewater. *Journal of Water Process Engineering*, 67, 106126.

Sharma, S., Sharma, K., & Grover, S. (2024). Real-time data analysis with smart sensors. In Gulati, S. (Ed.) *Application of Artificial Intelligence in Wastewater Treatment* (pp. 127–153). Springer Nature.

Singh, N. K., Yadav, M., Singh, V., Padhiyar, H., Kumar, V., Bhatia, S. K., & Show, P. L. (2023). Artificial intelligence and machine learning-based monitoring and design of biological wastewater treatment systems. *Bioresource Technology*, 369, 128486.

Skoumpopoulou, D., Toliyat, S. M. H., Ojra, A., Shokri, A., & Hu, S. (2024). Challenges of achieving digital transformation in manufacturing firms: The case of predictive maintenance and spare part inventory management. *Journal of Manufacturing Technology Management*, 36, 159–178.

Stengel, J., Aboagye, E., Le, P., DeNafo, M., Snyder, D., Nelson, N., & Yenkie, K. (2024). Systematic development of a machine learning-based asset management tool for wastewater pipeline networks. *ACS ES&T Water*, 4(12), 5555–5565.

Udo, W. S., Kwakye, J. M., Ekechukwu, D. E., & Ogundipe, O. B. (2024). Optimizing wind energy systems using machine learning for predictive maintenance and efficiency enhancement. *Journal of Renewable Energy Technology*, 28(3), 312–330.

VanEyk, L., & Heyns, P. S. (2025). A framework to define, design and construct digital twins in the mining industry. *Computers & Industrial Engineering*, 200, 110805.

Velásquez, D., Vallejo, P., Toro, M., Odriozola, J., Moreno, A., Naveran, G., Giraldo, M., Maiza, M., & Sierra, B. (2024). EDAR 4.0: Machine learning and visual analytics for wastewater management. *Sustainability*, 16(9), 3578.

Viveros, P., Miqueles, L., Mena, R., & Kristjanpoller, F. (2021). Opportunistic strategy for maintenance interventions planning: A case study in a wastewater treatment plant. *Applied Sciences*, 11(22), 10853.

Wang, A. J., Li, H., He, Z., Tao, Y., Wang, H., Yang, M., & Ren, N. (2024). Digital twins for wastewater treatment: A technical review. *Engineering*, 36, 21–35.

Wang, Y., Cheng, Y., Liu, H., Guo, Q., Dai, C., Zhao, M., & Liu, D. (2023). A review on applications of artificial intelligence in wastewater treatment. *Sustainability*, 15(18), 13557.

Wu, W., Pang, A., & Yang, W. (2024). Heterogeneous sensor fault detection for networked systems based on a graph transformer. *IEEE Sensors Journal*, 24(4), 5266–5278.

Xu, W. L., Wang, Y. J., Wang, Y. T., Li, J. G., Zeng, Y. N., Guo, H. W., Liu, H., Dong, K. L., & Zhang, L. Y. (2024). Application and innovation of artificial intelligence models in wastewater treatment. *Journal of Contaminant Hydrology*, 267, 104426.

Yao, J. F., Yang, Y., Wang, X. C., & Zhang, X. P. (2023). Systematic review of digital twin technology and applications. *Visual Computing for Industry, Biomedicine, and Art*, 6(1), 10.

Yazdi, M. (2024). Maintenance strategies and optimization techniques. In Yazdi, M. (Ed.) *Advances in Computational Mathematics for Industrial System Reliability and Maintainability* (pp. 43–58). Springer.

Yin, Y., Kou, C., Jia, S., Lu, L., Yuan, X., & Luo, Y. (2024). PCDMD: Physics-constrained dynamic mode decomposition for accurate and robust forecasting of dynamical systems with imperfect data and physics. *Computer Physics Communications*, 304, 109303.

Yousif, J. H., & Abdalgader, K. (2022). Experimental and mathematical models for real-time monitoring and auto watering using IoT architecture. *Computers*, 11(1), 7.

Zha, D., Bhat, Z. P., Lai, K. H., Yang, F., Jiang, Z., Zhong, S., & Hu, X. (2025). Data-centric artificial intelligence: A survey. *ACM Computing Surveys*, 57(5), 1–42.

Zhang, B., Wang, P., Liu, G., Ma, Z., & Zhao, T. (2024). AHU sensor fault diagnosis in various operating conditions based on a hybrid data-driven model combined energy consumption. *Journal of Building Engineering*, 87, 109028.

Zhao, X., Wang, W., Wen, L., Chen, Z., Wu, S., Zhou, K. & Wu, C. (2023). Digital twins in smart farming: An autoware-based simulator for autonomous agricultural vehicles. *International Journal of Agricultural and Biological Engineering*, 16(4), 184–189.

Zhu, M., Wang, J., Yang, X., Zhang, Y., Zhang, L., Ren, H., & Ye, L. (2022). A review of the application of machine learning in water quality evaluation. *Eco-Environment & Health*, 1(2), 107–116.

8 Future Prospects
AI Innovations for Enhanced Water Management

Sundram Singh, Swati Sangolgi, and Tapas Ranjan Das

8.1 INTRODUCTION

Living organisms and various ecosystems on the Earth require a clean, safe, and abundant supply of fresh water for their survival and support. Both natural and human interference have a major impact on water availability. The report of the WHO points out that about 3.6 million deaths occur every year due to diseases caused by drinking contaminated water worldwide. Bacteria such as *Escherichia coli* (*E. coli*), *Vibrio cholerae*, *Salmonella* spp., *Pseudomonas aeruginosa,* and protozoans like *Giardia lamblia, Entamoeba histolytica, Cryptosporidium* spp., *Cyclospora cayetanensis,* and viruses like hepatitis A virus, hepatitis E virus, Rotavirus, and Norovirus are involved in the contamination of water resources. These microbes and viruses affect the health of people and cause waterborne diseases like dysentery, cholera, hepatitis A, giardiasis, and typhoid fever. The situations in developing and underdeveloped countries, including Nigeria, Bangladesh, Kenya, and Haiti, are even worse. Factors affecting water quality in a rural area are very different from an urban area. This is due to differences in the environmental conditions, population density, and type of human activity. In rural areas, water quality is often affected by agricultural runoff, including pesticides, fertilizers, and animal waste.

Conversely, urban areas face challenges such as industrial discharge, improper waste disposal, and higher volumes of domestic and municipal wastewater. In addition, water bodies near urban areas often experience pollution from stormwater runoff, which carries oil, chemicals, and debris from the city streets. The types of wastewater impacting water quality are given in Figure 8.1, and each type presents unique challenges that require tailored management strategies to ensure water quality and sustainability.

With machine learning (ML) and deep learning (DL) with AI innovations, much work has been done on object recognition (Joy et al., 2023). A modification of the hydrological cycle worldwide, mainly by an increase in population and changes in agricultural and industrial activities, has degraded the water quality everywhere on Earth. Increased urbanization and climatic trends put extreme pressure on water resources. Water is a multifaceted resource; it is supplied from multiple sources for various purposes, so it has to be assessed and managed well at multiple consumption levels by the interested parties. AI can contribute

DOI: 10.1201/9781003600718-11

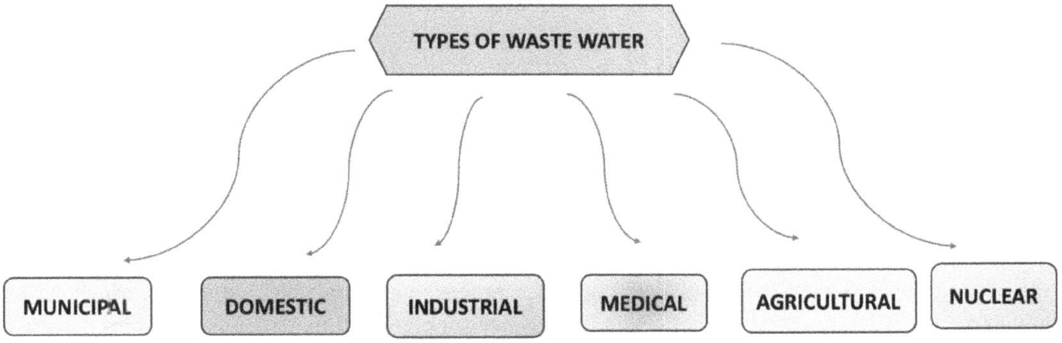

FIGURE 8.1 Types of wastewater.

to the saving of water resources through reducing contamination, efficient water supply, improving the quality of water, and better protection of humans against natural disasters such as floods and droughts. AI is transforming water management methods through optimizing water usage and improving water infrastructure. AI involves a lot of computing algorithms to formulate intelligent models that could predict and automate processes that would otherwise be done by a human being in the absence of AI. Highly complex models such as artificial neural networks (ANNs), support vector machines (SVMs), and hybrids have contributed to the advancement of water resources management. With the complexity that has led to water distribution, new AI methods turned out to be important to predict uncertainties, such as the new patterns of rain and new soil moisture, where original models had underperformed. An AI is said to improve the ability to mitigate the risk of disasters more closely related to extraordinary meteorological events like flash floods and dry spells because of its ability to work with many sets of data. One of the notable examples of the application of AI would be the case of the Yangtze River Basin. In such areas, risk assessment and flood mitigation are done using AI-based systems, which prove extremely valuable in protecting these areas. Water management through AI allows for making it more efficient, precise, and sustainable. It becomes vital to help us manage water resources better and with more responsibility. It can even imitate and sometimes even surpass the human thinking capacity by being swift in making logical judgments with steps that lead to definite objectives, for instance, multi-step predictions and pattern recognition. AI also gets along much better with dealing with big data and complex models easily (Kao et al., 2021). The application of AI techniques has considerably increased in various scientific and engineering fields since the mid-20th century. Interdisciplinary research is one of the newer causes for changes in AI advancement, leading to considerable changes in technological innovation, engineering science, water resource management, and medical diagnostics (Blake et al., 2021).

8.2 ARTIFICIAL INTELLIGENCE RESEARCH

Artificial intelligence (AI) imitates human brain activity through machines and is becoming essential in many areas like education, healthcare, production, and agriculture. It uses models like ant colony, particle swarm, DL, genetic algorithm, and others to improve efficiency and reduce human effort (Ni et al., 2015).

In the context of water management, AI can track water usage by continuously learning, predicting outcomes, optimizing processes, and suggesting new operations, surpassing the performance of pre-programmed tasks, unlike supervisory control and data acquisition (SCADA) systems (Fu et al., 2024). It also helps in optimizing water systems, leak detection, water quality diagnosis, and more (Boyle et al., 2022; Lunani, 2018; Doorn, 2021; Garza-Díaz & Sandoval-Solis, 2022). Research shows that AI improves leak detection and water quality control by identifying previous data patterns that are beyond human intervention

(Baird et al., 2019; Zhou et al., 2023; Cantos et al., 2020; Fitchett et al., 2020; Yazdekhasti et al., 2020; Dawood et al., 2020; Zhang et al., 2022; Fu et al., 2024). Moreover, its data handling and learning capabilities establish it as a valuable tool for managing and optimizing drinking water treatment processes (Fan et al., 2018; Alam et al., 2022; Dogo et al., 2019; Grbčić et al., 2021; Mboweni et al., 2021; Li et al., 2021).

By using data analytics, DL models, regression models, and algorithms, AI facilitates precise water allocation and management within cities according to their changing needs of the people. AI technology can also diagnose water quality, make decisions, and optimize operations, potentially establishing universal process analysis and predictive modeling platforms. According to an article on Analytics Insight, "The Promise of Artificial Intelligence in Water Management" (2021), accessed on September 12, 2023, AI offers significant potential in transforming water management practices, which mentions AI to be advantageous for governmental agencies handling water supplies and management systems for making cost-effective and sustainable water infrastructures that can adapt to various unforeseen situations and other environmental conditions that can lead to water scarcity.

8.3 HISTORICAL BACKGROUND OF AI IN WATER MANAGEMENT

AI-based evaluation and management of water resources have come a long way over the past few decades. As water challenges grew more complex and traditional systems struggled to adapt, AI technologies emerged as a more flexible and effective solution. The rapid growth of AI applications in water management began in the early 2000s as researchers sought to address these evolving challenges (Krysanova et al., 2018).

AI implementation in water resource management strategies started during the early 1990s, with one of the first AI techniques in hydrology being ANNs. These networks helped us understand complex relationships between weather data and water outcomes, like river flow and groundwater levels. Initial studies showed that ANNs could predict river flow better than traditional models, especially in data-scarce areas, by identifying and understanding the nonlinear connections between different factors (Imrie et al., 2000).

Throughout the 2000s, AI technology kept advancing, which resulted in the development of more sophisticated tools and techniques. The new tools are expected to improve traditional models by more effectively managing the uncertainties in water dynamics and processes, such as fluctuating rainfall patterns, soil moisture levels, and land use changes, which noticeably affect water requirements.

8.4 LEVERAGING AI FOR SUSTAINABLE AND EFFICIENT WATER RESOURCE MANAGEMENT

Water resource management involves tasks such as conserving, harvesting, planning, and distributing water to customers. Conventional methods often struggle with water scarcity, pollution, water wastage and resource allocation, and the impacts of climate change. These challenges are often compounded by agricultural sectors, industrial sources, mining, and urban runoff, which generally contribute to water pollution (Berthet et al., 2021).

AI technologies provide certain advantages to these and can greatly enhance water quality monitoring, resource allocation, and decision-making procedures. For example, it can predict water demand, identify real-time contaminants, and optimize distribution infrastructure for a more efficient and sustainable water supply.

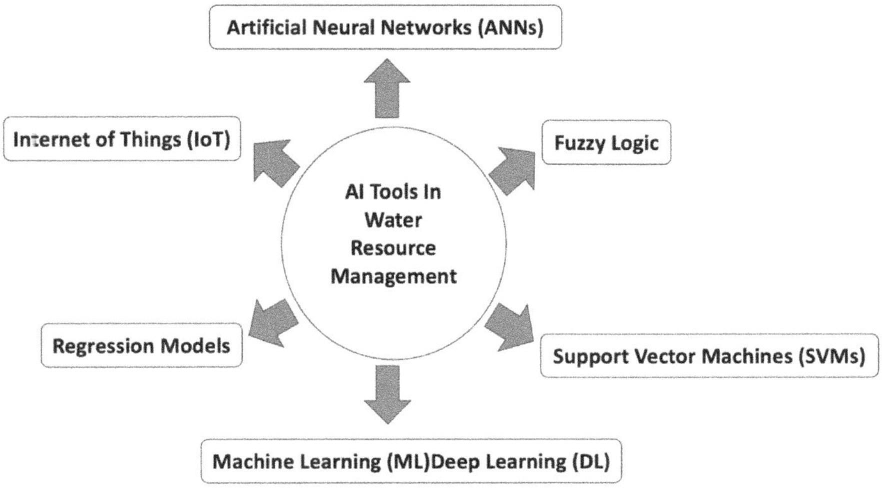

FIGURE 8.2 Various tools of AI in water management resources.

AI tools and method implementation vary with locations, specifying the need, requirement, and challenges that need to be accounted for before coming to the management and optimization practices. A list of AI tools used in water resource management is given in Figure 8.2, which contributes to managing different water challenges while promoting efficiency and sustainability. ANNs excel in detecting patterns in complex datasets and thus are useful for predictive modeling of water demand, quality, and distribution systems. Internet of Things (IoT) sensors, when integrated with AI, enables real-time monitoring of water levels, quality, offering solutions to smart irrigation and water leakage detection. ML, which consists of a vast set of algorithms, including regression models and SVM, is employed to forecast water usage and predict contaminants (Vasquez et al., 2014). DL is highly effective in analyzing large datasets for wastewater treatment (WWT) and optimization of water distribution networks. Fuzzy logic Systems aid in handling uncertainty and variability in water resource allocation. A baseline research study of the location will provide insight into the required model to be applied, which can solve all the water-related issues effectively.

In agriculture, AI-driven systems improve water access, efficient use, and sustainability, whereas for industries and urban areas, AI can manage water supply during shortages and suggest water-saving appliances and tariffs. Technologies like AI, IoT, ML, and DL create smart models to greatly improve water management practices globally, making them more sustainable and resilient (Alam et al., 2022; Singh et al., 2020). Combining AI with IoT creates smart models to handle complex water management challenges. AI tools help in many areas of water management. They are used for treating wastewater, controlling water pollution, helping in agriculture, optimizing water usage, and monitoring water quality in real time. Effective water management in agriculture is crucial for economic growth. For example, Liu et al. (2025) developed an IoT and data analytics-based system that improved agricultural water use efficiency. Predictive models help detect water infrastructure problems early, predict and prevent system failures, and reduce water losses.

8.4.1 Improving Clean Water Management through AI and IoT Innovations

AI and IoT innovations can contribute to better management of clean drinking water. These improvements involve monitoring systems, optimized water distribution networks, and ensuring a safe, consistent supply. Tools like remote sensing, adaptive management strategies, and information system integration address challenges related to water allocation.

IoT-based water systems minimize water wastage by monitoring data and adjusting the water flow according to weather and agricultural requirements. For example, Taiwan's integrated management approach employs a SCADA system for resource management, supply monitoring, quality evaluation, and leak detection.

Similarly, Mexico's "PUMAGUA" initiative improves water quality by implementing a smart network for water sources, reducing consumption and wastage. Another example is from South Korea, where they use data for water analysis and rain forecasts with various ground-level parameters in their Hydro Intelligent Toolkit.

A case study from Portugal highlights the benefits of constructing smart water grids, which have shown increased efficiency and sustainability of water management systems. Technologies for monitoring and controlling water loss have resulted in cost cutting, reduction in CO_2 emissions, and improved the city's global efficiency ranking from 20th to 5th.

8.4.2 Applications of AI in Different Water Management Approaches

8.4.2.1 Wastewater recycling and its management by various AI tools

Predicting real-time parameters for WWT presents significant challenges. Some researchers, including Li et al. (2022), have developed advanced methods for forecasting the treatment process of wastewater using anaerobic bioreactors (AnBRs). This approach involves monitoring six metrics like reactor temperature, environment, chemical oxygen demand (COD), and flux, and eight metrics for the evaluation of the treatment process, like effluent COD, pH, biogas composition, and redox potential.

To analyze these inputs and outputs, several DL models were used. The statistical analysis performed by these models was remarkably accurate, closely matching the real measurements on the AnBR. For instance, convolutional neural network (CNN) predicted results with an accuracy rate of 97.44%, with each calculation completed in less than a second. These findings underscore the effectiveness of DL techniques in accurately predicting AnBR performance.

Wastewater treatment plants(WWTPs) are being used to remove contaminants, pollutants, and other impurities, thereby mitigating environmental pollution caused by human activities. However, these plants also generate a large amount of residual matter and emit significant amounts of greenhouse gases. This necessitates further optimization to reduce their harmful effects on the environment and enhance sustainability.

8.4.2.2 Advancements in deep learning and machine learning for water quality assessment

Advancements in technology have made it easier to determine water quality, which is crucial as global water pollution continues to rise. DL models excel in analyzing time-oriented data and perform better in measuring water quality parameters. DL and ML models are being used to accurately predict localized water demand, supply patterns, and potential contaminants. These models can also optimize water treatment processes, improving efficiency and leading to cost reductions in terms of operation and management.

Latest research on bidirectional long short-term memory (LSTM) models for evaluating water quality of Yamuna River in New Delhi, focusing on training, and filling in missing data to accurately predict biochemical oxygen demand (BOD) and COD levels with minimum error rates.

Smart sensors further contribute to water quality measurements by collecting and preprocessing data. Advanced techniques like Geometric Synthetic Minority Over-sampling Technique (G-SMOTE) are used for feature extraction, while a multiclass deep learning model (MDLNN) has achieved outstanding results, including a high accuracy rate of 99.34% and a very low validation loss of 0.0415%.

These advancements highlight the potential of DL for predicting water quality, emphasizing the need for models that are both accurate and efficient to ensure safe and sustainable water use.

8.4.2.3 Revolutionizing rainwater and irrigation with AI innovations

Rainfall prediction is important for water harvesting and management. Bhattacharyya et al. (2021), in their study on rainwater management, used various AI models to predict rainfall in Andhra Pradesh, India. Rainwater harvesting systems are inherently complex, needing many inspections, but computer vision-based solutions can automate many of these processes. For example, a Seasonal Autoregressive Integrated Moving Average (SARIMA)-based rolling forecasting model is used in predicting rainfall, while a Mask Region-based Convolutional Neural Network (Mask R-CNN) segmentation model calculates how much rainwater a rooftop can collect.

In agriculture, using too much water can cause shortages. To address this, DL models are employed to classify soil texture and determine precise water requirements in irrigation. AI plays a significant role in optimizing irrigation scheduling by processing sensor data and comprehending the evolving dynamics of weather patterns, soil properties, and plant conditions during cultivation time (Talaviya et al., 2020; Qin et al., 2023). Additionally, advancements in sensing technologies, the widespread IoT revolution, and ML applications in smart agriculture have profoundly influenced irrigation water conservation, plant development, and crop yield (Mohammed et al., 2023b; Mohammed et al., 2022a, 2022b). These innovations contribute to efficient irrigation management and minimize electricity consumption and overall expenses (Abioye et al., 2020). Key parameters such as plant health, climatic variables, and soil moisture levels are critical for reducing water usage in irrigation without compromising crop productivity. Tools like automated weather stations and IoT-driven control systems have been developed to meet these needs (Yartu et al., 2022).

Advanced DL models are also revolutionizing irrigation management. Raei et al. (2022) developed a model by combining a U-shaped network (U-Net) with a residual network with 34 layers (ResNet-34) to identify irrigation control networks across a specific region, achieving efficient segmentation. Furthermore, unmanned aerial vehicles (drones), integrated with speed measurement systems, LSTM models, and component analysis, are used to predict evaporation rates and manage irrigation processes effectively. The integration of DL models such as U-net and CNN enables high-resolution mapping of irrigation systems. The firefly algorithm further optimizes AI-based systems, enhancing water irrigation predictions.

AI tools and technologies hold several water management applications in agriculture and allied sectors, as illustrated in Figure 8.3. They can be allocated for optimizing irrigation frequency based on crop needs and weather conditions. The use of DL models for the analysis of crop health and the detection of waterborne diseases in early stages of infections provides necessary data for timely interventions (Rathore et al., 2019). Neural network algorithms like recurrent neural network (RNN), deep neural network (DNN), and ANN offer faster predictions, essential for forecasting and examining patterns over time. The linear regression (LR) approach is also widely used in prediction and decision-making, such as modeling relationships between variables to address agricultural problems (Mohammed et al., 2023b; Sapankevych & Sankar, 2009).

Other applications of AI include surface water management in agricultural fields, monitoring of water quality, WWT, reusing water resources, efficient water delivery, rainwater collection, and storage. A study by Mohammed et al. (2023a) explored the optimum irrigation water and solar photovoltaic (PV) energy requirements for date palm irrigation using microirrigation systems. Furthermore, they developed four ML models using meteorological data and information on date palm age. These were LR, support vector regression (SVR), LSTM, neural network, and extreme gradient boosting (XGBoost) to predict water and energy requirements.

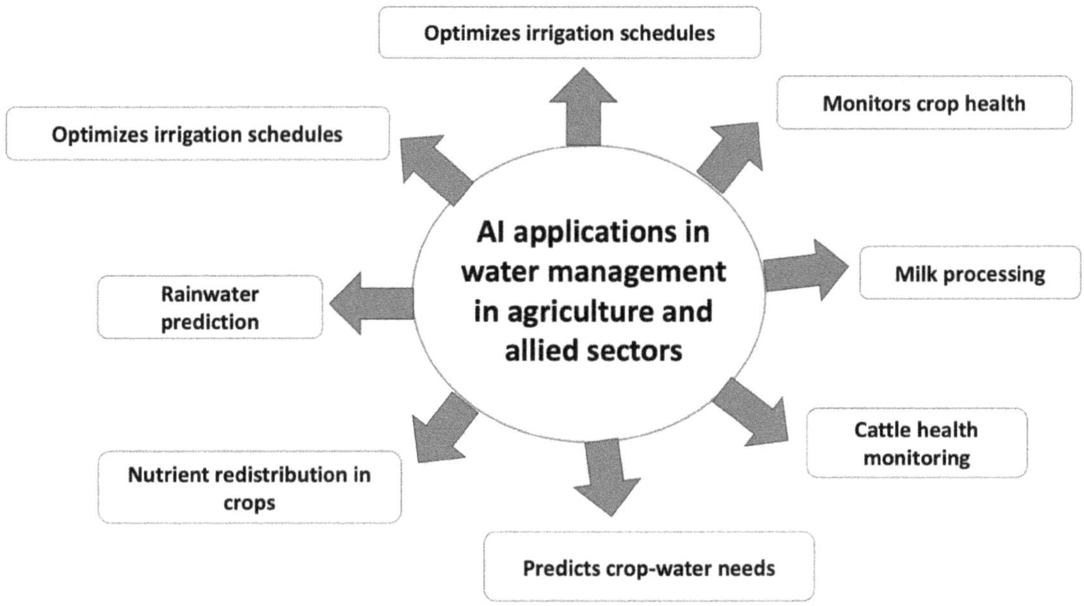

FIGURE 8.3 AI applications in water management in agriculture and allied sectors.

8.5 AI IN AUTOMATION OF WATER MANAGEMENT SYSTEMS

Cities and localities with less rainfall and low water tables often face greater challenges with water availability throughout the year. Ensuring clean water and supporting good public health in such regions is essential. Water management systems are socio-technical, meaning the demand for water by consumers varies. Every day, manual water operating systems are required to make thousands of decisions, ranging from pump operation to emergency response.

Sensors with real-time monitoring in association with IoT devices can continuously collect data on the quality of water, water levels, and its usage. Integrating data received from various sources can later be subjected to cumulative monitoring, ensuring more precise decision-making and resource allocation (Tang et al., 2023).

Natural disasters further complicate water availability and distribution. Predictive tools that leverage historical data on flood, drought, and tsunami patterns can integrate with water management systems to provide early warnings. This combination can act as a potent tool to mitigate critical water-related situations. By utilizing AI, authorities can address water scarcity more effectively through timely alerts, enabling swift and informed decision-making.

The automation of water operating systems is essential to tackle these challenges. AI-powered solutions can transform water management by optimizing resources, improving resilience against natural calamities, and ensuring sustainable water distribution. While AI holds great promise for revolutionizing water-related operations, several limitations must also be considered, such as data dependency and high operational costs. For a detailed breakdown of the benefits and challenges inherent in these innovations, refer to Table 8.1.

TABLE 8.1 Advantages and limitations of AI in drinking water operations

SECTION	*ADVANTAGES*	*BARRIERS AND CONSTRAINTS*
Automation	Automating complex systems	Personnel and financial barriers due to high costs of AI deployment
Integration	Integrating with other technologies	Need for ongoing expertise, either in-house or through consultants
Cost savings	Saving money through efficient operations	Data volume and quality issues, scalability, and repeatability challenges
Improved hydraulics and water quality	Enhancing hydraulics and water quality	Preference for existing technologies like SCADA systems (Sarni et al., 2019)
Water conservation	Efficient water conservation	Concerns over the long-term risks of AI investments
Time/Labor savings	Saving time and labor	Requirement for good human operators to manage drinking water systems; AI to support but not replace human decision-making in crucial areas
Unexpected benefits	Benefits experienced were often different from initial expectations (Fu et al., 2024)	Need for AI researchers and vendors to define the scope of AI's usefulness and reassure operators of their fundamental role (Walski, 2023)

8.5.1 Addressing Growing Complexity in Water Distribution

An explosion in population growth led to an increase in the complexity of operational decisions in water systems. According to Wallis-Lage (2020), more than 90% of North American water utilities use SCADA systems. Despite these advancements, a blog by Blue Conduit highlights a major concern that over 22 million people in the United States get drinking water from toxic lead pipes and pipes with rust, which carry harmful heavy metals with water. AI can help to overcome this by creating multiple checkpoints for the detection of rust buildup within the pipelines, along with sensors for the detection of harmful chemicals and pollutants within the supply water. Additionally, predictive models enable efficient planning for pipeline replacements, enhancing the overall reliability of water systems (Khan et al., 2021).

Technological improvements also support improved monitoring and frequent water loss audits. These innovations have encouraged customers to demand greater accountability for water losses, lower costs, better energy management, and faster service restoration. By leveraging AI and other technologies, water utilities can address these demands while navigating the growing complexity of water distribution systems.

8.5.2 AI Solutions for Water Utilities

Recent studies show that AI has great potential for improving water utilization (Fu et al., 2024). AI uses ML to learn from new data, going beyond what humans can program. This adaptability allows AI to predict and optimize novel operations as conditions evolve.

One of AI's key applications is in forecasting and managing water demand. It plays a vital role in drinking water operations by identifying bacterial hotspots and optimizing energy consumption through efficient pump scheduling (Helmbrecht et al., 2017). Additionally, AI is also used in the calibration of automated hydraulic-driven models for efficient infrastructure planning (Beh et al., 2015) and supporting large-scale water-related resource management. These AI-driven solutions empower water utilities to enhance operational efficiency, reduce costs, and ensure sustainable management of water resources.

8.5.3 Challenges and Future Directions in AI Adoption

Water utilities do not use AI as much as industries like energy and transport (Hadjimichael et al., 2016; Doorn, 2021). However, generative AI will soon be incorporated into customer service, staff training, and report generation. The concept of the digital water adoption curve helps understand how AI and other digital technologies progress in water management, moving from basic to higher transformational stages.

To maximize its impact, AI tools must be integrated into comprehensive water management systems and need to be extended to regions that are affected the most by water management problems, such as rural as well as urban areas, to provide them with standard, clean, and safe drinking water. This approach can help ensure fair access to clean and safe drinking water and close gaps in water management (Park et al., 2021).

8.6 AI IN MONITORING SURFACE WATER QUALITY

Three-fourths of the Earth's surface consists of water, out of which only 1% can be utilized for human consumption (Al Aani et al., 2019). With a rapid and ever-growing population and increasing industrialization, the level of harmful pollutants in oceans, rivers, and lakes is increasing (Adelodun et al., 2020). Moreover, surface water is most susceptible to pollution from urban runoffs, untreated wastewater, and excessive nutrient loading from agricultural activities, leading to eutrophication and algal blooms (Reckhow & Chapra, 1999). Climate change, such as erratic rainfall patterns, prolonged droughts, and rising temperatures, also decreases water availability and exacerbates water quality degradation. All these factors make the monitoring of surface water quality incessant (Yu et al., 2019).

8.6.1 Improving Water Quality Monitoring

Besides traditional approaches, new work is being done on IoT-based technologies (Geetha & Gouthami, 2016) and AI. AI tools like fuzzy logic, genetic algorithms, and ANNs are fast, affordable, and capable of near real-time monitoring, prediction, and assessment (Yetilmezsoy et al., 2011). These advanced technologies streamline and improve the accuracy of water quality management. The use of AI in monitoring and adjusting improper water pressure and dynamic flow control throughout the water supply in real time can be useful for minimizing pipe bursts and leakages. Automated reporting and standards monitoring to ensure water quality standards at regular intervals can easily be achieved. Moreover, AI can also be used to add essential micronutrients and remove harmful chemicals and heavy metals to ensure a safe and clean water supply.

Additionally, AI's capability to monitor water quality in real time allows for early detection of contaminants and optimization of water treatment processes. This helps in maintaining consistent water safety and preventing potential health risks.

8.6.2 AI Applications and Trends

AI applications in the assessment of water quality were initiated in the early 1990s and have been used for water reservoirs (Allawi et al., 2018) and other water bodies (Zaji & Bonakdari, 2019; Ceccaroni et al., 2018). These technologies have become indispensable for identifying trends, similarities, and research gaps in the field. Most of the research on the use of ANNs for quality monitoring of surface water comes from Iran and Southeast Asia, where limited freshwater makes water quality very important (Al Aani et al., 2019). New technologies, sewer, and water quality sensors, along with AI, may help us to better understand our assets and measure their performance.

8.7 DISINFECTION PRACTICES AND DBP FORMATION: ANN MODELS AND TREATMENT APPROACHES

Chemical disinfectants like chlorine and chlorine-based compounds such as sodium hypochlorite, calcium hypochlorite, and chloramine not only kill microorganisms but also tend to react with organic matter and Br^- ions to create by-products known as disinfection by-products (DBPs) like haloacetic acids (HAAs) and trihalomethanes (THMs), which can lead to health risks (Hamidin et al., 2008).

To treat Natural Organic Matter (NOM) and DBP precursors, three major processes are used, which are as follows:

1. Traditional methods such as flocculation, coagulation, media filtration, and sedimentation.
2. Adsorption techniques using carbon-based adsorbents, like activated charcoal.
3. Advanced nanofiltration methods for precise purification (Chen et al., 2007).

For a clear visual representation of these processes, see Figure 8.4.

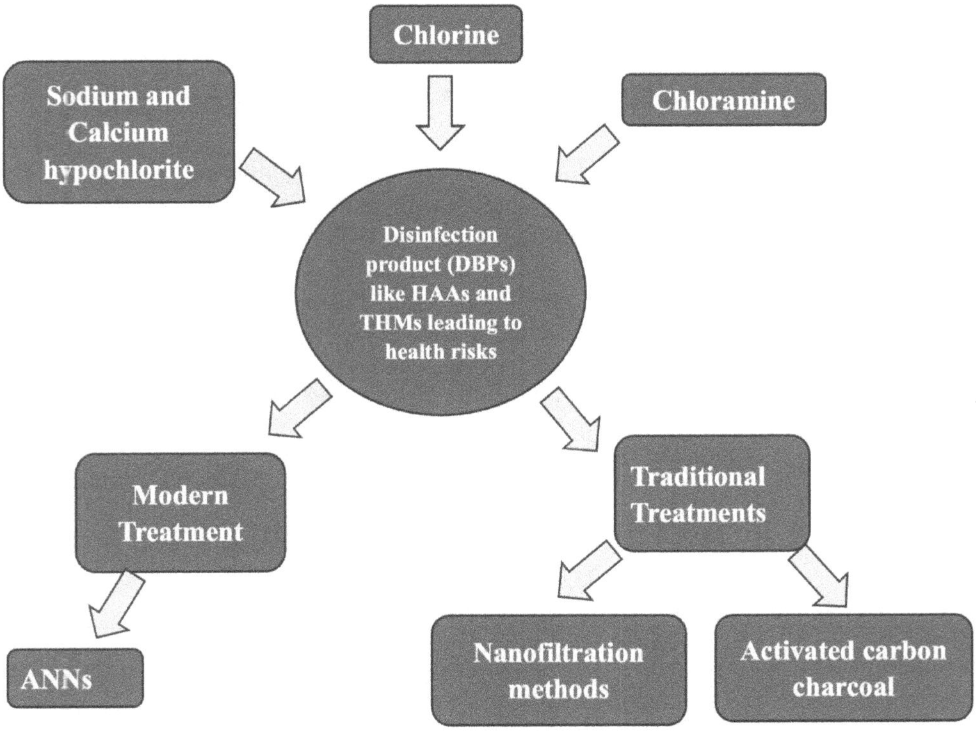

FIGURE 8.4 Various methods used for the treatment of NOM and DBP precursors.

8.7.1 Predicting DBP Formation with Neural Networks (NNs)

DBP formation involves a complex chemical reaction. Empirical models are used to predict DBP levels by considering factors like dissolved organic carbon (DOC) concentration, ultraviolet absorbance, concentration of bromide ions (Br^-), chlorination dosage, temperature, and pH.

ANNs, with their ability to handle complex and noisy data, have emerged as a useful tool for predicting the DBPs like THMs, HAAs, and Total Organic Halides (TOX) in both raw and treated water (Milot et al., 2002). Reliable ANN models are created using feed-forward and back propagation methods, often implemented with the MATLAB neural network toolbox.

8.7.2 Tailoring ANN Models for Different Water Treatment Methods

Principal component analysis (PCA) has been utilized for determining whether only a single neural network model could predict DBP concentrations for all water types or whether specialized models are needed. PCA results showed that different ANN models are required for various water treatment methods, like raw water, conventional treatment, granular activated carbon (GAC) adsorption, and nanofiltration.

It has been shown that ANNs could predict DBP concentrations, such as TOX, THM4, and HAA6, accurately, with a high regression coefficient of more than 0.75. Testing revealed that ANN predictions were accurate even with varying data sizes and input parameters.

Moreover, ANNs excel at identifying influential input variables. For raw water, DOC was the key factor for DBP formation. In contrast, for GAC-treated water and nanofiltration, Br^- ion concentration was more significant. Chlorine dose was consistently identified as a key parameter for reducing DBP concentrations (Chang et al., 2002).

8.7.3 Advantages of ANNs for DBP Formation Prediction

ANNs can predict the formation of DBPs across the United States using minimal training data. By examining the relationship between water quality parameters and chlorination conditions, ANNs can be used as a robust and efficient approach to understanding changes in DBP formation.

These models are invaluable for process optimization, control, and assessing changes in DBP formation, offering a cost-effective alternative to pilot-scale testing. However, it is important to apply ANNs carefully beyond the derived water quality parameters. Even though ANNs do not provide detailed mechanistic insights, they are useful for optimizing the operation and planning to meet required drinking water quality standards (Hua & Reckhow, 2008).

8.8 ARTIFICIAL INTELLIGENCE IN WASTEWATER TREATMENT

Water resources needed for mankind, various developmental and production processes, are increasingly threatened by pollutants from improper human interventions, as well as natural processes. Every day, a huge quantity of toxic pollutants is released as mostly untreated wastewater (Al Aani et al., 2019).

Treating wastewater helps to reduce the environmental impacts of depleting resources, save energy, and promote water reuse and sustainable development. Innovative technologies are needed to make WWT more efficient, reduce costs, and lower energy consumption (Chen et al., 2007). In WWT, AI models predict and optimize processes, showing excellent capabilities in process design, water quality monitoring, parameter optimization, and performance prediction, ensuring more effective and sustainable solutions.

8.8.1 AI Models Used in Wastewater Treatment

Models using AI and traditional methods are important for creating effective WWT systems and reusing sewage. AI models, such as those incorporating the IoT and SVM, facilitate to prediction and optimization of WWT (Ullah et al., 2017). These models address pollutants like dyes, agricultural wastes, microbes, drugs, pesticides, insecticides, and heavy metals while also predicting the presence of contaminants in wastewater, aiding treatment processes.

However, most of these applications are based on lab research and need a lot of data for accurate prediction and assessment (Nourani et al., 2018). AI models can accurately adapt to providing water of different qualities from wastewater through different treatment methods according to the needs. For instance, water intended for industrial applications serves as a coolant, cleaning solvent, or energy production medium, whereas water for agricultural use should retain essential micronutrients to support plant and crop sustainability. These varying requirements necessitate the adoption of effective AI-driven strategies to meet diverse needs.

8.8.2 Data Acquisition for Water Quality Monitoring

Continuous water quality monitoring sensors provide vast amounts of data, which are required for AI models in WWT. These sensors have been developed for assessment, rapid monitoring, as well as for automating WWT processes. Data acquired from sensors such as pH, dissolved oxygen (DO), temperature, turbidity, heavy metals (Oliveira et al., 2017), etc., can be statistically analyzed by computational models for carrying out effective water regulation procedures. Critical input parameters for AI models in WWT include BOD, COD, pH, DO, flow rate, temperature, and the initial concentration of pollutants. Effluent parameters play a vital role in evaluating the effectiveness and overall performance of WWT.

8.8.3 AI Methods in Performance Prediction

It is also crucial to predict the WWT performance to ensure efficiency and compliance with standards. Nourani et al. (2018) applied AI-based ensemble models to WWT planning, demonstrating the effectiveness of models like adaptive neuro-fuzzy inference systems (ANFIS) and NN ensembles. Similarly, Bisaria et al. (2022) simulated chlorpyrifos adsorption using ANN and ANFIS, with the validity of the effectiveness experimentally validated. Yang et al. (2022) proposed a model known as PCA-Nonlinear Autoregressive model with Exogenous inputs (NARX) for predicting residual quality. This model outperformed static ANN models in prediction accuracy. Additionally, reinforcement learning (RL)enables management systems to learn and optimize their effectiveness gradually with the help of received feedback. Furthermore, decision trees and random forests have proven effective in making predictions based on data-driven decision rules.

8.9 INNOVATIONS IN WATER MANAGEMENT: AI-DRIVEN CHATGPT SOLUTIONS

Water is an indispensable resource essential for sustainable development, animal habitat, and human life (Emenike et al., 2022). However, the water crisis around the world is marked by improper water usage, misuse, and pollution and thus requires major changes (Du Plessis, 2019). Conventional water treatment methods are not sufficient in providing real-time observations, utilizing resources efficiently, or minimizing environmental impacts, which leads to several problems, like a lack of quick responses to quality changes of water, inefficient use of energy resources leading to excessive energy consumption, operative techniques prone to human error, and high maintenance costs. As a result, innovative approaches are essential for developing effective water management strategies. In recent years, AI tools, specifically ChatGPT, have transformed multiple industries by increasing automation, providing real-time support and quality control, predictive maintenance of equipment, documentation, and compilation of large data. When applied to water management, AI-driven ChatGPT solutions can handle, examine, and process large datasets for improved decision-making. By combining DL models with ChatGPT's precise language processing ability, new understanding can be derived, and water management strategies can be optimized accordingly (Holzinger et al., 2023). AI-driven ChatGPT solutions can also enhance water resource management through the use of advanced data analytics and predictive technology. These solutions assist water authorities in optimizing water systems and predicting water availability and stressful situations.

8.10 FUTURE PROSPECTS

Developing AI tools and models for building efficient water infrastructures needs collaboration among researchers, developers, and water management authorities to ensure solutions from AI models for increasing the effectiveness of current management systems (Du Plessis, 2019; Emenike et al., 2022). National as well as global initiatives are required to utilize AI technology for effective water management and solving water-related issues and should also collaborate on water-related issues to provide integrated solutions. Moreover, the development of guidelines and ethical standards for regulating AI tools in water resource management within the country will provide a competitive advantage and responsible usage.

Upcoming research should emphasize the advancement in energy-efficient AI models and sustainable AI technology. AI models must continuously evolve to meet present-day water management needs with a sustainable approach. AI is a promising tool for WWT, offering advanced methods for evaluation, prediction, control, and automation. AI techniques have also enhanced control and automation in WWTPs, enabling operators to improve facility management and control (Mannina et al., 2019). Multiobjective optimization and integrated distributed expert systems have further enhanced WWTP stability and reduced costs (Dai et al., 2016).

AI-based technologies have transformed WWT processes, making it possible to handle complex systems, maximize equipment efficiency, and maintain sustainable water quality (Ullah et al., 2017). Further research and development of AI applications will optimize performance and address emerging issues. Additionally, new techniques and models for water conservation, harvesting, and management are emerging, supporting global water availability and sustainability. AI tools like ML, neural networks pave the way for future water resource conservation. Applications utilizing advanced neural network models and predictive algorithms hold significant promise for monitoring and managing water quality, ensuring efficient and sustainable water resource management.

8.11 CONCLUSION

AI and related technologies, as discussed like neural networks, IoT, DL, and predictive algorithms, are believed to have considerable potential to transform water management systems to be sustainable and more resilient in the face of present-day water-related challenges and those that lie ahead. These advanced technologies can improve water management systems by simplifying complexities in various sectors, from the treatment of wastewater, agricultural wastes, and industrial water to ensure water usage optimization and real-time monitoring.

AI-driven technologies in agriculture improve water use efficiency and reduce the use of valuable water resources. This contributes not only to environmental sustainability but also to the economic growth of the country. Beyond that, AI applications in water management strengthen industries toward low-cost products and services.

Despite AI's potential in managing complex water distribution systems, it is still underutilized. Researchers should integrate their work with daily water operators and clearly explain the benefits of AI and discuss its potential. Areas of discussion may involve precise leak detection, automated optimization, and quick responses. AI-incorporated systems, like ChatGPT and DL models, can optimize resource use and solve water issues efficiently. These technologies can provide water management authorities with valuable data about live monitoring of water quality, along with capabilities in active decision-oriented management and precision algorithms to protect water resources. However, one major concern is the high water usage associated with large AI systems, which necessitates ongoing research and development to address sustainability challenges. The future of water management lies in leveraging AI technologies to provide innovative solutions while ensuring continuous advancements in current systems to meet increasing water needs.

REFERENCES

Abioye, E. A., Abidin, M. S. Z., Mahmud, M. S. A., Buyamin, S., AbdRahman, M. K. I., Otuoze, A. O., Ramli, M. S. A., & Ijike, O. D. (2020). IoT-based monitoring and data-driven modelling of drip irrigation system for mustard leaf cultivation experiment. *Information Processing in Agriculture*, 8, 270–283.

Adelodun, B., Ajibade, F. O., Ibrahim, R. G., Bakare, H. O., & Choi, K. S. (2020). Snowballing transmission of COVID-19 (SARS-CoV-2) through wastewater: Any sustainable preventive measures to curtail the scourge in low-income countries? *Science of the Total Environment*, 742, 140680.

Al Aani, S., Bonny, T., Hasan, S. W., & Hilal, N. (2019). Can machine language and artificial intelligence revolutionize process automation for water treatment and desalination? *Desalination*, 458, 84–96.

Alam, G., Ihsanullah, I., Naushad, M., & Sillanpää, M. (2022). Applications of artificial intelligence in water treatment for optimization and automation of adsorption processes: Recent advances and prospects. *Chemical Engineering Journal*, 427(130011), 130011.

Allawi, M. F., Jaafar, O., Mohamad Hamzah, F., Mohd, N. S., Deo, R. C., & El-Shafie, A. (2018). Reservoir inflow forecasting with a modified coactive neuro-fuzzy inference system: A case study for a semi-arid region. *Theoretical and Applied Climatology*, 134, 545–563.

Baird, G. M., Hatler, D., & Carpenter, P. (2019). How cost effective is machine learning/AI applied to leak detection and pipe replacement prioritization? In *Pipelines 2019* (pp. 274–283). Reston, VA: American Society of Civil Engineers.

Beh, E. H., Maier, H. R., & Dandy, G. C. (2015). Adaptive, multiobjective optimal sequencing approach for urban water supply augmentation under deep uncertainty. *Water Resources Research*, 51, 1529–1551.

Berthet, A., Vincent, A., & Fleury, P. (2021). Water quality issues and agriculture: An international review of innovative policy schemes. *Land Use Policy*, 109, 105654.

Bhattacharyya, D., Swathi, K., Rao, N. T., & Ku-mari, N. M. J. (2021). Long term prediction of rainfall in Andhra Pradesh with deep learning. *The Journal of Medical, Pharmaceutical, and Allied* Sciences, 10, 3132–3137.

Bisaria, K., Singh, R., Gupta, M., Mathur, A., & Dixit, A. (2022). Novel acoustic-activated alkali-functionalized Trapa bispinosa peel biochar for green immobilization of chlorpyrifos from wastewater: Artificial intelligence modelling and experimental validation. *Biomass Conversion and Biorefinery*, 14, 1–20.

Blake, R. W., Mathew, R., George, A., & Papakostas, N. (2021). Impact of artificial intelligence on engineering: Past, present, and future. *Procedia CIRP*, 104, 1728–1733.

Boyle, C., Ryan, G., Bhandari, P., Law, K. M., Gong, J., & Creighton, D. (2022). Digital transformation in water organizations. *Journal of Water Resources Planning and Management*, 148(7), 03122001.

Cantos, W. P., Juran, I., & Tinelli, S. (2020). Machine-learning–based risk assessment method for leak detection and geolocation in a water distribution system. *Journal of Infrastructure Systems*, 26(1), 04019039.

Ceccaroni, L., Velickovski, F., Blaas, M., Wernand, M. R., Blauw, A., Subirats, L., ... & Aubrecht, C. (2018). Artificial intelligence and earth observation to explore water quality in the Wadden Sea. *Earth Observation Open Science and Innovation. ISSI Scientific Report Series*, 15, 311–320.

Chang, F. J., Chang, L. C., & Huang, H. L. (2002). Real-time recurrent learning neural network for stream-flow forecasting. *Hydrological Processes*, 16(13), 2577–2588.

Chen, C., Zhang, X. J., He, W. J., & Han, H. (2007). Simultaneous control of microorganisms and disinfection by-products by sequential chlorination. *Biomedical and Environmental Sciences*, 20(2), 119.

Dai, X., Chen, C., Yan, G., Chen, Y., & Guo, S. (2016). A comprehensive evaluation of re-circulated bio-filter as a pretreatment process for petroleum refinery wastewater. *Journal of Environmental Sciences*, 50, 49–55.

Dawood, T., Elwakil, E., Novoa, H. M., & Delgado, J. F. G. (2020). Artificial intelligence for the modeling of water pipes deterioration mechanisms. *Automation in Construction*, 120, 103398.

Dogo, E. M., Salami, A. F., Nwulu, N. I., & Aigbavboa, C. O. (2019). Blockchain and internet of things-based technologies for intelligent water management system. In Al-Turjman, F. (Ed.) *Artificial Intelligence in IoT* (pp. 129–150). Springer, Cham.

Doorn, N. (2021). Artificial intelligence in the water domain: Opportunities for responsible use. *Science of the Total Environment*, 755, 142561.

Du Plessis, A. (Ed.) (2019). Current and future water scarcity and stress. In *Water as an Inescapable Risk: Current Global Water Availability, Quality and Risks with a Specific Focus on South Africa* (pp. 13–25). Springer, Cham.

Emenike, E. C., Adeniyi, A. G., Omuku, P. E., Okwu, K. C., & Iwuozor, K. O. (2022). Recent advances in nano-adsorbents for the sequestration of copper from water. *Journal of Water Process Engineering*, 47, 102715.

Fan, M., Hu, J., Cao, R., Ruan, W., & Wei, X. (2018). A review on experimental design for pollutants removal in water treatment with the aid of artificial intelligence. *Chemosphere*, 200, 330–343.

Fitchett, J. C., Karadimitriou, K., West, Z., & Hughes, D. M. (2020). Machine learning for pipe condition assessments. *Journal - American Water Works Association*, 112(5), 50–55.

Fu, T., Qiu, Z., Yang, X., & Li, Z. (2024). The impact of artificial intelligence on green technology cycles in China. *Technological Forecasting and Social Change*, 209, 123821.

Garza-Díaz, L. E., & Sandoval-Solis, S. (2022). Changes in the stability landscape of a river basin by anthropogenic droughts. *Water*, 14(18), 2835.

Geetha, S., & Gouthami, S. J. S. W. (2016). Internet of things enabled real-time water quality monitoring system. *Smart Water*, 2, 1–19.

Grbčić, L., Kranjčević, L., & Družeta, S. (2021). Machine learning and simulation-optimization coupling for water distribution network contamination source detection. *Sensors*, 21, 1157.

Hadjimichael, A., Morera, S., Benedetti, L., Flameling, T., Corominas, L., Weijers, S., & Comas, J. (2016). Assessing urban wastewater system upgrades using integrated modeling, life cycle analysis, and shadow pricing. *Environmental Science & Technology*, 50(23), 12548–12556.

Hamidin, N., Yu, Q. J., & Connell, D. W. (2008). Human health risk assessment of chlorinated disinfection by-products in drinking water using a probabilistic approach. *Water Research*, 42(13), 3263–3274.

Helmbrecht, J., Pastor, J., & Moya, C. (2017). Smart solution to improve water-energy nexus for water supply systems. *Procedia Engineering*, 186, 101–109.

Holzinger, A., Keiblinger, K., Holub, P., Zatloukal, K., & Müller, H. (2023). AI for life: Trends in artificial intelligence for biotechnology. *New Biotechnology*, 74, 16–24.

Hua, G., & Reckhow, D. A. (2008). DBP formation during chlorination and chloramination: Effect of reaction time, pH, dosage, and temperature. *Journal-American Water Works Association*, 100(8), 82–95.

Imrie, C. E., Durucan, S., & Korre, A. (2000). River flow prediction using artificial neural networks: Generalisation beyond the calibration range. *Journal of Hydrology*, 233(1–4), 138–153.

Joy, C., Sundar, G. N., & Narmadha, D. (2023). Artificial intelligence based test systems to resist waterborne diseases by early and rapid identification of pathogens: A review. *SN Computer Science*, 4, 1620.

Kao, I. F., Liou, J. Y., Lee, M. H., & Chang, F. J. (2021). Fusing stacked autoencoder and long short-term memory for regional multistep-ahead flood inundation forecasts. *Journal of Hydrology*, 598, 126371.

Khan, F., Yarveisy, R., & Abbassi, R. (2021). Risk-based pipeline integrity management: A road map for the resilient pipelines. *Journal of Pipeline Science and Engineering*, 1(1), 74–87.

Krysanova, V., Donnelly, C., Gelfan, A., Gerten, D., Arheimer, B., Hattermann, F., & Kundzewicz, Z. W. (2018). How the performance of hydrological models relates to credibility of projections under climate change. *Hydrological Sciences Journal*, 63(5), 696–720.

Li, G., Ji, J., Ni, J., Wang, S., Guo, Y., Hu, Y., Liu, S., Huang, S. F., & Li, Y. Y. (2022). Application of deep learning for predicting the treatment performance of real municipal wastewater based on one-year operation of two anaerobic mem-brane bioreactors. *Science of the Total Environment*, 813, 151920.

Li, L., Rong, S., Wang, R., & Yu, S. (2021). Recent advances in artificial intelligence and machine learning for nonlinear relationship analysis and process control in drinking water treatment: A review. *Chemical Engineering Journal*, 405, 126673.

Liu, X., Zhao, Z., & Rezaeipanah, A. (2025). Intelligent and automatic irrigation system based on internet of things using fuzzy control technology. *Scientific Reports*, *15*(1), 14577.

Lunani, M. (2018). Artificial intelligence for water and wastewater: Friend or foe? *Opflow*, 44(6), 6–7.

Mannina, G., Rebouças, T. F., Cosenza, A., Sànchez-Marrè, M., & Gibert, K. (2019). Decision support systems (DSS) for wastewater treatment plants: A review of the state of the art. *Bioresource Technology*, 290, 121814.

Mboweni, I. V., Abu-Mahfouz, A. M., & Ramotsoela, D. T. (2021, October). A machine learning approach to intrusion detection in water distribution systems: A review. In *IECON 2021–47th Annual Conference of the IEEE Industrial Electronics Society*, Toronto, ON (pp. 1–7). IEEE.

Milot, J., Rodriguez, M. J., & Sérodes, J. B. (2002). Contribution of neural networks for modeling trihalomethanes occurrence in drinking water. *Journal of Water Resources Planning and Management*, 128(5), 370–376.

Mohammed, M., El-Shafie, H., & Munir, M. (2023a). Development and validation of innovative machine learning models for predicting date palm mite infestation on fruits. *Agronomy*, 13, 494.

Mohammed, M., Hamdoun, H., & Sagheer, A. (2023b). Toward sustainable farming: Implementing artificial intelligence to predict optimum water and energy requirements for sensor-based micro irrigation systems powered by solar PV. *Agronomy*, 13, 1081.

Mohammed, M., Munir, M., & Ghazzawy, H. S. (2022a). Design and evaluation of a smart ex vitro acclimatization system for tissue culture plantlets. *Agronomy*, 13, 78.

Mohammed, M., Riad, K., & Alqahtani, N. (2022b). Design of a smart IoT-based control system for remotely managing cold storage facilities. *Sensors*, 22, 4680.

Ni, J., Wu, L., Fan, X., & Yang, S. X. (2016). Bioinspired intelligent algorithm and its applications for mobile robot control: A survey. *Computational Intelligence and Neuroscience,* 2016, 3810903.

Nourani, V., Elkiran, G., & Abba, S. I. (2018). Wastewater treatment plant performance analysis using artificial intelligence: An ensemble approach. *Water Science and Technology*, 78(10), 2064–2076.

Oliveira, S. C., Abreu, M. A., & Santos, L. L. (2017). Application of artificial neural networks in estimating heavy metal concentrations in river sediments. *Journal of Environmental Management*, 193, 315–324.

Park, Y., Shin, S. H., & Lee, D. (2021). Artificial intelligence for drinking water quality management: A review. *Current Opinion in Environmental Science & Health*, 23, 100297.

Qin, H., Zhang, Y., Zhu, J., & Xu, J. (2023). Multi-objective optimization for irrigation scheduling using artificial intelligence techniques. *Computers and Electronics in Agriculture*, 207, 107601.

Raei, E., Asanjan, A. A., Nikoo, M. R., Sadegh, M., Pourshahabi, S., & Adamowski, J. F. (2022). A deep learning image segmentation model for agricultural irrigation system classification. *Computers and Electronics in Agriculture*, 198, 106977

Rathore, M. M., Son, H., Ahmad, A., & Paul, A. (2019). Machine learning techniques for the assessment of water quality and prevention of waterborne diseases. *Future Generation Computer Systems*, 92, 723–735.

Reckhow, K. H., & Chapra, S. C. (1999). Modeling excessive nutrient loading in water quality management. *Water Science and Technology*, 39(3), 25–32.

Sapankevych, N., & Sankar, R. (2009). Time series prediction using support vector machines: A survey. *IEEE Computational Intelligence Magazine*, 4, 24–38.

Sarni, W., White, C., Webb, R., Cross, K., & Glotzbach, R. (2019). Digital water: Industry leaders chart the transformation journey. *International Water Association and Xylem Inc*, 43.

Singh, R., Ramasamy, S., & Thirumal, V. (2020). Impact of artificial intelligence in water management: A global scenario. *Materials Today: Proceedings*, 39, 583–590.

Talaviya, T., Shah, D., Patel, N., Yagnik, H., & Shah, M. (2020). Implementation of artificial intelligence in agriculture for optimisation of irrigation and application of pesticides and herbicides. *Artificial Intelligence in Agriculture*, 4, 58–73. https://doi.org/10.1016/j.aiia.2020.04.002

Tang, C., Wu, M., Chen, S., & Du, H. (2023). Artificial intelligence and IoT integration for real-time monitoring and control of urban water networks. *Journal of Cleaner Production*, 372, 133943.

The Promise of Artificial Intelligence in Water Management. (2021). Accessed: September 12, 2023: https://www.analyticsinsight.net/the-promise-of-artificial-intelligence-in-water-management/

Ullah, R., Saeed, A., & Ali, R. (2017). Smart systems in water treatment using neural networks: A new age in wastewater management. *Environmental Technology & Innovation*, 8, 193–199.

Vasquez, M. I., Fatta-Kassinos, D., & Kalogerakis, N. (2014). Emerging contaminants in water and wastewater: Advanced treatment technologies. *Critical Reviews in Environmental Science and Technology*, 44(6), 1237–1270.

Wallis-Lage, C. (2020). Report reflects critical role of data and planning in water infrastructure resilience. *Journal AWWA*, 112(3), 38–46.

Walski, T. (2023). Find big leaks that are hard to find. *Opflow*, *49*(9), 8–9.

Yang, Y., Kim, K.-R., Kou, R., Li, Y., Fu, J., Zhao, L., & Liu, H. (2022). Prediction of effluent quality in a wastewater treatment plant by dynamic neural network modeling. *Process Safety and Environmental Protection*, 158, 515–524.

Yartu, M., Cambra, C., Navarro, M., Rad, C., Arroyo, Á., & Herrero, Á. (2022). Humidity forecasting in a potato plantation using time-series neural models. *Journal of Computational Science*, 59, 101547.

Yazdekhasti, S., Piratla, K. R., Sorber, J., Atamturktur, S., Khan, A., & Shukla, H. (2020). Sustainability analysis of a leakage-monitoring technique for water pipeline networks. *Journal of Pipeline Systems Engineering and Practice*, 11(1), 04019052.

Yetilmezsoy, K., Turkdogan-Aydinol, I., Gunay, A., & Ozis, I. (2011). Post-treatment of poultry slaughterhouse wastewater and appraisal of the economic outcome. *Environmental Engineering & Management Journal (EEMJ)*, 10(11), 1635–1645.

Yu, C. X., Yin, X. N., Yang, Z. F., & Dang, Z. (2019). Sustainable water resource management of regulated rivers under uncertain inflow conditions using a noisy genetic algorithm. *International Journal of Environmental Research and Public Health*, 16, 868.

Zaji, A. H., & Bonakdari, H. (2019). Robustness lake water level prediction using the search heuristic-based artificial intelligence methods. *ISH Journal of Hydraulic Engineering*, 25(3), 316–324.

Zhang, J., Lian, Z., Zhou, Z., Song, Z., Liu, M., & Yang, K. (2022). Leakage detection in a buried gas pipeline based on distributed optical fiber time-domain acoustic wave signal. *Engineering Failure Analysis*, 141, 106594.

Zhou, J., Lin, H., Li, S., Jin, H., Zhao, B., & Liu, S. (2023). Leakage diagnosis and localization of the gas extraction pipeline based on SA-PSO BP neural network. *Reliability Engineering & System Safety*, 232, 109051.

AI and the Future of Water Scarcity

9

Machine Learning for Efficient Resource Management

G. R. Manjunathareddy and H. R. Raveesha

9.1 INTRODUCTION TO WATER SCARCITY

Water scarcity is a critical challenge of the 21st century, intensified by a growing global population, urbanization, industrialization, and changing consumption patterns (Kumari et al., 2021). With the global population exceeding 8 billion and freshwater demand escalating unsustainably, traditional water management systems struggle under the pressure of inefficient resource allocation, aging infrastructure, and unpredictable climate change impacts (Evaristo et al., 2023). This chapter explores how artificial intelligence (AI) and machine learning (ML) can transform water management by improving water distribution efficiency, enhancing forecasting accuracy, and offering innovative solutions to combat water scarcity.

Several interconnected factors are contributing to the strain on freshwater systems:

Climate Change: Altered rainfall patterns, rising temperatures, extreme weather events, and droughts exacerbate water supply uncertainty, reducing freshwater availability due to increased evaporation and less predictable precipitation.

Population Growth: As urban populations expand and living standards rise, water demand increases, putting immense pressure on existing systems, particularly in densely populated regions (Pointet, 2022).

Inefficient Water Usage: Outdated infrastructure and inefficient irrigation systems lead to significant water losses, further straining available resources (Serafeim et al., 2024).

Agricultural and Industrial Demand: Agriculture is responsible for 70% of global freshwater consumption, while industrial sectors, including manufacturing and energy production, also contribute heavily to water depletion, often over-extracting from water sources (Musie and Gonfa, 2023).

DOI: 10.1201/9781003600718-12

TABLE 9.1 Applications of AI and ML in water management

AREA OF APPLICATION	*AI/ML TECHNOLOGY*	*KEY BENEFITS*	*REAL-WORLD EXAMPLES*
Optimizing water distribution	Predictive modeling, optimization algorithms	Improved water allocation based on real-time demand	AI-powered systems in smart cities. e.g., Barcelona (Salomons and Housh, 2020).
		Reduced water waste through accurate forecasting	AI-based water distribution systems in Israel (Ferrari et al., 2023).
Refining irrigation systems	ML, computer vision	Efficient water usage in agriculture	AI-driven irrigation in farms. e.g., California vineyards (Wei et al., 2024).
		Reduced over-irrigation, minimizing water loss	Precision agriculture in India using AI for water-saving (Oudah et al., 2024).
Forecasting water demand	Data analytics, time-series forecasting	Accurate prediction of water needs for planning	Predictive models in drought-prone areas like Cape Town (Onyeuwaoma et al., 2024).
		Adaptive resource management in fluctuating climates	Water demand prediction systems in Australia (Sanchez et al., 2020).
Leak detection and maintenance	AI-driven sensors, anomaly detection models	Early detection of leaks and system inefficiencies	Leak detection in urban water systems, e.g., Singapore (Farah and Shahrour 2024).
		Reduced water loss, improving system sustainability	Leak detection systems in Mexico City (Barrientos-Torres et al., 2023).
Monitoring water quality	ML, remote sensing	Real-time water quality monitoring	AI-based water quality monitoring in the UK (Sagan et al., 2020).
		AI-based water quality monitoring in the UK	AI-driven water quality monitoring in East Africa (Nallakaruppan et al., 2024)
Predicting water shortages	AI-based predictive analytics	Proactive response to potential shortages	AI systems predicting water scarcity in South Africa (Niyongabo et al., 2024).
		Improved drought management and crisis mitigation	AI in drought management in California and Australia (Nandgude et al., 2023).
Sustainability and conservation	Machine learning, optimization algorithms	Enhanced decision-making for sustainable resource use	AI in sustainable farming practices. e.g., precision irrigation in India (Abioye et al., 2022).
		Cost-effective resource management	AI applications in sustainable water use in Chile (Nti et al., 2023).

Given these challenges, AI and ML have emerged as powerful tools to optimize water management (Table 9.1). These technologies can enhance decision-making by monitoring water usage, predicting future demand, and adjusting allocations dynamically. AI models can also identify emerging issues and detect risks early, enabling proactive measures to avoid crises. In a changing world, AI and ML offer adaptive and responsive solutions for more sustainable water management, helping mitigate water scarcity.

9.2 UNDERSTANDING AI AND ML IN WATER MANAGEMENT

To grasp the role of AI and ML in water management, it is vital to realize the fundamental concepts behind these technologies and how they operate. AI involves the development of systems or machines that can carry out tasks typically requiring human intelligence. These tasks include decision-making, problem-solving, learning, and adaptation, all driven by the analysis of data inputs (Collins et al., 2021). AI can encompass a wide spectrum, ranging from simple systems that follow predefined rules to highly advanced neural networks capable of recognizing intricate patterns in vast datasets (Sarker, 2022). Essentially, AI mimics human cognitive functions, enabling machines to "learn" from experience and improve their performance over time. This broad capability of AI allows it to be applied to various fields, including water management, where it can optimize resource use, predict water demand, and improve infrastructure efficiency.

9.2.1 ML in Water Resource Management

ML is a subset of AI that focuses on enabling systems to learn from data and enhance their performance autonomously over time without requiring explicit programming (França et al., 2021). Unlike traditional systems that rely on predefined rules and manual instructions, machine learning algorithms (MLAs) are designed to identify patterns, make predictions, and adjust based on new information. By analyzing large datasets, ML systems can reveal hidden insights, support data-driven decision-making, and optimize results across various applications (Sarker, 2021). In water resource management, ML can be particularly beneficial, as it can help forecast demand, detect defects in infrastructure, and improve efficiency by continuously adapting to changing environmental and operational conditions.

ML can be categorized into three main types, each offering unique methods for analyzing and using data:

9.2.2 Supervised Learning

Supervised learning is a type of ML where the model is trained using a labeled dataset, meaning that the correct output or target values are already known for each input. The model's task is to identify patterns within the data and establish a relationship between the inputs and their corresponding outputs (Wang et al., 2023). Once trained, the model can use these learned relationships to predict new, unseen data.

For instance, supervised learning can be applied in water demand forecasting by analyzing historical data, such as past water consumption, temperature variations, and precipitation levels (Li et al., 2024). The model can predict future water demand by learning how these variables correlate with water usage, providing valuable resource planning and management insights. Through this process, supervised learning helps to optimize water distribution by anticipating shifts in demand based on identifiable patterns.

9.2.3 Unsupervised Learning

Unsupervised learning is an ML approach in which the model receives data that lacks labeled outputs. In this scenario, the algorithm uncovers underlying patterns, groupings, or structures within the data without knowing the desired results. This method is particularly useful for discovering hidden insights or identifying trends that may not be immediately obvious.

For example, unsupervised learning can be applied in water management to detect anomalies in water consumption patterns across different regions. It can also uncover previously unknown trends, such as excessive water wastage in urban areas or agricultural practices (Essamlali et al., 2024). By identifying these hidden patterns, unsupervised learning helps optimize water usage and guide more efficient resource allocation without predefined outcomes or labels.

9.2.4 Reinforcement Learning

Reinforcement learning (RL) is a type of ML where a model learns to make decisions by interacting with an environment and receiving feedback as rewards or penalties (Vaish et al., 2021). Unlike other learning methods, RL is based on trial and error, with the model continually adjusting its strategy to maximize long-term rewards. This process enables the model to improve its performance over time, making it highly effective for tasks that require dynamic decision-making and optimization.

In water resource management, RL can enhance irrigation networks or water distribution systems. By continuously adjusting based on real-time environmental conditions, resource availability, and demand, the system learns to make better decisions for water allocation. For example, it can optimize water usage by balancing conservation efforts with cost-reduction objectives. The system receives feedback on how well its actions align with these goals, allowing it to refine its approach and improve water management strategies (Chen et al., 2021). Ultimately, RL in water management can help automate tasks, improve decision-making, and forecast outcomes, contributing to more efficient and sustainable water use.

9.3 THE ROLE OF AI IN WATER DISTRIBUTION SYSTEMS

Urban water distribution systems face significant challenges, including aging infrastructure, water losses, and uneven resource distribution, exacerbated by rapid urbanization and growing demand (Adeyeye et al., 2020). These issues strain existing systems and hinder utilities' ability to ensure reliable and equitable access to water, especially in densely populated areas.

AI and ML offer powerful solutions to improve water distribution efficiency, reliability, and sustainability. Using predictive models and adaptive management strategies, AI enables smarter decision-making, optimizing operations, reducing waste, and improving real-time water management (Bolón-Canedo et al., 2024).

Key areas where AI can make a difference include predictive maintenance, real-time monitoring, water quality management, and demand forecasting (Table 9.1). AI can identify infrastructure weaknesses and predict failures, minimizing costly repairs (Adeyeye et al., 2020). It also helps monitor water quality by integrating sensor data throughout the network, ensuring a safe water supply (Bolón-Canedo et al., 2024). Furthermore, AI models can forecast water demand, enabling better resource allocation and preventing shortages and oversupply.

By strengthening water distribution systems, AI promotes efficient resource use and supports sustainable water management, ensuring a reliable water supply for future generations.

a. **Leak Detection and Prevention**

One of the most significant challenges in urban water systems is water loss due to leaks for up to 30%–40% of the total water supply in some cities, which is both wasteful and costly

(Ahopelto and Vahala, 2020). Traditional leak detection methods are often slow, reactive, and expensive, relying on manual inspections and physical monitoring, which may not be efficient enough to catch leaks early (Mohd Yussof and Ho, 2022)

AI can revolutionize leak detection by utilizing data from pressure monitors, flow meters, and acoustic sensors distributed throughout the water distribution network (Romero-Ben et al., 2023). AI algorithms can analyze this data in real time to detect inconsistencies or irregularities in water flow that might indicate the presence of a leak (Mahdi et al., 2024). Furthermore, machine learning models (MLMs) can predict potential leak locations by analyzing historical data and identifying patterns or anomalies (Fan et al., 2021). This predictive capability allows water utilities to take proactive measures, such as performing pre-emptive maintenance or replacing vulnerable infrastructure, before a significant failure occurs. By shifting from reactive to proactive leak management, AI can dramatically reduce water loss, saving resources and improving the overall efficiency of the water distribution system (Romero et al., 2023).

b. Smart Water Distribution

AI can also play a crucial role in optimizing the allocation and distribution of water across urban areas, ensuring that water resources are used as efficiently as possible (Kamyab et al., 2023). In cities with complex and often outdated water infrastructure, balancing water supply to different neighborhoods, sectors, or buildings can be particularly difficult during peak demand periods or when the system is under stress, such as during droughts or water shortages (Avazpour et al., 2023).

AI-powered systems can integrate real-time data from sensors embedded throughout the water distribution network to dynamically adjust water flow and pressure, ensuring that all areas receive the required water while minimizing waste (Krishnan et al., 2024). These intelligent systems can respond to changes in water demand, such as fluctuations caused by weather patterns, population movement, or consumption patterns. By continuously analyzing this data, AI systems can optimize water allocation to reduce waste, prevent over-supply, and prioritize critical infrastructure, such as hospitals, fire stations, or emergency services, during water shortages (Rashid et al., 2024).

In addition to optimizing general water distribution, AI can help ensure that water is directed more efficiently to areas that need it most, reducing inequalities in water access and improving overall system sustainability. This level of optimization and flexibility in real-time water distribution is impossible with traditional manual methods, making AI an invaluable tool for urban water management (Mana et al., 2024).

c. Infrastructure Maintenance and Management

Water distribution systems, consisting of pumps, valves, pipes, and treatment plants, require regular maintenance to function efficiently (Ahmad et al., 2020). Aging infrastructure poses a significant challenge, as many components are reaching the end of their lifespan. Unexpected breakdowns can cause service disruptions, costly repairs, and water supply loss, negatively impacting both customers and operational efficiency (Ford et al., 2022).

AI-driven predictive maintenance systems address this challenge by allowing water utilities to monitor infrastructure conditions through data from embedded sensors in pumps, valves, and other critical components (Ucar et al., 2024). MLA analyzes historical data to identify patterns and predict potential failures. These insights enable utilities to perform maintenance before breakdowns occur, reducing unplanned outages and extending the life of infrastructure (Asadi, 2024).

This proactive approach also enhances resource efficiency and reduces downtime. Utilities can schedule repairs during off-peak hours, minimizing service disruptions and lowering emergency repair costs (Murtaza et al., 2024). By preventing critical failures, AI and ML contribute to sustainable water management practices, minimizing waste and conserving resources (Alsabt et al., 2024).

9.4 MACHINE LEARNING FOR IRRIGATION SYSTEMS

Agriculture is the largest global consumer of water, with irrigation responsible for around 70% of total freshwater withdrawals. Despite its critical role, traditional irrigation practices are often inefficient, leading to significant water wastage (Ingrao et al., 2023). Factors such as over-irrigation, poor water management, and a lack of real-time monitoring can contribute to the depletion of water resources, diminished crop yields, and environmental damage (Ahmed et al., 2023). However, AI and ML present innovative solutions to tackle these challenges by optimizing irrigation practices, improving water usage efficiency, and boosting agricultural productivity. Below are some of the key applications of AI and ML in enhancing irrigation systems (Wei et al., 2024):

a. **Precision Agriculture**: ML is revolutionizing precision agriculture by optimizing water usage through data from weather forecasts, soil moisture levels, and crop types. Sensors in the soil measure factors like moisture, temperature, and humidity, providing real-time data processed by AI systems. These systems analyze the data to determine the precise water requirements for crops, considering variables like the crop's growth stage, soil type, and current weather conditions (Javaid et al., 2023).

   This method helps prevent under-irrigation, which can lead to water waste and crop stress, and over-irrigation, which unnecessarily depletes water resources. Precision irrigation conserves water and improves crop yields by optimizing water delivery. It ensures that plants receive water under ideal conditions, promoting healthier growth and enhancing productivity. With ML-driven precision agriculture systems, farmers can make informed decisions that balance water conservation with maintaining robust crop health and high yields (Benti et al., 2024).

b. **Weather Prediction and Forecasting:**

   Weather significantly influences irrigation decisions. Rainfall patterns are often unpredictable and vary from season to season, making irrigation planning challenging (Tarraf et al., 2024). MLAs can enhance weather forecasting by analyzing historical climate data and weather conditions to predict future rainfall patterns more accurately (Abdel Azeem and Dev, 2024).

   AI-powered weather prediction systems can forecast rainfall for specific agricultural regions, allowing farmers to adjust irrigation schedules accordingly (Hachimi et al., 2023). For example, if the system predicts rain, farmers can reduce or delay irrigation to conserve water and prevent over-irrigation. Alternatively, if a dry spell is forecasted, the system can prompt farmers to increase irrigation to ensure crops receive the necessary water during critical growth periods (Sangha et al., 2023).

   Farmers can optimize water use based on real-time and forecasted conditions by integrating weather predictions with ML-driven irrigation systems. Minimizes the need for manual adjustments, reduces water consumption, and prevents crop stress caused by inconsistent irrigation (Abioye et al., 2022).

c. **Automated Irrigation Systems:**

   Automated irrigation systems powered by ML and AI are transforming agricultural water management. Traditional irrigation methods rely on fixed schedules or manual adjustments, leading to inefficiency and water wastage. In contrast, AI-driven systems use real-time data from soil moisture sensors, weather stations, and other environmental factors to determine the exact timing and amount of water needed (Morchid et al., 2024).

   These smart irrigation systems continuously monitor soil moisture and only activate irrigation when required, ensuring water is used efficiently. By triggering irrigation only when moisture levels fall below a set threshold, AI systems help prevent over-irrigation and reduce waste (Obaideen et al., 2022). Additionally, they can integrate weather forecasts to adjust irrigation schedules in response to expected rainfall or temperature changes, further enhancing water efficiency (Bouali Et-Taibi et al., 2024).

In water-scarce regions, these systems can prioritize irrigation for the most water-sensitive areas or adjust based on crop type and specific needs. By quickly reacting to shifts in moisture levels or environmental conditions, AI-powered irrigation systems minimize water waste, support healthier crops, and boost agricultural productivity. This approach conserves water and ensures crops receive optimal hydration at the right time.

9.5 FORECASTING WATER DEMAND USING MACHINE LEARNING

Forecasting water demand is a crucial element of effective water resource management. It allows water utilities to plan for future supply needs, prevent shortages, and optimize the distribution of water resources (Bouali Et-Taibi et al., 2024; Younis et al., 2023). Water demand forecasting has traditionally been based on historical usage data and simplistic models, which often fail to consider the numerous variables influencing consumption patterns. However, with the integration of ML and AI, water demand forecasting has become more advanced, precise, and adaptable. These technologies enable utilities to analyze various factors, from weather patterns to demographic changes, allowing for more accurate predictions and better resource management (Essamlali et al., 2024). The following sections explore how AI and ML are transforming the way water demand is forecasted:

a. **Historical Data and Trend Analysis:**

Historical data and trend analysis are crucial for water demand forecasting, and ML models outperform traditional methods by uncovering hidden patterns in water usage. These models analyze large datasets from sources like water meters, utility records, and consumption reports, identifying trends that might not be immediately apparent (Thompson and Dickenson, 2021).

ML models can also account for external factors such as population growth, weather patterns, and urban development, helping to predict their impact on future water demand. By analyzing long-term trends, AI systems offer more accurate forecasts, considering factors like economic shifts and changing social behaviors influencing water use (Morain et al., 2024).

One of the key advantages of ML in historical data analysis is its ability to learn from large amounts of data and continuously improve predictions. These models refine their accuracy as more data is processed, allowing for more reliable short-term and long-term forecasts (Tien et al., 2022). This continuous improvement helps water utilities make better decisions, ensuring efficient resource management.

b. **Real-Time Demand Prediction:**

One of the most vital uses of ML in water demand forecasting is real-time prediction. Water demand can fluctuate significantly due to temperature, humidity, economic activity, and seasonal changes (Tańczyk et al., 2022; Niknam et al., 2022). ML models can accurately predict short-term water consumption by integrating real-time data like water usage, weather conditions, and smart meter readings (Kavya et al., 2023).

This capability allows water utilities to adjust their distribution strategies dynamically, ensuring resources are allocated efficiently and immediate demand is met without waste. For instance, during a heatwave, real-time predictions can help utilities anticipate a surge in demand as people and businesses increase water use for cooling. AI systems can then optimize water distribution to prevent overburdening the system and maintain an even flow of resources. This flexibility is crucial in water shortage, reducing waste and ensuring system reliability during peak demand (Dubey and Kumar, 2025).

Real-time forecasting also enables utilities to make necessary operational adjustments, such as modifying pressure levels or activating backup supplies when needed. By continuously monitoring and predicting demand fluctuations, ML models help utilities manage water distribution more effectively and ensure the system remains resilient under varying conditions.

c. **Scenario Modeling:**

ML is crucial for scenario modeling in water utilities, allowing them to simulate different demand scenarios under various conditions. This helps utilities prepare for uncertainties like population growth, climate change, and policy changes (Chakraborty et al., 2021). By creating hypothetical scenarios, utilities can assess the impact of urbanization, water conservation policies, and environmental shifts on water demand, aiding in decisions about infrastructure, conservation, and emergency plans (Dinar, 2024).

For example, AI can model how urbanization and population growth affect future water demand and predict the outcomes of water conservation policies (Dinar, 2024; Kim et al., 2023). It can also factor in climate-related changes, such as altered rainfall patterns, to help utilities plan for extreme weather and long-term shifts (Abbass et al., 2022).

This proactive approach enables utilities to anticipate future needs, such as additional storage or improved infrastructure, ensuring more sustainable and effective water management.

d. **Integration of Socio-Economic Factors:**

Water demand is influenced not only by weather conditions and population growth but also by various socio-economic factors, including income levels, industrial development, urbanization, and shifts in consumption habits (Wang et al., 2021). MLAs can integrate these socio-economic variables into their predictions, offering a more comprehensive and accurate forecast of future water needs.

For example, in areas experiencing rapid urbanization, lifestyle changes—such as increased use of water-intensive appliances, commercial developments, and higher living standards—can drive up water demand, even during drought. Similarly, fluctuations in income levels or industrial activity can lead to greater water demand in certain sectors (Lebu et al., 2024). AI models can track these socio-economic trends and adjust forecasts, accordingly, helping water utilities plan more effectively for consumption in both urban and industrial areas (Vekaria and Sinha, 2024).

By incorporating real-time data on economic and demographic changes, MLAs enable water utilities to respond more swiftly to shifts in demand. For instance, if a new manufacturing facility opens in a region, AI systems can predict the increased water demand associated with that development and adjust water distribution strategies to accommodate the new consumption load (Zhu et al., 2022).

9.6 MONITORING WATER QUALITY WITH AI AND ML

Ensuring access to safe and clean drinking water is crucial for public health, as contamination can lead to various health issues, including waterborne diseases (Shayo et al., 2023). While essential, traditional water quality testing methods are often limited by their scope, labor intensity, and time-consuming nature (Essamlali et al., 2024). In contrast, AI and ML offer transformative capabilities that significantly enhance water quality monitoring through real-time, continuous, and more accurate assessments.

By leveraging advanced data analytics, AI and ML systems can detect contamination early, predict potential water quality problems, and even automate water treatment processes to maintain safe drinking water (Sheik et al., 2024). These technologies enable continuous monitoring through data from various sensors deployed in water sources, such as temperature measurements, pH levels, turbidity, and chemical composition. This creates a dynamic and comprehensive view of water quality. Consequently, AI and ML improve water quality assessments' accuracy and allow quicker responses to emerging issues.

The following sections will explore how AI and ML are revolutionizing water quality monitoring, providing more efficient, reliable, and proactive management of water resources.

a. **Sensor Networks for Real-Time Monitoring:**

A significant advantage of AI and ML in water quality monitoring is their ability to integrate and analyze data from extensive sensor networks deployed across various water bodies, including rivers, lakes, reservoirs, and distribution systems (de Camargo et al., 2023). These sensors continuously measure key water quality parameters such as pH, turbidity, chemical concentrations, temperature, dissolved oxygen, and microbial contamination (Geetha et al., 2024).

By processing this data in real-time, ML algorithms enable continuous water quality monitoring, overcoming the limitations and delays of traditional manual sampling (Forhad et al., 2024). AI systems can detect subtle changes in water quality and offer early warnings of potential issues. For instance, a sudden increase in turbidity levels may signal the presence of pollutants or the onset of algal blooms, prompting immediate investigation and remediation efforts (Almuhtaram et al., 2021).

This real-time data processing enhances water utilities' ability to respond quickly and effectively to water quality concerns. In contamination cases, AI systems can promptly alert operators, guiding them on necessary actions such as adjusting treatment processes, shutting down water intakes, or notifying local authorities and residents about potential risks (Miller et al., 2023). This proactive approach helps ensure safe drinking water while reducing contamination risks.

b. **Predictive Analytics for Water Quality Trends:**

AI and ML are essential tools in water quality management. They utilize historical data alongside environmental factors such as rainfall, pollution levels, industrial activities, and water quality trends (Ramovha et al., 2024). This integration allows AI systems to predict potential water quality degradation, enabling proactive steps to prevent contamination.

For example, by analyzing past data, AI can predict increased pollutants following heavy rainfall or nearby industrial operations. These predictions help water utilities adjust treatment processes and prepare treatment plants for possible water quality issues (Alprol et al., 2024). Additionally, AI can forecast seasonal trends, like the increased risk of algae blooms in warmer months, guiding decision-makers to adjust water management practices accordingly (Atton Beckmann et al., 2025).

Furthermore, predictive analytics helps identify potential hotspots for water quality problems, enabling utilities to prioritize monitoring and preventative actions. By anticipating challenges, AI and ML enable more informed, data-driven decisions, ensuring safer, cleaner water while minimizing risks to public health.

c. **Automated Water Treatment:**

AI is pivotal in optimizing and automating water treatment processes, significantly improving water quality management (Pérez-Beltrán, 2024). Water treatment plants (WTPs) commonly use chemicals like chlorine or ozone to disinfect water, but the required doses vary depending on water quality. Incorrect dosing can lead to inefficiencies, excessive chemical use, or insufficient treatment.

AI systems can continuously monitor water quality in real-time and adjust chemical dosing based on detected contaminant levels (Liu et al., 2024). For example, if microbial contamination rises, AI can automatically increase the chlorine dosage to maintain safe water standards (Gámiz et al., 2020). Conversely, if the water quality improves, the system can lower the chemical levels, reduce waste, and minimize environmental harm.

Through automation, AI enhances the efficiency and sustainability of water treatment, resulting in utility cost savings while ensuring safe, high-quality water (Narayanan et al., 2024).

d. **Multi-Sensor Fusion for Comprehensive Water Quality Monitoring:**

AI improves water quality monitoring by integrating data from various sensor sources, creating a more accurate and comprehensive understanding of water health (Olawade et al., 2024). Sensors placed at different depths in water bodies measure parameters like pH, turbidity, and dissolved oxygen, while AI fuses these data streams to identify complex relationships between indicators (Jan et al., 2021).

AI can also combine sensor data with satellite imaging and remote sensing, allowing large-scale monitoring. For example, satellite images can detect algae blooms, which AI can cross-reference with in situ data to assess contamination severity (Park et al., 2024). This integrated approach enables more effective tracking and response to water quality threats.

By continuously analyzing these diverse data sources, AI can identify subtle patterns that might be missed, helping authorities prevent health risks and environmental damage through proactive, efficient water management.

e. **Early Warning Systems for Water Pollution:**

AI and ML-powered early warning systems (EWS) are essential in areas at risk from industrial pollution, agricultural runoff, or natural disasters. These systems analyze environmental data, such as emissions, runoff patterns, and weather forecasts, to predict water contamination (Essamlali et al., 2024). For example, AI can assess the impact of pollutants from factories or mines, helping authorities take preventive actions like shutting down water intake or adjusting treatment processes (Rautela and Goyal, 2024).

AI also predicts the effects of excess fertilizers or chemical spills, allowing utilities to prepare by activating filtration systems or specialized treatments (Mousavi and Hakemi, 2024). By offering early detection, these systems enable proactive measures to protect water quality and public health, improving water management and reducing contamination risks.

9.7 CASE STUDIES: AI APPLICATIONS IN WATER-SCARCE REGIONS

AI and ML technologies are increasingly deployed in regions facing significant water scarcity, providing innovative solutions to manage and optimize water resources (Jayakumar et al., 2024). These case studies illustrate how different organizations are harnessing the power of AI to improve water efficiency, enhance sustainability, and overcome challenges related to water distribution and conservation.

a. **Israel: Precision Irrigation and Water Management:**

Israel has long been at the forefront of water conservation technologies. The country has developed an integrated water management system that includes desalination, wastewater reuse, and highly efficient irrigation systems. AI and ML in Israel's agricultural sector have significantly boosted (Mishra et al., 2023).

In Israel's farming sector, precision irrigation systems powered by AI and ML water usage predict exactly how much each crop requires based on environmental factors like soil moisture, weather forecasts, and crop type (Rathor et al., 2024). MLM can analyze historical data on weather patterns, soil conditions, and crop needs to adjust irrigation schedules dynamically, ensuring that water is applied only when and where (Elbasi et al., 2023).

This system has led to remarkable improvements in water efficiency, reducing overall crop yields. The success of AI-driven precision irrigation in Israel is an example of how data-driven approaches can optimize agricultural water use and boost food security in water-scarce regions.

b. **Singapore: Smart Water Grid:**

Singapore, a city-state with limited natural freshwater resources, has made significant strides in water management through innovative technologies. The Public Utilities Board (PUB) of Singapore has implemented a "smart water grid" that leverages AI and ML to monitor and manage the city's water distribution network (Brears, 2021). This system incorporates data from thousands of sensors installed throughout the water infrastructure to track water flow and pressure levels and detect leaks in real-time.

The AI-driven system analyzes this data to optimize water distribution and prevent water loss, a common challenge in many urban areas due to aging infrastructure and leakage (Wu et al., 2024). MLAs can predict potential failure points within the network and trigger maintenance actions before a leak or break occurs, thereby preventing the loss of precious water resources (Medeiros et al., 2024).

In addition to leak detection, the smart water grid optimizes water flow throughout the city to ensure equitable distribution. This ensures that areas with higher demand, such as residential neighborhoods or key commercial zones, receive an adequate water supply while minimizing waste in lower-demand areas (Shiddiqi et al., 2024).

Singapore's smart water grid is an excellent example of how AI and ML can help urban centers better manage their water resources, ensuring long-term sustainability and reducing the environmental footprint of water distribution networks.

c. **California: Water Demand Forecasting:**

California has long struggled with recurring droughts and water scarcity. AI and ML to forecast water demand and improve resource management (Haider et al., 2024). Given its vast agricultural sector and rapidly growing urban areas, effective water demand forecasting is crucial for optimizing water use and ensuring that resources are allocated where they are needed most.

MLM in California predicts future water demand by analyzing weather patterns, population growth, historical usage trends, and agricultural needs (Emami et al., 2022). These models are variable of rainfall, temperature fluctuations, and the increasing demand driven by population growth and industrial development.

ML to predict fluctuations in water demand, water utilities can plan for potential shortages, optimize water distribution, and prioritize conservation efforts (Ou et al., 2023). For example, during drought or unusually high demand, these predictive models can help authorities allocate water more efficiently, ensuring that the most critical sectors, such as agriculture or municipal water supplies, are prioritized.

This approach has enabled California to manage its water resources more effectively, ensuring that water is used more efficiently in urban and agricultural contexts. ML in water demand forecasting represents a powerful tool for regions grappling with water scarcity to prepare for future challenges and adapt to changing environmental conditions (Jibrin et al., 2024).

9.8 CHALLENGES AND ETHICAL CONSIDERATIONS IN AI-DRIVEN WATER MANAGEMENT

While AI and ML hold tremendous potential for improving water management, several challenges need to be addressed:

a. **Data Privacy and Security:** The large-scale environmental and socio-economic data raise concerns about privacy and security. Ensuring that data is handled responsibly and securely is critical to gaining public trust and ensuring the success of AI-driven solutions.

b. **Data Accessibility:** Access to high-quality data is essential for the success of AI and ML applications. However, access to accurate and comprehensive data can be limited. Ensuring that data is accessible to all stakeholders is essential for equitable water management solutions.
c. **Equity and Inclusivity:** AI and ML solutions must benefit all communities, especially those most vulnerable to water scarcity. Ensuring technologies are accessible and inclusive is crucial for achieving fair and sustainable water management.

Case studies from various regions can illustrate the practical applications of AI and ML in addressing water scarcity. These real-world examples help and highlight the practical challenges and successes encountered in implementing AI-driven water management systems.

a. **Israel: Precision Irrigation and Water Management:**

   Israel has been a leader in water management, with AI and ML playing a key role in its success. One notable application is precision irrigation, which uses AI to optimize water use in agriculture. AI algorithms recommend irrigation practices based on crop type, soil conditions, and weather patterns. This system minimizes waste and ensures crops receive the right amount of water, helping Israel reduce water consumption despite severe water scarcity.
b. **Singapore: Smart Water Grid:**

   Singapore's PUB has developed one of the world's most advanced water management systems, using AI and ML to optimize water distribution and quality. The Smart Water Grid integrates data from sensors, water meters, and control systems to detect leaks, improve water quality, and ensure efficient distribution.

   The system continuously predicts areas requiring maintenance, such as forecasting pipe failures for proactive repairs. Real-time data helps optimize water flow, ensuring fair distribution across Singapore, even during droughts or shortages. A key innovation is data fusion from social media, weather forecasts, and satellite data.
c. **California: Water Demand Forecasting:**

   In California, ML is being used to address persistent droughts and rising water demand. Water utilities rely on ML models to predict demand and optimize resource allocation. The California Department of Water Resources uses these models to analyze historical consumption data, weather patterns, and demographic trends, helping forecast future water needs and identify potential shortages.

   Additionally, California has implemented AI-driven irrigation systems that adjust watering schedules based on real-time weather and soil moisture data, optimizing water use in agriculture while maintaining crop health.

   Despite these promising solutions, challenges such as data access, privacy, and ethical concerns must be addressed for AI and ML in water management.
d. **Data Ownership and Governance:** Data ownership and governance are key ethical concerns in using AI and ML for water management. Since water data comes from multiple sources—governments, private companies, and local communities—questions about ownership, sharing, and access where data may be limited; private companies often control valuable datasets, making it crucial to establish clear governance frameworks for data sharing, ownership, and security. These frameworks promote responsible data use, transparency, and fairness while safeguarding privacy. Transparent AI algorithms are essential to ensure fairness, particularly when AI-driven decisions affect local communities. Strong data governance fosters and ensures that AI technologies contribute positively to water management.
e. **Inclusivity in Water Management:** AI has the potential to improve water efficiency and quality, but its benefits must be accessible to all, including rural and underserved communities. For example, small-scale farmers in developing countries may lack access to advanced AI-driven irrigation systems, limiting the technology's reach.

AI applications should be tailored to local needs, ensuring affordability, scalability, and relevance. Additionally, capacity-building through training and resource support is crucial to help communities effectively utilize AI for water management.

9.9 THE FUTURE OF AI IN WATER RESOURCE MANAGEMENT

The potential for AI and ML to revolutionize water management is vast. As technology, AI will optimize water distribution, improve water quality, and forecast future demand. Key advancements on the horizon include:

Integration with the Internet of Things (IoT): IoT devices provide real-time data to AI systems, enabling even more precise and adaptive water management strategies.
AI in Climate Change Adaptation: AI models can help cities and countries adapt to the impacts of climate change, including more frequent droughts and floods.
Global Water Management Systems: Collaborative global platforms powered by AI will enable cross-border sharing of water data, facilitating better management of shared water resources.

The potential of AI and ML to address water scarcity will continue to grow as new technologies emerge and AI systems become more refined and powerful. In the coming years, we can expect to see the following advancements:

a. **Integration with Emerging Technologies:** AI will increasingly be integrated with other emerging technologies such as Blockchain for secure data management, IoT for real-time monitoring, and 5G networks for faster data transmission. These technologies will enable more efficient and responsive water management systems, providing real-time insights that can be used to prevent water crises before they occur.
b. **AI and Climate Change Mitigation:** AI will also play a key role in helping communities adapt to the effects of climate change. By analyzing vast amounts of environmental data, AI systems can help forecast the impacts of changing weather patterns, such as prolonged droughts or heavy rainfall, on water supply systems. This will enable cities and regions to develop adaptive strategies, such as changing water storage practices, adjusting irrigation methods, or implementing conservation measures.
c. **Collaborative Global Platforms:** In the future, AI may facilitate global collaboration in managing shared water resources. By integrating data across countries and regions, AI can help optimize transboundary water sources that cross national borders, such as rivers and lakes. Collaboration could help prevent conflicts over water resources and ensure that water is distributed fairly and sustainably across the globe.

9.10 CONCLUSION

AI and ML offer innovative solutions to water scarcity by improving water management in distribution, irrigation, demand forecasting, leak detection, and water quality monitoring. These technologies allow real-time data analysis, optimizing water use and reducing waste, especially in agriculture, where consumption is high. By predicting demand and identifying infrastructure issues, AI ensures efficient and safe water use, minimizing shortage risks.

Despite their potential, challenges persist, such as limited data access in developing regions and concerns over data privacy and security, particularly in vulnerable communities. Without proper infrastructure, the full benefits of AI and ML may not be realized.

As AI and ML advance, their ability to enhance water management grows, enabling smarter resource allocation and better crisis management. However, responsible implementation, with attention to ethical, privacy, and equity issues, is essential. International cooperation and thoughtful planning are crucial to making AI a key tool in combating water scarcity and fostering a sustainable future.

REFERENCES

Abbass, K., Qasim, M. Z., Song, H., Murshed, M., Mahmood, H. & Younis, I. (2022). A review of the global climate change impacts, adaptation, and sustainable mitigation measures. *Environmental Science and Pollution Research*, *29*, 42539–42559. https://doi.org/10.1007/s11356-022-19718-6

Abdel Azeem, M., & Dev, S. (2024). A performance and interpretability assessment of machine learning models for rainfall prediction in the Republic of Ireland. *Decision Analytics Journal*, *12*, 100515. https://doi.org/10.1016/j.dajour.2024.100515

Abioye, E. A., Hensel, O., Esau, T. J., Elijah, O., Abidin, M. S. Z., Ayobami, A. S., Yerima, O., & Nasirahmadi, A. (2022). Precision irrigation management using machine learning and digital farming solutions. *AgriEngineering*, *4*(1), 70–103. https://doi.org/10.3390/agriengineering4010006

Adeyeye, K., Gibberd, J., & Chakwizira, J. (2020). Water marginality in rural and peri-urban communities. *Journal of Cleaner Production*, *273*, 122594. https://doi.org/10.1016/j.jclepro.2020.122594

Ahmad, S., Jia, H., Chen, Z., Li, Q., & Xu, C. (2020). Water-energy nexus and energy efficiency: A systematic analysis of urban water systems. *Renewable and Sustainable Energy Reviews*, *134*, 110381. https://doi.org/10.1016/j.rser.2020.110381

Ahmed, Z., Gui, D., Murtaza, G., Yunfei, L., & Ali, S. (2023). An overview of smart irrigation management for improving water productivity under climate change in drylands. *Agronomy*, *13*(8), 2113. https://doi.org/10.3390/agronomy13082113

Ahopelto, S., & Vahala, R. (2020). Cost-benefit analysis of leakage reduction methods in water supply networks. *Water*, *12*(1), 195. https://doi.org/10.3390/w12010195

Almuhtaram, H., Kibuye, F. A., Ajjampur, S., Glover, C. M., Hofmann, R., Gaget, V., Owen, C., Wert, E. C., & Zamyadi, A. (2021). State of knowledge on early warning tools for cyanobacteria detection. *Ecological Indicators*, *133*, 108442. https://doi.org/10.1016/j.ecolind.2021.108442

Alprol, A. E., Mansour, A. T., Ibrahim, M. E. E.-D., & Ashour, M. (2024). Artificial intelligence technologies revolutionizing wastewater treatment: Current trends and prospects. *Water*, *16*(2), 314. https://doi.org/10.3390/w16020314

Alsabt, R., Alkhaldi, W., Adenle, Y. A., & Alshuwaikhat, H. M. (2024). Optimizing waste management strategies through artificial intelligence and machine learning: An economic and environmental impact study. *Cleaner Waste Systems*, *8*, 100158. https://doi.org/10.1016/j.clwas.2024.100158

Asadi, Y. (2024). Employing machine learning in water infrastructure management: Predicting pipeline failures for improved maintenance and sustainable operations. *Industrial Artificial Intelligence*, *2*, 8. https://doi.org/10.1007/s44244-024-00022-w

Atton Beckmann, D., Werther, M., Mackay, E. B., Spyrakos, E., Hunter, P., & Jones, I. D. (2025). Is more data always better? – Machine learning forecasting of algae based on long-term observations. *Journal of Environmental Management*, *373*, 123478. https://doi.org/10.1016/j.jenvman.2024.123478

Avazpour, B., Osmond, P., & Corkery, L. (2023). The dynamic challenges of mainstreaming water sensitive cities in our built environment: Lessons from Australia. *Cities*, *143*, 104615. https://doi.org/10.1016/j.cities.2023.104615

Barrientos-Torres, D., Martinez-Ríos, E. A., Navarro-Tuch, S. A., Pablos-Hach, J. L., & Bustamante-Bello, R. (2023). Water flow modeling and forecasting in a water branch of Mexico City through ARIMA and transfer function models for anomaly detection. *Water*, *15*(15), 2792. https://doi.org/10.3390/w15152792

Benti, N. E., Chaka, M. D., Semie, A. G., Warkineh, B., & Soromessa, T. (2024). Transforming agriculture with machine learning, deep learning, and IoT: Perspectives from Ethiopia: Challenges and opportunities. *Discovery Agriculture*, *2*, 63. https://doi.org/10.1007/s44279-024-00066-7

Bolón-Canedo, V., Morán-Fernández, L., Cancela, B., & Alonso-Betanzos, A. (2024). A review of green artificial intelligence: Towards a more sustainable future. *Neurocomputing*, *599*, 128096. https://doi.org/10.1016/j.neucom.2024.128096

Bouali Et-taibi, M., Abid, M. R., Boufounas, E.-M., Morchid, A., Bourhnane, S., Abu Hamed, T., & Benhaddou, D. (2024). Enhancing water management in smart agriculture: A cloud and IoT-based smart irrigation system. *Results in Engineering*, *22*, 102283. https://doi.org/10.1016/j.rineng.2024.102283

Brears, R. C. (2021). Smart digital water management and managing customers of the future. In *Water Resources Management: Innovative and Green Solutions* (pp. 107–130). De Gruyter. https://doi.org/10.1515/9783110685640-007

Chakraborty, D., Alam, A., Chaudhuri, S., Başağaoğlu, H., Sulbaran, T., & Langar, S. (2021). Scenario-based prediction of climate change impacts on building cooling energy consumption with explainable artificial intelligence. *Applied Energy*, *291*, 116807. https://doi.org/10.1016/j.apenergy.2021.116807

Chen, M., Cui, Y., Wang, X., Xie, H., Liu, F., Luo, T., Zheng, S., & Luo, Y. (2021). A reinforcement learning approach to irrigation decision-making for rice using weather forecasts. *Agricultural Water Management*, *250*, 106838. https://doi.org/10.1016/j.agwat.2021.106838

Collins, C., Dennehy, D., Conboy, K., & Mikalef, P. (2021). Artificial intelligence in information systems research: A systematic literature review and research agenda. *International Journal of Information Management*, *60*, 102383. https://doi.org/10.1016/j.ijinfomgt.2021.102383

de Camargo, E. T., Spanhol, F. A., Slongo, J. S., da Silva, M. V. R., Pazinato, J., de Lima Lobo, A. V., Coutinho, F. R., Pfrimer, F. W. D., Lindino, C. A., Oyamada, M. S., & Martins, L. D. (2023). Low-cost water quality sensors for IoT: A systematic review. *Sensors*, *23*(9), 4424. https://doi.org/10.3390/s23094424

Dinar, A. (2024). Challenges to water resource management: The role of economic and modeling approaches. *Water*, *16*(4), 610. https://doi.org/10.3390/w16040610

Dubey, P., & Kumar, M. (2025). Integrating explainable AI with federated learning for next-generation IoT: A comprehensive review and prospective insights. *Computer Science Review*, *56*, 100697. https://doi.org/10.1016/j.cosrev.2024.100697

Elbasi, E., Zaki, C., Topcu, A. E., Abdelbaki, W., Zreikat, A. I., Cina, E., Shdefat, A., & Saker, L. (2023). Crop prediction model using machine learning algorithms. *Applied Sciences*, *13*(16), 9288. https://doi.org/10.3390/app13169288

Emami, M., Ahmadi, A., Daccache, A., Nazif, S., Mousavi, S.-F., & Karami, H. (2022). County-level irrigation water demand estimation using machine learning: A case study of California. *Water*, *14*(12), 1937. https://doi.org/10.3390/w14121937

Essamlali, I., Nhaila, H., & El Khaili, M. (2024). Advances in machine learning and IoT for water quality monitoring. *Heliyon*, *10*(6), e27920. https://doi.org/10.1016/j.heliyon.2024.e27920

Evaristo, J., Jameel, Y., Tortajada, C., & others. (2023). Water woes: The institutional challenges in achieving SDG 6. *Sustain Earth Reviews*, *6*, 13. https://doi.org/10.1186/s42055-023-00067-2

Fan, X., Zhang, X., & Yu, X. B. (2021). Machine learning model and strategy for fast and accurate detection of leaks in the water supply network. *Journal of Infrastructure Preservation and* Resilience, 2, 10 (2021). https://doi.org/10.1186/s43065-021-00021-6

Farah, E., & Shahrour, I. (2024). Water leak detection: A comprehensive review of methods, challenges, and future directions. *Water*, *16*(20), 2975. https://doi.org/10.3390/w16202975

Ferrari, E., Verda, D., Pinna, N., & Muselli, M. (2023). Optimizing water distribution through explainable AI and rule-based control. *Computers*, *12*(6), 123. https://doi.org/10.3390/computers12060123

Ford, G.L., Ancell, G., Hill, E. S., Levine, J., Reali, C., Rijks, E., & Sanchis, G. (2022). Management of aged infrastructure. In Ancell, G., et al. (Eds.) *Power System Assets. CIGRE Green Books*. Springer, Cham. https://doi.org/10.1007/978-3-030-85514-7_4

Forhad, H. M., Uddin, M. R., Chakrovorty, R. S., Ruhul, A. M., Faruk, H. M., Kamruzzaman, S., Sharmin, N., Molla Jamal, A. H. M. S. I., Haque, M. M., & Morshed, A. K. M. M. (2024). IoT-based real-time water quality monitoring system in water treatment plants (WTPs). *Heliyon*, *10*(23), e40746. https://doi.org/10.1016/j.heliyon.2024.e40746

França, R. P., Monteiro, A. C. B., Arthur, R., & Iano, Y. (2021). An overview of deep learning in big data, image, and signal processing in the modern digital age. In Piuri, V., Raj, S., Genovese, A., & Srivastava, R. (Eds.) *Hybrid Computational Intelligence for Pattern Analysis: Trends in Deep Learning Methodologies* (pp. 63–87). Academic Press, London. https://doi.org/10.1016/B978-0-12-822226-3.00003-9

Gámiz, J., Grau, A., Martínez, H., & Bolea, Y. (2020). Automated chlorine dosage in a simulated drinking water treatment plant: A real case study. *Applied Sciences*, *10*(11), 4035. https://doi.org/10.3390/app10114035

Geetha, T. S., Gandhimathi, G., Chellaswamy, C., & Thiruvalar Selvan, P. (2024). Comprehensive river water quality monitoring using convolutional neural networks and gated recurrent units: A case study along the Vaigai River. *Journal of Environmental Management*, *365*, 121567. https://doi.org/10.1016/j.jenvman.2024.121567

Hachimi, C. E., Belaqziz, S., Khabba, S., Sebbar, B., Dhiba, D., & Chehbouni, A. (2023). Smart weather data management based on artificial intelligence and big data analytics for precision agriculture. *Agriculture*, *13*(1), 95. https://doi.org/10.3390/agriculture13010095

Haider, S., Rashid, M., Tariq, M. A. U. R., & Nadeem, A. (2024). The role of artificial intelligence (AI) and ChatGPT in water resources, including its potential benefits and associated challenges. *Discover Water*, *4*, 113. https://doi.org/10.1007/s43832-024-00173-y

Ingrao, C., Strippoli, R., Lagioia, G., & Huisingh, D. (2023). Water scarcity in agriculture: An overview of causes, impacts, and approaches for reducing the risks. *Heliyon*, *9*(8), e18507. https://doi.org/10.1016/j.heliyon.2023.e18507

Jan, F., Min-Allah, N., & Düştegör, D. (2021). IoT-based smart water quality monitoring: Recent techniques, trends, and challenges for domestic applications. *Water*, *13*(13), 1729. https://doi.org/10.3390/w13131729

Javaid, O., Haleem, A., Haleem Khan, I., & Suman, R. (2023). Understanding the potential applications of artificial intelligence in the agriculture sector. *Advanced Agrochem*, *2*(1), 15–30. https://doi.org/10.1016/j.aac.2022.10.001

Jayakumar, D., Bouhoula, A., & Al-Zubari, W. K. (2024). Unlocking the potential of artificial intelligence for sustainable water management, focusing on operational applications. *Water*, *16*(22), 3328. https://doi.org/10.3390/w16223328

Jibrin, A. M., Al-Suwaiyan, M., Aldrees, A., Dan'azumi, S., Usman, J., Abba, S. I., Yassin, M. A., Scholz, M., & Sammen, S. S. (2024). Machine learning predictive insight of water pollution and groundwater quality in the Eastern Province of Saudi Arabia. *Scientific Reports*, *14*, 20031. https://doi.org/10.1038/s41598-024-70610-4

Kamyab, H., Khademi, T., Chelliapan, S., SaberiKamarposhti, M., Rezania, S., Yusuf, M., Farajnezhad, M., Abbas, M., Jeon, B. H., & Ahn, Y. (2023). The latest innovative avenues for utilizing artificial intelligence and big data analytics in water resource management. *Results in Engineering*, *20*, 101566. https://doi.org/10.1016/j.rineng.2023.101566

Kavya, M., Mathew, A., Shekar, P. R., & Sarwesh, P. (2023). Short-term water demand forecast modeling using artificial intelligence for smart water management. *Sustainable Cities and Society*, *95*, 104610. https://doi.org/10.1016/j.scs.2023.104610

Kim, D., Choi, S., Kang, S., & Noh, H. (2023). A study on developing an AI-based water demand prediction and classification model for the Gurye Intake Station. *Water*, *15*(23), 4160. https://doi.org/10.3390/w15234160

Krishnan, A., Sundaram, T., Nagappan, B., Devarajan, Y., & Bhumika. (2024). Integrating artificial intelligence in nanomembrane systems for advanced water desalination. *Results in Engineering*, *24*, 103321. https://doi.org/10.1016/j.rineng.2024.103321

Kumari, U., Swamy, K., Gupta, A., Karri, R. R., & Meikap, B. C. (2021). Global water challenge and future perspective. In Dehghani, M. H., Karri, R., & Lima, E. (Eds.) *Green Technologies for the Defluoridation of Water* (pp. 197–212). Elsevier. https://doi.org/10.1016/B978-0-323-85768-0.00002-6

Lebu, S., Lee, A., Salzberg, A., & Bauza, V. (2024). Adaptive strategies to enhance water security and resilience in low- and middle-income countries: A critical review. *Science of the Total Environment*, *925*, 171520. https://doi.org/10.1016/j.scitotenv.2024.171520

Li, Z., Wang, G., Lin, D., & Mashhadi, A. (2024). A hybrid approach is used for accurate water demand prediction using socio-economic and climatic factors with ELM optimization. *Heliyon*, *10*(3), e25028. https://doi.org/10.1016/j.heliyon.2024.e25028

Liu, J., Long, Y., Zhu, G., & Hursthouse, A. S. (2024). Application of artificial intelligence in the management of coagulation treatment engineering system. *Processes*, *12*(9), 1824. https://doi.org/10.3390/pr12091824

Mahdi, N. M., Jassim, A. H., Abulqasim, S. A., Basem, A., Ogaili, A. A. F., & Al-Haddad, L. A. (2024). Leak detection and localization in water distribution systems using advanced feature analysis and an artificial neural network. *Desalination and Water Treatment*, *320*, 100685. https://doi.org/10.1016/j.dwt.2024.100685

Mana, A. A., Allouhi, A., Hamrani, A., Rehman, S., el Jamaoui, I., & Jayachandran, K. (2024). Sustainable AI-based production agriculture: Exploring AI applications and implications in agricultural practices. *Smart Agricultural Technology*, *7*, 100416. https://doi.org/10.1016/j.atech.2024.100416

Medeiros, V. d. S., dos Santos, M. D., & Brito, A. V. (2024). Case study for predicting failures in water supply networks using neural networks. *Water*, *16*(10), 1455. https://doi.org/10.3390/w16101455

Miller, M., Kisiel, A., Cembrowska-Lech, D., Durlik, I., & Miller, T. (2023). IoT in water quality monitoring: Are we here? *Sensors*, *23*(2), 960. https://doi.org/10.3390/s23020960

Mishra, S., Kumar, R., & Kumar, M. (2023). Use of treated sewage or wastewater as irrigation water for agricultural purposes: Environmental, health, and economic impacts. *Total Environment Research Themes*, *6*, 100051. https://doi.org/10.1016/j.totert.2023.100051

Mohd Yussof, N. A., & Ho, H. W. (2022). Review of water leak detection methods in smart building applications. *Buildings*, *12*(10), 1535. https://doi.org/10.3390/buildings12101535

Morain, A., Ilangovan, N., Delhom, C., & Anandhi, A. (2024). Artificial intelligence for water consumption assessment: State of the art review. *Water Resources Management*, *38*, 3113–3134. https://doi.org/10.1007/s11269-024-03823-x

Morchid, A., Alblushi, I. G. M., Khalid, H. M., El Alami, R., Sitaramanan, S. R., & Muyeen, S. M. (2024). High-technology agriculture system to enhance food security: A concept of smart irrigation system using Internet of Things and cloud computing. *Journal of the Saudi Society of Agricultural Sciences*, *24*(2), 123–130. https://doi.org/10.1016/j.jssas.2024.02.001

Mousavi, S. R., & Hakemi, A. (2024). Artificial intelligence and insecticides in pest control. IntechOpen. https://doi.org/10.5772/intechopen.1006987

Murtaza, A. A., Saher, A., Zafar, M. H., Moosavi, S. K. R., Aftab, M. F., & Sanfilippo, F. (2024). A paradigm shift for predictive maintenance and condition monitoring from Industry 4.0 to Industry 5.0: A systematic review, challenges, and case study. *Results in Engineering*, *24*, 102935. https://doi.org/10.1016/j.rineng.2024.102935

Musie, W., & Gonfa, G. (2023). Freshwater resource, scarcity, water salinity challenges, and possible remedies. *Heliyon*, *9*(8), e18685. https://doi.org/10.1016/j.heliyon.2023.e18685

Nallakaruppan, M. K., Gangadevi, E., Shri, M. L., Balusamy, B., Bhattacharya, S., & Selvarajan, S. (2024). Reliable water quality prediction and parametric analysis using explainable AI models. *Scientific Reports*, *14*, 7520. https://doi.org/10.1038/s41598-024-56775-y

Nandgude, N., Singh, T. P., Nandgude, S., & Tiwari, M. (2023). Drought prediction: A comprehensive review of different drought prediction models and adopted technologies. *Sustainability*, *15*(15), 11684. https://doi.org/10.3390/su151511684

Narayanan, D., Bhat, M., Samuel Paul, N. R., Khatri, N., & Saroliya, A. (2024). Artificial intelligence-driven advances in wastewater treatment: Evaluating techniques for sustainability and efficacy in global facilities. *Desalination and Water Treatment*, *320*, 100618. https://doi.org/10.1016/j.dwt.2024.100618

Niknam, A., Zare, H. K., Hosseininasab, H., Mostafaeipour, A., & Herrera, M. (2022). A critical review of short-term water demand forecasting tools: What method should I use? *Sustainability*, *14*(9), 5412. https://doi.org/10.3390/su14095412

Niyongabo, A., Zhang, D., Guan, Y., Wang, Z., Imran, M., Nicayenzi, B., Guyasa, A. K., & Hatungimana, P. (2024). Predicting urban water consumption and health using artificial intelligence techniques in Tanganyika Lake, East Africa. *Water*, *16*(13), 1793. https://doi.org/10.3390/w16131793

Nti, E. K., Cobbina, S. J., Attafuah, E. E., Senanu, L. D., Amenyeku, G., Gyan, M. A., Forson, D., & Safo, A.-R. (2023). Water pollution control and revitalization using advanced technologies: Uncovering artificial intelligence options towards environmental health protection, sustainability, and water security. *Heliyon*, *9*(7), e18170. https://doi.org/10.1016/j.heliyon.2023.e18170

Obaideen, K., Yousef, B. A. A., AlMallahi, M. N., Tan, Y. C., Mahmoud, M., Jaber, H., & Ramadan, M. (2022). An overview of smart irrigation systems using IoT. *Energy Nexus*, *7*, 100124. https://doi.org/10.1016/j.nexus.2022.100124

Olawade, D. B., Wada, O. Z., Ige, A. O., Egbewole, B. I., Olojo, A., & Oladapo, B. I. (2024). Artificial intelligence in environmental monitoring: Advancements, challenges, and future directions. *Hygiene and Environmental Health Advances*, *12*, 100114. https://doi.org/10.1016/j.heha.2024.100114

Onyeuwaoma, N., Sivakumar, V., & Bade, M. (2024). Modeling drought in South Africa: Meteorological insights and predictive parameters. *Environmental Monitoring and Assessment*, *196*(965). https://doi.org/10.1007/s10661-024-13009-y

Ou, Z., He, F., Zhu, Y., Lu, P., & Wang, L. (2023). Analysis of driving factors of water demand based on explainable artificial intelligence. *Journal of Hydrology: Regional Studies*, *47*, 101396. https://doi.org/10.1016/j.ejrh.2023.101396

Oudah, M., Al-Naji, A., Al-Janabi, T. Y., Namaa, D. S., & Chahl, J. (2024). Automatic irrigation system based on computer vision and an artificial intelligence technique using Raspberry Pi. *Automation*, *5*(2), 90–105. https://doi.org/10.3390/automation5020007

Park, J., Patel, K., & Lee, W. H. (2024). Recent advances in algal bloom detection and prediction technology using machine learning. *Science of the Total Environment*, *938*, 173546. https://doi.org/10.1016/j.scitotenv.2024.173546

Pérez-Beltrán, C. H., Robles, A. D., Rodriguez, N. A., Ortega-Gavilán, F., & Jiménez-Carvelo, A. M. (2024). Artificial intelligence and water quality: From drinking water to wastewater. *TrAC Trends in Analytical Chemistry*, *172*, 117597. https://doi.org/10.1016/j.trac.2024.117597

Pointet, T. (2022). The United Nations World Water Development Report 2022 on groundwater, a synthesis. *LHB, 108*(1). https://doi.org/10.1080/27678490.2022.2090867
Ramovha, N., Chadyiwa, M., Ntuli, F., & Sithole, T. (2024). The potential of stormwater management strategies and artificial intelligence modeling tools to improve water quality. *Water Resources Management, 38*, 3527–3560. https://doi.org/10.1007/s11269-024-03841-9
Rashid, A. B., & Kausik, M. A. K. (2024). AI revolutionizing industries worldwide: A comprehensive overview of its diverse applications. *Hybrid Advances, 7*, 100277. https://doi.org/10.1016/j.hybadv.2024.100277
Rathor, A. S., Choudhury, S., Sharma, A., Nautiyal, P., & Shah, G. (2024). Empowering vertical farming through IoT and AI-driven technologies: A comprehensive review. *Heliyon, 10*(15), e34998. https://doi.org/10.1016/j.heliyon.2024.e34998
Rautela, K. S., & Goyal, M. K. (2024). Transforming air pollution management in India with AI and machine learning technologies. *Scientific Reports, 14*, 20412. https://doi.org/10.1038/s41598-024-71269-7
Romero-Ben, L., Alves, D., Blesa, J., Cembrano, G., Puig, V., & Duviella, E. (2023). Leak detection and localization in water distribution networks: Review and perspective. *Annual Reviews in Control, 55*, 392–419. https://doi.org/10.1016/j.arcontrol.2023.03.012
Sagan, V., Peterson, K. T., Maimaitijiang, M., Sidike, P., Sloan, J., Greeling, B. A., Maalouf, S., & Adams, C. (2020). Monitoring inland water quality using remote sensing: Potential and limitations of spectral indices, bio-optical simulations, machine learning, and cloud computing. *Earth-Science Reviews, 205*, 103187. https://doi.org/10.1016/j.earscirev.2020.103187
Salomons, E., & Housh, M. (2020). Practical real-time optimization for energy-efficient water distribution systems operation. *Journal of Cleaner Production, 275*, 124148. https://doi.org/10.1016/j.jclepro.2020.124148
Sanchez, G. M., Terando, A., Smith, J. W., García, A. M., Wagner, C. R., & Meentemeyer, R. K. (2020). Forecasting water demand across a rapidly urbanizing region. *Science of the Total Environment, 730*, 139050. https://doi.org/10.1016/j.scitotenv.2020.139050
Sangha, L., Shortridge, J., & Frame, W. (2023). The impact of nitrogen treatment and short-term weather forecast data in irrigation scheduling of corn and cotton on water and nutrient use efficiency in humid climates. *Agricultural Water Management, 283*, 108314. https://doi.org/10.1016/j.agwat.2023.108314
Sarker, I. H. (2021). Machine learning: Algorithms, real-world applications, and research directions. *SN Computer Science, 2*, 160. https://doi.org/10.1007/s42979-021-00592-x
Sarker, I. H. (2022). AI-based modeling: Techniques, applications and research issues towards automation, intelligent and smart systems. *SN Computer Science, 3*, 158. https://doi.org/10.1007/s42979-022-01043-x
Serafeim, A. V., Fourniotis, N. T., Deidda, R., Kokosalakis, G., & Langousis, A. (2024). Leakages in water distribution networks: Estimation methods, influential factors, and mitigation strategies: A comprehensive review. *Water, 16*(11), 1534. https://doi.org/10.3390/w16111534
Shayo, G. M., Elimbinzi, E., Shao, G. N., & Fabian, C. (2023). The severity of waterborne diseases in developing countries and the effectiveness of ceramic filters for improving water quality. *Bulletin of the National Research Centre, 47*, 113. https://doi.org/10.1186/s42269-023-01088-9
Sheik, A. G., Kumar, A., Sharanya, A. G., Amabati, S. R., Bux, F., & Kumari, S. (2024). Machine learning-based monitoring and design of managed aquifer rechargers for sustainable groundwater management: Scope and challenges. In *Environmental Science and Pollution Research*. https://doi.org/10.1007/s11356-024-35529-3
Shiddiqi, A. M., Za'in, C., Lathifah, A., Ahmad, T., & Purwitasari, D. (2024). GA-sense: Sensor placement strategy for detecting leaks in water distribution networks based on time series flow and genetic algorithm. *MethodsX, 12*, 102612. https://doi.org/10.1016/j.mex.2024.102612
Tańczyk, J., Kajewska-Szkudlarek, J., Lipiński, P., & Rychlikowski, P. (2022). Improving short-term water demand forecasting using evolutionary algorithms. *Scientific Reports, 12*, 13522. https://doi.org/10.1038/s41598-022-17177-0
Tarraf, B., Brun, F., Raynaud, L., Roux, S., Zhang, Y., Davadan, L., & Deudon, O. (2024). Assessing the impact of weather forecast uncertainties in crop water stress model predictions. *Agricultural and Forest Meteorology, 349*, 109934. https://doi.org/10.1016/j.agrformet.2024.109934
Thompson, K. A., & Dickenson, E. R. V. (2021). Using machine learning classification to detect simulated increases of de facto reuse and urban stormwater surges in surface water. *Water Research, 204*, 117556. https://doi.org/10.1016/j.watres.2021.117556
Tien, P. W., Wei, S., Darkwa, J., Wood, C., & Calautit, J. K. (2022). Machine learning and deep learning methods for enhancing building energy efficiency and indoor environmental quality. *Energy and AI, 10*, 100198. https://doi.org/10.1016/j.egyai.2022.100198
Ucar, A., Karakose, M., & Kırımça, N. (2024). Artificial intelligence for predictive maintenance applications: Key components, trustworthiness, and future trends. *Applied Sciences, 14*(2), 898. https://doi.org/10.3390/app14020898

Vaish, R., Dwivedi, U. D., Tewari, S., & Tripathi, S. M. (2021). Machine learning applications in power system fault diagnosis: Research advancements and perspectives. *Engineering Applications of Artificial Intelligence*, *106*, 104504. https://doi.org/10.1016/j.engappai.2021.104504

Vekaria, D., & Sinha, S. (2024). aiWATERS: An artificial intelligence framework for the water sector. *AI in Civil Engineering*, *3*(6). https://doi.org/10.1007/s43503-024-00025-7

Wang, Q., Wang, X., Liu, Y., & Li, R. (2021). Urbanization and water consumption at national- and subnational-scale: The roles of structural changes in economy, population, and resources. *Sustainable Cities and Society*, *75*, 103272. https://doi.org/10.1016/j.scs.2021.103272

Wang, Y., Cui, Z., & Ke, R. (2023). Machine learning basics. In Wang, Y., Cui, Z., & Ke, R. (Eds.) *Machine Learning for Transportation Research and Applications* (pp. 25–40). Elsevier. https://doi.org/10.1016/B978-0-32-396126-4.00008-4

Wei, H., Xu, W., Kang, B., Eisner, R., Muleke, A., Rodriguez, D., deVoil, P., Sadras, V., Monjardino, M., & Harrison, M. T. (2024). Irrigation with artificial intelligence: Problems, premises, promises. *Human-Centered Intelligent Systems*, *4*, 187–205. https://doi.org/10.1007/s44230-024-00072-4

Wu, Y., Liu, S., & Kapelan, Z. (2024). Addressing data limitations in leakage detection of water distribution systems: Data creation, data requirement reduction, and knowledge transfer. *Water Research*, *267*, 122471. https://doi.org/10.1016/j.watres.2024.122471

Younis, H. I., Kizhisseri, M. I., & Mohamed, M. M. (2023). Forecasting future water demands for sustainable development in Al-Ain City, United Arab Emirates. *Water*, *15*(21), 3800. https://doi.org/10.3390/w15213800

Zhu, M., Wang, J., Yang, X., Zhang, Y., Zhang, L., Ren, H., Wu, B., & Ye, L. (2022). A review of the application of machine learning in water quality evaluation. *Eco-Environment & Health*, *1*(2), 107–116. https://doi.org/10.1016/j.eehl.2022.06.001

PART IV

AI Revolutionizing Water Treatment

Automation and Optimization in Water Treatment

10

Neha Raghuwanshi, Shagun Sharma, and Chaitanya Vijay Kale

10.1 INTRODUCTION TO AUTOMATION IN WATER TREATMENT

10.1.1 Overview of Water Treatment Systems

Because water is highly essential to our daily existence, there is an increasing need to maintain and enhance its quality. Our precious water supplies are being contaminated by both point and non-point sources. The primary causes of water pollution include home, industrial, and agricultural practices, as well as other environmental and global shifts. In many parts of the world, the ground and surface water are tainted and unfit for human consumption. The world's population is expected to exceed 7.9 billion people by 2020, which might lead to a severe freshwater shortage.

This has led to efforts to compare different water treatment and recycling methods. There have also been attempts to implement a strategy for recycling and water treatment. By examining the technologies' performance, sludge output, lifespan, and mode of operation, a comparison has been provided. This article aims to offer recommendations for choosing the precise technology or their combinations for different applications (Gupta et al., 2012).

10.1.2 The Role of Automation in Enhancing Efficiency

Industrial operations have been completely changed by software automation, which has led to notable increases in productivity and efficiency. Beyond simple mechanization, it has a wide range of applications in contemporary industry that improve operational processes, maximize resource utilization, and guarantee high-quality results. Automation technologies are now essential to attaining operational excellence and competitive advantage because they simplify repetitive operations, reduce human error, and facilitate real-time data-driven decision-making (Ajiga et al., 2024).

DOI: 10.1201/9781003600718-14

By automating time-consuming, repetitive activities, it increases efficiency and frees up human operators to perform more strategic and imaginative duties. In industrial processes, automation guarantees constant quality and accuracy, minimizes errors, and decreases the need for manual intervention. Lower operating costs, less downtime, and more effective resource use result in significant cost savings. These advantages are further enhanced by the incorporation of cutting-edge technologies like artificial intelligence (AI), Internet of Things (IoT), and robotics, which allow industries to maintain high levels of efficiency, quickly respond to market needs, and adapt to changing conditions (Silva, 2023).

10.2 THE FUNDAMENTALS IN PROCESS OPTIMIZATION

10.2.1 Defining Process Optimization in Water Treatment

Wastewater treatment includes wastewater collection, processing, and reuse. Unwanted performance in the wastewater collecting network, one of the crucial infrastructures, can result in a variety of issues. Predicting the variables in wastewater treatment plants (WWTPs) is crucial since wastewater treatment is a process that depends on a variety of physical and chemical elements. Conversely, urban treated wastewater has restrictions but is a suitable resource for a variety of applications, including industry and agriculture. When these resources are misused, there may be health and environmental hazards, which can lead to social unrest and occasionally financial difficulties. According to the study, wastewater treatment has several advantages, including enhancing soil quality, supplying plant nutrients, and reducing the cost of using freshwater for irrigation. Nonetheless, there are concurrently detrimental effects on society, environment, and people. All these topics were examined methodically and critically, and they were also marked for future research directions (Macedonio & Drioli, 2019).

10.2.2 Objectives of Processing Optimization

The process of treating water involves a variety of physical, chemical, physicochemical, and biological procedures with the goal of removing or minimizing contaminants or undesirable aspects of the water. Water treatment makes water suitable for its intended end use by eliminating impurities and unwanted components or lowering their concentration. Both drinking and irrigation usage are made possible by this treatment, which is essential to human health. Automation can guarantee that treated water satisfies legal requirements and is fit for human consumption. Leaks, pressure changes, and equipment failures are examples of possible risks that automation can identify and react to. Automation can increase production, decrease resource waste, and streamline procedures. Energy expenses, chemical use, and the environmental effects of water treatment can all be decreased with automation. Also, it can save money and use resources better in several ways. First, by adjusting chemical doses and energy use based on real-time needs, automated systems cut waste and use resources more efficiently (Al-Juaidi et al., 2014).

10.2.3 Benefits of Optimized Processes

Smart water quality monitoring systems (SWQMSs) have been developed in recent years as a result of IoT and software technology advancements like cloud computing, software as a service (SaaS), and service-oriented architecture (SOA). These systems integrate knowledge, technology, and functionality from fields such as biology, chemistry, nanoscience, and cognitive sciences with components and technologies from microsystems, which are miniature electric, mechanical, optical, and fluid devices. Thankfully, the difficulties posed by the wide range of IoT technologies, systems, and design principles can be addressed with the aid of IoT software platforms (Ullah et al., 2024).

10.3 TECHNOLOGIES ENABLING AUTOMATION AND OPTIMIZATION

10.3.1 Advanced Sensors and IoT in Treatment

A technology with multiple layers is called an IoT platform. It transfers information between cloud storage and a hardware device. At the moment, the IoT platform provides customers with an integrated feature that facilitates the development of software applications for hardware devices that are connected. It also handles scalability, data security, and cross-device compatibility. Sensor node technologies, such as wireless sensor networks, miniaturization, and nanotechnology, are important IoT-related technologies. Since IoT is associated with a lot of wireless sensor devices, it generates a lot of data, which IoT also processes. IoT is essentially made up of three layers (Li et al., 2020).

1. The application layer
2. The network layer
3. The perception layer

Sensors, cameras, RFID readers, and other data-gathering terminals are all part of the perception layer of IoT, which creates the data acquisition interface. To gather sensor data and to collaborate and integrate environments, the data acquisition interface is essential. Monitoring water quality can be used to keep an eye on the effluent treatment plant. One of the main IoT application areas is water quality monitoring, which uses sensors to measure pollution and calculate the water quality factor value. Using the GSM module, that sensor data collection can be sent to the IoT cloud server (Al-Juaidi et al., 2014; Ullah et al., 2024).

10.3.2 ML and AI in Process Optimization

We provide a quick overview of the pertinent AI principles and techniques to help guide the discussions in Section 10.3. AI and machine learning (ML) technologies are frequently utilized in Discrete Wavelet Transform (DWT). Four categories of problems—classification, regression, dimensionality reduction, and clustering—are frequently resolved by ML. Predictive models are created using ML techniques to address regression and classification issues using process data. Stated otherwise, a thorough application process for the ML approach consists of the following: (i) measuring the relationship between the system's input parameters and the desired output from a given data set, or the training process; and (ii) predicting the system's new output using the estimated nonlinear relationship. The goal of ML is to predict rather than estimate, which makes ML suitable for water quality prediction, algal bloom forecasting, and pipe network issue diagnosis (Kourou et al., 2014).

Principal component analysis (PCA), a dimensionality reduction technique, is primarily used to process high-dimensional data (such as fluorescence excitation–emission matrix), while artificial neural networks (ANNs), deep learning (DL), support vector machines (SVMs), and random forests (RFs) are used to handle nonlinear classification and regression analysis. Both ANNs and DL carry out distributed parallel information processing and mimic the behavioral traits of animal brain networks. DL differs from ANNs in that it often has a higher prediction accuracy and a more complicated structure. However, DL is more likely to overfit and requires more data to train the network. In contrast to the classic ANN, the SVM has stronger mathematical theory support, which enhances the interpretability of the model to a certain extent.

Furthermore, the SVM may be converted into a convex optimization problem, which guarantees the algorithm's global optimality and helps neural networks avoid the local minimum problem that they struggle to handle. Nevertheless, SVM models need a lot of training time and are challenging to train using big sample data. The ability of RFs to assess the significance of variables during classification or regression analysis is a key benefit over ANNs, DL, and SVMs. A more thorough statistical analysis is necessary, even though the weight of the ANN can also be used to assess the significance of the input parameters. Because of their strong ability to handle nonlinear relationships, the four ML techniques mentioned above are appropriate for building prediction models in DWT situations where the background substances are complex and their interactions are unknown. Examples of these situations include the analysis of Disinfection Biproducts (DBP) precursors, membrane fouling, and the prediction of contaminants in the source water. Additionally, the target requirement of the prediction model, the size of the current data collection, and the features of various ML approaches are taken into consideration while choosing and implementing the ML method (Chen et al., 2013; Kourou et al., 2014; Mo et al., 2023).

10.3.3 Automated Control Systems (SCADA, PLCs)

CPU, memory, input/output ports, relays, timers, counters, and data storage containers make up a PLC. The S7–300 modular Programmable Logic Controller (PLC) is utilized for mid-range and low-end functioning. It features 38 analog and 1,024 digital input/output ports, and it has 312–318 modules. For the project to run three DC pumps, one AC pump, one motor, and two solenoids, eight Digital Output (DO) ports are needed. Simatic Manager is a PLC programming program. The software SimaticwinCC is utilized for Supervisory Control and Data Acquisition (SCADA) programming. PLC languages are as follows:

1. Ladder diagram
2. Functional block diagram
3. Structural text
4. Instruction list
5. Sequential functional charts

Ladder diagrams are employed from the aforementioned techniques. First-rate process quality is a crucial requirement for a prosperous water sector. With the SIMATIC S7 process control system, for instance, totally integrated automation offers the ideal circumstances for this. This cutting-edge process control system serves as the foundation for safe and efficient process management and regulates every process in the plant. Remote or widely dispersed plants, like pumping stations, can be effectively managed and observed (Bratby, 2015).

The entire sewage water treatment process is monitored and managed via the SCADA screen. This screen allows the operator to send signals to all the field operations that are currently underway. There is one emergency stop key on the screen that can be utilized in an emergency. After receiving the pH data, the operator displays the value on the screen and distributes the purified water for either household or agricultural use based on the pH value.

10.4 AUTOMATION IN KEY WATER TREATMENT PROCESS

10.4.1 Coagulation and Flocculation

Both suspended and dissolved particles are present in surface and groundwater. To remove the suspended materials from the water, flocculation and coagulation are employed. Source, charge, particle size, shape, and density are all different for suspended particles. These elements are necessary for proper flocculation and coagulation applications. When two suspended particles in water get close to one another, their negative charges repel one another because they have the same kind of surface charge. In the absence of appropriate flocculation and coagulation, suspended solids will not clump together and settle out of the water; instead, they will stay in suspension (Ghernaout & Ghernaout, 2012).

Particle collision and floc growth are made possible by the sequential coagulation and flocculation processes. The next step after this is sedimentation. The flocculation step will fail if coagulation is not complete, and the sedimentation step will fail if flocculation is not complete. Particle size grows from submicroscopic microfloc to visible suspended particles by flocculation, a mild mixing stage. When microfloc particles collide, they form bonds that result in larger, more noticeable flocs known as pinflocs. With more collisions and interaction with additional organic or inorganic polymers (coagulants), the floc size keeps growing. In order to help bridge, bind, and strengthen the floc, as well as to add weight and speed up the settling process, high molecular weight polymers known as coagulant aids may be added to the floc after macroflocs have formed. Water is prepared for sedimentation when floc reaches its ideal size and strength. In earlier periods, Ancient Egyptians used a combination of coagulation with alum, which caused impurities to clump together and settle out (Bratby, 2015; Ghernaout & Ghernaout, 2012).

To counteract the negative charges on non-settleable solids (such as clay and organic materials that provide color), coagulant chemicals are introduced to water that have charges opposite to those of the suspended solids. The tiny suspended particles have the ability to adhere to one another once the charge is neutralized. These somewhat bigger particles, known as microflocs, are invisible to the human eye. The newly created microflocs should be surrounded by clear water. Otherwise, the charge of some of the particles and coagulation have not been neutralized. Additional coagulant chemicals can be required. Pollutants in colloidal form are frequently found in wastewater. In these situations, the colloidal suspension could include metal oxides, organic components, poisonous compounds that are insoluble, stable emulsions, and substances that cause turbidity. The elimination of turbidity from water is the main goal of the coagulation/flocculation process. The most significant physicochemical process used to clean water and wastewater was coagulation, which was accomplished both chemically and electrically (Narges et al., 2021).

Coagulation and flocculation are advantageous in ways other than eliminating turbidity from water. Numerous microorganisms that are suspended in water are eliminated by the technique, which can also be used to eliminate water's color. Compared to groundwater, surface water has far more turbidity and color. Surface water gathers up a lot of particles as it travels over the ground to streams, via streams, and finally through rivers. Because of this, surface water typically requires treatment by flocculation and coagulation, whereas groundwater typically requires aeration (Krupińska, 2020).

It was discovered that Brownian movement occurs in water when negatively charged surfaces repel one another, creating a stable dispersed suspension. The two main types of metal coagulants that are frequently utilized are those based on iron and those based on aluminum. This chapter discusses aluminum-based coagulants. Aluminum sulfate, also known as "alum," is arguably the most popular metal coagulant and has been used for decades to clean water. Simple metal coagulants (conventional) are widely used, primarily because they are inexpensive and have an easier application process. Nevertheless, they have a number of drawbacks, including the requirement to adjust pH either before or after treatment, sensitivity to temperature fluctuations, and the requirement for larger dosages because charge neutralization is typically insufficient, the sensitivity to sample-specific properties and composition, and the excessive production of sludge. According to research on flocculation, colloidal particle agglomeration forms microflocs that eventually transform into observable floc masses, leading to sensitivity to particular sample composition and properties and overproduction of sludge (Nicolaisen, 2003).

10.4.2 Filtration and Membrane Technologies

Membrane technology is frequently used to produce several types of water from brackish water, seawater, well water, and surface water. Membrane technology has recently expanded into the treatment of secondary and tertiary municipal wastewater as well as oil field-produced water. It is also utilized in industrial processes and in the treatment of industrial wastewater. To produce water of increasing purity and quality for a variety of uses, one membrane technique is frequently followed by another. Therefore, one kind of membrane may improve the performance of another to achieve objectives ranging from wastewater disposal to the creation of drinking water from unexpected sources. The ability to manage a region's whole water resources is thus made possible by membrane technology, which is particularly useful in places where natural water supplies are limited. The most popular membrane element configuration is the spiral wound one because of its high packing density and affordable cost. The scientific advancements in the field of novel membranes and application ideas for spiral wound membrane elements will be discussed in the following. The four well-recognized membrane technologies—microfiltration (0.01–10 microns), ultrafiltration (500–100,000 Da), nanofiltration (100–500 Da), and reverse osmosis (up to 100 Da)—are all covered by spiral wound components (Nicolaisen, 2003).

A feed spacer is put between the opposing membrane surfaces to complete the membrane package, which consists of two membrane sheets sandwiched together with an inserted permeate carrier. The permeate leaves the element through a perforated center tube, which is encircled by the membrane package. A suitable outer wrap is applied to a membrane element to secure its physical shape. Operational parameters are used to determine the physical and chemical characteristics of the different components, including the membrane. Limitations of the standard reverse osmosis elements include temperature (45°(*Z*), pH value (2-11), silt density index (less than 3 Slit Density Index (SDI)), chlorine (dichlorination required), and several other factors. These restrictions need to be reduced or eliminated for more intricate membrane applications, although they are suitable for traditional pure water applications. Spiral wound elements' construction, materials, and membranes have all benefited from the use of advanced materials. Elements with better operating characteristics and broader application regions are the outcome of this endeavor. With the exception of the requirement that new applications be based on a profitable basis for the user, a wide range of novel membrane applications have been made available and have no end in sight. Increased use of membrane technology in a variety of applications is necessary due to factors like water scarcity, environmental regulations, and the obvious reasoning behind reusing water rather than releasing it into the environment (Hand & Cusick, 2021).

10.4.3 Disinfection and Chemical Dosage Control

As an alternative to conventional chemical dosing disinfection techniques, electrochemical disinfection—a process in which chemical oxidants are produced *in situ* via redox reactions on the surface of an electrode—has gained more interest recently. Electrochemical disinfection shows promise for usage in resource-constrained settings as well as an adjunct to aging centralized systems since it eliminates the need for the transportation and storage of hazardous materials and can be expanded across centralized and distributed treatment contexts (Hand & Cusick, 2021).

The specific oxidant generated can affect the necessary dose across treatment settings and introduce chemical reaction pathways that compete with pathogen inactivation, even if the basic idea behind electrochemical disinfection by *in situ* synthesis of oxidants remains the same. Similarly, the oxidant requirement for disinfection might vary significantly depending on water composition in different treatment situations. This study examines three treatment contexts for electrochemical disinfection: distributed wastewater treatment, centralized wastewater treatment, and water treatment. Similar to conventional chemical disinfection procedures, electrochemical disinfection would take place in the water treatment setting following the coagulation/flocculation, sedimentation, and filtering phases, when applicable. The composition and quality of water to be treated depend on the source and season, even though the size of centralized water treatment facilities can vary.

Following primary and secondary biological treatment in a centralized treatment plant, electrochemical disinfection would also follow a similar approach to conventional chemical disinfection in the context of centralized wastewater treatment. Regardless of any physical, chemical, or biological pretreatment, electrochemical disinfection takes place in this study's dispersed wastewater treatment setting following waste dilution with flush water and other gray water sources. This situation is comparable to both new sanitation systems and more widely utilized septic systems in environments with limited resources around the world (Alam et al., 2021).

10.5 AI AND ML IN AUTOMATION LEARNING PROCESS

10.5.1 AI Tool Application in Water Treatment

Numerous studies have documented the use of AI approaches for water treatment process modeling and optimization, including the removal of contaminants from water. The adsorptive removal of metals, colors, organic compounds, nutrients, medicines, medications, pesticides, and Personal Care Products (PCPs) from water is accomplished by commonly used AI approaches. AI methods have proved successful in determining the correlation between water treatment factors. For instance, the initial concentration of the pollutant, the dose of the adsorbent, time, pH, agitation rate, and temperature are often utilized as input variables in the adsorption of pollutants. Removal effectiveness (%) and adsorption capacity are the primary output variables. The coefficient of determination (R2), mean squared error (MSE), sum of squared error (SSE), and root-mean-square error (RMSE) statistics are used to validate the outcomes that the models predict (Alam et al., 2021).

The findings of the experiments and the model generally agreed quite well. According to certain studies, AI will help remove multiple toxins from water at once. These results increase the possibility of using AI to increase the effectiveness of actual water treatment systems. In addition to batch adsorption, AI methods can be used to forecast the adsorbent's removal performance in column experiments (Ihsanullah, 2020).

10.5.2 Challenges and Prospects

Compared to traditional mathematical modeling, AI techniques have several advantages. They can be used to reduce experimental costs and forecast how different water treatment processes will perform. Still, a few restrictions have prevented these methods from being widely used in actual water treatment. The main disadvantage of AI tools like ANNs is their low reproducibility, which might lead to a locally optimal solution because of random weights and bias. Predicting the effectiveness of pollutant removal during the adsorption process can also be done by combining several AI tools. High accuracy and prediction can be attained using DL and deep ANNs (Cairone et al., 2024; Kumari et al., 2024; Paschen et al., 2019).

But for experimental training, testing, locating local minima, and overfitting, there must be enough data. Certain situations may also cause the process performance that AI tools forecast to differ from the actual outcomes. AI technologies might make incorrect predictions, for instance, if operating settings and water quality suddenly change. The prediction capabilities of AI tools need to be improved to be used in a variety of situations and to handle abrupt changes in input variables. The literature that is currently accessible shows that AI tools have performed quite well when it comes to modeling the batch adsorption process with a restricted range of data. It has not yet been investigated how AI technologies might be used in real-world wastewater treatment with a variety of data.

The selection and availability of data are significant obstacles that are pertinent to AI-based water treatment. To create datasets for system optimization and prediction, water utilities are hired to develop, collect, process, assess, and analyze data. Since random data selection has significant disadvantages, extra care must be exercised while choosing the training data for AI tools, an experimental design technique. However, in order to get an appropriate response, experimental design techniques (like Response Surface Methodology (RSM)) typically require a big input dataset. AI models can be trained using operational data from actual water treatment facilities, improving the accuracy of pollution removal predictions (Dalibard et al., 2025).

In addition to protecting the environment, AI technology has the potential to significantly lower operational costs and play a key role in sustainable wastewater treatment. In addition to forecasting the effectiveness of the water treatment process, AI tools can be used to integrate the entire water treatment process, including water discharge, transportation, sludge management, environmental implications, economics, and legislation. To successfully employ AI techniques in the water treatment domain, data collection from various water treatment processes is required. To maintain data integrity, however, extra caution should be used when gathering the data. When reporting the data, all relevant information should be included, including the location, process environment, dataset ontology, and data sources. Researchers, students, and engineers will find this material useful in reusing data in other experimental domains for future forecasting. AI gives the water sector a chance to improve and control water management and monitoring. To solve several issues in water treatment and management, like water quality, leak detection, and water process optimization, new AI-based algorithms must be developed in order to make wise decisions. Prediction accuracy can be improved by using hybrid AI algorithms, which lower operating and energy costs. To compare different AI-based standalone and hybrid water treatment methods and recommend the most effective ones for use in actual treatment procedures, a benchmark or framework should be created (Zhang et al., 2023).

10.6 CHALLENGES AND LIMITATIONS OF AUTOMATION IN WATER TREATMENT

10.6.1 Challenges in Automation in Water Treatment

The use of AI in water management, in particular, in wastewater systems, is a high-potential and challenging domain. Besides the technical requirements, there are conceptual and general challenges related to the characteristics of water systems that need to be considered when developing or deploying AI in these sectors. For example, some of the most crucial water parameters for environmental and health compliance (microbiology, chemical oxygen demand, biochemical oxygen demand, etc.) must be analyzed in laboratories, resulting in lower data quantities compared to sensor data, as well as complex temporal and statistical relationships. In addition, many AI techniques require large amounts of data to reflect a certain variability in the processes to build robust models. As treatment plants usually have stable processes, this requirement implies a long data collection time (Dalibard et al., 2025; IBM, 2024).

Insufficient data can increase the risk of overfitting a model, in which the model can produce good results within the boundaries of the training data but struggles to extrapolate results beyond those boundaries or underfitting when there are not enough data. Further challenges arise from the natural drift of sensors, which can decrease data credibility with implications for the model and its interpretation. This is currently the main barrier for advancements toward self-learning systems, the ultimate vision for AI (Ziembowicz & Kida, 2022).

Choosing the right ML model for a wastewater problem is a difficult task. Several factors need to be considered, such as data typology, size and quantity of the dataset, number of features, complexity of the problem, and evaluation criteria and performance measures. Transferability of one model to another setting can be difficult as most AI methods are built on data for that particular context (process, location, etc.)

and would need comprehensive reworking to work in a new setting. Finally, the interpretability of some models can be difficult as some techniques, such as ANNs, can appear non-transparent, which might lead to false conclusions or wrong causalities (Pandey et al., 2022)

10.6.2 Limitations of Automation in Water Treatment

10.6.2.1 *Limitations in the use of the classic Fenton process in the technology of water and wastewater treatment*

The key advantages that make the Fenton process superior to other processes are its ability to degrade recalcitrant organic pollutants and perform these processes at room temperature and atmospheric pressure, and the availability of required reagents. The Fenton process has the shortest reaction time among advanced oxidation processes and can be implemented and integrated into existing water and wastewater treatments, such as coagulation, filtration, and biological oxidation. Additionally, it is effective for the inactivation of pathogenic microorganisms (e.g., hepatitis A virus and SARS-CoV-2 virus) in various types of wastewaters. The popularity of the Fenton process in organic micropollutant degradation has been owing to the lack of complicated equipment and chemical requirements and the possibility of employing it in the treatment of numerous types of wastewater; however, Fenton oxidation has some limitations that limit its industrial application (Pliego et al., 2015).

The classic Fenton reaction still has some disadvantages, including the required low pH of the reaction mixture (approx. 3). There is agreement that an acid reaction is required for the effective use of Fenton's reagent, as H^+ ions are essential for the degradation of H_2O_2. The use of a higher pH causes the concentration of H^+ ions to be too low to generate a sufficiently large amount of OH radicals. The high efficiency and usefulness of advanced oxidation largely depend on the presence and amount of ·OH radicals, which initiate a series of chemical reactions. When fully mineralized, they decompose organic compounds into carbon dioxide, water, and inorganic ions. According to the available literature on the Fenton process, at pH>4, the decomposition of H_2O_2 may occur on the flocs of the formed Fe(III) hydroxide; therefore, an appropriate amount of OH radicals is not produced (Figure 10.1). However, below pH=2.5, an excess of Fe^{3+} ions remains in the reaction medium, which not only causes rapid decomposition of H_2O_2 in accordance with reaction 1.18, and subsequently by the Fenton reaction 1.19, but also accelerates the competitive reactions 1.20–1.23, which reduces the amount of ·OH radicals involved in the oxidation of substrates (Cairone et al., 2024).

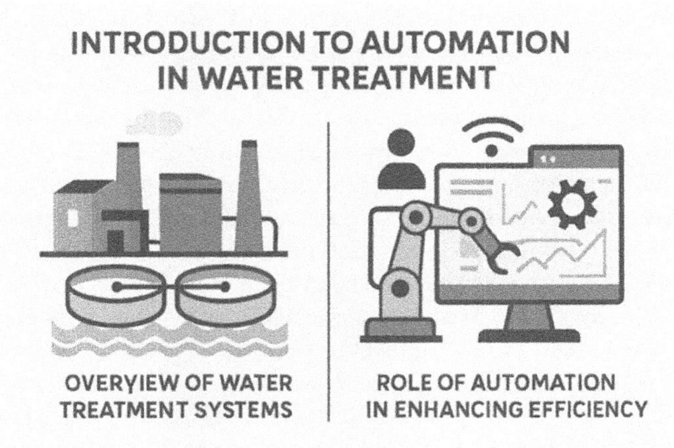

FIGURE 10.1 Key limitations of the classic Fenton process: underlying causes and potential solutions.

10.7 FUTURE TRENDS AND INNOVATION IN AUTOMATION AND PROCESS OPTIMIZATION

10.7.1 Role of Big Data and Cloud Computing in Water Treatment

Variations in influent wastewater pollutant concentrations, flow rates, process parameters, and ambient variables make wastewater treatment procedures intrinsically complex. These differences make it difficult to guarantee proper supervision and oversight of WWTP operations. Additionally, WWTPs use a lot of energy and materials. However, taking appropriate action can lead to large operational cost savings, notably in terms of energy and labor. To guarantee efficient energy use, Torregrossa et al. (2019) suggested an approach based on the analysis of daily data to identify plant energy inefficiencies. This methodology supports the design and operational stages of WWTPs. Monitoring operational conditions and optimizing WWTP performance can be achieved by analyzing both historical and real-time measurements. But a major obstacle still exists: insufficient raw data processing brought on by a shortage of equipment and/or specialized training among WWTP employees. Numerous studies have put forth novel methods based on big data analytics and cloud computing to track and evaluate wastewater treatment operations in response to these difficulties (Pandey et al., 2022).

Large datasets are gathered, processed, and analyzed as part of big data analytics, which helps operators manage these enormous volumes of data and obtain a better grasp of the state of operations. WWTP performance can be greatly improved by combining precise predictive models with big data analytics. To guarantee sufficient information and have precise and sophisticated predictive models, the raw data collected alone are insufficient. To get a high-quality dataset that can be used to extract trustworthy information, data preparation, processing, and analysis are crucial procedures. Additionally, in data-driven techniques, reducing measurement mistakes is essential. In WWTPs, calibrated sensors are usually used for online measurements. Even if the quality of these sensors has improved, a few issues, including solids deposition, biofilm formation, and calibration drift, can still affect real measurements. These factors impact data quality and, in turn, interpretation. If there are flaws in the measured data, the data analysis results will also have inaccuracies, and the information gathered might not accurately reflect the treatment process under study (Rashid & Kausik, 2024).

The ability to store data on cloud storage systems in addition to local storage devices facilitates the quick adoption of big data analytics. For data storage in a virtual environment that can be accessed remotely from any authorized device with an internet connection, Cloud Computing (CC) offers a versatile, effective, and affordable option. This method greatly accelerates and simplifies dataset exchange, which promotes teamwork and decision-making. Furthermore, cloud-based storage systems' data backup features reduce the chance of data loss from hardware failure. In WWTPs, CC can streamline data storage, facilitating effective effluent quality and process parameter monitoring and analysis without requiring the expenses of more conventional data collection techniques. Additionally, CC makes data accessible in real time, allowing operators to quickly make well-informed decisions. A CC-based data gathering system was incorporated into WWTPs. In order to track and evaluate treatment procedures, allowing for prompt access to the data gathered. In order to create a low-cost, easily adjustable, adaptable, and portable system that can continuously monitor and manage influent wastewater in a municipal WWTP and avoid adverse impacts on the treatment process, Salem et al. combined CC and IoT devices. In order to enable remote monitoring, this system gathers and sends sensor data on the temperature and pH of influent wastewater in real time to a cloud server. Additionally, it modifies inlet valves to stop wastewater with inappropriate properties from entering the plant and notifies operators of unusual events through SMS and alarms (e.g., industrial wastewater can be detected and directed to appropriate treatment plants) (Sony & Naik, 2019).

10.7.2 Advancements in AI and Automation Technologies

Existing WWTPs continue to use conventional biological wastewater treatment, especially the Active Server Pages (ASP). AI-based methods have shown promise in increasing ASP efficiency, lowering energy consumption, and improving effluent quality. Furthermore, AI-based monitoring has demonstrated efficacy in precisely recognizing WWTP flaws, including toxic shocks, sludge thickening, and sensor malfunctions, and has made it possible to pinpoint their causes, offering important insights to avert associated issues. This method included recirculated sludge flow to the aeration tank, influent temperature and flow rate, contaminant concentrations in the influent and treated effluent, and total energy consumption. The impact of various elements (and combinations of input variables) on the variance of sludge settleability in an ASP was examined by Zheng et al. using ANNs (Futia & Vetrò, 2020; Luu, 2024).

They put up a strategy for choosing suitable input variables in order to forecast these kinds of fluctuations. By preventing sludge thickening, a major issue in ASP, this method can be used to optimize the process. By modifying operating parameters, including nitrate and activated sludge recirculation rates and aeration intensity, these models enabled enhancing effluent quality and lowering the ASP's energy consumption. This offered an alternative to traditional process control based on operator expertise. In comparison to conventional regulation, where aeration is based on a preset dissolved oxygen level, they were able to reduce air consumption by 16%. In a pilot-scale WWTP, Wang et al. (2023) used an online intelligent management approach based on the integration of an Multi layer perceptron (MLP) regression model with Intelligent Attack Surface Managements (ASMs) to achieve aeration energy savings of almost 49% when compared to the manual operating mode (Sharma et al., 2021; Sundaresan et al., 2021).

Because membrane-based technologies have the potential to deliver enhanced wastewater treatment, they have garnered increased interest among treatment approaches. By enabling sustainable resource recovery and safe wastewater reuse, these technologies reduce adverse environmental effects. Membrane methods for wastewater treatment have been effectively optimized with the use of AI modeling. To aid in decision-making, s Im et al. suggested AI models for forecasting process parameters (water flux, membrane fouling, and removal efficiencies of Dissolved organic carbon (DOC), Total Nitrogen (TN), and Total Potassium (TP)) in a Forward Osmosis (FO) membrane system. Based on a sizable dataset gathered from earlier research, these models demonstrated efficacy in forecasting how operating factors will affect the osmotic membrane's ability to remove micropollutants. To optimize the removal of every micropollutant, this method also offered information for enhancing the system's operation and design. AI-based methods have been shown to be sufficient for optimizing membrane fabrication processes, in addition to enabling process optimization and control. This has resulted in the creation of membranes with beneficial properties for wastewater treatment (Sharma et al., 2021).

Considering water quality parameters including pH, turbidity, and color, AI has been used to forecast the ideal coagulant dosages for the coagulation–flocculation process. This strategy works well for maximizing the use of chemicals and has a major positive impact on social, environmental, and economic factors. Additionally, AI modeling has been utilized to optimize electrochemical processes like electro-oxidation, electrocoagulation, electro-Fenton, and electrodialysis that are employed in the treatment of water and wastewater.

10.7.3 Role of Blockchain and Robotics and Drones in Water Treatment

The facts on wastewater management are unclear because businesses all across the world provide inaccurate or missing information. Unified data, such as that provided by numerous government organizations, however, show that wastewater production management is insufficient. But keeping an eye on WWTPs is a challenging undertaking. The introduction of the Industrial IoT (IIoT)has made it possible to monitor treatment plants. One of the process's fundamental challenges is sensor-generated data storage. When treating wastewater, a variety of sensors, including titrimetric and biosensors, are used to collect data on different factors. The data are then stored in a database for analysis and monitoring. This information is highly vulnerable to hacker manipulation or misunderstanding by third parties (Satilmisoglu et al., 2024).

Blockchain technology is currently the subject of research because it may be able to solve the issue. Research examining several aspects of blockchain technology and IIoT for treatment plants has been published in recent years. The main areas of substantial study for resolving important IoT problems are the development of robust frameworks and the incorporation of blockchain technology into IIoT applications. It is suggested that blockchain technology can be used to purify wastewater produced by homes and businesses. To assess and track the trash being produced, they suggested a comprehensive method consisting of several steps. The system's ability to monitor water consumption and wastewater output in real time was one of its main advantages (Latif et al., 2021).

Chemical-sensing payloads for use in atmospheric chemistry, industrial pollution monitoring, environmental law enforcement, agriculture, chemical enterprises, and wastewater treatment facilities can now be installed on small drones. In an effort to reduce the expenses and hazards involved with walkover surveys with handheld detectors, WWTPs are now testing the use of drones. This is done in an effort to reduce the frequency of walkover surveys that need to be conducted. At WWTPs, routine odor monitoring is carried out to assess the effectiveness of odor abatement systems, identify escape emissions, and predict and reduce off-site odor effects (Burgués et al., 2021).

Compounds with offensive scents cause issues for human settlements even at low quantities. By scanning portions of the infrastructure of WWTPs for issues like cracks or a host of other issues, drones are utilized to ensure worker safety. Drones can typically record high-quality videos in a relatively short period of time. A maintenance manager may decide how to allocate available resources and take preemptive action before a disaster occurs if a drone finds serious problems. Research has been done on the effectiveness of deploying drones and robotics equipped with electrochemical sensors for detecting odorous compounds like ammonia or hydrogen sulfide to monitor odor in oil refineries and WWTPs Casino et al. (2018).

10.8 CONCLUSION

In conclusion, the integration of automation and process optimization in water treatment is transforming the sector, significantly enhancing the efficiency and sustainability of water management systems. The exploration of various technologies, from advanced sensors and IoT to ML and AI, has highlighted their crucial role in improving the reliability, cost-effectiveness, and quality of water treatment processes. The application of automated systems, including SCADA and PLCs, has streamlined operations and minimized human error, while also facilitating real-time monitoring and decision-making.

Key areas such as coagulation, flocculation, filtration, membrane technologies, disinfection, and chemical dosage control have greatly benefited from automation, resulting in enhanced precision and consistency. AI and ML tools further optimize these processes by enabling predictive maintenance, adaptive control, and continuous learning, ensuring the systems are always operating at their peak efficiency.

Looking toward the future, the role of big data, cloud computing, blockchain, and even robotics and drones is expected to grow, creating a more connected, transparent, and automated water treatment ecosystem. These innovations will provide water treatment plants with deeper insights, greater automation capabilities, and the potential for more dynamic and scalable operations.

Ultimately, as the demand for clean water continues to rise globally, the future of automation in water treatment holds great promise for providing sustainable and efficient solutions. The convergence of AI, automation, and emerging technologies will be pivotal in meeting the challenges of water scarcity, contamination, and operational efficiency, shaping a future where clean water is accessible to all.

REFERENCES

Ajiga, N. D., Okeleke, N. P. A., Folorunsho, N. S. O., & Ezeigweneme, N. C. (2024). The role of software automation in improving industrial operations and efficiency. *International Journal of Engineering Research Updates*, 7(1), 022–035. https://doi.org/10.53430/ijeru.2024.7.1.0031

Alam, G., Ihsanullah, I., Naushad, M., & Sillanpää, M. (2021). Applications of artificial intelligence in water treatment for optimization and automation of adsorption processes: Recent advances and prospects. *Chemical Engineering Journal*, 427, 130011. https://doi.org/10.1016/j.cej.2021.130011

Al-Juaidi, A. E., Kaluarachchi, J. J., and Mousa, A. I. (2014). Hydrologic-economic model for sustainable water resources management in a coastal aquifer. *Journal of Hydrologic Engineering*, 19, 04014020.

Bratby, J. (2015). *Coagulation and Flocculation in Water and Wastewater Treatment:* Second Edition. IWA Publishing. *Water Intelligence Online*, *5*(0), 9781780402321. https://doi.org/10.2166/9781780402321

Burgués, J., Esclapez, M. D., Doñate, S., Pastor, L., & Marco, S. (2021). Aerial mapping of odorous gases in a wastewater treatment plant using a small drone. *Remote Sensing*, 13(9), 1757. https://doi.org/10.3390/rs13091757

Cairone, S., Hasan, S. W., Choo, K., Lekkas, D. F., Fortunato, L., Zorpas, A. A., Korshin, G., Zarra, T., Belgiorno, V., & Naddeo, V. (2024). Revolutionizing wastewater treatment toward circular economy and carbon neutrality goals: Pioneering sustainable and efficient solutions for automation and advanced process control with smart and cutting-edge technologies. *Journal of Water Process Engineering*, 63, 105486. https://doi.org/10.1016/j.jwpe.2024.105486

Casino, F., Dasaklis, T. K., & Patsakis, C. (2018). A systematic literature review of blockchain-based applications: Current status, classification, and open issues. *Telematics and Informatics*, 36, 55–81. https://doi.org/10.1016/j.tele.2018.11.006

Chen, J., Li, G., & Wu, S. (2013). Assessing the potential of support vector machine for estimating daily solar radiation using sunshine duration. *Energy Conversion and Management*, 75, 311–318. https://doi.org/10.1016/j.enconman.2013.06.034

Dalibard, A., Kriem, L. S., Beckett, M., Scherle, S., Yeh, Y., & Schließmann, U. (2025). Challenges and requirements of AI-based waste water treatment systems. *At - Automatisierungstechnik*, 73(1), 40–49. https://doi.org/10.1515/auto-2024-0023

Futia, G., & Vetrò, A. (2020). On the integration of knowledge graphs into deep learning models for a more comprehensible AI: Three challenges for future research. *Information*, 11(2), 122. https://doi.org/10.3390/info11020122

Ghernaout, D., & Ghernaout, B. (2012). Sweep flocculation as a second form of charge neutralisation: A review. *Desalination and Water Treatment*, 44(1–3), 15–28. https://doi.org/10.1080/19443994.2012.691699

Gupta, V. K., Ali, I., Saleh, T. A., Nayak, A., & Agarwal, S. (2012). Chemical treatment technologies for waste-water recycling: An overview. *RSC Advances*, 2(16), 6380. https://doi.org/10.1039/c2ra20340e

Hand, S., & Cusick, R. D. (2021). Electrochemical disinfection in water and wastewater treatment: Identifying impacts of water quality and operating conditions on performance. *Environmental Science & Technology*, 55(6), 3470–3482. https://doi.org/10.1021/acs.est.0c06254

IBM. (2024, December 11). *Overfitting vs. Underfitting*. Ibm.com. https://www.ibm.com/think/topics/overfitting-vs-underfitting

Ihsanullah, I. (2020). MXenes (two-dimensional metal carbides) as emerging nanomaterials for water purification: Progress, challenges and prospects. *Chemical Engineering Journal*, 388, 124340. https://doi.org/10.1016/j.cej.2020.124340

Kourou, K., Exarchos, T. P., Exarchos, K. P., Karamouzis, M. V., & Fotiadis, D. I. (2014). Machine learning applications in cancer prognosis and prediction. *Computational and Structural Biotechnology Journal*, 13, 8–17. https://doi.org/10.1016/j.csbj.2014.11.005

Krupińska, I. (2020). Aluminium drinking water treatment residuals and their toxic impact on human health. *Molecules*, 25(3), 641. https://doi.org/10.3390/molecules25030641

Kumari, S., Chowdhry, J., & Garg, M. C. (2024). AI-enhanced adsorption modeling: Challenges, applications, and bibliographic analysis. *Journal of Environmental Management*, 351, 119968. https://doi.org/10.1016/j.jenvman.2023.119968

Latif, S., Idrees, Z., Huma, Z. E., & Ahmad, J. (2021). Blockchain technology for the industrial Internet of Things: A comprehensive survey on security challenges, architectures, applications, and future research directions. *Transactions on Emerging Telecommunications Technologies*, *32*(11). https://doi.org/10.1002/ett.4337

Li, L., Rong, S., Wang, R., & Yu, S. (2020). Recent advances in artificial intelligence and machine learning for nonlinear relationship analysis and process control in drinking water treatment: A review. *Chemical Engineering Journal*, 405, 126673. https://doi.org/10.1016/j.cej.2020.126673

Luu, V. (2024, Octobre 5). *10 Remarkable artificial intelligence applications in 2024. Bestarion*. Consulté 4 avril 2025, à l'adresse. https://bestarion.com/10-remarkable-artificial-intelligence-applications-in-2024/

Macedonio, F., & Drioli, E. (2019). Membrane distillation development. In Galanakis, C. M. and Agrafioti, E. (Eds.) *Sustainable Water and Wastewater Processing* (pp. 133–159). Elsevier eBooks. https://doi.org/10.1016/b978-0-12-816170-8.00005-3

Mo, F., Querejeta, M. U., Hellewell, J., Rehman, H. U., Rezabal, M. I., Chaplin, J. C., Sanderson, D., & Ratchev, S. (2023). PLC orchestration automation to enhance human–machine integration in adaptive manufacturing systems. *Journal of Manufacturing Systems*, 71, 172–187. https://doi.org/10.1016/j.jmsy.2023.07.015

Narges, S., Ghorban, A., Hassan, K., & Mohammad, K. (2021). Prediction of the optimal dosage of coagulants in water treatment plants through developing models based on artificial neural network fuzzy inference system (ANFIS). *Journal of Environmental Health Science & Engineering*, 19, 1543–1553.

Nicolaisen, B. (2003). Developments in membrane technology for water treatment. *Desalination*, 153(1–3), 355–360. https://doi.org/10.1016/s0011-9164(02)01127-x

Pandey, S., Twala, B., Singh, R., Gehlot, A., Singh, A., Montero, E. C., & Priyadarshi, N. (2022). Wastewater treatment with technical intervention inclination towards smart cities. *Sustainability*, 14(18), 11563. https://doi.org/10.3390/su141811563

Paschen, U., Pitt, C., & Kietzmann, J. (2019). Artificial intelligence: Building blocks and an innovation typology. *Business Horizons*, 63(2), 147–155. https://doi.org/10.1016/j.bushor.2019.10.004

Pliego, G., Zazo, J. A., Garcia-Muñoz, P., Muñoz, M., Casas, J. A., & Rodriguez, J. J. (2015). Trends in the intensification of the Fenton process for wastewater treatment: An overview. *Critical Reviews in Environmental Science and Technology*, 45(24), 2611–2692. https://doi.org/10.1080/10643389.2015.1025646

Rashid, A. B., & Kausik, M. A. K. (2024). AI revolutionizing industries worldwide: A comprehensive overview of its diverse applications. *Hybrid Advances*, 7, 100277. https://doi.org/10.1016/j.hybadv.2024.100277

Satilmisoglu, T. K., Sermet, Y., Kurt, M., & Demir, I. (2024). Blockchain opportunities for water resources management: A comprehensive review. *Sustainability*, 16(6), 2403. https://doi.org/10.3390/su16062403

Sharma, N., Shamkuwar, M., Kumaresh, S., Singh, I., & Goje, A. (2021). Introduction to blockchain and distributed systems: Fundamental theories and concepts. In Krishnan, S., Balas, V. E., Harold Robinson, Y., & Kumar, R. (Eds.) *Blockchain for Smart Cities* (pp. 183–210). Elsevier eBooks. https://doi.org/10.1016/b978-0-12-824446-3.00002-8

Silva, J. A. (2023). Wastewater treatment and reuse for sustainable water resources management: A systematic literature review. *Sustainability*, 15(14), 10940. https://doi.org/10.3390/su151410940

Sony, M., & Naik, S. (2019). Key ingredients for evaluating Industry 4.0 readiness for organizations: A literature review. *Benchmarking an International Journal*, 27(7), 2213–2232. https://doi.org/10.1108/bij-09-2018-0284

Sundaresan, S., Kumar, K. S., Kumar, T. A., Ashok, V., & Julie, E. G. (2021). Blockchain architecture for intelligent water management system in smart cities. In Krishnan, S., Balas, V. E., Harold Robinson, Y., & Kumar, R. (Eds.) *Blockchain for Smart Cities* (pp. 57–80). Elsevier eBooks. https://doi.org/10.1016/b978-0-12-824446-3.00006-5

Torregrossa, D., Castellet-Viciano, L., & Hernández-Sancho, F. (2019). A data analysis approach to evaluate the impact of the capacity utilization on the energy consumption of wastewater treatment plants. *Sustainable Cities and Society*, 45, 307–313. https://doi.org/10.1016/J.SCS.2018.11.036

Ullah, I., Adhikari, D., Su, X., Palmieri, F., Wu, C., & Choi, C. (2024). Integration of data science with the intelligent IoT (IIoT): Current challenges and future perspectives. *Digital Communications and Networks*, 11, 280–298. https://doi.org/10.1016/j.dcan.2024.02.007

Zhang, S., Jin, Y., Chen, W., Wang, J., Wang, Y., & Ren, H. (2023). Artificial intelligence in wastewater treatment: A data-driven analysis of status and trends. *Chemosphere*, 336, 139163. https://doi.org/10.1016/j.chemosphere.2023.139163

Ziembowicz, S., & Kida, M. (2022). Limitations and future directions of application of the Fenton-like process in micropollutants degradation in water and wastewater treatment: A critical review. *Chemosphere*, 296, 134041. https://doi.org/10.1016/j.chemosphere.2022.134041

Energy Optimization and Environmental Impact Reduction through AI

11

Ayoub Belcaid and Ayoub Ouarga

LIST OF ABBREVIATIONS

AI	Artificial intelligence
ANN	Artificial neural network
BOD	Biological oxygen demand
CCD	Central composite design
CNN	Convolutional neural network
COD	Chemical oxygen demand
CPC	Cassava peel carbon
DO	Dissolved oxygen
DT	Decision tree
Fe_3O_4	Magnetite
GA	Genetic algorithm
GRNN	General regression neural network
MB	Methylene Blue
MG	Malachite green
ML	Machine learning
MLP	Multilayer Perceptron
MSE	Mean square error
NARX	Nonlinear autoregressive exogenous
NP	Nanoparticles
RBF	Radial basis function
RMSE	Root mean square error
RSM	Response surface methodology
SM	Suspended matter

DOI: 10.1201/9781003600718-15

SVM	Support vector machine
TDS	Total dissolved solids
WW	Wastewater
WWT	Wastewater treatment
α-Fe_2O_3	Hematite
γ-Fe_2O_3	Maghemite
γ-Fe_2O_3-NP-CPC	Maghemite nanoparticle loaded on CPC biochar
$\chi 2$	Reduced chi-square

11.1 INTRODUCTION

With rapid urbanization and industry growth, huge quantities of pollution and toxic effluents are released every day directly into the environment without treatment, which negatively impacts the entire ecosystem including the fauna, flora, and soil, especially water (Anastopoulos et al., 2019; Hou et al., 2020; Rose et al., 2023). Water is a valuable and an important resource for all living beings in our planet earth, and it must be protected and well preserved for the coming generations (Kallawar et al., 2024). However, access to clean water becomes difficult due to pollution, lack of rain, and global warming. Contaminated water is a major issue all over the world, which needs effective and efficient solutions for the well-being of the environment and humans (Patel et al., 2025). Therefore, many wastewater (WW) techniques were developed to resolve such issues, for instance, biological water treatment, physicochemical treatment, and advanced treatment techniques (Karuppaiyan et al., 2023). Adsorption shows its efficiency and high-level performance for pollution removal from water; furthermore, it is an eco-friendly and economic technique with no waste generation, using biomaterials (Priya et al., 2019; Belcaid et al., 2023).

Among various materials, nanocomposites based on metal oxides and biochar originating from agricultural waste are considered a sustainable solution for pollution removal from WW (Okut et al., 2025), combined with artificial intelligence (AI)-driven predictive modeling (Belcaid et al., 2024).

Compared to traditional mathematical models, the integration of AI algorithms demonstrates high performance in optimizing the wastewater treatment (WWT) process and energy consumption (Zhang et al., 2023). Indeed, AI stands out as a new wave and an efficient tool to control the key influencing factors on WWT and energy optimization. Furthermore, machine learning (ML) using predictive models can reduce laboratory costs, workload, chemicals, and energy, space and time requirements (Heddam, 2024; Behera et al., 2024). That is why AI models, namely artificial neural network (ANN), have been adopted in this chapter (Garg et al., 2024).

This chapter aims to (i) highlight the crucial role of AI and ML in WWT applications and energy optimization, especially in the adsorption process for toxic pollutants removal by nanomaterials, (ii) investigate the efficiency of AI to predict WWT process and energy optimization via well-developed models such as ANN, central composite design (CCD), response surface methodology (RSM), support vector machine (SVM), genetic algorithm (GA), and decision tree (DT), and (iii) finally propose a perspective on challenges and research needs of AI on WWT and water sustainability (Mao et al., 2024).

This chapter gives a fresh perspective for water and WW researchers about the importance of using AI for efficient water treatment and energy optimization.

11.2 METHODS AND TECHNIQUES AVAILABLE FOR WWT

Two techniques of WWT are used across the world: biological and physical-chemical techniques (Fetisoa et al., 2024). Each method involves various and efficient processes that successfully remove pollution from water.

WW is generated from different sources, mainly urban or domestic activities and industries, for example textile, paint, mining, food, biomedical, medical, petrochemical, paper, cement, battery, fertilizer, and agricultural industries (Marcińczyk et al., 2022); these industries release huge quantities of toxic pollutants such as dyes, heavy metals, suspended solids, bacteria, pathogens, abrasive slurries, oil, grease, colloids, nutrients, sediments, and organic matter (Patel et al., 2025).

The selection of an adequate treatment technique relies on the kind of contaminants in WW, besides the WW characteristics, composition, and volume. These factors are determinants for the choice of a suitable treatment process (Kallawar et al., 2024).

11.2.1 Biological Wastewater Techniques

Organic pollution and nutrients, including phosphorus and nitrogen generated by domestic activities (Rout et al., 2021), are removed by biological WWT. This method depends on the metabolic activities of micro-organisms and encompasses large techniques, namely activated sludge, biological films, biological disks, lagoon systems, fungi treatment, algae treatment, biological membranes, and biological filters. Most of the techniques are aerobic WWT; indeed, micro-organisms use oxygen for the degradation process of organic pollutants (Fetisoa et al., 2024).

In contrast, anaerobic techniques occur in a media deprived of oxygen (lack of oxygen). This means that microorganisms transform organic pollutants into biomass, dioxide of carbon and methane (CH_4) with no requirement of oxygen. This kind of microorganisms are rarely used (Okut et al., 2025), and in such treatment, we can find anaerobic membrane bioreactor, biofilm reactor, hydrolysis, and methanation process.

11.2.2 Physical and Chemical Wastewater Treatment Techniques

Industrial WW effluents are different and more difficult to treat than domestic ones, due to their chemical properties and complex and distinctive composition (Altowayti et al., 2019). The most common chemical and physical techniques include coagulation–flocculation, filtration, sedimentation, flotation, adsorption, oxidation, precipitation, ion exchange, membrane filtration, and membrane separation, besides electrocoagulation (Okut et al., 2025) (Figure 11.1).

Table 11.1 provides comprehensive insights into all existing WW techniques associated with their typical pollution and the source of each effluent.

Standard WWT plants use both physical-chemical and biological techniques. Such systems are typically organized into five sequential phases: pretreatment, primary treatment, secondary treatment, ternary treatment, and affinity treatment (or advanced treatment) (Fetisoa et al., 2024; Nagpal et al., 2024).

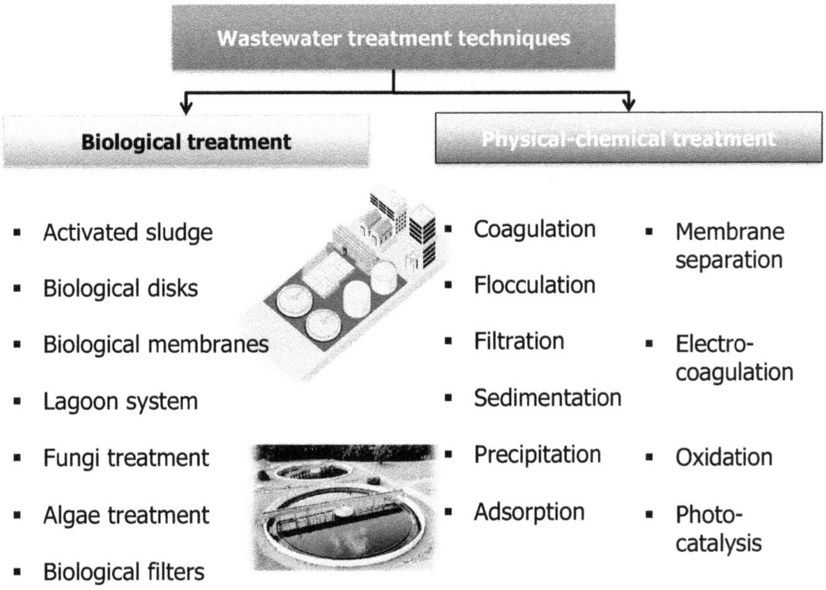

FIGURE 11.1 Different wastewater treatment techniques.

TABLE 11.1 Various wastewater treatment techniques

	INDUSTRY	*CONTAMINANTS*	*TREATMENT TECHNIQUES*	*REFERENCES*
Biological	Meat processing	Organic matter, pathogens, bacteria, suspended solids, oils, grease.	Activated sludge, biological membranes, coagulation flocculation, flotation, and adsorption.	Patel et al. (2025)
	Food processing	Organic matter, pathogens, bacteria, suspended solids, oils, grease, food additives, nutriments.	Activated sludge, biological filters, coagulation flocculation, flotation, and adsorption.	Nishat et al. (2023)
	Sugar industry	Organic matter, nutriments, suspended solids.	Activated sludge, coagulation flocculation, flotation, and adsorption.	Kallawar et al. (2024)
	Domestic wastewater	Organic matter, pathogens, bacteria, suspended solids, oils, grease, phosphor, nitrate, nitrite, azote, nitrogen, BOD, COD, phosphorous.	Activated sludge, biological disks, coagulation flocculation, flotation, adsorption, lagoon system, biological membranes, membrane separation, and oxidation.	Garg et al. (2024)
Physical-chemical	Textile	Dyes, heavy metals, suspended matter.	Adsorption, precipitation, oxidation process, photocatalysis, nanofiltration, and filtration.	Belcaid et al. (2024)
	Hospital	Pharmaceuticals, pathogeneses, bacteria, disinfectant, pigments, dyes, radioactive materials, blood, hazardous chemicals and materials, dissolved inorganic matter.	Adsorption, oxidation, photocatalysis, and anaerobic bioreactor.	Fetisoa et al. (2024)

(Continued)

TABLE 11.1 (*Continued*) Various wastewater treatment techniques

INDUSTRY	*CONTAMINANTS*	*TREATMENT TECHNIQUES*	*REFERENCES*
Paper industry	Acids, organic matter, dyes, heavy metals, sulfates.	Flotation, coagulation–flocculation, oxidation, precipitation, adsorption, and photocatalysis.	Patel et al. (2025)
Cement	Alkalinity, suspended solids, chemicals, organic compounds, sediments.	Flotation, flocculation, and sedimentation.	Okut et al. (2025)
Battery	Heavy metals, acidity, alkalinity, chemicals, hazardous chemicals.	Adsorption, oxidation, and precipitation.	Shin et al. (2024)
Mineral	Heavy metals, radioactive materials, suspended matter, sediments, hazardous and toxic chemicals, iron, zinc, copper, cadmium.	Filtration, membrane filtration, and adsorption.	Matebese et al. (2024)
Fertilizer	Nutrients, suspended solids, organic matter, heavy metals, biodegradable compounds, ammonia compounds, phosphates.	Oxidation, photocatalysis, coagulation, flocculation, biofilms, and precipitation.	Marcińczyk et al. (2022)

The pretreatment stage relies on physical and chemical methods including WW screening to eliminate small-diameter suspended and particulate matter from drainage systems. It also ensures the protection of the downstream equipment, including the piping system of the plant from damages and clogging caused by bigger objects such as paper, plastics, metal, dead animals, and rags (Matebese et al., 2024). In addition, using coagulation for rapid sedimentation of particular matters, the efficiency of the pretreatment can be improved by using advanced techniques, including the separation of liquid–liquid-like oils and greases from water by a flotation process, alongside separation methods of solid–liquid, such as electrocoagulation, electro-oxidation, and coalescence filtration, leading to cleaner pretreated water. The separation (liquid–liquid or solid–liquid) often takes place in circular clarifiers (Nagpal et al., 2024).

The primary treatment uses iron oxides, manganese dioxide, and iron chloride for the flocculation and coagulation processes, as well as precipitation to remove the suspended solids left after the pretreatment stage and to ensure clean water from suspensions. The primary treatment takes place in circular clarifiers (Nishat et al., 2023). Then, the secondary treatment or the biological treatment comes after the physical-chemical treatment for organic pollution removal using bacterial activities. We can find two types of bacteria in WWT plants: free bacteria that can move easily into clarifiers, with a high degree for organic pollution degradation; this kind of treatment is called activated sludge; the second type of bacteria are fixed on biofilms or biofilters; the polluted water pass vertically through the biofilters besides biological discs and get out clean from pollution. Often, in the secondary treatment, we can find rectangular clarifiers. Nevertheless, the biological treatment is not sufficient to eliminate micropollutants and bad smells from WW; that is why it is important to proceed with a final treatment (affinity or ternary treatment) such as chlorination, membrane filtration, activated carbon filtration, ozonation, and UV treatment (de Araujo et al., 2022).

11.3 ADSORPTION

Adsorption is among the best techniques for the separation of pollution, in the form of either liquid or gas, in the field of chemical and environmental engineering. It is a treatment technique that adopts surface phenomena; the pollutants are attached to the surface of solid material, resulting in the appearance of numerous chemical reactions. Adsorption is also the ability of a certain number of solid, organic, mineral, or inorganic materials to concentrate the molecules of an aqueous solution on their surface, thus allowing their separation and recovery according to various chemical or physical processes. The solid material is called an adsorbent, and the adsorbed solution is called an adsorbate (Borah et al., 2015).

Adsorption technology is widely used in various fields for the treatment and separation of liquids and gas pollution, including the petrochemical industry, the pharmaceutical industry, the textile industry, the food industry, the mining industry, and the environmental sector (WWT, drinking water treatment) (Wang et al., 2022).

11.3.1 Adsorption Isotherm for Single System

For a single system, the sorption study is performed by fitting data to three models: Langmuir, Freundlich, and Redlich–Peterson. The Langmuir model is applied for monolayer sorption onto a homogenous surface with identical active sites; on the other hand, the Freundlich model is appropriate for multilayer sorption onto a heterogeneous surface, while the Redlich–Peterson model is a combination of Langmuir and Freundlich models. The three models can be expressed as follows, respectively:

$$Q_e = \frac{Q_{\max} K_L C_e}{1 + K_L C_e} \tag{11.1}$$

$$Q_e = K_F C_e^{1/n} \tag{11.2}$$

$$Q_e = \frac{K_R C_e}{1 + \alpha C_e^{\beta}} \tag{11.3}$$

Q_e (mg/g) signifies the sorption amount at equilibrium, $Q_{\max}$ (mg/g) is the maximum sorption capacity, C_e (mg/L) represents the dye concentration at equilibrium, the terms K_L, n are Langmuir constant, alpha is freundlich contant, while β is Redlich–peterson constant. This can be corrected as Langmuir (K_L, n), Freundlich (α) and Redlich–Peterson constants (β); when β value is equal to 1, the Redlich–Peterson model is reduced to the Freundlich model; when β is equal to 0, the model is reduced to the Langmuir model. Besides, n is an indicator of sorption intensity and measures nanomaterial surface heterogeneity; if the value of $1/n$ gets closer to 0, it means that the surface is heterogeneous.

R_L is a dimensionless parameter related to the Langmuir model and determines whether sorption is favorable ($0 < R_L < 1$) or unfavorable ($R_L > 1$). This parameter is defined by the following equation:

$$R_L = \frac{1}{1 + K_L C_0} \tag{11.4}$$

11.3.2 Adsorption Isotherm for the Binary System

The sorption isotherms for the two systems are presented by two models: extended Langmuir and extended Freundlich, expressed as follows, respectively:

$$Q_{ei} = \frac{Q_{\max i} K_{Li} C_{ei}}{1 + \sum_{j=1}^{2} K_{Lj} C_{ej}} \tag{11.5}$$

For $i,j = 1,2$ with $i \neq j$

$$Q_{e1} = \frac{Q_{\max 1} K_{L1}}{1 + K_{L1} C_{e1} + K_{L2} C_{e2}} \tag{11.6}$$

$$Q_{e2} = \frac{Q_{\max 2} K_{L2}}{1 + K_{L1} C_{e1} + K_{L2} C_{e2}} \tag{11.7}$$

$$Q_{ei} = \frac{K_{fi} C_{ei}^{\left(\frac{1}{n_i}\right) + x_i}}{C_{ei}^{x_i} + Y_i C_{ej}^{z_i}} \tag{11.8}$$

For $i,j = 1,2$ with $i \neq j$

$$Q_{e1} = \frac{K_{f1} C_{ei}^{\left(\frac{1}{n_1}\right) + x_1}}{C_{e1}^{x_1} + Y_1 C_{e2}^{z_1}} \tag{11.9}$$

$$Q_{e2} = \frac{K_{f2} C_{e2}^{\left(\frac{1}{n_2}\right) + x_2}}{C_{e2}^{x_2} + Y_2 C_{e1}^{z_2}} \tag{11.10}$$

$Q_{e.}$ and $Q_{\max i}$ are the equilibrium and maximum adsorption capacities (mg/g), respectively, C_{ei} and C_{ej} (mg/L) are the initial concentrations at equilibrium, n is the intensity parameter, and the terms K_L, K_F, x_i, y_i, and z_i represent the model constants.

11.3.3 Adsorption Thermodynamic (Energy)

To determine the spontaneity of the sorption process, three thermodynamic parameters are employed: Gibbs free energy change (ΔG) (kJ/mol), entropy change (ΔS) (kJ/mol.K), and enthalpy change (ΔH) (kJ/mol) calculated using van't Hoff plot.

$$\Delta G = -RT \ln K_d \tag{11.11}$$

$$K_d = \frac{Q_e}{C_e} \tag{11.12}$$

$$\ln K_d = \frac{-\Delta H}{RT} + \frac{\Delta S}{R} \tag{11.13}$$

$$\Delta G = -T\Delta S + \Delta H \tag{11.14}$$

R is the ideal gas constant (8.314 J/mol.K), K_d is a constant (L/mg), and T is the temperature (K).

11.4 TYPES OF ADSORPTIONS

There are two main categories of adsorption: physisorption and chemisorption. The difference between these two types is the energy emitted and brought into play by the material during the adsorption process.

11.4.1 Physisorption

Physisorption results from the forces of attraction between the material and the adsorbate and involves low-energy forces, and the adsorption energy is estimated between 10 and 15 Kcal/mol, with a low adsorption heat.

Physisorption is perfectly reversible, so desorption is a complete operation because of the low energy involved, and it takes place without modifying the molecular structure of the material (Wang et al., 2022).

11.4.2 Chemisorption

Chemisorption is a chemical interaction characterized by an exchange of electrons and the creation of ionic or covalent bonds between the functional groups of the adsorbent and the adsorbate. This type of adsorption is done in a monolayer or monomolecular layer. Chemisorption is characterized by strong interactions between the surface of the adsorbent and the molecules of the adsorbate, high selectivity. Moreover, it involves forces with high energy level in the order of 100 Kcal/mol besides high heat of adsorption (Hou et al., 2020).

11.5 TYPES OF ADSORBENTS

11.5.1 Activated Carbon

Activated carbon is a porous solid material in powder form, with highly developed adsorption capacity. Activated carbons have a carbonaceous black color without odor or taste, insoluble in aqueous solutions. They can be used in WW purification, drinking water, WW deodorization, adsorption treatment of toxic gases, and other industrial uses. Activated carbon is produced either from a solid organic material of plant origin rich in carbon, such as the peel of agricultural waste, banana peels, potato peels, coconuts, and olive stones, or from a solid material of mineral origin such as clay, peat, silica gels, alumina, and coal (Anastopoulos et al., 2019).

11.5.2 Nanomaterial

Nanotechnology is the science that studies infinitely small elements at the atomic scale, including characterization, design, production, and demonstration of applications and systems controlling adsorbent size at the nanoscale. It is a field of research and development of materials and structures with a nanometric size ranging from 1 to 100 nm. Nanotechnology encompasses a range of multidisciplinary sciences with chemical, physical, and biological concepts (Qu et al., 2023).

The interest in nanomaterials is due to their structural properties, which make them a very interesting research topic with several technical, economic, and environmental advantages. Nanomaterials are utilized in several industrial fields (chemistry, biology, and physics). Given their size and dimensions similar to several biological elements such as viruses, membranes, bacteria, and proteins, they are recommended to be studied and used in biological aspects to eliminate and treat several diseases.

11.5.3 Iron Oxide

Iron oxide-based nanoparticles exist in different forms, magnetite (Fe_3O_4), hematite (α-Fe_2O_3), and maghemite (γ-Fe_2O_3) being the most used. These three forms of iron oxides are the most abundant in the soil and have a common property—magnetism. This magnetic property is influenced by particle size (Qu et al., 2023).

11.5.4 Nanocomposites

Nanocomposites result from the combination of nanoparticles and organic material. The incorporation of nanoparticles into the organic material gives it several properties, such as mechanical strength, flexibility, and rigidity, besides thermal and magnetic properties.

We can find various nanocomposites: manganese dioxide nanoparticles, iron oxide nanoparticles, and zinc oxide nanoparticles loaded on biochar originating from agricultural waste (Lin & Chen, 2020).

11.6 ENVIRONMENTAL BENEFITS OF AI AND ML IN WASTEWATER TREATMENT

Most WWT plants adopt traditional methods such as laboratory experiments, data analysis, and inspection procedures to interpret results of water quality evaluation. Unfortunately, these methods have some inconveniences, including high costs, high errors, slow response time, high energy consumption, gas emissions, and waste generation. AI and ML are efficient tools to optimize, monitor, predict, and discover uncertainty with errors in complicated real-time systems. AI/ML can model the reduction of environmental impact through the optimization of sorption process parameters for the uptake of toxic dyes from WW. This includes monitoring and reducing pollution parameters such as biochemical oxygen demand (BOD), chemical oxygen demand (COD), total dissolved solids (TDS), microbial activity, nutrients, heavy metals, and dyes. Synthetic dyes are among the organic pollutants that have garnered significant attention due to their harmful effect on the environmental ecosystem. They affect light penetration to water, thus affecting the microbiological activity (Soliman et al., 2019). Employing AI models (ANN, GA, MLP, and others) is efficient to predict and optimize the treatment of WW containing toxic dyes. AI algorithms can also be used to forecast the removal of numerous inorganic pollutants, such as heavy metals, salts, nitrates, minerals, phosphates, and chlorides. These pollutants eliminate the microorganisms needed for the biological water treatment in WWT plants.

On the other hand, excessive nutrient discharges into the environment lead to water contamination and eutrophication, thus adversely affecting the aquatic system. That is why ANN and SVM have been used to predict the removal of phosphorus and nitrogen from WW, aiming for excellent and stable water quality (Nagpal et al., 2024).

11.7 AI IN WASTEWATER TREATMENT

AI is a similar process to human intelligence via machines; AI requires ML, which is a part of AI. ML generally uses a machine to resolve a problem and a complex system via data obtained from multiple resources. The fundamental principle of ML is to employ algorithms and equations to create a relationship between the input and output data, to make a decision for an optimal solution (Gupta et al., 2024).

ML consists of three stages; training, validation, and testing. The training stage is for adjusting all the parameters of the suitable models for the system, including the input data and their interactions; the validation stage is the identification stage of the best representative model that can be used to explore the adequate solution for the system (Gupta et al., 2024; Garg et al., 2024); finally, the chosen optimum model is tested, and its results are compared with an experimental model. Based on this comparison and error parameters, the developed model can be utilized for prediction.

11.8 ENERGY OPTIMIZATION IN WASTEWATER PLANTS USING AI

AI is also used for energy optimization in WW plants, specifically in the aeration process (biological treatment), pumping requirements, decantation, chemical dosing (coagulation and flocculation, chlorination), reagents, material waste, filtration, membrane filtration, fluids' speed, fluids' flow rate, flotation process (air requirement), oxidation, membrane cleaning, and adsorption process. In such a strategy, we can increase the performance of the WW plant besides the reduction in gas emissions. The application of AI helps in reducing energy consumption by 32% and air needs by 15% without affecting the effluent quality (Learning et al., 2023). Other studies similarly confirmed the successful use of AI in reducing energy consumption in WW plants. For example, Zhang et al. (2023) confirmed the reduction of energy in the range of 10%–25% using an ANN model. AI anticipates potential equipment failures and enables proactive maintenance of all equipment. Further in the energy management system, AI identifies the peak demand periods and regulates operations to optimize equipment, process, and reagent usages during these periods (Zhang et al., 2023).

11.8.1 Biogas Production Control by AI in Wastewater Treatment Plant

WW plants generate a significant quantity of biogases used in electricity production, such as CH_4 and carbon dioxide (CO_2). The gaseous emission can be controlled, predicted, and estimated using AI algorithms. Herein, some studies reported the utilization of AI in WW plant gas emission. Mihály et al. (2023) studied the application of an ANN-based model to predict CH_4 production; Adibimanesh et al. (2023) reported the use of both ANN and SVM models separately in CH_4 and CO_2 generation. Moreover, other studies combined ANN with GA to efficiently control CH_4 production in a UASB (Upflow Anaerobic Sludge Blanket) bioreactor (Mihály et al., 2023).

11.8.2 AI and ML Models in Energy Consumption of Pumps and Effluent Flow Rate

Among AI algorithms, the ANN model achieved excellent accuracy (less than 3% error) in the energy consumption of pumps and effluent flow rate. The ANN model not only enhanced sewage pumping

TABLE 11.2 Example of AI applications in WWT and energy optimization

WATER TREATMENT APPLICATIONS	*EMPLOYED AI/ ML*	*DETAILS*	*REFERENCES*
Organic matter degradation using microalgae	DT algorithms	Microalgae cultivation enhances wastewater treatment using different classes of algae; the optimization of the cultivation parameters was done using AI	Oruganti et al. (2023)
Degradation and the removal of phosphor and azote (nutriments) from wastewater	ANN and SVR	Prediction of nutrient removal by identifying the key controlling parameters and their interaction using ANN and SVR algorithms	Cairone et al. (2024)
CO_2 production in WWT plant	GA associated with ANN	ANN and GA were both used to enhance CO_2 productivity by biomass, such as light intensity, pH, and temperature	Sanjeevi et al. (2024)
Municipal wastewater	GA, ANN, DT	Optimization of biological WWT parameters, BOD, COD, DO, TDS, pH, P, N	Sanjeevi et al. (2024)
Textile wastewater	AdaBoost	COD, NaCl concentration, and time were controlled by the AdaBoost AI algorithm	Cairone et al. (2024)
Combined municipal and industrial WWT plant	RBF and ANN	Influent BOD, COD, DO, pH, NH_4^+, N, P, and TDS	Cairone et al. (2024)
Membrane bioreactors for municipal wastewater treatment	CNN, Dense net	pH, BOD, COD, biogases, DO production, methane, CO_2	N. K. Singh et al. (2023)
Bed biofilm reactor for municipal wastewater treatment	ANN	Predicting nutrient removal such as P, N, and effluent concentration	N. K. Singh et al. (2023)
Coagulation flocculation for industrial wastewater treatment	ANN, GA, RSM, MLP, GRNN	Controlling the coagulant and flocculant doses such as sodium sulfate and ferrous sulfate	Safeer et al. (2022)
Gas emission in the nitrification reactor	GA, NARX	Controlling the CO_2 production by GA and NARX models	Mihály et al. (2023)

systems but also maintained optimal pumping performance with minimal energy consumption (Zhang et al., 2024). The use of AI shows potential for minimizing energy consumption in the range of 10%–25% in WW pumping operations (Table 11.2).

11.9 DIFFERENT AI AND ML MODELS

11.9.1 Artificial Neural Network (ANN)

ANN consists of neurons, which are the basic elements besides layers. Generally, three main layers are used: the first layer introduces inputs or independent data (X) into the model; the second or hidden layer processes the data; and the third layer shows outputs or responses of the ANN model (Y). Each layer contain neurons. Moreover, in the ANN, there is bias (b) that transforms information among neurons and links the neurons in each layer. The weights are the connection between links. Besides, each neuron is related to activation or transformation functions (linear and nonlinear functions) that convert inputs into outputs (response). Sigmoid activation function is used for hidden neurons (Sanjeevi et al., 2024), while Purelin linear function is used for neurons in the first and output layers. These functions are expressed as follows:

Purelin linear function

$$f(x) = x \tag{11.15}$$

Sigmoid function

$$f(x) = \frac{1}{1+e^{-x}} \tag{11.16}$$

The output response (neuron) is expressed as follows:

$$a_i^n = F_n\left(\sum_j W_i^n a_j^{n-1} + b_i^n\right) \tag{11.17}$$

where

a^n: the neuron corresponding to layer n,
b^n: the bias vector of layer n,
W^n: the weight connection between the layer $(n - 1)$ and n,
F_n: the activation function of layer n.

To validate the significance of the developed model and compare the predicted values with experimental values, several measurement criteria were performed such as (R^2), (R^2_{adj}), RMSE, and MSE, which are expressed as follows:

$$R^2 = 1 - \sum_{i=1}^{n} \frac{(y_{i,\text{pred}} - y_{i,\text{exp}})^2}{(y_{\text{ave,exp}} - y_{i,\text{exp}})^2} \tag{11.18}$$

$$R^2_{\text{adj}} = 1 - \left((1 - R^2)\frac{n-1}{n-K-1}\right) \tag{11.19}$$

$$\text{RMS} = \sqrt{\frac{\sum_{i=1}^{n} (y_{i,\text{pred}} - y_{i,\text{exp}})^2}{n}} \tag{11.20}$$

$$\text{MSE} = \frac{1}{n}\sum_{i=1}^{n} (y_{i,\text{pred}} - y_{i,\text{exp}})^2 \tag{11.21}$$

where n is the total number of tests, $y_{i,\text{pred}}$ and $y_{i,\text{exp}}$ are the predicted and experimental values, respectively, y_{ave} is the average of experimental values, and K is the number of input variables.

11.9.2 CCD and RSM

CCD is a statistical and mathematical approach applied to assess the interactions between the input parameters and the related response and to minimize the costs of experiments, besides defining the optimal process conditions in removing pollution from WW. Hence, the commonly adopted equation for CCD is described as follows:

Factors	Ci	Ads. Dose	T°
A	-1	-1	-1
B	+1	-1	-1
C	-1	+1	-1
D	+1	+1	-1
E	-1	-1	+1
F	+1	-1	+1
G	-1	+1	+1
H	+1	+1	+1

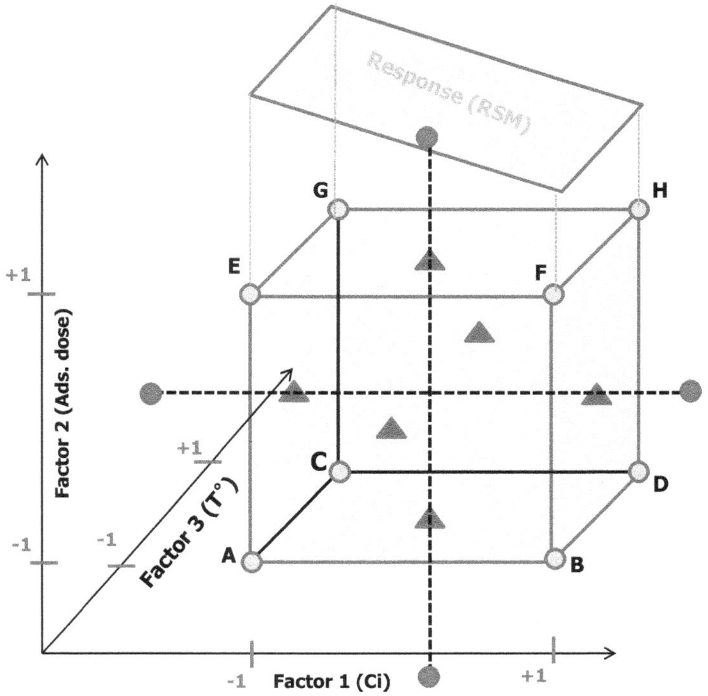

FIGURE 11.2 Central composite design (CCD) model.

$$Y = a_0 + \sum_{i=1}^{k} a_i X_i + \sum_{i<j}^{k} a_{ij} X_i X_j + \sum_{i=1}^{k} a_{ii} X_i^2 + \varepsilon \tag{11.22}$$

where Y represents the response, X_i and X_j stand for independent variables or factors, a_0, a_i, a_{ii}, and a_{ij} are the zero-order, linear (first-order), quadratic (or second-order), and interactive coefficients, respectively, and ε indicates the experiment error.

CCD is often associated with RSM (S. Singh et al., 2024) to plan and model experimental tests of all process factors (Figure 11.2).

11.9.3 Support Vector Machine

SVM is an ML algorithm for classification and regression analysis. It uses a nonlinear mapping function to transform input data into coordinates in a high-dimensional space (n-dimensional space) where linear classification can be performed. Most of the input data in real experiments are not linear. SVM makes these data linear and separable (Figure 11.3). To minimize error and optimize data manipulation, SVM utilizes many kernel functions, including polynomial kernel, RBF kernel, and sigmoid kernel.

11.9.4 Genetic Algorithm in Wastewater Treatment

GA is based on the natural concept of genetic evolution through selection, crossover, and mutation processes. The first step is the selection operation, which is an essential step in identifying the best individuals among a variety of bacterial populations by measuring their fitness in the case of biological WWT process. This step is followed by the crossover step, which involves the combination of two different genes

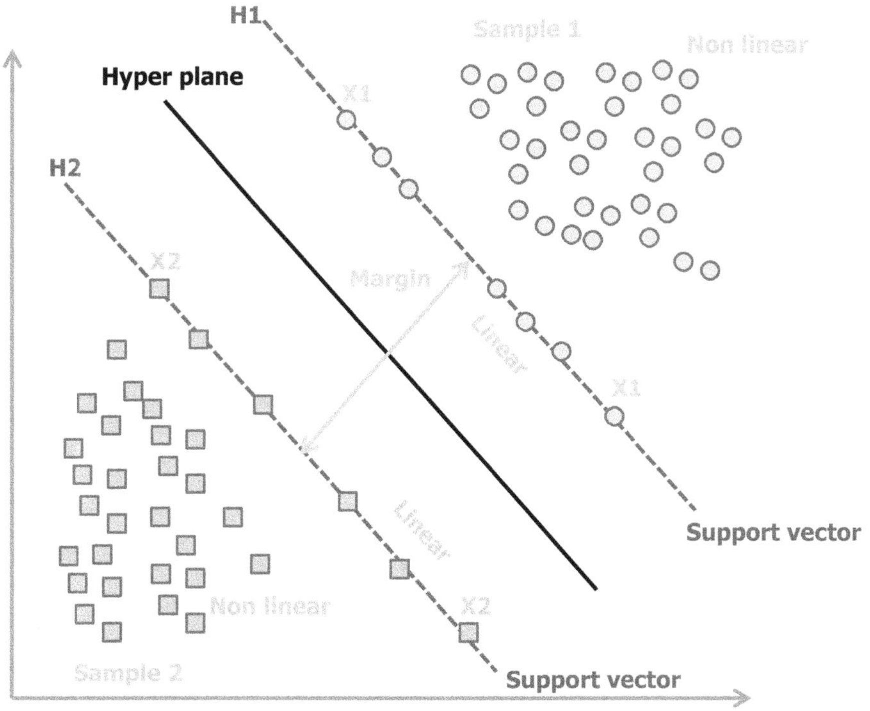

FIGURE 11.3 SVM model.

from the selected individuals (two different parents) to produce a new efficient bacteria with new genes. The final step is the mutation operation, which aims to integrate new genes in some bacterial population randomly (Figure 11.4). Eventually, the three steps are completely achieved till we find the best bacteria with high biological performance also can easily and efficiently reduce the BOD, and COD and remove the organic matter from WW (Salari et al., 2024).

11.9.5 Decision Tree

DT adopts the tree concept. In this model, all data are set into a tree structure and then subdivided into smaller sub-datasets to reduce variability. DT is a simple algorithm that is easy to implement. However, the model becomes complex and overfits as the tree grows deeper. To mitigate and limit tree depths (Learning et al., 2023), techniques such as regularization and pruning are used.

11.10 APPLICATION OF AI FOR DYE ADSORPTION IN WWT

AI is an efficient tool for optimizing and predicting dye sorption from WW. It is a key in designing and controlling adsorption, leading to an advanced WW system aligned with sustainable development principles.

This section examines the application of AI in MB and MG sorption, with an emphasis on (i) the experimental design of adsorption using ANN and (ii) modeling results.

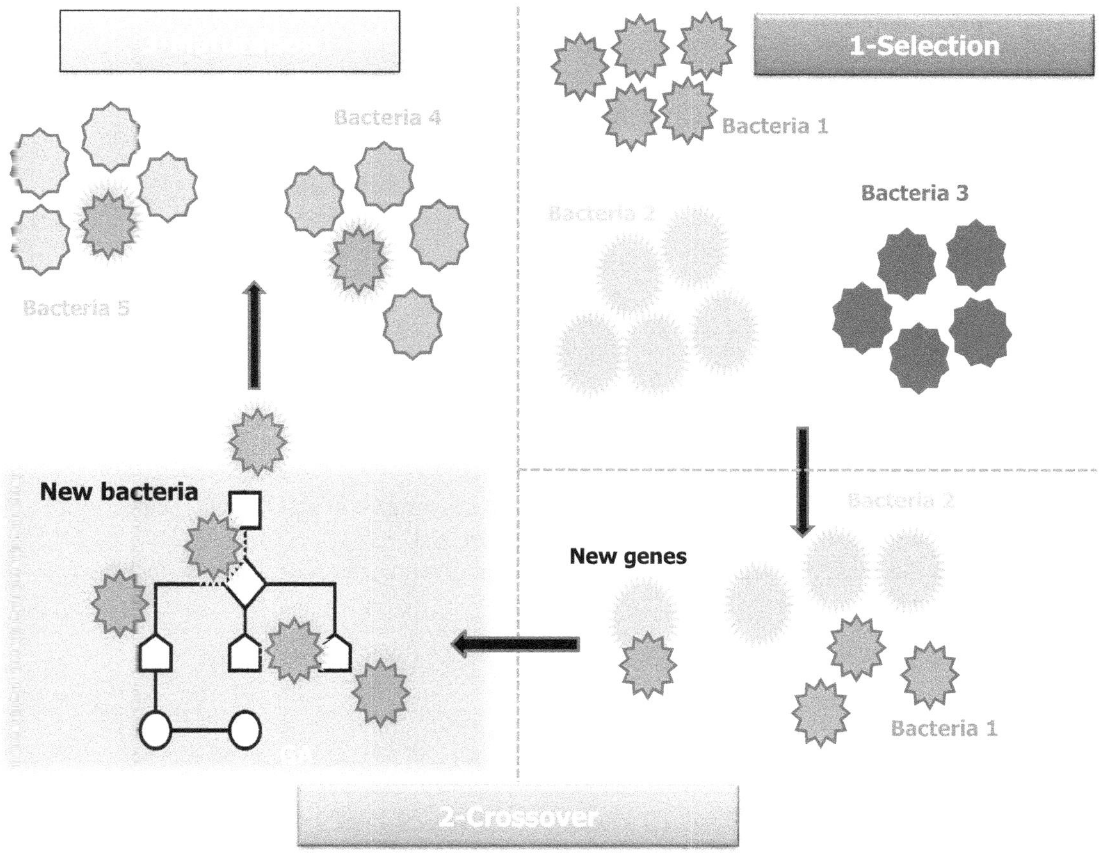

FIGURE 11.4 Genetic algorithm (GA) model.

11.10.1 Adsorption Measurements

Adsorption tests were carried out using a synthetic solution of MB and MG dyes in single (MB, MG), binary (MB+MG), and ternary (MB+MG+Rhodamine B (RB)) systems. The influence of various parameters was studied; indeed, in a single system, the initial concentration effect was determined using a constant mass of 20 mg from γ-Fe_2O_3-NP-CPC nanomaterial mixed separately with a known concentration of each dye ranging from 10 to 30 mg/L in 200-mL Erlenmeyer flasks, under continuous agitation of 300 rpm in a double jacket beaker at room temperature for 3 hours. The time effect was then examined at a concentration of 20 mg/L of MB and MG mixed separately with 20 mg of the nanomaterial in 200-mL Erlenmeyer flasks at a temperature of 25°C and a time ranging from 5 to 180 minutes. The mixture was agitated at 300 rpm, while the temperature effect was achieved at a range of 15°C, 20°C, 25°C, 30°C, and 35°C, keeping the other parameters fixed (nanomaterial dose 30 mg, time 3 hours, agitation 300 rpm). Finally, the pH effect (2–8) and adsorbent mass effect (5, 10, 15, 20, 25 and 30 mg) were completely studied in single dye system.

In the binary system (MB+MG), the effect of initial concentration was investigated by mixing 20 mg of the nanomaterial with different concentrations of dye solutions: 5+5, 7.5+7.5, 10+10, 15+15 mg/L, in a 200-mL Erlenmeyer flask under continuous stirring of 300 rpm for 3 hours. The temperature effect was achieved at various temperatures: 5°C, 10°C, 15°C, 20°C, 25°C, 30°C and 35°C, using a fixed adsorbent mass 5 mg and a dye concentration of 10+10 mg/L mixed in a 200-mL flask under magnetic stirring of 300 rpm for 3 hours.

The dye concentrations were measured using a UV–visible spectrophotometer.

In a single system, MB and MG concentrations were calculated by measuring their molar absorptivity at 663 and 621 nm, respectively, and Q_e (mg/g) of MB and MG adsorbed onto γ-Fe_2O_3-NPs-CPC nanomaterial at equilibrium and at time t Q_t (mg/g). Adsorption efficiency (%) η was calculated by the following equations:

$$\eta = \frac{(C_0 - C_e)\times 100}{C_0} \tag{11.23}$$

$$Q_e = \frac{(C_0 - C_e)\times V}{W} \tag{11.24}$$

$$Q_t = \frac{(C_0 - C_t)\times V}{W} \tag{11.25}$$

where C_0 and C_e (mg/L) are the initial concentration and concentration at equilibrium, respectively, V (L) is the solution volume, and W (g) is the mass of the nanomaterial.

Concerning the binary mixture, the concentrations of components A and B were measured by the following equations:

$$C_A = \frac{K_{B2}A_1 - K_{B1}A_2}{K_{A1}K_{B2} - K_{A2}K_{B1}} \tag{11.26}$$

$$C_B = \frac{K_{A1}A_2 - K_{A2}A_1}{K_{A1}K_{B2} - K_{A2}K_{B1}} \tag{11.27}$$

where C_A and C_B are the concentrations of components A and B; K_{A1}, K_{B1}, K_{A2}, and K_{B2} are the calibration constants for components A and B at maximum absorbance wavelengths λ_{1max} and λ_{2max}; and A_1 and A_2 represent the absorbance at wavelengths λ_1 and λ_2, respectively.

Q_{ei} (mg/g) of (i) dye adsorbed onto γ-Fe_2O_3-NPs-CPC nanomaterial at equilibrium and at time t, Q_{ti} (mg/g), and η_i (%) were calculated by the following equations:

$$\eta_i = \frac{(C_{0i} - C_{ei})\times 100}{C_{0i}} \tag{11.28}$$

$$Q_{ei} = \frac{(C_{0i} - C_{ei})\times V}{W} \tag{11.29}$$

$$Q_{ti} = \frac{(C_{0i} - C_{ti})\times V}{W} \tag{11.30}$$

where C_{0i} and C_{ei} (mg/L) are the initial concentration and concentration at equilibrium, respectively, of (i) dye in the binary mixture, V (L) is the solution volume, and W (g) is the amount of the nanomaterial.

11.10.2 Experimental Design of Adsorption Process Using AI

11.10.2.1 A-ANN

ANN was investigated for predicting MB and MG removal by γ-Fe_2O_3-NP-CPC nanomaterial. The network utilized in this research compromises five inputs (neurons): pH ($X1$), contact time ($X2$), initial

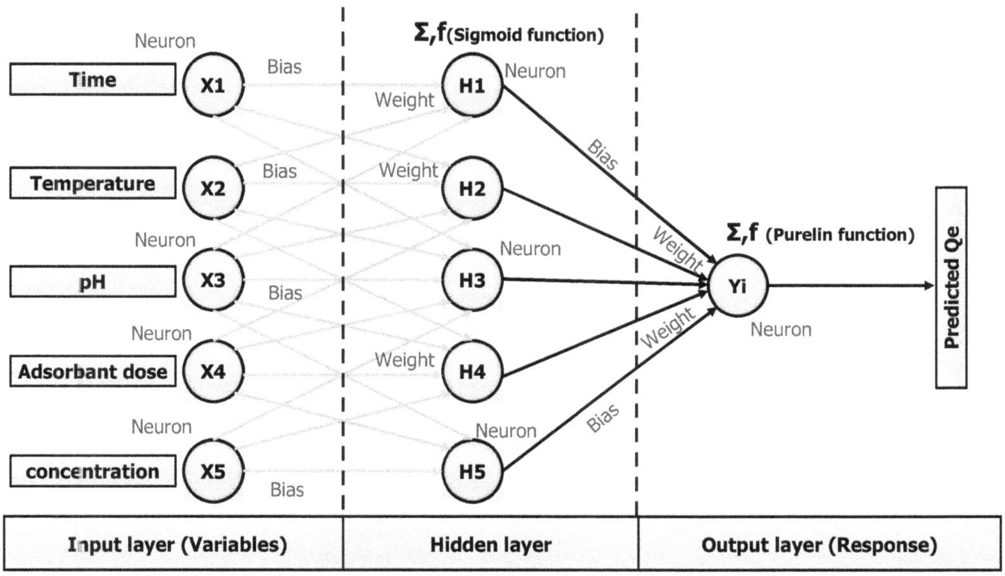

FIGURE 11.5 ANN model architecture for MB and MG sorption. (Belcaid et al., 2024.)

concentration ($X3$), adsorbent dose ($X4$), and temperature ($X5$) in the input player; five neurons in the second layer; and one neuron in the output layer (Y), which is Q_e (mg/g). Figure 11.5 shows the architecture of the constructed model for dye sorption through ANN.

11.11 OPTIMIZATION RESULTS

11.11.1 Adsorption Isotherm for Single Dye System

Table 11.3 demonstrates that the Langmuir model presents the best isotherm model with a very high R^2 (0.97) and a lower chi-square χ^2 (0.47) compared to the Freundlich model, indicating monolayer sorption on the nanomaterial surface. The experimental values of $Q_{\max}$ for MB and MG removal were 290.80 mg/g and 338.98 mg/g, respectively; besides, the maximum sorption capacities $Q_{\max}$ of MB and MG dyes were 535.27 and 440.79 mg/g, respectively (Ray et al., 2019). These values were close to the experimental ones.

11.11.2 Adsorption Isotherm for Binary Dye System

The extended Langmuir and extended Freundlich parameters are summarized in Table 11.4. It is clear from this table that the extended Langmuir model describes well the isotherm sorption in the binary dye

TABLE 11.3 Adsorption isotherm of MB and MG in a single system

	FREUNDLICH MODEL				*LANGMUIR MODEL*			
	K_F (MGL/G)	*N*	*X^2*	*R^2*	*Q_{MAX} (MG/G)*	*K_L (L/MG)*	*X^2*	*R^2*
MB	312.11	1.68	0.47	0.95	535.27	138.51	0.31	0.97
MG	300.53	2.56	0.58	0.95	440.79	219.83	0.13	0.94

TABLE 11.4 Adsorption isotherm of MB and MG in binary systems

	EXTENDED FREUNDLICH MODEL (FOR BINARY PARAMETERS)							EXTENDED LANGMUIR MODEL (FOR BINARY PARAMETERS)				
	X_I	Y_I	Z_I	K_{FI}	N_I	X^2	R^2	Q_{MAX} (MG/G)	K_{L1} (L/MG)	K_{L2} (L/MG)	X^2	R^2
MB (MB+MG)	1.39	0.43	0.59	436.9	3.55	10.35	0.98	406.81	0.017	1.57	0.59	0.99
MG (MG+MB)	0.31	1.21	0.042	312.49	2.69	0.29	0.90	182.59	5.84	2.67	0.37	0.93

TABLE 11.5 Adsorption thermodynamic of MB and MG

PARAMETERS	ΔH_{ADS}	ΔS_{ADS}										
UNITS	(KJ/MOL)	(KJ/MOL)	ΔG_{ADS} (KJ/MOL)					Q_E (MG/G)				
Temperature			15	25	35	45	50	15	25	35	45	50
MB	22.38	0.11	−1.29	−1.41	−1.72	−1.85	−2.23	287.33	290.85	297.12	298.12	299.50
MG	30.62	0.15	−1.27	−1.36	−1.41	−1.52	−1.56	285.95	289.12	290.27	293.11	293.52

system with R^2 values of 0.98 and 0.93 for MB(MB+MG) and MG(MB+MG), respectively; therefore, the sorption process was chemisorption in nature. Furthermore, Q_{max} capacities were 406.81 and 182.59 mg/g for MB(MB+MG) and MG(MB+MG), respectively.

11.11.3 Adsorption Thermodynamics for Single Dye System (Energy)

Table 11.5 shows the values of ΔH and ΔS, which are estimated from the slopes and intercepts of ln K_d versus $1/T$, besides the values of ΔG. It is clear that ΔH and ΔS values were positive, while ΔG were negative, demonstrating an endothermic and a spontaneous adsorption mechanism for both dyes with an increase in randomness at the solid/liquid interface during the sorption of MB and MG dyes onto γ-Fe_2O_3-NP-CPC nanomaterial.

11.11.4 Modeling with ANN

a. **Modeling with ANN for Single Dye Sorption:** The ANN model applied in our research had one input layer with five variables corresponding to temperature, reaction time, initial dye concentration, adsorbent dose, and pH. The hidden layer includes five neurons and one output represented by the adsorbed amount Q_e of MB and MG in the third layer, with sigmoid function as transfer function. The predicted values of MB and MG removal using the ANN model are shown in Table 11.6, which are much closer to the experimental values. Moreover, the R^2 and R^2_{adj} values for both dyes in Tables 11.8 and 11.10 were close to 1 and were 0.95 for MB and 0.97 for MG. Furthermore, the RMSE values of MB and MG were equal to 9.56 and 7.19, respectively, indicating the significance of the ANN model to predict the removal of the dye by γ-Fe_2O_3-NP-CPC nanomaterial in a single dye system.

   Table 11.9 demonstrates the applicability of the ANN model for the sorption of MB and MG at various pH values.
b. **Modeling with ANN for Binary Dye Sorption:** Tables 11.7 and 11.8 show higher R^2 values for both binary mixtures, of the order of 0.98 and 0.99 for MB (MB+MG) and MG (MG+MB), respectively, with lower values of RMSE in the order of 6.17 for MB (MB+MG) and 0.71 for

TABLE 11.6 ANN results of MB and MG sorption in a single system

	INDEPENDENT VARIABLES INPUTS					Q_E (MG/G) OF MB OUTPUT		Q_E (MG/G) OF MG OUTPUT	
NO. TESTS	TIME (MINUTES)	CONCENTRATION (MG/L)	PH	T (°C)	ADSORBENT DOSE (MG)	EXPERIMENTAL VALUES (MG/G)	PREDICTED VALUES (MG/G)	EXPERIMENTAL VALUES (MG/G)	PREDICTED VALUES (MG/G)
1	180	10	6.25	25	0.02	97.69430052	122.66	99.81068289	116.54
2	180	15	6.25	25	0.02	147.1243523	131.63	148.8640974	133.09
3	180	20	6.25	25	0.02	196.4507772	183.25	197.1737661	184.79
4	180	22.5	6.25	25	0.02	220.3108808	225.47	219.1311697	220.91
5	180	25	6.25	25	0.02	244.5854922	255.13	243.5226504	251.49
6	180	27.5	6.25	25	0.02	269.5336788	268.77	268.0493577	271.25
7	180	30	6.25	25	0.02	290.8031088	273.81	290.2772143	281.89

TABLE 11.7 ANN results of MB and MG sorption in binary dye system

	INDEPENDENT VARIABLES INPUTS					Q_E (MG/G) OF MB (MB+MG) OUTPUT		Q_E (MG/G) OF MG (MG+MB) OUTPUT	
NO. TESTS	TIME (MINUTES)	C_I (MG/L)	PH	T (°C)	ADSORBENT DOSE (MG)	EXPERIMENTAL VALUES (MG/G)	PREDICTED VALUES (MG/G)	EXPERIMENTAL VALUES (MG/G)	PREDICTED VALUES (MG/G)
1	180	5+5	4	25	0.02	49.31555594	61.50	49.31555594	49.37
2	180	7.5+7.5	4	25	0.02	73.81956849	72.61	73.81956849	74.02
3	180	10+10	4	25	0.02	97.79138401	96.51	97.79138401	97.63
4	180	15+15	4	25	0.02	146.6864229	147.57	146.6864229	107.48
5	180	5+15	4	25	0.02	49.68514706	46.20	145.6840174	147.17
6	180	10+10	4	25	0.02	97.79138401	101.38	96.88129588	96.86
7	180	15+5	4	25	0.02	145.5892809	145.15	48.51758946	48.37
8	180	15+15	4	25	0.02	145.2516866	145.85	145.4852102	144.70

TABLE 11.8 ANN error parameters

			SAME CONCENTRATIONS		DIFFERENT CONCENTRATIONS	
	MB	*MG*	*MB(MB+MG)*	*MG(MG+MB)*	*MB(MB+MG)*	*MG(MG+MB)*
FACTORS	*ANN MODEL*	*ANN MODEL*	*ANN MODEL*	*ANN MODEL*	*ANN MODEL*	*ANN MODEL*
R^2	0.95	0.97	0.98	0.87	0.99	0.99
R^2_{adj}	0.94	0.96	0.97	0.86	0.98	0.98
RMSE	9.56	7.19	6.17	19.60	2.52	0.71
MSE	91.54	51.83	38.08	384.30	6.39	0.84

TABLE 11.9 ANN results of MB and MG at various pH ranges

	INDEPENDENT VARIABLES INPUTS					Q_E (MG/G) OF MB OUTPUT		Q_E (MG/G) OF MG OUTPUT	
NO. TESTS	*TIME (MINUTES)*	*CONCENTRATION (MG/L)*	*PH*	*T (°C)*	*ADSORBENT DOSE (MG)*	*EXPERIMENTAL VALUES (MG/G)*	*PREDICTED VALUES (MG/G)*	*EXPERIMENTAL VALUES (MG/G)*	*PREDICTED VALUES (MG/G)*
1	60	10	2	25	0.02	89.97	92.40	94.59	95.87
2	60	15	2	25	0.02	122.41	126.25	141.27	141.15
3	60	20	2	25	0.02	158.68	173.62	186.55	192.34
4	60	25	2	25	0.02	203.19	221.27	225.00	232.51
5	60	10	4	25	0.02	99.82	96.75	100.28	100.41
6	60	15	4	25	0.02	148.73	131.95	150.23	145.31
7	60	20	4	25	0.02	196.40	181.87	199.77	197.31
8	60	25	4	25	0.02	243.65	232.17	248.16	237.97
9	60	10	6	25	0.02	100.03	103.92	100.75	102.74
10	60	15	6	25	0.02	149.46	140.38	150.34	147.70
11	60	20	6	25	0.02	197.18	191.62	200.23	201.48
12	60	25	6	25	0.02	246.40	240.96	248.58	245.17
13	60	10	8	25	0.02	100.28	113.70	100.80	103.16
14	60	15	8	25	0.02	149.66	150.05	150.60	147.82
15	60	20	8	25	0.02	198.63	200.58	200.13	203.34
16	60	25	8	25	0.02	246.66	246.88	249.61	252.04

TABLE 11.10 ANN error parameters at various pH ranges

FACTORS	*MB*	*MG*
pH	ANN model	ANN model
R^2	0.96	0.99
R^2_{adj}	0.95	0.98
RMSE	19.66	8.35
MSE	386.70	69.82

MG (MG+MB); on the other hand, a good agreement is observed between the experimental and the predicted values; thus, the constructed ANN model has good ability to predict the sorption of the dyes in binary mixture.

We can consider that the ANN model shows good performance in predicting dye sorption at any range of given pH and various initial concentrations (Tables 11.9 and 11.10).

11.12 FUTURE PERSPECTIVES AND CHALLENGES

The selection and the exact combination of adequate WW techniques are very essential for water treatment. Therefore, it is crucial to conduct an in-depth analysis of many input data collected from the non-treated water to make a final choice. These data include BOD, COD, turbidity, conductivity, TDS, SM, pH, T, volume, flow rate, water speed, time, organic matter, microorganism, adsorbent dose, dissolved oxygen (DO), nitrite, nitrate, phosphor, azote, ammonium, heavy metal, and dyes. All these parameters must be controlled and well analyzed to conclude with a satisfactory treatment technique. Recently, AI and ML algorithms have shown a novel WWT route by analyzing multiple variables (input data) for training and validation and predicting the treatment performance (outputs) without experimental analysis costs, chemical, and energy consumption.

The difficulty in implementing these AI and ML models is the availability of huge amounts of data. The performance of AI models increases with the increase in data; therefore, obtaining this quantity of information is time-consuming and needs a lot of experimentation in real-time settings. Furthermore, in developing countries, AI and ML are less used in environmental applications such as the WWT sector, and the present scenario suggests that more experience in AI and ML applications at a laboratory scale is required in water treatment to better comprehend the complex relationship between WWT, and input data, hidden layers and neurons (Garg et al., 2024).

The most recent studies use one single AI model to predict and optimize their system. However, the combination of various models can provide even better results with minimum errors, compared to using one system. Integrated models (e.g., ANN coupled with CCD and RSM) should be the subject of future research works.

11.13 CONCLUSION

WWT technology necessitates a large amount of data, such as reactor design, bacterial selection, bacterial cultivation, pollution parameters, type of pollution, water characteristics, and energy consumption. These data must be controlled and well organized for an adequate treatment process. This chapter discussed AI algorithms such as ANN, CCD, RSM, SVM, DT, and GA as intelligent tools for water quality monitoring and energy optimization. In addition, an ANN model was developed to predict and optimize the sorption of MB and MG. It demonstrated high predictive performance with minimal errors (RMSE and MSE) and presented an architecture composed of one hidden layer, five inputs, and one output, using sigmoid and Purelin functions applied to the hidden and output layers, respectively.

Future research in water treatment should develop AI models and use them in WW plants. Using one AI model is not efficient; hybrid and combined models are recommended to better control, optimize, and monitor WW plants.

This chapter presented AI as an outstanding solution to optimize the treatment parameters and also as a new and valuable tool to cut down the time, laboratory costs, and human efforts needed for pollution removal from WW. AI can deal with nonlinear problems. It helps and improves water quality to easily meet the environmental standards in an intelligent, eco-friendly, autonomous, and cost-effective way.

REFERENCES

Adibimanesh, B., Polesek-Karczewska, S., Bagherzadeh, F., Szczuko, P., & Shafighfard, T. (2023). Energy consumption optimization in wastewater treatment plants: Machine learning for monitoring incineration of sewage sludge. *Sustainable Energy Technologies and Assessments*, *56*(January), 103040. https://doi.org/10.1016/j.seta.2023.103040

Altowayti, W. A. H., Allozy, H. G. A., Shahir, S., Goh, P. S., & Yunus, M. A. M. (2019). A novel nanocomposite of aminated silica nanotube (MWCNT/Si/NH_2) and its potential on adsorption of nitrite. *Environmental Science and Pollution Research*, *26*(28), 28737–28748. https://doi.org/10.1007/s11356-019-06059-0

Anastopoulos, I., Pashalidis, I., Hosseini-Bandegharaei, A., Giannakoudakis, D. A., Robalds, A., Usman, M., Escudero, L. B., Zhou, Y., Colmenares, J. C., Núñez-Delgado, A., & Lima, É. C. (2019). Agricultural biomass/waste as adsorbents for toxic metal decontamination of aqueous solutions. *Journal of Molecular Liquids*, *295*. https://doi.org/10.1016/j.molliq.2019.111684

Behera, S. K., Karthika, S., Mahanty, B., Meher, S. K., Zafar, M., Baskaran, D., Rajamanickam, R., Das, R., Pakshirajan, K., Bilyaminu, A. M., & Rene, E. R. (2024). Application of artificial intelligence tools in wastewater and waste gas treatment systems: Recent advances and prospects. *Journal of Environmental Management*, *370*(September), 122386. https://doi.org/10.1016/j.jenvman.2024.122386

Belcaid, A., Beakou, B. H., Bouhsina, S., & Anouar, A. (2023). New insights on manganese dioxide nanoparticles loaded on cellulose-based biochar of cassava peel for the adsorption of three cationic dyes from wastewater. *International Journal of Biological Macromolecules*, *241*(January), 124534. https://doi.org/10.1016/j.ijbiomac.2023.124534

Belcaid, A., Beakou, B. H., Bouhsina, S., & Anouar, A. (2024). Biosorption of cobalt and chromium from wastewater using manganese dioxide and iron oxide nanoparticles loaded on cellulose-based biochar: Modeling and optimization with machine learning (artificial neural network). *International Journal of Biological Macromolecules*, *282*(P3), 136855. https://doi.org/10.1016/j.ijbiomac.2024.136855

Borah, L., Goswami, M., & Phukan, P. (2015). Adsorption of methylene blue and eosin yellow using porous carbon prepared from tea waste: Adsorption equilibrium, kinetics and thermodynamics study. *Journal of Environmental Chemical Engineering*, *3*(2), 1018–1028. https://doi.org/10.1016/j.jece.2015.02.013

Cairone, S., Hasan, S. W., Choo, K. H., Li, C. W., Zarra, T., Belgiorno, V., & Naddeo, V. (2024). Integrating artificial intelligence modeling and membrane technologies for advanced wastewater treatment: Research progress and future perspectives. *Science of the Total Environment*, *944*(March), 173999. https://doi.org/10.1016/j.scitotenv.2024.173999

de Araujo, C. M. B., Ghislandi, M. G., Rios, A. G., da Costa, G. R. B., do Nascimento, B. F., Ferreira, A. F. P., da Motta Sobrinho, M. A., & Rodrigues, A. E. (2022). Wastewater treatment using recyclable agar-graphene oxide biocomposite hydrogel in batch and fixed-bed adsorption column: Bench experiments and modeling for the selective removal of organics. *Colloids and Surfaces A: Physicochemical and Engineering Aspects*, *639*(November 2021), 128357. https://doi.org/10.1016/j.colsurfa.2022.128357

Fetisoa, M. R., Acheampong, A., Bondzie-Quaye, P., Bashir, M. A., & Huang, Q. (2024). Synergistic strategies for optimizing microalgae-based wastewater treatment: A comprehensive review. *Journal of Water Process Engineering*, *68*(July), 106409. https://doi.org/10.1016/j.jwpe.2024.106409

Garg, M. C., Kumari, S., & Agarwal, S. (2024). The integration of artificial intelligence in advanced wastewater treatment systems. In Garg, M. C. (Ed.) *The AI cleanse: Transforming Wastewater Treatment Through Artificial Intelligence. Springer Water: Vol. Part F*3340. Springer Nature, Switzerland. https://doi.org/10.1007/978-3-031-67237-8_1

Gupta, P., Bhardwaj, G., Dubey, S., Tayal, T., Sengupta, A., & Narad, P. (2024). AI-enabled process optimization for sustainable wastewater treatment solutions. In Garg, M. C. (Ed.) *The AI cleanse: Transforming Wastewater Treatment Through Artificial Intelligence. Springer Water: Vol. Part F*3340. Springer Nature, Switzerland. https://doi.org/10.1007/978-3-031-67237-8_6

Heddam, S. (2024). Predictive modeling and machine learning for optimal wastewater treatment performance. In Garg, M. C. (Ed.) *The AI cleanse: Transforming Wastewater Treatment Through Artificial Intelligence. Springer Water: Vol. Part F*3340. Springer Nature, Switzerland. https://doi.org/10.1007/978-3-031-67237-8_11

Hou, T., Yan, L., Li, J., Yang, Y., Shan, L., Meng, X., Li, X., & Zhao, Y. (2020). Adsorption performance and mechanistic study of heavy metals by facile synthesized magnetic layered double oxide/carbon composite from spent adsorbent. *Chemical Engineering Journal*, *384*(October 2019), 123331. https://doi.org/10.1016/j.cej.2019.123331

Kallawar, G., Thakare, N., Bonde, S., Barai, D., Bhanvase, B. A., Sonawane, A., Sonawane, S. H., & Manickam, S. (2024). Exploring sonochemical synthesis for photocatalyst nanocomposites in water and wastewater treatment: An in-depth review. *Journal of Cleaner Production*, *485*(October), 144279. https://doi.org/10.1016/j.jclepro.2024.144279

Karuppaiyan, J., Jeyalakshmi, R., Kiruthika, S., Wadaan, M. A., Khan, M. F., & Kim, W. (2023). A study on the role of surface functional groups of metakaolin in the removal of methylene blue: Characterization, kinetics, modeling and RSM optimization. *Environmental Research*, *226*(March), 115604. https://doi.org/10.1016/j.envres.2023.115604

Learning, M., Energy, B., & Prediction, C. (2023). Unlocking the potential of wastewater treatment: Machine learning based energy consumption prediction. *Water*, *15*(13), 2349.

Lin, M., & Chen, Z. (2020). A facile one-step synthesized epsilon-MnO_2 nanoflowers for effective removal of lead ions from wastewater. *Chemosphere*, *250*, 126329. https://doi.org/10.1016/j.chemosphere.2020.126329

Mao, Z., Li, X., Zhang, X., Li, D., Lu, J., Li, J., & Zheng, F. (2024). Optimization of effluent quality and energy consumption of aeration process in wastewater treatment plants using artificial intelligence. *Journal of Water Process Engineering*, *63*(May), 105384. https://doi.org/10.1016/j.jwpe.2024.105384

Marcińczyk, M., Ok, Y. S., & Oleszczuk, P. (2022). From waste to fertilizer: Nutrient recovery from wastewater by pristine and engineered biochars. *Chemosphere*, *306*(April). https://doi.org/10.1016/j.chemosphere.2022.135310

Matebese, F., Mosai, A. K., Tutu, H., & Tshentu, Z. R. (2024). Mining wastewater treatment technologies and resource recovery techniques: A review. *Heliyon*, *10*(3), e24730. https://doi.org/10.1016/j.heliyon.2024.e24730

Mihály, N. B., Luca, A. V., Simon-Várhelyi, M., & Cristea, V. M. (2023). Improvement of air flowrate distribution in the nitrification reactor of the waste water treatment plant by effluent quality, energy and greenhouse gas emissions optimization via artificial neural networks models. *Journal of Water Process Engineering*, *54*(March). https://doi.org/10.1016/j.jwpe.2023.103935

Nagpal, M., Siddique, M. A., Sharma, K., Sharma, N., & Mittal, A. (2024). Optimizing wastewater treatment through artificial intelligence: Recent advances and future prospects. *Water Science and Technology*, *90*(3), 731–757. https://doi.org/10.2166/wst.2024.259

Nishat, A., Yusuf, M., Qadir, A., Ezaier, Y., Vambol, V., Ijaz Khan, M., Ben Moussa, S., Kamyab, H., Sehgal, S. S., Prakash, C., Yang, H. H., Ibrahim, H., & Eldin, S. M. (2023). Wastewater treatment: A short assessment on available techniques. *Alexandria Engineering Journal*, *76*, 505–516. https://doi.org/10.1016/j.aej.2023.06.054

Okut, N., Hamzat, A. K., Rajakaruna, R. A. D. N. V., & Asmatulu, E. (2025). Agricultural wastewater treatment and reuse technologies: A comprehensive review. *Journal of Water Process Engineering*, *69*(November 2024), 106699. https://doi.org/10.1016/j.jwpe.2024.106699

Oruganti, R. K., Biji, A. P., Lanuyanger, T., Show, P. L., Sriariyanun, M., Upadhyayula, V. K. K., Gadhamshetty, V., & Bhattacharyya, D. (2023). Artificial intelligence and machine learning tools for high-performance microalgal wastewater treatment and algal biorefinery: A critical review. *Science of the Total Environment*, *876*(March), 162797. https://doi.org/10.1016/j.scitotenv.2023.162797

Patel, S. K., Shukla, S. C., Natarajan, B. R., Asaithambi, P., Dwivedi, H. K., Sharma, A., Singh, D., Nasim, M., Raghuvanshi, S., Sharma, D., Sen, S., Dubey, S., & Prajapati, A. K. (2025). State of the art review for industrial wastewater treatment by electrocoagulation process: Mechanism, cost and sludge analysis. *Desalination and Water Treatment*, *321*(September 2024), 100915. https://doi.org/10.1016/j.dwt.2024.100915

Priya, Sharma, A. K., Kaith, B. S., Tanwar, V., Bhatia, J. K., Sharma, N., Bajaj, S., & Panchal, S. (2019). RSM-CCD optimized sodium alginate/gelatin based ZnS-nanocomposite hydrogel for the effective removal of biebrich scarlet and crystal violet dyes. *International Journal of Biological Macromolecules*, *129*, 214–226. https://doi.org/10.1016/j.ijbiomac.2019.02.034

Qu, J., Che, N., Niu, G., Liu, L., Li, C., & Liu, Y. (2023). Iron/manganese binary metal oxide-biochar nano-composites with high adsorption capacities of Cd^{2+}: Preparation and adsorption mechanisms. *Journal of Water Process Engineering*, *51*(November 2022), 103332. https://doi.org/10.1016/j.jwpe.2022.103332

Ray, J., Jana, S., Mondal, B., & Tripathy, T. (2019). Enhanced and rapid adsorptive removal of toxic organic dyes from aqueous solution using a nanocomposite of saponified polymethyl acrylate grafted dextrin with embedded nanosilica. *Journal of Molecular Liquids*, *275*, 879–894. https://doi.org/10.1016/j.molliq.2018.11.126

Rose, P. K., Kumar, R., Kumar, R., Kumar, M., & Sharma, P. (2023). Congo red dye adsorption onto cationic amino-modified walnut shell: Characterization, RSM optimization, isotherms, kinetics, and mechanism studies. *Groundwater for Sustainable Development*, *21*(March), 100931. https://doi.org/10.1016/j.gsd.2023.100931

Rout, P. R., Zhang, T. C., Bhunia, P., & Surampalli, R. Y. (2021). Treatment technologies for emerging contaminants in wastewater treatment plants: A review. *Science of the Total Environment*, *753*, 141990. https://doi.org/10.1016/j.scitotenv.2020.141990

Safeer, S., Pandey, R. P., Rehman, B., Safdar, T., Ahmad, I., Hasan, S. W., & Ullah, A. (2022). A review of artificial intelligence in water purification and wastewater treatment: Recent advancements. *Journal of Water Process Engineering*, *49*(July), 102974. https://doi.org/10.1016/j.jwpe.2022.102974

Salari, M., Alahabadi, A., Rahmani-Sani, A., Miri, M., Yazdani-Aval, M., Lotfi, H., Saghi, M. H., Rastegar, A., Sepehr, M. N., & Darvishmotevalli, M. (2024). A comparative study of response surface methodology and artificial neural network based algorithm genetic for modeling and optimization of EP/US/GAC oxidation process in dexamethasone degradation: Application for real wastewater, electrical energy consump. *Chemosphere*, *349*(November 2023), 140832. https://doi.org/10.1016/j.chemosphere.2023.140832

Sanjeevi, R., Sathvara, P. B., Tripathi, S., & Anuradha, J. (2024). The AI cleanse: Revolutionizing the future of wastewater treatment with AI and machine learning. In Garg, M. C. (Ed.) *The AI cleanse: Transforming Wastewater Treatment Through Artificial Intelligence.* Springer Water: Vol. Part F3340. Springer Nature, Switzerland. https://doi.org/10.1007/978-3-031-67237-8_10

Shin, J. H., Kim, S. H., Yoo, C. H., Lee, H. J., Nguyen, B. T. D., Lee, G. G., Kim, J. F., & Lee, J. S. (2024). Valorization of battery manufacturing wastewater: Recovery of high-value metal ions through reaction-enhanced membrane cascade. *Chemical Engineering Journal*, *493*(May), 152247. https://doi.org/10.1016/j.cej.2024.152247

Singh, N. K., Yadav, M., Singh, V., Padhiyar, H., Kumar, V., Bhatia, S. K., & Show, P. L. (2023). Artificial intelligence and machine learning-based monitoring and design of biological wastewater treatment systems. *Bioresource Technology*, *369*(December 2022), 128486. https://doi.org/10.1016/j.biortech.2022.128486

Singh, S., Mahanty, B., Gujjala, L. K. S., & Dutta, K. (2024). Optimized phenol degradation and lipid production by *Rhodosporidium toruloides* using response surface methodology and genetic algorithm-optimized artificial neural network. *Chemosphere*, *363*(July), 142971. https://doi.org/10.1016/j.chemosphere.2024.142971

Soliman, N. K., Moustafa, A. F., Aboud, A. A., & Halim, K. S. A. (2019). Effective utilization of Moringa seeds waste as a new green environmental adsorbent for removal of industrial toxic dyes. *Journal of Materials Research and Technology*, *8*(2), 1798–1808. https://doi.org/10.1016/j.jmrt.2018.12.010

Wang, G., Yang, R., Liu, Y., Wang, J., Tan, W., Liu, X., Jin, Y., & Qu, J. (2022). Adsorption of Cd(II) onto *Auricularia auricula* spent substrate biochar modified by CS_2: Characteristics, mechanism and application in wastewater treatment. *Journal of Cleaner Production*, *367*(June), 132882. https://doi.org/10.1016/j.jclepro.2022.132882

Zhang, S., Jin, Y., Chen, W., Wang, J., Wang, Y., & Ren, H. (2023). Artificial intelligence in wastewater treatment: A data-driven analysis of status and trends. *Chemosphere*, *336*(May), 139163. https://doi.org/10.1016/j.chemosphere.2023.139163

Zhang, S., Wang, J., & Ren, H. (2024). Introduction to artificial intelligence in optimization of wastewater treatment processes. In Garg, M. C. (Ed.) *The AI cleanse: Transforming Wastewater Treatment Through Artificial Intelligence.* Springer Water: Vol. Part F3340. Springer Nature, Switzerland. https://doi.org/10.1007/978-3-031-67237-8_2

PART V

AI Combatting Contaminants for Safe Drinking Water

Detection and Mitigation of Microbial Threats Using AI

12

Shubhrajit Sarkar and Ankita Mazumder

12.1 INTRODUCTION

Microbial threat detection involves the identification of pathogens by microbiological techniques, including microbial culture, imaging, and molecular analysis, which are considered a crucial aspects in the public healthcare realm. There are certain issues with conventional techniques, including improper sample collection, handling, and storage, intricacies in microbial culture, erroneous identification, and difficulties in antimicrobial susceptibility tests. These conventional approaches also demand more manpower and a longer time duration, which eventually leads to the delay in treatment (Franco-Duarte et al., 2019). In the present era, microbiologists need to handle and characterize extensive datasets of biological and chemical data by adopting effective contemporary technologies. Apart from conventional laboratory activities and field experiments, the development of proficient skills to utilize contemporary computational methods is highly essential for extracting and analyzing meaningful insights from these extensive data collections.

In this context, the evolution of artificial intelligence (AI) has brought immense innovations in the domain of microbial diagnosis by offering highly precise, accurate, and current outcomes. The pervasive domain of microbiology is presently at the peak of a revolutionary period because of the amalgamation of AI in healthcare systems. With the integration of AI, the analysis of voluminous data, the recognition of hidden patterns within datasets, and the consequent clinical diagnostic process have become quite rapid. The application of AI is quintessential in today's world to detect disease at a very early stage, advance medical treatment, personalize and customize the treatment, and also detect and monitor epidemic outbreak. With the implementation of a diversified array of AI-based techniques, including machine learning (ML), object detection, and computer vision, AI exhibits enough potential to transform the healthcare field. Nevertheless, ethical concerns are associated with the case of AI-enabled microbial diagnosis. Important ethical considerations are privacy protection of patients, addressing algorithmic biases, maintenance of data security, facilitation of transparency, and ensuring equalized accessibility, which need to be carefully taken care of. For ensuring patients' safety, personnel surveillance is essential. With the

DOI: 10.1201/9781003600718-17

continual stringency in ethical regulations and policies for the implementation of AI in public healthcare and safety, AI is expected to be a potent component in delivering accurate, personalized diagnoses without hampering patients' rights and accessibility to healthcare services (Ali, Ahmed, & Aslam, 2023). This chapter will comprehensively cover the insights into the utilization of AI in microbial threat detection and consequent mitigation.

12.2 MICROBIAL DIAGNOSIS FOR DETECTING MICROBIAL THREATS

Identification of microbial threats (microorganisms) by adopting microbial diagnosis is an imperative aspect of the healthcare domain as it facilitates detecting the causing agents of various diseases and infections. The procedure adopted for microbe detection primarily involves the cultivation of fungi and bacteria on specific growth media or the isolation of viruses by cell culture, and the characterization of the infection-causing agents through genetic, antigenic, or biochemical methods (Buchan & Ledeboer, 2014). In an alternate manner, infectious diseases can be scrutinized through detecting a distinguished immune response. One of the most usual modes of such detection is the scrutiny of antibodies that are being developed within the body during the timeline of the illness. Moreover, in some cases, infection-causing agents are visualized within infected tissues, which may direct toward microbial diagnosis depending upon the distinguished morphological characteristics. At least, such visualization will facilitate categorizing the type of microorganism that is responsible for causing specific infections. Traditional well-established methodologies for microbial threat diagnosis are already there. However, these methods involve extensive, lengthy duration and are labor-intensive. On the contrary, modern techniques demand cost-effective and rapid methodologies for the primary classification of microbes as fungal and bacterial isolates. Furthermore, contemporary analytical methods necessitate detailed microbial profiling, primarily based on molecular techniques like 16S rRNA technique, polymerase chain reaction (PCR) technique, etc. The finding of such methods is quite important for the development of effective curative approaches (Franco-Duarte et al., 2019). Microbial threat diagnoses include multiple steps, starting with sample collection, followed by their transport and proper storage (Gori, Callea, Alberani, & Orlando, 2014). This fundamental step is critical in the diagnostic procedure as the accuracy of consecutive tests is greatly dependent on appropriate collection, handling, and storage of clinical samples (Miller et al., 2018).

12.3 CHALLENGES OF CONVENTIONAL MICROBIAL DIAGNOSIS TECHNIQUES

Microbial diagnosis is a fundamental aspect of contemporary microbiology and the medical field. However, it faces some intricacies during the entire process of diagnosis. One of the primary challenges is regarding the collection and transport of clinical specimens (Rueda et al., 2022). The diagnostic procedure begins with the collection of clinical samples. Nevertheless, it is utterly arduous to ensure the timely and proper collection, handling, and consequent transportation of those samples to the testing laboratory. Inappropriate handling, storage, and delay in transport may increase the risk of the integrity and viability of microbes present within the collected sample, resulting in erroneous results of diagnosis (Shrestha & Pokharel, 2020). Consequently, a fundamental conventional diagnostic technique involves culturing in appropriate media under optimized conditions, followed by the isolation of microorganisms. This procedure requires an elongated time duration to derive the findings of culture. Also, inappropriate microbial cultivation enhances the chances of false negative findings. Furthermore, utilization of excessive antibiotic drugs in clinical practice may influence the growth and subsequent isolation of distinct microorganisms (Dietvorst, Vilaplana, Uria, Marco, & Muñoz-Berbel, 2020).

Proper identification and classification of microorganisms through microbial diagnosis face another serious intricacy. Conventional techniques involving biochemical tests and microscopic observations are not capable of consistently carrying out the correct identification because of the confusing and deceptive morphological attributes or any unusual behavior portrayed by specific microorganism specimens (Zhang et al., 2022). Another challenging aspect is to enumerate the susceptibility level of bacterial isolates to antibiotic drugs for selecting a proper treatment strategy. Conventional techniques employed for susceptibility testing, including the disk diffusion method, are highly labor-intensive and time-consuming processes with poor reproducibility efficacy. Moreover, in the case of a few antimicrobials, such as colistin and daptomycin cases, the disk diffusion method cannot be adopted. Furthermore, the risk of interpretation errors is high in the semi-automated measurement of zone diameters. Traditional techniques also necessitate a combination of highly qualified and skilled persons, specially designed equipment, and substantial resources for accurate diagnosis of medical conditions. Here, these methods involve high chances of human errors in settings with restricted resources because of the dependency on skilled personnel. Therefore, it is quite crucial to introduce novel, user-friendly, and cost-effective diagnostic techniques (Gajic et al., 2022; Kontsevaya et al., 2024).

Conventional diagnostic techniques often create an unwanted delay in the primary phase of medical treatment or therapy, resulting in the uncontrolled progress of the illness (Peri, Stewart, Hume, Irwin, & Harris, 2021). In such scenarios, empirical medical therapies may increase the chances of antimicrobial resistance (AMR) (Ventola, 2015). Moreover, the probability of false positive results and cross-contamination makes the conventional diagnosis more formidable. As conventionally executed, microbial threat diagnosis may not deliver real-time data to effectively surveil epidemiology because of the higher chances of cross-contamination. This may eventually provide false positive findings from serological examination (Asgharzadeh et al., 2020). The inadequacy of standard diagnostic procedures and quality-ensuring measures in medical testing laboratories may lead to improper results that are less consistent, accurate, and trustworthy.

12.4 OVERVIEW OF AI AND ML

The exploration of the concept of artificial brains has been initiated since the early 1900s. However, the term 'artificial intelligence' was first coined in the Dartmouth Conference by John McCarthy in the year 1955, which aims at simulating and replicating the intelligence and problem-solving quality of human beings along with the formulation and development of several theoretical techniques, methods, and applied systems. AI is basically the capability of any computer system or computer-regulated robot to execute tasks that are typically associated with the intelligence of humans. In simple terms, AI consists of both hardware and software methods that can imitate or replicate human behavior, thinking ability, and problem-solving ability. Moreover, AI systems are categorized based on the level of intelligence involved with the system as compared to a human being. Two major categories of AI are strong AI, or general AI, and weak AI, or weak AI. The contemporary practical applications employ soft or weak AI, which exhibits the capability of solving individual problems along with a satisfactory level of accuracy of the provided solutions. On the other hand, huge research efforts are concentrated on strong AI or general AI (Goertzel, 2014). AI involves various significant scientific domains, including natural language processing (NLP), computer vision, ML, planning, speech recognition, expert systems, and robotics. The prime domains and sub-domains of AI are schematically depicted in Figure 12.1.

In the case of AI utilization, a vast number of applications are developed based on the methods of ML. The concept of ML was initially introduced by Arthur Samuel at Bell Labs, IBM, and Stanford

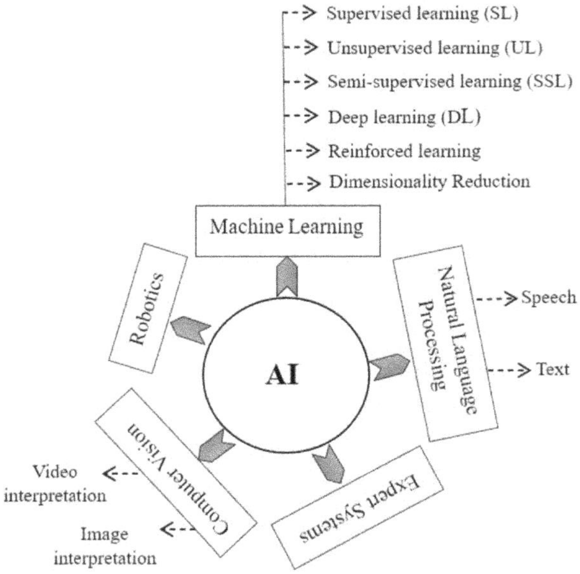

FIGURE 12.1 Schematic representation of the prime domains and subdomains of artificial intelligence (AI).

in the year 1959, which is fundamentally a specialized section of AI. It is developed with the primary objective of discerning features from an extensive and diversified dataset. In ML methods, algorithms are adopted for analyzing wide sets of data, followed by the automated identification of specific patterns within the dataset and consequently leveraging these identified patterns to make predictions and undertake decisions in real-life scenarios (Mukhamediev, Popova, et al., 2022). In the case of traditional software programming, programmers manually write a definite code to be consequently followed by the computer for a specified task. On the contrary, ML methods adopt algorithms for training large sets of data and thereby prepare themselves to carry out real-life complex tasks by learning from the datasets themselves. The suitability of ML is better for handling complex and dynamic tasks requiring continued adaptation and comprehensive analysis. Nowadays, the application of ML is prominent in diversified fields including speech emotions (Kerkeni et al., 2019), speech recognition (Yang, Yao, Liao, & Huang, 2024), agriculture (Attri, Awasthi, & Sharma, 2023), robotics (Kim et al., 2021), astronomy (Ball & Brunner, 2010), geological processes (Besimbayeva, Khmyrova, Tutanova, Flindt, & Sharafutdinov, 2023), environmental studies (Spence et al., 2018), petrographic studies (Kuchin, Mukhamediev, & Yakunin, 2020), construction (Lai, Qiu, Feng, Chen, & Fan, 2016), economics (Zhong, Chen, Dan, & Rezaeipanah, 2022), chemistry (Artrith et al., 2021), computational biology (Zitnik et al., 2019), medicine (Miotto, Wang, Wang, Jiang, & Dudley, 2018), mining and exploration (Mukhamediev, Kuchin, Amirgaliyev, Yunicheva, & Muhamedijeva, 2022), etc. ML is currently recognized as a powerful tool to analyze data and solve complex scientific issues.

At present, there are thousands of ML algorithms for the purpose of prediction, pattern identification, or recommendation. The model selection indicates selecting the best algorithm for any specified problem to be handled. This is fundamentally carried out by the research aim (such as infectious disease diagnosis, phenotyping, recommendation of clinical treatment, prediction of risk, etc.), the researchers' insight about the appropriate ML algorithms that will best fit the purpose, and constraints evolving from the available datasets (dimensionality and volume of datasets). Based on the learning method and goal of the algorithm, ML methods may be broadly categorized into the following six sectors: deep learning (DL), unsupervised learning (UL), supervised learning (SL), semi-supervised learning (SSL), reinforcement learning (RL), and dimensionality reduction (Goertzel, 2014; Mukhamediev, Popova, et al., 2022). Here, one crucial point to be noted is that the circumference for most of these above-mentioned algorithms is not stringent, and each algorithm may be hypothetically categorized into more than one subgroup. Among them, UL algorithms typically carry out the job of segregating the datasets of unlabeled objects into intersecting or isolated groups through the adoption of an automatic methodology depending on the attributes of those unlabeled objects, without any influence or supervision of human beings. UL proclaims the hidden patterns along with the imbalances and anomalies within the dataset. The adoption of UL algorithms is common in such cases where data without labeling and specific results are only accessible. The analogies and resemblances of observations within the dataset are evaluated for categorizing the data into different groups having no previous labeling.

The SL methods are usually preferred for solving regression or classification issues. Such techniques are adopted in the case of the allocation of a definite group of explicitly labeled objects in a possibly infinite dataset. It is adopted to predict definite results from a wide set of predictors. The datasets are primarily labeled, and then, training of algorithms with those labeled data is being carried out. SL develops an automated system that can evaluate whether specific items of interest (for example, clinical features of a patient) will belong to a definite class (presence of a particular disease) or not. However, SSL is basically a blend of UL and SL. In a few instances, the expense of labeling datasets becomes comparatively very high because of the engagement of qualified and expert manpower. As a consequence, if labeling is lacking for most of the datasets but available only in a few data, SSL methods may then be utilized to construct the model. RL algorithm is another type of ML technique in which software is trained for taking the best decision from a collection of probable actions through interaction with its environment. The algorithm, commonly referred to as the agent, collects feedback in terms of either punishments or rewards. Feedback collection occurs depending on the variations in the environment induced by the preceding action. In the case of the selection of each future action, both the instantaneous rewards received from an action and the possible future rewards based on that specific recent action will play pivotal roles. Let us understand with an example from the field of medicine. Here, the actions may be assumed as the collection of feasible treatment strategies, the environment as the patient's physiological function, and the rewards as the status of the health conditions of that patient. RL algorithms usually perform self-learning through the adoption of a trial-and-error method for enumerating the best decision-making action considering the possible uncertain conditions (Woodman & Mangoni, 2023).

DL models and neural networks may be adopted with both unlabeled and labeled datasets (for instance, clustering with unlabeled datasets or image analysis with labeled datasets). In the case of DL methods, the hidden characteristics in the provided large data arrays are proclaimed through the adoption of neural networks involving a vast number of hidden layers and networks of a deep special architecture. Several thousands of hidden layers exist, and each individual layer learns unique features from the preceding layers (Hinton, 2018). In such a context, the transfer learning (TL) concept is frequently considered. This TL concept signifies the enhancement of a learner from one specific domain through the transfer of information from an associated domain. Furthermore, ML algorithms may be provisionally classified into traditional and contemporary ones (Mukhamediev, Kuchin, et al., 2022). The traditional SL models are support vector machine (SVM), k-nearest-neighbor (k-NN), decision tree (DT), logistic regression (LR), and feed-forward artificial neural networks (ANNs). In the case of UL models, traditional ones include principal component analysis (PCA) and k-means. However, isometric mapping (ISOMAP), t-distributed stochastic neighbor embedding (t-SNE), locally linear embedding (LLE), multidimensional scaling (MDS), and kernel principal component analysis (KPCA) are considered contemporary UL models. The modern DL, RL, SL, and SSL models involve the ensemble methods (random forest (RF), boosting, etc.), long short-term memory (LSTM), recurrent neural networks (RNNs), convolutional neural networks (CNNs), deep feed forward neural networks (DFFNNs), etc. (Mukhamediev, Popova, et al., 2022; Woodman & Mangoni, 2023).

Among all these above-mentioned domains of ML, DL is the fastest evolving subdomain and is considered the most progressive ML method for a wide range of applications (LeCun, Bengio, & Hinton, 2015). DL engages deep neural networks, in simple terms, the networks comprising multiple (two or more) hidden layers. The prime benefit of deep architectures is associated with the capability of solving such tasks that utilize the end-to-end method. This technique reduces the necessity of initial data processing. This is because an image or signal vector is provided as input to the network, followed by the independent identification of the regularities by the network linking the input to the target variable. The network itself executes the complex, laborious process of choosing the compelling features. The functioning of this network has largely simplified the tasks of the researchers. Nevertheless, these benefits can be availed only if a sufficiently vast quantity of training datasets is available and the neural network architecture is correctly selected. Amidst the varied available architectures, three fundamental architectures are the standard feed-forward neural network (FFNN), recurrent neural network (RNN), and convolutional neural network (CNN). In the domain of novel drug development, DL-based algorithms play pivotal roles. A bulk quantity of data (both chemical and biological data) can be proficiently analyzed with these DL algorithms, thereby facilitating proper identification of drug targets, accurate prediction of drug characteristics, and expedition of invention of novel therapeutic drugs. Virtual screening is a prime aspect of the utilization of DL algorithms in drug discovery and development. DL-based models can precisely predict the likeliness of binding of a particular molecule to a specified target, by permitting proper training on vast datasets of molecules and their inherent characteristics. This in turn reduces the resource requirement and time duration involved in searching for potent therapeutic drugs (Mukhamediev, Popova, et al., 2022).

12.5 AI INTEGRATION IN MICROBIAL DIAGNOSIS

As discussed previously in Section 12.3, conventional microbial diagnosis methods are associated with some intricacies during the entire diagnostic procedure, starting from specimen collection to microbial species identification and susceptibility testing. In this context, the integration of AI in the microbial diagnostic process has substantially remodeled the sector by delivering faster and highly accurate dependable results as compared to traditional methods. AI shows high proficiency in the comprehensive analysis of data, which comprises different tasks like identification of microbial pathogens, ascertainment of novel microbial species, and prediction of drug resistance. ML, being a domain of AI, helps accelerate the procedure of analyzing genetic data, which eventually leads to increased speed, accuracy, and reliability of the technique (Sunuwar & Azad, 2021). Traditional microbial threat diagnosis faces difficulties all along the procedure, initiating from the collection of specimens to the conduction of susceptibility testing. These intricacies may result in delivering erroneous outcomes, mostly because of the improper treatment and handling of the collected samples. Moreover, the delay involved in the diagnostic procedure is a key contributing component to the rise of AMR (Ventola, 2015). The amalgamation of AI provides dependable, accurate, and expedited results through the adoption of ML and genetic analysis. AI algorithms can identify distinct patterns and anomalies swiftly and accurately from a vast number of microbial datasets, which is quintessential for the proper detection of infectious threats at a very early stage, followed by the consequent development of adequate restrictive measures (Agbehadji, Awuzie, Ngowi, & Millham, 2020). AI algorithms execute predictive modeling for predicting microbial activity through utilizing previous microbiological data, which facilitates providing anticipatory warnings against specific severe disease outbreaks, embracing patterns in AMR, and optimizing effective treatment strategies. AI may also play a significant role in the customization and optimization of disease treatment regimens by accounting for microbiological profiles, patient's unique characteristics, and hereditary history. This eventually helps derive much more efficient, successful, and tailored treatment strategies. ML models, like DNA sequencers, execute predictions regarding the genomic sequences of microbes (including viruses, bacteria, and fungi), facilitating clinicians to properly select efficient treatment regimens. AI promotes automation, thereby resulting in reduced turnaround times and improved patient care (Májek, Lüftinger, Beisken, Rattei, & Materna, 2021). In short, AI has the capability of entirely revolutionizing microbial diagnosis by contributing rapid, more accurate, and efficient techniques that can alleviate the challenges associated with traditional microbial threat diagnosis and improve patient treatment and care (as shown in Figure 12.2).

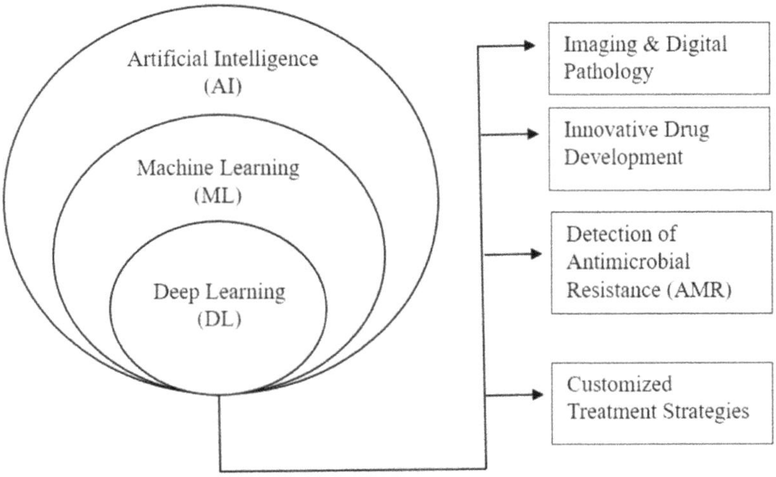

FIGURE 12.2 Role of artificial intelligence (AI) in revolutionizing microbial diagnosis.

12.6 ANTIMICROBIAL RESISTANCE

AMR in microbes, specifically in the case of bacteria, signifies a worldwide health emergency, induced by the expeditious evolution of drug-resistant microbes and the extensively lengthy time period for the development of advanced antimicrobial drugs. For tackling this health crisis, AI, specifically in the form of ML, exhibits a promising approach for impeding the escalation of AMR. The integration of AI provides a significant scope for effective prediction and identification of AMR in bacterial species. Additionally, the synergistic implementation of laboratory testing and ML methods has huge potential to accelerate the invention of novel antimicrobial drugs (Melo, Maasch, & de la Fuente-Nunez, 2021). This synergy strategy will not only help in deriving innovative solutions for resisting AMR but also commit to lowering the misuse of antimicrobial drugs through proper and accurate microbial threat identification.

In the past few years, several research studies have manifested the significant proficiency of AI in terms of resisting AMR. Notable applications of AI algorithms depending on sequencing have been observed in research studies on AMR that showcased satisfactory outcomes. A valuable strategy is to collect clinical data for formulating support systems to undertake appropriate clinical decisions. This approach is expected to aid medical practitioners in monitoring and impeding AMR escalation for improving the judicious utilization of antimicrobial drugs. Furthermore, extensive usage of AI algorithms is currently observed in the synthesis of new antimicrobial drugs and also in the exploration of synergistic drug combinations (Jukič & Bren, 2022). Tackling critical aspects, including the detection of infection-causing agents at an initial stage, differentiating between non-infectious and infectious agents, and taking proper clinical decisions are integral factors in resisting AMR.

For instance, a detailed study was carried out by Yelin et al. in which an extensive dataset of 10 years of timeline comprising more than 0.7 million cases of community-gathered urinary tract infections (UTIs) was investigated. This study manifested the significant contribution of demographic attributes, the prior records of urine cultures, and the previous history of the patient's utilization of antibiotic drugs in the development of AMR. Following the outcomes of these analyses, an ML-based AMR prediction model was developed, which can recognize the specifics of pathogens highly responsible for causing UTI along with their distinct AMR patterns (Yelin et al., 2019). Another research study explored multivariate associations utilizing several ML-based models, such as SVM, LR, and RF, with the primary motive of categorizing drug resistance against eight different anti-tuberculosis drugs and identifying multidrug resistance.

From the results, it has been manifested that these adopted ML models outperformed pre-existing techniques, specifically in the case of sensitivity toward the classification of resistance. Amidst the models considered in this study, the SVM model analyzing a dataset comprised of 1,839 tuberculosis samples was found to be the top-performing one (Yang et al., 2018). In a similar manner, another interesting research work on tuberculosis resistance study was conducted, in which 11 different drugs were analyzed on the basis of a larger dataset consisting of 13,402 samples taken from 16 different countries. Here, the LR model outperformed other ML-based algorithms. Henceforth, it may be concluded that the variation in the performance of ML algorithms is due to dissimilar training data (Kouchaki et al., 2019). The selection of the most appropriate ML-based algorithm is highly influenced by the distinct properties and range of the training dataset being used.

ML is also adopted for predicting a specific pathogen's resistance to a definite antimicrobial. Different AI-based algorithms are primarily developed with genomes of pathogens and their corresponding susceptibility profile or resistance to a definite antimicrobial, followed by the ultimate creation of a model. This model will eventually undertake the prediction of the AMR status in another new but alike genome set. Also, the model may be further tuned for enhancing the efficacy of prediction.

The mechanisms responsible for AMR are diversified. However, antimicrobial resistance genes (ARGs) are the typical driving agents that deliberate the antimicrobial effects in case of being expressed in microbial cells. Thus, the adoption of molecular approaches is an integral part of the research on AMR. In this context, the genomic sequencing and consequent detection of ARGs within either the sole microbe or whole microbial communities are the elementary aspects. Techniques that are used to detect and characterize AMR may vary from specific culturing and PCR techniques to high-throughput sequencing. Nevertheless, appropriate resources of such larger-scale extensive datasets are quintessential to carry out proper analysis. Presently, AMR databases are custom-made for small-scale datasets and genes' functional profiling utilizing depictive annotations. Such properties are not helpful in the analysis of extensive sequencing data, like datasets derived by the utilization of metagenomics for supervision. For tackling these constraints, two reliable databases, namely MEGARes and the Comprehensive Antibiotic Resistance Database (CARD), provide reliable peer-reviewed data regarding AMR determinants (Bonin et al., 2023). These databases contain valuable and reliable information about microbes, concise annotations depicting the working mechanism of the AMR determinants on a specific category of antimicrobial drugs, and analogous protein sequences. Such databases provide reliable resources to train AI algorithms for developing models competent enough to identify certain novel or already known AMRs. Additionally, these databases provide significant information regarding the minimum inhibitory concentration (MIC) of multidrug-resistant (MDR) microbes (Nguyen et al., 2019).

At present, two potent techniques are employed to detect AMR. The first technique utilizes antimicrobial susceptibility testing (AST), while the second one involves whole genome sequencing (WGS) to execute antimicrobial susceptibility testing (called WGS-AST) (Su, Satola, & Read, 2019). For improving the performance of AST methods, a synergistic strategy comprising flow cytometer antimicrobial susceptibility testing (FAST) and supervised ML was employed. The integrated ML-based method portrayed reliable outcomes in a short time period (<3 hours), manifesting a substantial enhancement in the speed of the method. In another research study by Lechowicz, Urbaniak, Adamus-Białek, and Kaca (2013), the timespan involved in doing AST was promisingly reduced to just 0.5 hours from 24 hours. Another remarkable progress was attained through the evolution of an infrared ray spectrometer method that amalgamates ANN with infrared (IR) spectroscopy (Stuart, 2015). The synergetic integration of AI-based IR-spectrometry (requiring 30 minutes) and the FAST technique (requiring 3 hours) reflects notable rapid performance in comparison with conventional AST techniques, which usually demand 24 hours.

In the domain of WGS to carry out AST, contemporary research works majorly rely on k-mer analysis. This method is typically evolved from the entire genome of specimens. Nevertheless, the datasets derived by k-mer analysis are majorly redundant and larger, which thereby makes these datasets illogical and impractical to be used directly in AI-dependent uses. For overcoming this intricacy, the transformation of k-mer dataset to a binary matrix adopting Rapid Annotation Using Subsystem Technology (RAST) server can be a viable option (Davis et al., 2016).

12.7 ARG IDENTIFICATION ALGORITHMS

Various algorithms are available to identify and detect ARGs. Among them, the Mykrobe predictor is a predominant one (Hunt et al., 2019), which is enabled with user-friendly, downloadable software with the capacity to rapidly process whole genome sequences and offer swift outcomes reflecting whether the microbial species in detection is resistant or susceptible to specific antimicrobial drugs. However, a major drawback associated with this ARG identification algorithm is its range of application, as it is designed categorically for bacterial species, including *Mycobacterium tuberculosis and Staphylococcus aureus* (Davis et al., 2016).

Another algorithm named DeepARG was introduced by Arango-Argoty et al. (2018), specifically for predicting ARGs utilizing both short-sequence reads and full-length open reading frames (ORFs) (Arango-Argoty et al., 2018). In spite of exhibiting greater sensitivity and specificity in the identification of ARGs, DeepARG faces challenges in the detection of rare, uncommon ARGs, such as those ARGs that show resistance toward sulfonamide and triclosan (Suzuki, Ogo, Koike, Takada, & Newman, 2015; Yazdankhah et al., 2006). However, this algorithm shows a promising future as an online tool to successfully identify novel ARGs. It demands supplementary training datasets comprised of less commonly known ARGs for increasing its comprehensive performance (Arango-Argoty et al., 2018).

In contrast to methods that depend on WGS and allied AMR metadata for predicting phenotypes of AMR, an alternative approach is to utilize short oligonucleotides, commonly termed as k-mers, for predicting AMR. This approach functions without any previous information on microbial AMR genes (ARGs) (Nguyen et al., 2018). In this aspect, typically adopted algorithms in the ML model are eXtreme Gradient Boosting (XGBoost), RAST tool kit (RASTtk), and Kover (Davis et al., 2016; Drouin et al., 2016; Nguyen et al., 2018). These algorithms generally use short k-mers or oligonucleotides, normally consisting of 10–40 bases. Among these algorithms, the Kover algorithm outperforms RASTtk in terms of better accuracy, but its setting up is complex and demands high expertise in both statistics and information technology for proper functioning (Drouin et al., 2016). Kover functions by considering the different weightage of scores of k-mer resistance and also the correlation between these k-mer combos and thereby delivers comprehensive outcomes of resistance. On the other hand, the XGBoost algorithm ascertains the presence of microbial ARGs through employing MIC data rather than showing dependency on the clinical breakpoint data for differentiating microbial species as resistant, intermediate, or susceptible ones. For instance, the XGBoost algorithm-based ML model showcased an overall accuracy greater than 90% for two experimentally different bacterial genera (named *Nontyphoidal salmonella and Klebsiella pneumoniae*). Nevertheless, the applicability of the XGBoost algorithm was majorly observed for the prediction of ARGs in gram-negative bacterial strains. In the case of gram-positive bacterial strains, its employability is still erratic (Nguyen et al., 2018). To address this limitation, the adoption of the SVM algorithm has caught attention in recent times for categorizing existing and novel ARGs in the case of both gram-positive and gram-negative bacteria. This SVM algorithm performs training by using pan-genome data and reference genomes (Sherman & Salzberg, 2020). The promising outcomes of the SVM algorithm have been found in predicting AMR genes in varied gram-negative bacteria (such as *Pseudomonas, Escherichia coli*) and gram-positive bacteria (such as *Mycobacterium tuberculosis, Staphylococcus aureus*, etc.) (Her & Wu, 2018; Hyun, Kavvas, Monk, & Palsson, 2020; Kavvas et al., 2018).

12.8 ANTIMICROBIAL PROTEINS (AMPS)

In the present era, the use of antimicrobial proteins (AMPs) as a potent replacement for antibiotic drugs has accumulated significant attention in the medical field. AMPs are nothing but proteins that are derived naturally in most organisms, including human beings, through inherent immune systems for combating microbial infectious diseases (Ioannou, Baliou, & Kofteridis, 2023). Nonetheless, the procedure of discovering novel antimicrobials is quite strenuous and time-oriented one, associated with the danger of identifying such compounds having no or low antimicrobial characteristics *in vivo*. To overcome this drawback, various antimicrobial peptide databases were successfully demonstrated by research fraternities (Lei et al., 2019; Zhang et al., 2023). These databases include Antiviral Peptides Database (AVPdb), Antimicrobial Peptide Database (APD), Collection of Anti-Microbial Peptides (CAMP), Data repository of antimicrobial peptides (DRAMP), BACTIBASE, etc., that are comprised of requisite information on different types of AMPs derived from diversified sources. With the use of these AMP databases, the mining and consequent prediction of AMPs have been significantly facilitated. The primary phase of designing AMPs requires the selection of input training data, comprised of both AMPs and non-AMPs databases. These databases provide elaborate detailing of protein properties for maintaining a well-organized representation of negative and positive inputs. Some of these AMP databases, such as Biofilm-active AMPs database (BaAMPs), APD, and CAMPR3, are accessible to the public. Most of the research works typically consider the exclusion of datasets for non-AMPs, which is attributed to the non-existence of experimentally verified databases comprised of non-AMPs (Yin et al., 2024). Consequently, the proper selection of peptide descriptors is a crucial aspect of the contemplated final outcome. The role of these peptide descriptors is pivotal in giving significance to particular structural features in the process of designing a protein's antimicrobial attributes. Commonly, there are four classes of descriptors, namely inductive, 2D Quantitative Structure-Activity Relationship (QSAR), 3D QSAR, and other descriptors like the number of residues present in between amino acids.

For discovering and synthesizing peptides, Fields et al. executed a comprehensive analysis of an automated system developed through the amalgamation of ML methods and biophysical selection of nominal bacteriocin fields (Fields et al., 2020). In this study, the ML system was trained employing a sequence-free prediction algorithm for designing and identification of bacteriocin-based peptides. In the primary stage, 20-mers (20 amino acid-peptide candidates) were created to evaluate, eventually generating 28,895 20-mer peptides. Consequently, the selection of 16 sequences was carried out for synthesis purposes, followed by the scrutiny of hemolytic, cytotoxicity, and antimicrobial activities. The results manifested that the peptides dependent on bacteriocin had portrayed substantial antimicrobial activities against bacterial strains like *Pseudomonas aeruginosa and Escherichia coli*. This study concluded the proficiency of the ML-based method as a promising strategy for synthesizing new bacteriocin candidates (Cardoso et al., 2020). Another approach was to utilize Increment of Diversity with Quadratic Discriminant (IDQD) analysis, which is basically a segment of ML methods and is designed specifically to undertake sequence pattern recognition. For instance, a research work by Feng et al. employed the integrated utilization of ML-based IDQD analysis and primary sequence information to categorize AMPs (Feng, Wang, & Yu, 2019). During the scientific analysis of specified characteristics within the sequences, IDQD develops a distinct scheme for recognizing and predicting patterns (Lu, Luo, Zhang, Chen, & Zhang, 2010). The outcomes of this study manifested an overall accuracy, specificity, and sensitivity of 86.02%, 92.79%, and 74.31%, respectively, in the identification of AMPs. The study concluded that this technique outperformed other currently existing automated methods adopted for categorizing AMPs (Feng et al., 2019). Again, Parvaiz et al. carried out a study in which ML was adopted for performing a comprehensive search of 700,000 compounds to categorize those having beta-lactamase inhibition characteristics. Only 74 compounds were detected and consequently projected for empirical validation. The result manifested that 11 and 7 compounds were identified as enhancers and inhibitors of beta-lactamase CMY-10, respectively. One compound exhibited promising potential as being identified as a β-lactamase inhibitor as well as a β-lactam enhancer (Parvaiz, Ahmad, Yu, MacKerell, & Azam, 2021).

More research emphasis on non-antimicrobial peptides is still quintessential for improving the judicial design of new effective antimicrobials. It is imperative to elaborate on the databases comprising information on non-AMPs that will facilitate AI for executing highly precise identification of AMR determinants. In short, AI has already achieved promising improvements in terms of the identification of AMR determinants and the discovery and design of novel antimicrobials through a considerable reduction of the timespan, efforts, and resources in comparison to conventional techniques.

12.9 AI IN EPIDEMIOLOGY

An epidemic usually shows some unique characteristics, including contagiousness, recurrence, epidemic nature, etc. The sudden emergence of contaminations triggered by formerly unidentified pathogens becomes a huge threat to both human health and food security systems. To address the medical complications involved with epidemiology, the prime focus of biomedical researchers is to identify and characterize pathogen microorganisms and also to predict and screen infectious diseases. In this context, ML (majorly DL)-based methods are mostly dominating in batch image classification, which leads to a significant decline in the time consumption and computational costs (Ghosh, Chakraborty, Kodamana, & Chakraborty, 2022). Research works focused on epidemiology generally investigate the mode of spread of severe infectious diseases and their evolution over the course of time within populations of varied locations, mostly under diversified environmental parameters. Investigations concentrated on those diseases induced by animal or plant viruses provide comprehensive data collection incorporating information on chromosome structure, characteristics, genetic diversity, gene expression, corresponding symptoms, and illnesses.

In the recent past, a worldwide pandemic was caused by severe acute respiratory syndrome coronavirus 2 (SARS-CoV-2) infection, which had already jeopardized the health system by causing more than five million deaths. The utilization of DL-based techniques in such scenarios is significant to identify novel compounds for combating the different virus variants and their consequent mutations. Choi et al. adopted an integrated strategy involving DL-based methods and molecular simulation-based screening methods (Choi et al., 2023). This research manifested the potential of the azithromycin drug to target the RNA polymerase. *In vitro* antiviral capability of this drug through inhibition of the replication of SARS-CoV-2 was thus affirmed (Zhang, Ye, Xi, Juhas, & Li, 2021).

In a study, an innovative k-mer based technique named VirFinder was explored by Ren, Ahlgren, Lu, Fuhrman, and Sun (2017) to identify sequences of prokaryotic viruses from diverse metagenomic datasets, which manifested the rapid screening of pathogenic proviruses at the genetic level (Ren et al., 2017). The synergistic usage of ML and hyperspectral imaging (HSI) has evolved as a novel innovative methodological procedure for image detection of viruses. The use of HSI provides high-dimensional data, which is often comprised of redundant information. In such a scenario, the role of ML is pivotal, which facilitates reducing dimensionality through evaluating a specific effective wavelength range by data preprocessing. Remarkably, DL outperforms other conventional ML methods as it permits automated extraction of features, which eventually leads to reduced computational cost, time, and effort. Additionally, it reduces the dependency on other specialized expertise for feature extraction.

12.10 ETHICAL CONSIDERATIONS AND LIMITATIONS

The amalgamation of AI with microbial diagnostic methods is rapidly advancing, offering an unparalleled scope for enhancing healthcare reports. It is possible to collect and store online healthcare datasets through genetic testing, bioinformatic companies, and different social media. However, this is associated with the inevitable risk of hacking these databases and consequently utilizing them for unethical and unauthorized purposes. These ethical concerns circumscribe different topics, which include data security, privacy of patients, algorithmic bias, proper accessibility, and general responsibility of healthcare practitioners (Jeyaraman, Balaji, Jeyaraman, & Yadav, 2023).

Ensuring the privacy of patients is the fundamental ethical issue involved with the utilization of AI in microbial threat detection. As these AI- and ML-based methods largely depend on vast datasets that may be comprised of sensitive health-oriented information, it is pivotal to ensure the privacy of patients. As per ethical guidelines, ahead of collecting, storing, and utilizing the personal health data of patients for any AI-involved diagnosis, informed consent or approval should be obtained from them. In this context, it is strictly required to anonymize the stored datasets for preventing the disclosure of the identity of patients and to ensure proper and ethical utilization of patients' information (Padmapriya & Parthasarathy, 2024). Additionally, in recent times, another major issue has been regarding cybersecurity and data security. On account of the enhanced dependency on cloud-based data storage systems and interconnected systems, chances of breaching of data security are also gradually increasing. It is strongly essential to ethically ensure the complete protection of patients' information against any unauthorized access and cyberattacks. Both AI developers and healthcare organizations should enable sufficient encryption methods, cybersecurity protections, and conformity with privacy rules for ensuring the fidelity and confidentiality of patient datasets (Javaid, Haleem, Singh, & Suman, 2023). There must be transparency between clinical practitioners and patients regarding the working strategy of AI-enabled microbial threat diagnosis. Thus, ethical standards focus on the development and advancement of transparent AI-based diagnostic models that may deliver logical justifications very rapidly for their incurred predictions.

In some instances, AI may also be involved with a few structural barriers, including cultural diversity, gender, language variation, or racial distinction that may impact the inclusion of a specific patient in the data. Automatic scheduling software or customer service can result in discrepancies. As a result, it is quintessential that AI model developers and healthcare units must include robust methods to recognize and manage the above-discussed biases. Another important aspect is to morally ensure proper access to AI-enabled microbial threat diagnostic tools for people from all sections of society. In this context, healthcare units should emphasize narrowing the technological gap for underprivileged people to get appropriate access to novel AI diagnosis strategies (Naik et al., 2022). AI has the capability to further enhance the proficiency of diagnosis, but obviously involving the role of human beings (clinical professionals). Keeping a proper balance between human intelligence and AI-enabled automation is thus critical for ensuring ethics and patient security. Merits and demerits of AI-integrated microbial diagnosis are presented in Table 12.1.

TABLE 12.1 Merits and demerits of integration of AI in microbial threat diagnosis

MERITS	*DEMERITS*
Rapid diagnosis with enhanced precision, accuracy, and minimum probability of false positives for improved diagnosis outcomes.	AI-enabled diagnostic procedures are often cost-intensive and difficult to install, which can impose intricacies for small-scale healthcare units.
It is possible to detect intricate hidden patterns and details that human beings may skip, resulting in improved and precise identification of pathogenic specimens and their inherited characteristics.	Unbiased and high-quality datasets are quintessential for the learning and training of AI, as biased or incomplete datasets may result in erroneous diagnoses and medication strategy.
It may provide assistance to academic fields for taking justified decisions through the detection of hidden patterns to achieve promising scientific discoveries.	In a few instances, AI-based decision-making processes are complicated to understand by healthcare practitioners, which thereby induces trust issues regarding the recommendations given by AI.
Customized treatment strategy may be developed using patient data, which may deliver improved outcomes and reduce adverse impacts.	AI-enabled methods may deliver inaccurate recommendations because of the biases acquired from the training data collections.
With the use of AI, substantial reduction in manual labor requirement and associated expense is possible.	For proper monitoring of AI-based operations and interpretation of outcomes, qualified AI experts are essential.
AI-based processes are continuously evolving to stay ahead and tackle novel infectious diseases.	Healthcare professionals do not always show eagerness in the adoption of AI-enabled technologies because of the fear of job loss or displacement.
As human errors are restricted to a great extent, AI provides consistent and genuine results, which makes patient treatment safe.	Ethical concerns and privacy protection issues are prevalent during the handling of patient personal data utilized for diagnosis.

12.11 THE FUTURE PROSPECT OF AI IN MICROBIAL THREAT DETECTION

AI-based microbial threat diagnostic tools will have a promising future with the continuous advancement of ML and computational abilities. With the evolvement of AI-based diagnostic systems, disease management has undergone a huge transformation through rapid identification of infection-causing pathogens, even antimicrobial drug-resistant and novel microbial strains. It is expected that eventually patients may be able to identify infectious diseases and monitor their personal health accordingly in real time (Kumar, Koul, Singla, & Ijaz, 2023). Personalized treatments and precision medicinal strategies for individual patients depending on their own microbiota of patients are some promising future aspects of AI. The role of AI in new vaccine development, epidemiological surveillance, and AMR will gradually expand, increasing human's capability of responding effectively toward severe diseases (Arora et al., 2021). Moreover, AI is anticipated to make a significant contribution to the comprehensive in-depth understanding of the human microbiome, emphasizing interactive mechanisms between microbes and hosts. Complex relationships existing in microbial ecosystems may also be explored with the synergistic combination of AI analysis and multi-omics data. Enhanced automation powered by AI-based robotics may simplify regular laboratory work, increasing the efficacy and lowering the chances of human mistakes as the amalgamation of AI will quicken specimen analysis and consequent outcome interpretation. As regulatory norms and ethical considerations are continuously wavering, the utilization of AI-based methods in diagnosis will be a prime component in the future healthcare domain, and the key domains are reflected in Figure 12.3. It is expected to derive substantial transformations in this field by obtaining accurate and personalized medical solutions, ensuring patients' privacy without breaching their personal information, and providing proper accessibility of advanced healthcare techniques to underprivileged sections (Gerke, Minssen, & Cohen, 2020).

12.12 SUMMARY

The strategies for microbial threat detection are continuously evolving with the amalgamation of AI into the field of diagnostics. In conclusion, AI exhibits promising avenues to enhance the in-depth understanding of microbial ecosystems and their complex interactive mechanisms with hosts. Throughout this

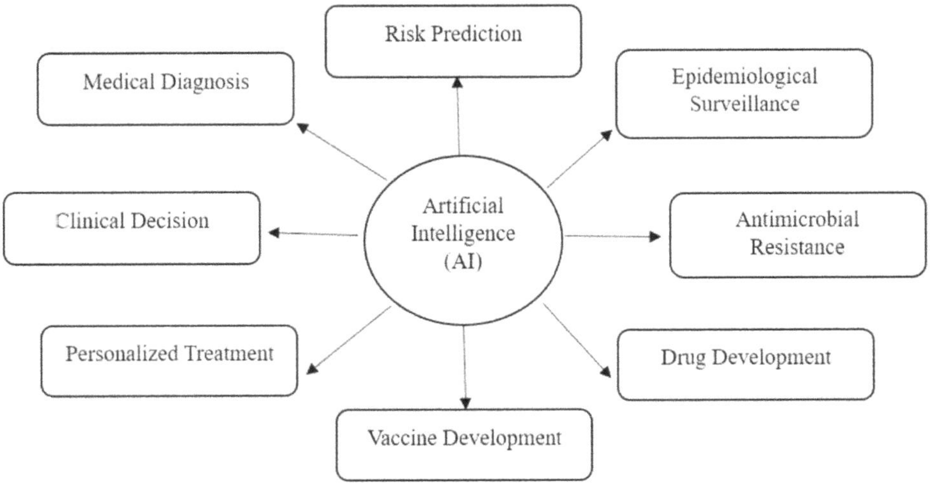

FIGURE 12.3 Key domains of artificial intelligence (AI) in future healthcare diagnostics.

chapter, various applications of AI in microbial diagnosis are explored, starting from predictive modeling to epidemic surveillance, highlighting the huge transformational potential of AI in microbiological research and development. AI shows its proficiency in the accurate and rapid detection of infection-causing pathogens, AMR detection at an initial stage, customized treatments, advanced diagnostic procedures, and detection of epidemic outbreaks. Therefore, with the evolution of AI, healthcare and public health organizations have considerably benefited. Nevertheless, as AI is becoming highly active in the process of medical decision-making, some intricate issues emerge. Intricacies include ethical concerns (confidentiality of patient information), data security, data quality, biases in algorithms, accessibility, transparency, and human intelligence in the supervision process. Careful handling of these issues is quintessential for ensuring effective, responsible, and safe utilization of AI tools in the microbiology sector.

REFERENCES

Agbehadji, I. E., Awuzie, B. O., Ngowi, A. B., & Millham, R. C. (2020). Review of big data analytics, artificial intelligence and nature-inspired computing models towards accurate detection of COVID-19 pandemic cases and contact tracing. *International Journal of Environmental Research and Public Health*, 17(15), 5330. https://doi.org/10.3390/ijerph17155330

Ali, T., Ahmed, S., & Aslam, M. (2023). Artificial intelligence for antimicrobial resistance prediction: Challenges and opportunities towards practical implementation. *Antibiotics*, 12(3), 523. https://doi.org/10.3390/antibiotics12030523

Arango-Argoty, G., Garner, E., Pruden, A., Heath, L. S., Vikesland, P., & Zhang, L. (2018). DeepARG: A deep learning approach for predicting antibiotic resistance genes from metagenomic data. *Microbiome*, 6(1), 23. https://doi.org/10.1186/s40168-018-0401-z

Arora, G., Joshi, J., Mandal, R. S., Shrivastava, N., Virmani, R., & Sethi, T. (2021). Artificial intelligence in surveillance, diagnosis, drug discovery and vaccine development against COVID-19. *Pathogens*, 10(8), 1048. https://doi.org/10.3390/pathogens10081048

Artrith, N., Butler, K. T., Coudert, F.-X., Han, S., Isayev, O., Jain, A., & Walsh, A. (2021). Best practices in machine learning for chemistry. *Nature Chemistry*, 13(6), 505–508. https://doi.org/10.1038/s41557-021-00716-z

Asgharzadeh, M., Ozma, M. A., Rashedi, J., Poor, B. M., Agharzadeh, V., Vegari, A., … & Kafil, H. S. (2020). False-positive *Mycobacterium tuberculosis* detection: Ways to prevent cross-contamination. *Tuberculosis and Respiratory Diseases*, 83(3), 211–217. https://doi.org/10.4046/trd.2019.0087

Attri, I., Awasthi, L. K., & Sharma, T. P. (2023). Machine learning in agriculture: A review of crop management applications. *Multimedia Tools and Applications*, 83(5), 12875–12915. https://doi.org/10.1007/s11042-023-16105-2

Ball, N. M., & Brunner, R. J. (2010). Data mining and machine learning in astronomy. *International Journal of Modern Physics D*, 19(07), 1049–1106. https://doi.org/10.1142/S0218271810017160

Besimbayeva, O. G., Khmyrova, E. N., Tutanova, M. S., Flindt, N., & Sharafutdinov, R. R. (2023). Modern data analysis technologies used for geomechanical monitoring: Review. *Kompleksnoe Ispol'zovanie Mineral'nogo Syr'â/Complex Use of Mineral Resources/Mineraldik Shikisattardy Keshendi Paidalanu*, 326(3), 5–15. https://doi.org/10.31643/2023/6445.23

Bonin, N., Doster, E., Worley, H., Pinnell, L. J., Bravo, J. E., Ferm, P., … & Boucher, C. (2023). MEGARes and AMR++, v3.0: An updated comprehensive database of antimicrobial resistance determinants and an improved software pipeline for classification using high-throughput sequencing. *Nucleic Acids Research*, 51(D1), D744–D752. https://doi.org/10.1093/nar/gkac1047

Buchan, B. W., & Ledeboer, N. A. (2014). Emerging technologies for the clinical microbiology laboratory. *Clinical Microbiology Reviews*, 27(4), 783–822. https://doi.org/10.1128/CMR.00003-14

Cardoso, M. H., Orozco, R. Q., Rezende, S. B., Rodrigues, G., Oshiro, K. G. N., Cândido, E. S., & Franco, O. L. (2020). Computer-aided design of antimicrobial peptides: Are we generating effective drug candidates? *Frontiers in Microbiology*, 10. https://doi.org/10.3389/fmicb.2019.03097

Choi, S. Y., Choi, A., Baek, S.-E., Ahn, J. Y., Roh, Y. H., & Kim, J. H. (2023). Effect of multimodal diagnostic approach using deep learning-based automated detection algorithm for active pulmonary tuberculosis. *Scientific Reports*, 13(1), 19794. https://doi.org/10.1038/s41598-023-47146-0

Davis, J. J., Boisvert, S., Brettin, T., Kenyon, R. W., Mao, C., Olson, R., … & Stevens, R. (2016). Antimicrobial resistance prediction in PATRIC and RAST. *Scientific Reports*, 6(1), 27930. https://doi.org/10.1038/srep27930

Dietvorst, J., Vilaplana, L., Uria, N., Marco, M.-P., & Muñoz-Berbel, X. (2020). Current and near-future technologies for antibiotic susceptibility testing and resistant bacteria detection. *TrAC Trends in Analytical Chemistry*, 127, 115891. https://doi.org/10.1016/j.trac.2020.115891

Drouin, A., Giguère, S., Déraspe, M., Marchand, M., Tyers, M., Loo, V. G., ... & Corbeil, J. (2016). Predictive computational phenotyping and biomarker discovery using reference-free genome comparisons. *BMC Genomics*, 17(1), 754. https://doi.org/10.1186/s12864-016-2889-6

Feng, P., Wang, Z., & Yu, X. (2019). Predicting antimicrobial peptides by using increment of diversity with quadratic discriminant analysis method. *IEEE/ACM Transactions on Computational Biology and Bioinformatics*, 16(4), 1309–1312. https://doi.org/10.1109/TCBB.2017.2669302

Fields, F. R., Freed, S. D., Carothers, K. E., Hamid, M. N., Hammers, D. E., Ross, J. N., ... & Lee, S. W. (2020). Novel antimicrobial peptide discovery using machine learning and biophysical selection of minimal bacteriocin domains. *Drug Development Research*, 81(1), 43–51. https://doi.org/10.1002/ddr.21601

Franco-Duarte, R., Černáková, L., Kadam, S., S. Kaushik, K., Salehi, B., Bevilacqua, A., ... & Rodrigues, C. F. (2019). Advances in chemical and biological methods to identify microorganisms: From past to present. *Microorganisms*, 7(5), 130. https://doi.org/10.3390/microorganisms7050130

Gajic, I., Kabic, J., Kekic, D., Jovicevic, M., Milenkovic, M., Mitic Culafic, D., ... & Opavski, N. (2022). Antimicrobial susceptibility testing: A comprehensive review of currently used methods. *Antibiotics*, 11(4), 427. https://doi.org/10.3390/antibiotics11040427

Gerke, S., Minssen, T., & Cohen, G. (2020). Ethical and legal challenges of artificial intelligence-driven healthcare. In *Artificial Intelligence in Healthcare* (pp. 295–336). Elsevier. https://doi.org/10.1016/B978-0-12-818438-7.00012-5

Ghosh, D., Chakraborty, S., Kodamana, H., & Chakraborty, S. (2022). Application of machine learning in understanding plant virus pathogenesis: Trends and perspectives on emergence, diagnosis, host-virus interplay and management. *Virology Journal*, 19(1), 42. https://doi.org/10.1186/s12985-022-01767-5

Goertzel, B. (2014). Artificial general intelligence: Concept, state of the art, and future prospects. *Journal of Artificial General Intelligence*, 5(1), 1–48. https://doi.org/10.2478/jagi-2014-0001

Gori, E., Callea, E., Alberani, F., & Orlando, L. (2014). Microbial monitoring and methods of sample collection: A GITMO survey (Gruppo Trapianto di Midollo Osseo). *Ecancermedicalscience*, 8. https://doi.org/10.3332/ecancer.2014.421

Her, H.-L., & Wu, Y.-W. (2018). A pan-genome-based machine learning approach for predicting antimicrobial resistance activities of the *Escherichia coli* strains. *Bioinformatics*, 34(13), i89–i95. https://doi.org/10.1093/bioinformatics/bty276

Hinton, G. (2018). Deep learning: A technology with the potential to transform health care. *JAMA*, 320(11), 1101. https://doi.org/10.1001/jama.2018.11100

Hunt, M., Bradley, P., Lapierre, S. G., Heys, S., Thomsit, M., Hall, M. B., ... & Iqbal, Z. (2019). Antibiotic resistance prediction for *Mycobacterium tuberculosis* from genome sequence data with Mykrobe. *Wellcome Open Research*, 4, 191. https://doi.org/10.12688/wellcomeopenres.15603.1

Hyun, J. C., Kavvas, E. S., Monk, J. M., & Palsson, B. O. (2020). Machine learning with random subspace ensembles identifies antimicrobial resistance determinants from pan-genomes of three pathogens. *PLOS Computational Biology*, 16(3), e1007608. https://doi.org/10.1371/journal.pcbi.1007608

Ioannou, P., Baliou, S., & Kofteridis, D. P. (2023). Antimicrobial peptides in infectious diseases and beyond: A narrative review. *Life*, 13(8), 1651. https://doi.org/10.3390/life13081651

Javaid, M., Haleem, A., Singh, R. P., & Suman, R. (2023). Towards insighting cybersecurity for healthcare domains: A comprehensive review of recent practices and trends. *Cyber Security and Applications*, 1, 100016. https://doi.org/10.1016/j.csa.2023.100016

Jeyaraman, M , Balaji, S., Jeyaraman, N., & Yadav, S. (2023). Unraveling the ethical enigma: Artificial intelligence in healthcare. *Cureus*. https://doi.org/10.7759/cureus.43262

Jukič, M., & Bren, U. (2022). Machine learning in antibacterial drug design. *Frontiers in Pharmacology*, 13. https://doi.org/10.3389/fphar.2022.864412

Kavvas, E. S., Catoiu, E., Mih, N., Yurkovich, J. T., Seif, Y., Dillon, N., ... & Palsson, B. O. (2018). Machine learning and structural analysis of Mycobacterium tuberculosis pan-genome identifies genetic signatures of antibiotic resistance. *Nature Communications*, 9(1), 4306. https://doi.org/10.1038/s41467-018-06634-y

Kerkeni, L., Serrestou, Y., Raoof, K., Mbarki, M., Mahjoub, M. A., & Cleder, C. (2019). Automatic speech emotion recognition using an optimal combination of features based on EMD-TKEO. *Speech Communication*, 114, 22–35. https://doi.org/10.1016/j.specom.2019.09.002

Kim, D., Kim, S.-H., Kim, T., Kang, B. B., Lee, M., Park, W., ... & Jo, S. (2021). Review of machine learning methods in soft robotics. *PLoS One*, 16(2), e0246102. https://doi.org/10.1371/journal.pone.0246102

Kontsevaya, I., Cabibbe, A. M., Cirillo, D. M., DiNardo, A. R., Frahm, N., Gillespie, S. H., ... & Goletti, D. (2024). Update on the diagnosis of tuberculosis. *Clinical Microbiology and Infection*, 30(9), 1115–1122. https://doi.org/10.1016/j.cmi.2023.07.014

Kouchaki, S., Yang, Y., Walker, T. M., Sarah Walker, A., Wilson, D. J., Peto, T. E. A., ... & Clifton, D. A. (2019). Application of machine learning techniques to tuberculosis drug resistance analysis. *Bioinformatics*, 35(13), 2276–2282. https://doi.org/10.1093/bioinformatics/bty949

Kuchin, Y. I., Mukhamediev, R. I., & Yakunin, K. O. (2020). One method of generating synthetic data to assess the upper limit of machine learning algorithms performance. *Cogent Engineering*, 7(1), 1718821. https://doi.org/10.1080/23311916.2020.1718821

Kumar, Y., Koul, A., Singla, R., & Ijaz, M. F. (2023). Artificial intelligence in disease diagnosis: A systematic literature review, synthesizing framework and future research agenda. *Journal of Ambient Intelligence and Humanized Computing*, 14(7), 8459–8486. https://doi.org/10.1007/s12652-021-03612-z

Lai, J., Qiu, J., Feng, Z., Chen, J., & Fan, H. (2016). Prediction of soil deformation in tunnelling using artificial neural networks. *Computational Intelligence and Neuroscience*, 2016, 1–16. https://doi.org/10.1155/2016/6708183

Lechowicz, L., Urbaniak, M., Adamus-Białek, W., & Kaca, W. (2013). The use of infrared spectroscopy and artificial neural networks for detection of uropathogenic Escherichia coli strains' susceptibility to cephalothin. *Acta Biochimica Polonica*, 60(4), 713–718. https://www.ncbi.nlm.nih.gov/pubmed/24432322

LeCun, Y., Bengio, Y., & Hinton, G. (2015). Deep learning. *Nature*, 521(7553), 436–444. https://doi.org/10.1038/nature14539

Lei, J., Sun, L., Huang, S., Zhu, C., Li, P., He, J., ... & He, Q. (2019). The antimicrobial peptides and their potential clinical applications. *American Journal of Translational Research*, 11(7), 3919–3931. https://www.ncbi.nlm.nih.gov/pubmed/31396309

Lu, J., Luo, L., Zhang, L., Chen, W., & Zhang, Y. (2010). Increment of diversity with quadratic discriminant analysis: An efficient tool for sequence pattern recognition in bioinformatics. *Open Access Bioinformatics*, 2, 89–96. https://doi.org/10.2147/OAB.S10782

Májek, P., Lüftinger, L., Beisken, S., Rattei, T., & Materna, A. (2021). Genome-wide mutation scoring for machine-learning-based antimicrobial resistance prediction. *International Journal of Molecular Sciences*, 22(23). https://doi.org/10.3390/ijms222313049

Melo, M. C. R., Maasch, J. R. M. A., & de la Fuente-Nunez, C. (2021). Accelerating antibiotic discovery through artificial intelligence. *Communications Biology*, 4(1), 1050. https://doi.org/10.1038/s42003-021-02586-0

Miller, J. M., Binnicker, M. J., Campbell, S., Carroll, K. C., Chapin, K. C., Gilligan, P. H., ... & Yao, J. D. (2018). A guide to utilization of the microbiology laboratory for diagnosis of infectious diseases: 2018 Update by the Infectious Diseases Society of America and the American Society for Microbiologya. *Clinical Infectious Diseases*, 67(6), e1–e94. https://doi.org/10.1093/cid/ciy381

Miotto, R., Wang, F., Wang, S., Jiang, X., & Dudley, J. T. (2018). Deep learning for healthcare: Review, opportunities and challenges. *Briefings in Bioinformatics*, 19(6), 1236–1246. https://doi.org/10.1093/bib/bbx044

Mukhamediev, R. I., Kuchin, Y., Amirgaliyev, Y., Yunicheva, N., & Muhamedijeva, E. (2022). Estimation of filtration properties of host rocks in sandstone-type uranium deposits using machine learning methods. *IEEE Access*, 10, 18855–18872. https://doi.org/10.1109/ACCESS.2022.3149625

Mukhamediev, R. I., Popova, Y., Kuchin, Y., Zaitseva, E., Kalimoldayev, A., Symagulov, A., ... & Yelis, M. (2022). Review of artificial intelligence and machine learning technologies: Classification, restrictions, opportunities and challenges. *Mathematics*, 10(15), 2552. https://doi.org/10.3390/math10152552

Naik, N., Hameed, B. M. Z., Shetty, D. K., Swain, D., Shah, M., Paul, R., ... & Somani, B. K. (2022). Legal and ethical consideration in artificial intelligence in healthcare: Who takes responsibility? *Frontiers in Surgery*, 9. https://doi.org/10.3389/fsurg.2022.862322

Nguyen, M., Brettin, T., Long, S. W., Musser, J. M., Olsen, R. J., Olson, R., ... & Davis, J. J. (2018). Developing an in silico minimum inhibitory concentration panel test for Klebsiella pneumoniae. *Scientific Reports*, 8(1), 421. https://doi.org/10.1038/s41598-017-18972-w

Nguyen, M., Long, S. W., McDermott, P. F., Olsen, R. J., Olson, R., Stevens, R. L., ... & Davis, J. J. (2019). Using machine learning to predict antimicrobial MICs and associated genomic features for nontyphoidal *Salmonella*. *Journal of Clinical Microbiology*, 57(2). https://doi.org/10.1128/JCM.01260-18

Padmapriya, S. T., & Parthasarathy, S. (2024). Ethical data collection for medical image analysis: A structured approach. *Asian Bioethics Review*, 16(1), 95–108. https://doi.org/10.1007/s41649-023-00250-9

Parvaiz, N., Ahmad, F., Yu, W., MacKerell, A. D., & Azam, S. S. (2021). Discovery of beta-lactamase CMY-10 inhibitors for combination therapy against multi-drug resistant Enterobacteriaceae. *PLoS One*, 16(1), e0244967. https://doi.org/10.1371/journal.pone.0244967

Peri, A. M., Stewart, A., Hume, A., Irwin, A., & Harris, P. N. A. (2021). New microbiological techniques for the diagnosis of bacterial infections and sepsis in ICU including point of care. *Current Infectious Disease Reports*, 23(8), 12. https://doi.org/10.1007/s11908-021-00755-0

Ren, J., Ahlgren, N. A., Lu, Y. Y., Fuhrman, J. A., & Sun, F. (2017). VirFinder: A novel k-mer based tool for identifying viral sequences from assembled metagenomic data. *Microbiome*, 5(1), 69. https://doi.org/10.1186/s40168-017-0283-5

Rueda, Z. V., Aguilar, Y., Maya, M. A., López, L., Restrepo, A., Garcés, C., ... & Vélez, L. A. (2022). Etiology and the challenge of diagnostic testing of community-acquired pneumonia in children and adolescents. *BMC Pediatrics*, 22(1), 169. https://doi.org/10.1186/s12887-022-03235-z

Sherman, R. M., & Salzberg, S. L. (2020). Pan-genomics in the human genome era. *Nature Reviews Genetics*, 21(4), 243–254. https://doi.org/10.1038/s41576-020-0210-7

Shrestha, L. B., & Pokharel, K. (2020). Standard operating procedure for specimen collection, packaging and transport for diagnosis of SARS-COV-2. *Journal of Nepal Medical Association*, 58(228). https://doi.org/10.31729/jnma.5260

Spence, M. A., Blanchard, J. L., Rossberg, A. G., Heath, M. R., Heymans, J. J., Mackinson, S., ... & Blackwell, P. G. (2018). A general framework for combining ecosystem models. *Fish and Fisheries*, 19(6), 1031–1042. https://doi.org/10.1111/faf.12310

Stuart, B. (2015). Infrared spectroscopy. In *Kirk-Othmer Encyclopedia of Chemical Technology* (pp. 1–18). Wiley. https://doi.org/10.1002/0471238961.0914061810151405.a01.pub3

Su, M., Satola, S. W., & Read, T. D. (2019). Genome-based prediction of bacterial antibiotic resistance. *Journal of Clinical Microbiology*, 57(3). https://doi.org/10.1128/JCM.01405-18

Sunuwar, J., & Azad, R. K. (2021). A machine learning framework to predict antibiotic resistance traits and yet unknown genes underlying resistance to specific antibiotics in bacterial strains. *Briefings in Bioinformatics*, 22(6). https://doi.org/10.1093/bib/bbab179

Suzuki, S., Ogo, M., Koike, T., Takada, H., & Newman, B. (2015). Sulfonamide and tetracycline resistance genes in total- and culturable-bacterial assemblages in South African aquatic environments. *Frontiers in Microbiology*, 6. https://doi.org/10.3389/fmicb.2015.00796

Ventola, C. L. (2015). The antibiotic resistance crisis: Part 1: Causes and threats. *P & T: A Peer-Reviewed Journal for Formulary Management*, 40(4), 277–283. https://www.ncbi.nlm.nih.gov/pubmed/25859123

Woodman, R. J., & Mangoni, A. A. (2023). A comprehensive review of machine learning algorithms and their application in geriatric medicine: Present and future. *Aging Clinical and Experimental Research*, 35(11), 2363–2397. https://doi.org/10.1007/s40520-023-02552-2

Yang, Y., Niehaus, K. E., Walker, T. M., Iqbal, Z., Walker, A. S., Wilson, D. J., ... & Clifton, D. A. (2018). Machine learning for classifying tuberculosis drug-resistance from DNA sequencing data. *Bioinformatics*, 34(10), 1666–1671. https://doi.org/10.1093/bioinformatics/btx801

Yang, Y., Yao, F., Liao, M., & Huang, Y. (2024). Ubranch conformer: Integrating up-down sampling and branch attention for speech recognition. *IEEE Access*, 12, 144684–144697. https://doi.org/10.1109/ACCESS.2024.3471183

Yazdankhah, S. P., Scheie, A. A., Høiby, E. A., Lunestad, B.-T., Heir, E., Fotland, T. Ø., ... & Kruse, H. (2006). Triclosan and antimicrobial resistance in bacteria: An overview. *Microbial Drug Resistance*, 12(2), 83–90. https://doi.org/10.1089/mdr.2006.12.83

Yelin, I., Snitser, O., Novich, G., Katz, R., Tal, O., Parizade, M., ... & Kishony, R. (2019). Personal clinical history predicts antibiotic resistance of urinary tract infections. *Nature Medicine*, 25(7), 1143–1152. https://doi.org/10.1038/s41591-019-0503-6

Yin, K., Xu, W., Ren, S., Xu, Q., Zhang, S., Zhang, R., ... & Li, R. (2024). Machine learning accelerates de novo design of antimicrobial peptides. *Interdisciplinary Sciences: Computational Life Sciences*, 16(2), 392–403. https://doi.org/10.1007/s12539-024-00612-3

Zhang, J., Li, C., Rahaman, M. M., Yao, Y., Ma, P., Zhang, J., ... & Grzegorzek, M. (2022). A comprehensive review of image analysis methods for microorganism counting: From classical image processing to deep learning approaches. *Artificial Intelligence Review*, 55(4), 2875–2944. https://doi.org/10.1007/s10462-021-10082-4

Zhang, K., Teng, D., Mao, R., Yang, N., Hao, Y., & Wang, J. (2023). Thinking on the construction of antimicrobial peptide databases: Powerful tools for the molecular design and screening. *International Journal of Molecular Sciences*, 24(4), 3134. https://doi.org/10.3390/ijms24043134

Zhang, Y., Ye, T., Xi, H., Juhas, M., & Li, J. (2021). Deep learning driven drug discovery: Tackling severe acute respiratory syndrome Coronavirus 2. *Frontiers in Microbiology*, 12. https://doi.org/10.3389/fmicb.2021.739684

Zhong, Y., Chen, L., Dan, C., & Rezaeipanah, A. (2022). A systematic survey of data mining and big data analysis in internet of things. *The Journal of Supercomputing*, 78(17), 18405–18453. https://doi.org/10.1007/s11227-022-04594-1

Zitnik, M., Nguyen, F., Wang, B., Leskovec, J., Goldenberg, A., & Hoffman, M. M. (2019). Machine learning for integrating data in biology and medicine: Principles, practice, and opportunities. *Information Fusion*, 50, 71–91. https://doi.org/10.1016/j.inffus.2018.09.012

AI-Powered Decision Support Systems for Sustainable Wastewater Management 13

Addressing Nutrient Cycling and Heavy Metal Contamination

Qudrat Ullah, Muhammad Waqar, Sankar Sangeetha Vasanthkumar, and Thanet Khomphet

LIST OF ABBREVIATIONS

ABM	Agent-Based Model
AI	Artificial Intelligence
ANFIS	Adaptive Neuro-Fuzzy Inference System
ANN	Artificial Neural Network
As	Arsenic
BOD	Biological Oxygen Demand
BP-ANN	Backpropagation Artificial Neural Network
BPNN	Backpropagation Neural Network
Cd	Cadmium
COD	Chemical Oxygen Demand
DO	Dissolved Oxygen

DOI: 10.1201/9781003600718-18

FF	Feedforward
GA	Genetic Algorithm
Hg	Mercury
ML	Machine Learning
MLP-ANN	Multilayer Perceptron Artificial Neural Network
MPC	Model Predictive Control
NH_4^+	Ammonium Ion
NRMSE	Normalized Root Mean Square Error
Pb	Lead
PDSS	Powered Decision Support System.
PGPRs	Plant Growth-Promoting Rhizobacteria
PO_4^{3-}	Phosphate Ion
R^2	Coefficient of Determination
RSM	Response Surface Methodology
SVM	Support Vector Machine
TDS	Total Dissolved Solids
TMDLs	Total Maximum Daily Loads
TN	Total Nitrogen
TSS	Total Suspended Solids
WW	Wastewater
WWM	Wastewater Management
WWT	Wastewater Treatment
WWTPs	Wastewater Treatment Plants

13.1 INTRODUCTION

Wastewater management (WWM) is full of challenges, needing more solutions with sustainable approaches to the environment and the population's health. The global generation of wastewater (WW) is on the rise; it is expected to reach 1.3 billion tons per day by 2030, which puts additional pressure on the old and existing WWM facilities as well as natural systems (Wurtsbaugh, Paerl, & Dodds, 2019). Conventional techniques of liquid waste management are unable to remove many of the contaminants, especially nitrogen and phosphorus, which cause eutrophication in receiving water bodies. Additionally, discharges of heavy metals from industries and through urban runoff increase the risk to human life and the well-being of other species (Zhang et al., 2020). These challenges are especially difficult because they require continuous automatic monitoring systems capable of adaptation for responding to variations in the characteristics of WW and environmental factors.

Nutrient recycling and heavy metal pollution mechanisms are essential for reducing the impact of WW on the environment and health. Nutrient control is a critical factor in preventing increased nutrient concentrations in water bodies that lead to the generation of unwanted algal blooms that, in turn, reduce the oxygen concentration and are unfavorable to aquatic life (Briffa, Sinagra, & Blundell, 2020; Lan, Liu, Hu, & Zhu, 2024). Additionally, many hazardous metals and nonmetals, toxic to humans and fauna, for example, lead, cadmium, and mercury, are also long-lasting pollutants capable of accumulating in food chains (Upadhyay, Rani, Kumar, Mythili, & Jain, 2023). The need for the reuse and recycling of nutrients and the elimination of heavy metals through advanced sustainable WWM is core to the principles of a circular economy. Sustainable WWM aligns with the principles of a circular economy by promoting the reuse and recycling of resources. Figure 13.1 depicts this concept through a closed-loop cycle involving recycling, consuming, and manufacturing. In the context of wastewater treatment (WWT), this cycle reflects the recovery of nutrients like nitrogen and phosphorus for reuse, the sustainable management of

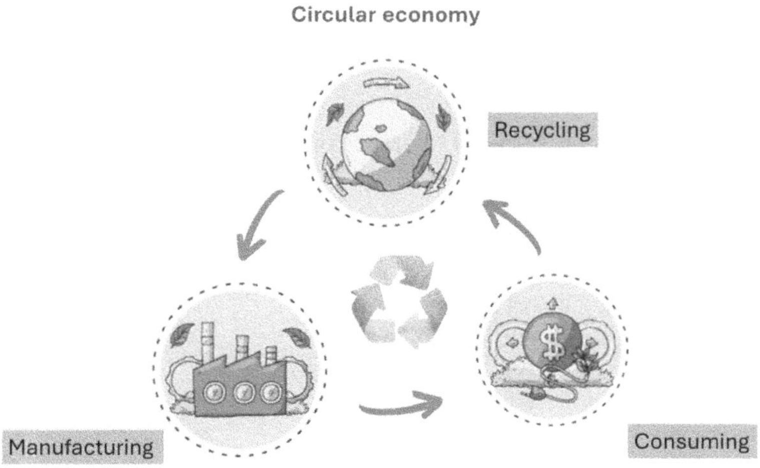

FIGURE 13.1 Circular economy cycle in sustainable wastewater management (WWM).

treated water for consumption, and the manufacturing of byproducts, all of which reduce environmental impact and support long-term ecological balance. These sustainability goals call for these advanced treatment technologies, together with adequate monitoring systems in place.

Machine learning (ML) is now widely adopted as a means of improving decision support systems in the field of sewage management while providing challenging new methods in nutrient recycling and heavy metal removal. ML algorithms, predictive analytics, and other artificial intelligence (AI) enhancements analyze real-time data, which are available in large volumes from the WWT processes and help achieve higher levels of prediction concerning pollutant removal efficiency (Sahu, Kaur, Singh, & Arya, 2023). For example, it has been established that an artificial neural network (ANN) could predict water quality management parameters with accuracies, likely R^2 values, of between 0.64 and 1, which has boosted the working effectiveness of treatment plants (Hassanien, Abouelmagd, Mahmoud, & Darwish, 2023). When applied to DSS, AI helps WWT centers manage their operations efficiently, minimizes costs, and improves the system's capacity to deal with ecological issues, toward supporting environmentally friendly WW handling.

13.2 AI TECHNOLOGIES IN WASTEWATER MANAGEMENT

13.2.1 Overview of AI Applications

AI technologies have found their way into WWM strategies as effective tools that can revolutionize both the shape and efficiency of WWM systems. Among the most common AI applications is the prediction model, which uses past data and trained algorithms to estimate pollutant removal efficiency (Baarimah et al., 2024). For example, AI models found that predictive efficiency (R^2) for most pollutants is between 0.64 and 1.00, which can be considered a highly accurate estimate of treatment effectiveness (Cechinel et al., 2024). Such capability lets operators set treatment parameters and achieve the intended levels of organic and inorganic pollutants in aspects of permitted regulation and safety to the environment.

Besides predictive analysis, probably the most important use of AI in the domain of WWT is real-time data monitoring of water quality indicators. Conventional water quality parameters, which include pH, chemical oxygen demand (COD), biological oxygen demand (BOD), turbidity, total dissolved solids (TDS), and conductivity, which are critical to water quality, can be evaluated through sophisticated advanced sensors integrated with AI algorithms (Ghosh, Adhikary, & Sau, 2024). The combination of these technologies allows wastewater treatment plants (WWTPs) to adapt effectively to changes in influent characteristics, thus keeping treatment conditions constant. For instance, real-time data analysis could cause immediate adjustments in aeration and chemical addition, which are vital in attaining good treatment efficiency and conforming to the stipulated environmental requirements (Olmo Bau, 2024). This aggressive form of monitoring dramatically improves the operational readiness of WWT facilities.

Another important application area of AI technologies is fault detection in treatment processes. This will help WWTPs leverage the ML methodologies on the operational data to pick out deviations and probable equipment failures before they occur. Such a predictive maintenance function is crucial to improving system availability and keeping down the maintenance expenses associated with failures. Literature proves that fault detection systems facilitated by AI considerably decrease the maintenance expenses by reducing to 30%, which improves the reliability of treatment processes in general (Gawde et al., 2023; Goriparthi, 2024). Moreover, the function of real-time fault identification means that the accuracy of facilities to sustain peak performance is enhanced, particularly when faced with different modes of operation, hence enhanced treatment.

13.2.2 Predictive Accuracy of AI Models (R^2 Values)

To evaluate the performance of AI-based predictive models, the coefficient of determination (R^2) is used. It quantifies how well the predicted values ($\hat{y}_i$) align with the actual observed values (y_i), offering a measure of the model's accuracy. The equation is expressed as:

$$R^2 = 1 - \frac{\sum (y_i - \hat{y}_i)^2}{\sum (y_i - \bar{y})^2}$$

where:

y_i: observed values.
$\hat{y}_i$: predicted values by the model.
$\bar{y}$: mean of the observed values.
R^2: coefficient of determination (ranges from 0 to 1, with higher values indicating better predictive accuracy).

This equation is widely used to assess the performance of AI models in WWT, such as nitrogen and phosphorus removal predictions.

13.2.2.1 *For example*

- ANN predictive accuracy for nitrogen removal: $\boldsymbol{R^2 = 0.989 - 0.997}$ (Du et al., 2021).
- SVM for phosphorus prediction: $\boldsymbol{R^2 = 0.93}$ (Manu & Thalla, 2017).
- Hybrid AI models for nutrient removal efficiencies: $\boldsymbol{R^2 > 0.98}$ (Elbeltagi et al., 2022).

13.2.3 Benefits of AI Integration

It is most important to note that the use of AI technologies in the process of WWM brings about key advantages such as cost optimization and smooth efficiency. Through computerizing action plans for the treatment process and incorporating methods of data forecasting and monitoring, WWTPs can experience significant cost savings. For example, AI applications in optimizing aeration procedures to decrease power usage by 20%–30% would lead to notable cost savings, but maintain the accomplishment of environmental standards (Gu, Li, Yuan, & Yang, 2023). Furthermore, desired variations in characteristics of WW reflect opportunities for better resource management regarding fluctuation in appearances, which bring about advancements in the sustainability of WWM.

In addition, the effectiveness of operations that result from AI integration enhances the treatment's effectiveness and the environment. Regular and proactive outlooks allow for timely reaction to shifts in water quality, whereas the effectiveness of the treatment in compliance with WWTPs' regulations is consistently adapted. This versatility is pivotal to dealing with influential characteristics that vary in a way that affects treatment effectively. In addition, many AI implementations cover aspects of preprocessing data in WWTPs, big dataset collection, and analysis, resulting in better decision-making as well as more effective operational activities (Gupta et al., 2024). Due to the increasing demand for cost-effective, sustainable WWT, the application of AI in the treatment of water resources will provide a better impetus for enhanced water treatment systems.

In addition, it is now clear that the use of AI technologies when planning and carrying out WWT increases the operational efficiency of existing technologies and significantly reduces costs while significantly improving the results of work. The challenge that can be addressed using predictive analytics, real-time monitoring, and fault detection systems in WWTPs has provided sound insights into how these plants can operate within environments that are quite complex in terms of regulation (Therrien, Nicolaï, & Vanrolleghem, 2020). AI is a relatively young field, and it will take further research and development before its capabilities are fully realized to assist WWM in meeting future needs in the optimal and sustainable use of available resources.

The application of AI technologies in WWM involves different subsystems that can help improve the performance of the system. Issues such as pollutant removal efficiencies regarding WW can effectively be analyzed using ANNs, which are among the core AI subsystems (Hassanien et al., 2023). SVMs are well suited for handling small databases and result in good performance of classification, thus ensuring the identification of the ideal condition of the treatment (Manu & Thalla, 2017). Besides, integrating an ANN with another metaheuristic algorithm called genetic algorithm (GA) enhances the optimization since the model's parameters are adjusted to adapt to dynamic operational environments (Elbeltagi et al., 2022). They all contribute to enhancing the deterministic and functional capacities of WWTPs that address sustainable management goals.

13.3 NUTRIENT CYCLING IN WASTEWATER TREATMENT

13.3.1 Importance of Nutrient Management

Nutrient control plays an important role in the balance of aquatic systems of the biosphere and in stopping pollution of the environment. Excessive nutrient discharges, nitrogen and phosphorus, for instance, cause water quality concerns such as eutrophication, where water becomes over-enriched with algae, which depletes water's oxygen levels (Narvarte et al., 2023). They could also cause the death of the water animals and the deterioration of water, thus being dangerous to human and animal life. To be more specific, nutrient over-enrichment has increased the rate of biological phosphorus removal by 38% if enhanced through further treatment processes (Hu, Tong, & He, 2023). So, sound nutrition management measures are crucial to dealing with these impacts and sustaining water assets.

This makes nutrient management even more important than the rules and regulations that are being put into place in the management of WWT. Many areas have set high standards, providing strict control over the emissions of nutrients to prevent the phenomenon of eutrophication in the water. For instance, there was also observed increased practice application of the schemes of total maximum daily loads (TMDLs) focused on nitrogen and phosphate, which called for the application of additional refinement technologies to meet the legal constraints (Rai et al., 2024). The difficulty comes in removing the nutrients while at the same time retaining the efficiency and cost-effectiveness of WWTPs. Therefore, new technologies like AI have been considered a possible solution to strengthen nutrient management practices.

13.3.2 Artificial Intelligence Approaches to Nutrient Recycling

Literature review shows that there are various methods that AI presents in nutrient cycling in WWT, hence increasing nutrient removal effectiveness (Sahu et al., 2023). Another way is based on the creation of predictive models using such approaches as ML to make inferences on such data and predict nutrient removal performance under different operating conditions (Zaghloul & Achari, 2022). Such models have been established to predict nitrogen removal processes with reasonable measures of precision, as evidenced by R^2 values of between 0.989 and 0.997 (Du et al., 2021). By adopting such approaches, the WWTP operators are well placed to obtain a clear understanding of which strategies to adopt on the processes in scenarios where nutrient removal goals need to be achieved without using a lot of resources.

Apart from the implementation of equations for the prediction of nutrient cycling efficiency, real-time adjustments derived from AI analytics are also highly significant. A combination of real-time sensor technology and AI makes for long-term water quality, utilizing primary indicators, including total nitrogen (TN) and phosphorus. Real-time data also enable the establishment and application of control measures that counteract current conditions. This fact tells us that fluctuations in the nutrient concentration of inflowing WW can cause changes in the dynamics of the treatment process. For instance, AI systems can predict, determine, and control the required aeration rates and the chemical deposition rate in real time, which has been found to result in high nutrient removal efficiency as well as low operational costs (Gu et al., 2023; Singh et al., 2023). Some of such adaptive management strategies can not only ameliorate nutrient cycling but also assist in the general sustainability of the given WWT operations. AI applications in nutrient management effectively respond to some of the issues regarding relative variability and uncertainty of data in the scope of WWT. Using big data and complicated multivariate statistics, AI models can analyze patterns that define process improvement. This capability is particularly relevant within nutrient cycling due to direct variations in influent characteristics due to seasons and differences in practices like land usage. Consequently, the use of AI-supported approaches enables a better understanding and decoding of nutrient dynamics for the development of effective preventive measures and management practices, guaranteeing improvements in water quality and the condition of aquatic environments (Murmu et al., 2024). The efficiency of nutrient recycling techniques in AI subsystems reduces the impacts of WW through predictive and adaptive procedures. ANNs perform well in predicting nitrogen removal with R^2 values of between 0.989 and 0.997, as in anammox and partial nitridations (Du et al., 2021). SVM is effective in the prediction of phosphorus removal with a coefficient of determination of 0.93 and enables accurate decisions to be made on the changes of the parameters in the treatment process (Manu & Thalla, 2017). Aldrich artificial neural network–support vector machine (ANN-SVM) mixes the advantages of both methods and achieves an efficiency of nutrient cycling equal to 88% using real-time data on dissolved oxygen (DO), total suspended solids (TSS), and influent flow rates (Guo et al., 2018). These subsystems make it possible to manage nutrients in the WWTPs under different water quality situations, thus contributing to environmental conservation.

Table 13.1 provides a brief comparison of different AI models used for predicting nutrient removal efficiency for the WWT process. Using the coefficient of determination (R^2), the better the prediction accuracy of the models, the higher the nutrient removal outcome they forecast. It is shown that the choice of ANN as well as SVM yields high prediction accuracy ranging from 0.989 to 0.997, evidencing that the methods are useful for estimating nitrogen removal efficiencies. The agent-based model (ABM) also produces a good fit of 90% accuracy and records a 38% enhancement in biological phosphorus removal than

TABLE 13.1 Summary of AI models for nutrient removal efficiency

MODEL TYPE	*PREDICTION ACCURACY (R^2)*	*KEY FINDINGS*	*REFERENCES*
Artificial neural network (ANN)	0.989–0.997	Demonstrated high accuracy in predicting nitrogen removal during anammox and partial nitritation processes.	Du et al. (2021)
Support vector machine (SVM)	0.989–0.997	Effective in forecasting nitrogen concentration in wastewater effluent, comparable to ANN performance.	Wu et al. (2024)
Hybrid AI models	Not specified	Proposed to improve the predictive accuracy of pollutant removal efficiency and water quality parameters.	Wu and Wang (2022)
Agent-based model (ABM)	0.90	Outperformed traditional models with a 38% improvement in biological phosphorus removal rates.	Smith (2022)
Q-learning algorithm	Not specified	Developed for controlling aerobic and anaerobic processes, aiming for stable effluent quality.	Croll, Ikuma, Ong, and Sarkar (2023)

using conventional approaches. In addition, the potential of hybrid AI models is recognized, but concrete numbers of accuracy are not presented, suggesting further work to improve prediction. The adoption of Q-learning as part of the systems enhancements can be noted as a progressive step in the overall approach to nutrient removal, wherein adaptive management has been stressed to result in sustainable performance of WWT facilities. In total, it is possible to conclude that the application of AI technologies is effective in raising the level of nutrient management practices, which is extremely significant in minimizing the environmental consequences of WW discharge.

13.4 HEAVY METAL CONTAMINATION

13.4.1 Sources and Effects of Heavy Metals

Increased concentration of heavy metals in WW is an important environmental problem because these metals are toxic to human organisms and aquatic life. Some of the known heavy metals present in WW pollution are lead (Pb), cadmium (Cd), mercury (Hg), and arsenic (As), which are usually because to industrial discharge, mining, and distasteful waste disposal (Jagaba et al., 2024). For example, in the electroplating process, battery manufacturing, and metal finishing industries, WW seems to contain high levels of heavy metals (Zhang et al., 2024). When in water bodies, these metals accumulate in the tissues of living organisms, have toxic effects which are bio-magnified through the food chain, and affect humans through the consumption of contaminated fish and other marine products.

It turns out that heavy metals impact every cell of the human body and have a wide range of consequences for human health. Heavy metals can lead to different health complications, including neurological diseases, growth problems, different cancers, and so on. For instance, exposure to lead can badly affect young people's health and can cause numerous impairments resulting in developmental delays (Jomova et al., 2022; World Health Organization, 2023). Moreover, heavy metals pose a threat to such habitats by changing species abundance and overall species diversity. These metals are potentially toxic to aquatic life, affect the growth and/or reproduction of species at the population level, and result in disruptions in the balance of biotic and abiotic factors in an ecosystem. Thus, efficient measures should be taken to minimize the effects provoked by heavy metal pollution in WW.

13.4.2 Artificial Intelligence for Management of Heavy Metal

AI provides unique approaches to eliminating high concentrations of heavy metals in WWT. An example of applying this is the utilization of ML techniques to identify the removal efficiency of heavy metals over the treatment process. Such predictive models can be used to examine the patterns and associations of historical data retrieved from the WWTPs so that operational changes can be made. For example, researchers have shown that techniques based on ML provide good prediction capabilities; the coefficient of determination is higher than 0.90 for some of the heavy metal removal techniques (Yaqub, Eren, & Eyupoglu, 2021). By using these models, operators will be able to adjust treatment conditions, including chemical dosages and contact times, to achieve better removals of heavy metals.

Other than predictive modeling, the installation of sensor networks for continuous detection of the concentration of heavy metals in WW is important in the management of heavy metals (Leburu et al., 2024). It is composed of sensor networks that allow constant monitoring of pertinent attributes, including pH, conductivity, and concentration of certain heavy metals, to produce an early signal on contamination. Such data can be analyzed using AI to give insights that can be used to intervene when the level of heavy metals present in water reaches the industry-set threshold. For instance, when levels of lead or cadmium are high, the AI systems automatically control the treatment processes or warn operators that there are certain problems (Safeer et al., 2022). This added approach not only increases the efficiency of removing heavy metals but also helps avoid violations of ecological legislation and safeguards people's lives and water resources.

Acceptance and uncertainty in heavy metal management also redefine the role of AI technologies in the treatment of WW. Traditional ML models are particularly useful because they can incorporate comprehensive large datasets and complex statistical analysis to explain heavy metal behavior in WW; in addition, AI models can adapt ML to account for complex interactions between heavy metal and other WW constituents and fluctuations in influent composition (Jiang, Li, Song, & Wang, 2022; Yuan et al., 2023). This capability is useful, especially in responding to treatment processes, which may require other great and stronger management plans. In the future, as improvements in research continue in this field, heavy metal management with the assistance of AI solutions will further improve the sustainability and efficiency of WWT facilities and greatly improve the quality of water resources and the protection of the health of the population. Removal and management of heavy metals in waste sludge utilize AI subsystems for high effectiveness in the process and integration. Possible uptakes from 85% to 95% are realized by ANNs in simulating the adsorption characteristics of heavy metals like lead and cadmium (Ashkanani et al., 2024). Random forest (RF), with its removal efficiencies ranging from 82-RFC to 92-RFC, is efficient in managing multivariable interaction patterns in complex WWs (Ashkanani et al., 2024). Reinforcement learning provides additional improvement in the management by enabling the implementation of effective changes in real time, in response to possible variations in heavy metal concentrations (Ashkanani et al., 2024). All these subsystems enhance capabilities in WWTPs to reduce the effects of heavy metal pollutants in line with the objectives of protecting human health as well as aquatic life forms.

Several AI techniques have been used to remove heavy metals from WW; Table 13.2 presents a summary of various AI techniques, their effectiveness determined by the percentage of removal. The studies prove that both the hybrid AI models and ANN have the highest removal efficiencies, ranging from 90% to 95% and 85%, respectively, and therefore are sensitive options for enhancing the treatment processes of heavy metals. Decision tree and RF manage to show a good amount of percentage efficiency of ranges 80%–90% and 88%–93%, respectively; this shows that the two classifiers are reliable for heavy metal removal in perspective applications. However, fuzzy logic Systems yield the least removal efficiencies, which fall between 70% and 80%, and the authors noted that while it can be beneficial, it may not be as efficient as other AI classes. Collectively, these values show that the considered AI-driven methodologies can help to improve the effectiveness of heavy metal removal processes: the trend reveals that the hybrid and neural network methodologies, which are the focus of the present study, can be particularly beneficial when it comes to achieving high removal levels in the context of WWT.

TABLE 13.2 AI techniques for heavy metal removal

TECHNIQUE	*REMOVAL EFFICIENCY (%)*	*REFERENCES*
Artificial neural networks (ANN)	85–95	Ashkanani et al. (2024)
Support vector machines (SVM)	80–90	
Decision trees	75–85	
Random forest	82–92	
Genetic algorithms	78–88	
Fuzzy logic systems	70–80	
Hybrid AI models	90–95	
Reinforcement learning	85–90	
Ensemble learning	88–93	

13.5 DECISION SUPPORT SYSTEMS (DSS) POWERED BY AI

13.5.1 Framework for AI-Powered DSS

The incorporation of AI into DSS is a type of modern management approach to improving decision-making across multiple domains, especially in the field of WW. The key components of a functioning AI-based DSS have been identified as data management, model management, user interface, and knowledge management. Big data collection includes the necessary raw material for preparing the basis of AI models, including data collection and storage (Himeur et al., 2023; Soori, Jough, Dastres, & Arezoo, 2024). For example, WWTP data as input data feeds the AI system to predict subsequent pollutant efficiency for removals (Z. Ahmed, Mohamed, Zeeshan, & Dong, 2020; Bahramian, Dereli, Zhao, Giberti, & Casey, 2023). Decision-making and efficiency in managing WW are facilitated using multiple AI subsystems in the development of the AI-powered DSS for WWM. ANNs are based on predictive accuracy, which would show accurate fulfillment when predicting treatment results in nutrient and heavy metal removal processes (Himeur et al., 2023). SVMs help bring aspects by monitoring variations in water quality parameters, including COD or TDS, and maintain the stability of the system (Soori et al., 2024). Optimization through GAs adjusts parameters, is dynamically scalable, and adapts the properties needed for a DSS to effectively address various operational crises. The AI-powered decision support system (AI-PDSS) leverages advanced technologies to optimize WWT processes. As depicted in Figure 13.2, the system integrates ML modules such as ANNs and SVMs for predictive accuracy, alongside real-time monitoring through sensors that feed data into cloud-based AI platforms. This framework facilitates significant outcomes, including a 31.4% energy reduction through optimized aeration control, efficient nutrient removal (e.g., 80%–93% efficiency for nitrogen and phosphorus), and heavy metal reduction (e.g., 85%–95% removal of Pb and Cd using ANN/SVM control), contributing to sustainable WWM. At AgroFresh Solutions, Inc., this integration helps in the creation of a framework that enhances sustainable methods in WW handling and treatment.

13.5.2 Components of an Effective Decision Support System

It also means that any AI-powered DSS in WWM must include superior analytical tools that support the real-time analysis feature (Nagpal, Siddique, Sharma, Sharma, & Mittal, 2024). These tools help operators evaluate the efficiency of the treatment processes continuously and make decisions. For instance,

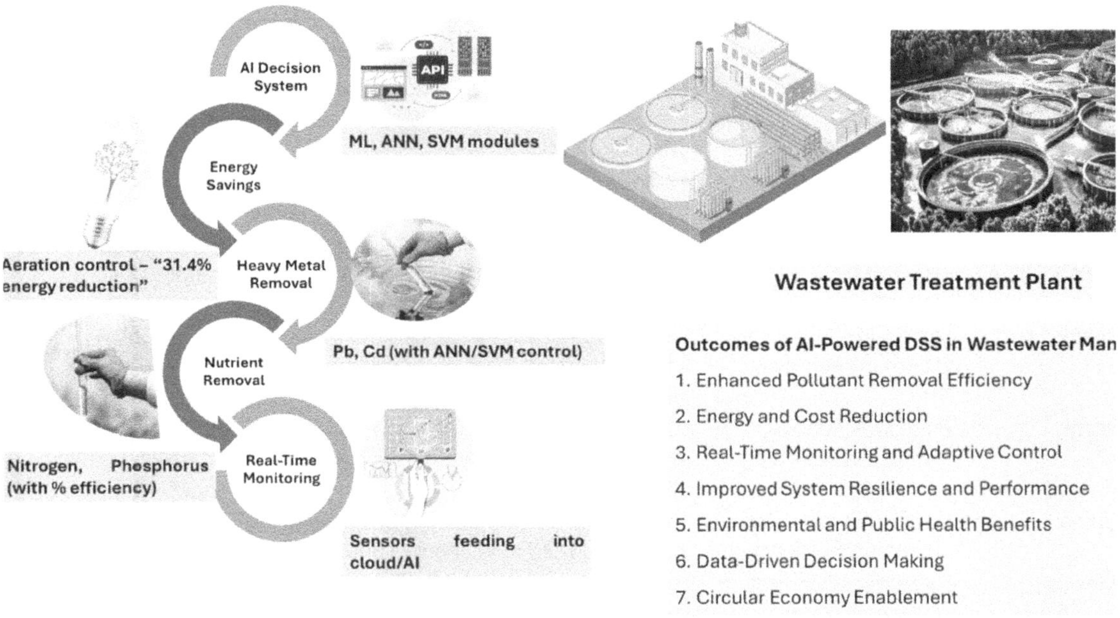

FIGURE 13.2 AI-powered decision support system (AI-PDSS) for wastewater management (WWM).

heavy metal removal efficiencies can be predicted by using AI models with reported values ranging from 85% to 95% for ANN and 80% to 90% for the SVM (Yuan et al., 2023). In addition, the integration of sensor networks to provide real-time data improves the performance of the system in changing parameters of water quality like BOD, COD, and TDS (U. Ahmed, Mumtaz, Anwar, Mumtaz, & Qamar, 2020). When these components are incorporated, an AI-PDSS can enhance the functionality of WWTPs, lower costs, and meet environmental regulations.

13.5.3 Usage and Case Studies

Several examples are also demonstrated to explain how the AI-embedded DSS could be helpful in WWM, as several of its cases have been applied to solve the problems successfully. For example, Maleki, Karimi-Jashni, and Mousavifard (2024) discussed how the use of ML models in the prediction of the removal efficiency of metals in a municipal WWTP. Another study concluded that the level of accuracy that was realized by the chosen ANN model was high, as pointed out by the R^2 value of 0.998 (Kasa & Mohd, 2024). This capability enabled operators to adjust the treatment conditions, which led to an appreciable decrease in the concentration of heavy metals in effluent releases. Likewise, Bahramian et al. (2023) investigated the applicability of composite ML approaches with many AI techniques for improving the predictive efficiency of pollutant removal. Another work showed that hybrid models could remove efficiency of more than 90%, and therefore, the use of the AI-powered DSS could enhance WWT systems (Sanjeevi, Sathvara, Tripathi, & Anuradha, 2024).

AI-integrated DSS not only improves decision-making in the field of WWM but also provides direction toward the sustainability of the environment. With the support of the most sophisticated approaches, like ML and real-time supervision, such systems make it possible to enhance the efficiency of treatment activities and minimize spending and adverse consequences for the environment. Nonetheless, there are challenges like the availability of high-quality data, how to combine numerous data sources, and how best to design interfaces that are easy to use. Subsequent research should target the following by

TABLE 13.3 Case studies of AI-powered DSS in wastewater management

LOCATION	*AI APPLICATION*	*OUTCOMES*
Detroit, Michigan	Data Mining for Aeration Optimization	Achieved a 31.4% reduction in energy consumption while maintaining effluent quality above standards (Asadi, Verma, Yang, & Mejabi, 2017).
Australia	AI-Powered Intelligence Platform	Enhanced operational efficiency, leading to significant cost savings and improved treatment outcomes (Nagpal et al., 2024).
Municipal WWTP, USA	Machine Learning for Heavy Metal Removal	ANN model achieved an R^2 value of 0.998, optimizing treatment conditions and reducing heavy metal concentrations (Rajaee, Khani, & Ravansalar, 2020).
Spain	Hybrid AI Models for Pollutant Removal	Removal efficiencies exceeded 90%, demonstrating the effectiveness of combined machine-learning techniques (Bahramian et al., 2023).
Brazil	Predictive Control with Deep Learning	Achieved a 16.7% decrease in electricity consumption compared to normal operating conditions (Filipe, Bessa, Reis, Alves, & Póvoa 2019).
China	Neural Networks for Pumping Efficiency	Average energy savings of approximately 10%, with a maximum reduction of 25% in energy consumption (Huang et al., 2020).
Mexico	Hierarchical Control Strategy	Reduced overall costs by 0.8% and decreased aeration energy by around 6% (Choudhury, Karki, Choudhury, Narad, & Sengupta, 2024).

imposing different approaches to data management and improving the integration of AI systems into an organization's framework. With the advancement of the field of AI, there will be a greater possibility for AI-powered DSS to change the practice of WWM to a more efficient and sustainable manner because of the growing environmental challenges.

The idealized form of numerous case studies that seek to show the use of AI-based DSS in several places is presented in Table 13.3. Here, the applications of AI are shown across a wide spectrum: data mining for aeration optimization in Detroit has led to a substantial 31.4% reduction in energy while maintaining effluent quality adequately. For example, in Australia, the adoption of the intelligence platform resulted in efficiency gains and a reduction of operational costs streamwise. ML applications for heavy metal removal in a real municipal WWTP in the USA presented outstanding predictive efficiency, in which an ANN model had an R^2 of 0.998, which resulted in the appropriate control of the treatment conditions. In addition, AI models used in Spain brought pollutant removal efficiency of more than 90%, and predictive control in Brazil lowered electricity utilization by 16.7%. The use of the neural network, an advanced technology used in China, revealed an average energy saving of about 10%, and a maximum energy saving of up to 25%. Finally, a hierarchical control strategy developed in Mexico contributed to the overall cost decrease of 0.8% and aeration energy reduction of approximately 6%. Given collectively, these cases all highlight AI-powered DSS as an essential enabling factor in renewing and improving the efficiency and sustainability of processes of WWT through varied positive consequences and impacts, revealing energy savings, cost optimization, and generally, better treatment rates.

13.5.3.1 Equations for predictive modeling

In WWT systems, the application of predictive modeling in establishing the effectiveness of different treatment processes is paramount. Presented in this chapter are common equations that have been used in predicting nutrient and heavy metal removal efficiencies in the WWT systems. These equations are used as basic equations by engineers and researchers to assess the efficiency of treatment technologies and control the parameters of operation.

13.6 EQUATION FOR PREDICTING NUTRIENT REMOVAL EFFICIENCY

The efficiency of nutrient removal in WWT can be quantified using the following equation:

$$E = \frac{C_{in} - C_{out}}{C_{in}} \times 100$$

where:

- E is the nutrient removal efficiency expressed in percentage.
- C_{in} is the influent concentration of the nutrient (e.g., nitrogen, phosphorus) entering the treatment system.
- C_{out} is the effluent concentration of the nutrient after treatment.

This equation provides a straightforward method for calculating the percentage of nutrients removed during the treatment process (Mojiri, Ohashi, Ozaki, Aoi, & Kindaichi, 2020; Narvarte et al., 2023; Zhao et al., 2018). By comparing the influent concentration (C_{in}) with the effluent concentration (C_{out}), the equation allows for the assessment of how effectively the treatment system is functioning. A higher value of E indicates a more efficient removal process, which is critical for meeting regulatory standards and protecting environmental water quality.

For instance, if a WWTP receives influent with a nitrogen concentration of 50 mg/L and discharges effluent with a nitrogen concentration of 10 mg/L, the nutrient removal efficiency can be calculated as follows:

$$E = \frac{50 - 10}{50} \times 100 = 80\%$$

This result indicates that 80% of the nitrogen has been effectively removed from WW, showcasing the treatment system's capability.

13.6.1 Equation for Heavy Metal Removal Prediction

The removal of heavy metals from WW is another critical aspect of treatment processes. The efficiency of heavy metal removal can be predicted using the following equation:

$$R = \frac{C_{in} - C_{final}}{C_{in}} \times 100$$

where:

- R is the removal rate of heavy metals expressed in percentage.
- C_{in} is the initial concentration of the heavy metal in the influence.
- C_{final} is the final concentration of the heavy metal after treatment.

This equation is essential for evaluating the effectiveness of various treatment methods in removing heavy metals, which are often toxic and pose significant environmental risks (Mojiri et al., 2020; Vakili et al., 2021; Zaferani, Emami, Amiri, & Binaeian, 2019). By calculating the removal rate (R), operators can determine how well the treatment system is performing in terms of heavy metal reduction.

For example, if a treatment facility receives WW with an initial lead concentration of 100 μg/L and achieves a final concentration of 10 μg/L after treatment, the removal rate can be calculated as follows:

$$R = \frac{100 - 10}{100} \times 100 = 90\%$$

This means that 90% of the lead ions have been removed from WW, and their treatment process shows how the intensity of heavy metal pollution can be alleviated.

The equations demonstrated in this chapter are basic equations for the predictive modeling of WWT. They give a numerical denominator upon which the effectiveness of nutrient and heavy metal removal mechanisms can be measured. While practicing the formulas listed above, it will help WWTPs achieve their goal and meet environmental standards, hence contributing to water conservation. These equations will lay the basic framework for creating further enhanced modeling techniques with evolving AI and ML technologies for optimizing the effectiveness of WWTPs.

13.7 CHALLENGES AND FUTURE DIRECTIONS

Some serious challenges exist while applying sophisticated predictive modeling approaches in the domain of WWM. The first of these is data quality and availability, which poses one of the greatest problems for a company. Normally, the data achievable from WWTPs can be inadequate, sparse, or low quality, which may compromise the accuracy of the prediction models. For instance, Bhagat et al. (2021) noted that the numerical value of the correlation coefficient for the RF and ANN model was larger than 0.99, which pertains to high accuracy figures, yet such an assertion is valid on quality input parameters alone. Lack of proper data hinders the working of any model and overfitting as well as underfitting of models may occur which badly hamper their applications in real-life scenarios. Thus, revisiting of data quality problems is urgently needed to increase the efficiency of the prediction models in WWTPs.

However, due to the variability of the operational conditions of different WWTPs, the second challenge regarding the practical use of predictive modeling arises from data quality. It has been found that influent characteristics, seasonal variation, and other operational practices can have profound effects on the treatment (Khurshid & Pani, 2023). For instance, Wang and Man (2021) explained that the behavior of an artificial neural network–genetic algorithm (ANN-GA) model designed for predicting the efficiency of a WWTP witnessed a shift due to the changes in the influent parameters such as pH, TSS, and COD. This non-stationarity requires the creation of rich, flexible models capable of tracking possible changes and correctly predicting treatment performance in different scenarios. As a result, healthcare facilities need to develop procedures for the collection of standardized data that would enable comparisons between facilities.

Regarding future work with the proposed framework, one promising avenue for research is the combination of AI models by employing different modes to enhance the predictive power of the models. ANN as well as SVM can be integrated with GA, which will complement or enhance the features of a single model for improving the efficacy of the predictive models in WWTPs. For instance, it was found that the combination of ANN with GA has produced a relatively low normalized root mean square error (NRMSE) (Elbeltagi et al., 2022). The research going forward should therefore involve testing other complex hybrid models of algorithms as well as fine-tuning those used in this study for enhanced qualitative performances.

In particular, the introduction of big data analytics and ML approaches in real-time predictions for model optimization provides a promising opportunity for WWTP improvement. Another advantage is that through real-time data that sensors and monitoring systems can provide in detail, researchers can create models born from real-time data and update them with new data as it is shown that operational conditions have changed. Besides, it improves the reliability of the forecast and provides timely decision-making

TABLE 13.4 Cost and energy savings through AI-driven wastewater treatments

AI APPLICATION	*REDUCTION IN COST (%)*	*ENERGY SAVINGS (%)*	*AI MODEL USED*	*OPTIMIZATION AREA*	*REFERENCE*
Aeration process	31.4	N/A	Data mining	Aeration-induced oxygen reduction	Nagpal et al. (2024)
Energy consumption	16.7	25	ANN	Sewage pumping systems	Esteves, Cardoso, Leitão, and Pires (2023)
Hybrid AI systems	30	2.2–6	Adaptive kernel models	Energy during variable weather	Zafar et al. (2022)
Operational costs	0.8	6	MPC and FF	Aeration and wastewater quality	Choudhury et al. (2024)
Cost-effectiveness in adsorption	15	10	ANN-GA	Adsorption optimization	Thakur and Dharmendra (2024)

while managing the treatment processes. According to Cardoso, Rodrigues, Gaspar, and Gomes (2021) and (Morris, Georgiou, Guenther, & Caucci, 2021), there is also the potential to identify specific shifts in influent characterizations, which brings better efficiency to WWT systems. Thus, future research should focus on the integration of real-time data analysis with hybrid AI models to compensate for the variability and the issues of data quality to make an advanced step toward more actual and sustainable WWM.

13.7.1 Cost and Energy Savings through AI-Driven Wastewater Treatments

Another multifaceted benefit of applying AI in the management of the water and WW working processes is cost and energy economization, as presented in Table 13.4, reflecting the diverse applications of AI in operation management brought significant savings to the operational costs and energy consumption in various treatment areas. For example, based on the analysis of factors influencing the efficiency of aeration, the introduction of data mining methods of AI has made it possible to save 31.4% of the costs, which indicates the role of AI in improving the effectiveness of traditional approaches (Qambar & Al Khalidy, 2022).

Besides costs, the use of AI models has also resulted in saving energy. ANN integration into the sewage pumping systems has led to a 25% energy savings, proving the practical applicability of AI in energy-focused usage (Mannina, Rebouças, Cosenza, Sànchez-Marrè, & Gibert, 2019). In addition, actual hybrid AI systems have demonstrated the ability to regulate power demand in conditions of fluctuating weather—the cost savings, while preserving the activity level, can exceed 30% (Mazhar et al., 2023). These insights affirm that AI technologies should not only be used to improve treatment outcomes but also should be used to address the management of WW.

Therefore, referring to the data shown in Table 13.4, it can be outlined that the application of AI in WWTs makes it possible to reach the goals of cost and energy-efficient approaches. This study shows how refinements of treatment procedures and minimization of service costs are achievable by AI applications, which greatly enhances the financial sustainability of WWTPs. As more and more people seek ways in which to manage water effectively and sustainably, AI is set to play more of a significant role in determining the future of efficient WWT.

Table 13.4 shows the comparison of cost and energy savings that have been realized by implementing AI models in WWT. It encompasses cost-cutting measures, energy use minimization, and enhancement of treatment procedures.

TABLE 13.5 Heavy metal adsorption efficiencies using various AI models

HEAVY METAL	*TREATMENT METHOD*	*AI MODEL*	*EFFICIENCY (%)*	*R^2 VALUE*	*SIGNIFICANT VARIABLES*	*REFERENCES*
Lead (Pb^{2+})	Adsorption on modified chitosan	MLP-ANN	92	0.995	pH, initial concentration	Zaferani et al. (2019)
Cadmium (Cd^{2+})	Adsorption on biochar	ANN and RF	78.1	0.948	Cation exchange capacity	Zhu, Wang, & Ok (2019)
Zinc (Zn^{2+})	Microfiltration-Complexation	BPNN	85	0.955	TSS, pH	Sekulić, Antanasijević, Stevanović, and Trivunac (2017)
Nickel (Ni^{2+})	Adsorption on Typha domingensis	ANFIS	31	0.972	Adsorption time, pH	Fawzy, Nasr, Adel, Nagy, and Helmi (2016)
Mixed metals	Biochar	ANN	90	0.99	Temperature, biochar properties	Mojiri et al. (2020)

13.7.2 Heavy Metal Adsorption Efficiencies Using Various AI Models

The accuracy of different AI algorithms in the adsorption efficiencies of heavy metals depends on the efficiency of the WWT process. From the results presented in Table 13.5, it is evident that AI models of various effectiveness in eradicating heavy metals, including lead, cadmium, zinc, and nickel. The MLP-ANN model is exceptionally promising, with 92% adsorption efficiency for Pb^{2+} using the modified chitosan as a treatment approach and $R^2=0.995$ (Gautam et al., 2023; Gouvêa, 2024). This means that the model equations predicted the removal efficiencies quite well with the corresponding experimental values, suggesting that the model can therefore be used quite effectively in practice.

The ANFIS model of the adsorption efficiency of Ni^{2+} is only 31%, but it has a good R^2 value of 0.972 (Salim, Oukhrib, & Zaidi, 2024). This shows that nickel removal is not as easy as other metals; it could be affected by factors such as adsorption time and pH (Vakili et al., 2021). The table also shows that the efficiency of the ANN and RF models for predicting the uptake of cadmium (Cd^{2+}) is a combined 78.1%, for which the cation exchange capacity is determined to be a major predictor variable. These conclusions underline the need to choose the correct AI models according to the type of heavy metal to be removed and the type of treatment used.

In general, the information in Table 13.5 indicates the prospects of the AI models for the improvement of the efficiency of heavy metal extraction from WW. These models assist in the proper prediction of the required processes, and given the nature of the treatment processes that go on in the WWT facilities, proper prediction of the same amount to the achievement of higher and better results in terms of environmental qualities as well as meeting regulatory standards. This paper has established that continued development of AI technologies in this field will lead to improvement of heavy metal remediation efforts and hence improvement of water management.

Table 13.5 presents the adsorption effectiveness of developed AI models for several heavy metals employing various treatments. It emphasizes aspects of the accuracy of the models and important findings on metal removal rates.

13.7.3 Performance Metrics of AI Models for Nutrient Removal

It is necessary to measure different AI models for nutrient removal in WW because the performance measure is essential to validate the models in real-world systems. As reflected in Table 13.6 below,

TABLE 13.6 Performance metrics of AI models for nutrient removal

PARAMETER	*AI MODEL*	*REMOVAL EFFICIENCY (%)*	*R^2 VALUE*	*KEY INPUTS*	*REFERENCES*
Total nitrogen (TN)	ANN-GA	90.9	0.990	Influent TN, pH, volumetric flow rate	Mojiri et al. (2020)
Ammonium (NH_4^+)	BP-ANN	80–85	0.989–0.997	Aeration rate, influent NH_4^+	Antwi et al. (2019)
Phosphorus (PO_4^{3-})	Agent-Based Model (ABM)	93.36	N/A	Biological phosphorus uptake rates	Zhao et al. (2018)
Nitrogen (general)	SVM	86	0.93	Influent pH, COD, total solids	Manu and Thalla (2017)
Nutrient cycling	ANN-SVM Hybrid	88	0.98	DO, TSS, influent flow rate	Guo et al. (2018)

it is seen that different AI models have varying effectiveness in removing key nutrients such as nitrogen and phosphorus that are important in water quality and avoiding water blooms. The table shows that with the help of the ANN-GA model, the removal efficiency of TN is very high, namely 90.9% with a high R^2 value of 0.990, revealing a good correlation of the predicted values with the real values. This implies that the ANN-GA model is most appropriate when influent TN, pH, and volumetric flow rate are well measured.

Furthermore, the analysis of data in detail shows that the ABM outcompetes all the models in phosphorus removal with an efficiency of 93.36%. While the R^2 value is not useful here, the model is based on biological phosphorus uptake rates that are particularly useful in biological treatment applications. For ammonium removal, the model also shows high efficiency, which is between 80% and 85%, and an R^2 value between 0.989 and 0.997 to prove its effectiveness under different operation conditions. These observations underscore the need to choose the right AI models relevant to given nutrient removal goals and operating conditions, thus offering better practice in handling WW.

Table 13.6 gives more information about the AI models used in nutrient removal from WW, particularly nitrogen and phosphorus. Some of the comparison is done by the performance indicators such as the R^2 values, removal efficiencies, and the areas of application. The table also contains major conclusions and sources to support the information provided in the paper.

13.8 CONCLUSION

In conclusion, the enhancement initiative of AI for sustainable WWM is a transformation opportunity for increasing efficiency and effectiveness in WWT. The literature has shown that ML algorithms and hybrid modeling methods are promising tools for enhancing pollutant elimination, water quality parameter surveillance, and prognostic analytics to support management choices. Nevertheless, to maximize these advantages, efforts need to be made to solve current problems like data quality, non-similarity of some operational conditions, etc., as well as using the further development of the model, which combines different approaches to the solution of similar problems. Hence, there is a need for a call for action that will support other research and implementation endeavors geared toward establishing standard data collection practices, improving model performance, and incorporating real-time analysis. It is therefore critical to encourage multidisciplinary research among scholars, practitioners, and policymakers to unleash the benefits of AI in sustainable WWM and in general.

REFERENCES

Ahmed, U., Mumtaz, R., Anwar, H., Mumtaz, S., & Qamar, A. M. (2020). Water quality monitoring: From conventional to emerging technologies. *Water Supply*, *20*(1), 28–45.

Ahmed, Z., Mohamed, K., Zeeshan, S., & Dong, X. (2020). Artificial intelligence with multi-functional machine learning platform development for better healthcare and precision medicine. *Database*, *2020*, baaa010.

Antwi, P., Zhang, D., Xiao, L., Kabutey, F. T., Quashie, F. K., Luo, W., Meng, J., & Li, J. (2019). Modeling the performance of single-stage nitrogen removal using anammox and partial nitritation (SNAP) process with backpropagation neural network and response surface methodology. *Science of the Total Environment*, *690*, 108–120.

Asadi, A., Verma, A., Yang, K., & Mejabi, B. (2017). Wastewater treatment aeration process optimization: A data mining approach. *Journal of Environmental Management*, *203*, 630–639.

Ashkanani, Z., Mohtar, R., Al-Enezi, S., Smith, P. K., Calabrese, S., Ma, X., & Abdullah, M. (2024). AI-assisted systematic review on remediation of contaminated soils with PAHs and heavy metals. *Journal of Hazardous Materials*, *468*, 133813.

Baarimah, A. O., Bazel, M. A., Alaloul, W. S., Alazaiza, M. Y., Al-Zghoul, T. M., Almuhaya, B., ... & Mushtaha, A. W. (2024). Artificial intelligence in wastewater treatment: Research trends and future perspectives through bibliometric analysis. *Case Studies in Chemical and Environmental Engineering*, *10*, 100926.

Bahramian, M., Dereli, R. K., Zhao, W., Giberti, M., & Casey, E. (2023). Data to intelligence: The role of data-driven models in wastewater treatment. *Expert Systems with Applications, 217*, 119453.

Bhagat, S. K., Pyrgaki, K., Salih, S. Q., Tiyasha, T., Beyaztas, U., Shahid, S., & Yaseen, Z. M. (2021). Prediction of copper ions adsorption by attapulgite adsorbent using tuned-artificial intelligence model. *Chemosphere*, *276*, 130162.

Briffa, J., Sinagra, E., & Blundell, R. (2020). Heavy metal pollution in the environment and their toxicological effects on humans. *Heliyon*, *6*(9), e04691.

Cardoso, B. J., Rodrigues, E., Gaspar, A. R., & Gomes, Á. (2021). Energy performance factors in wastewater treatment plants: A review. *Journal of Cleaner Production*, *322*, 129107.

Cechinel, M. A. P., Neves, J., Fuck, J. V. R., de Andrade, R. C., Spogis, N., Riella, H. G., ... & Soares, C. (2024). Enhancing wastewater treatment efficiency through machine learning-driven effluent quality prediction: A plant-level analysis. *Journal of Water Process Engineering*, *58*, 104758.

Choudhury, P. R., Karki, P., Choudhury, A., Narad, P., & Sengupta, A. (2024). Economic implications and cost–benefit analysis of ai in wastewater treatment. In Garg, M. C. (Ed.) *The AI Cleanse: Transforming Wastewater Treatment Through Artificial Intelligence: Harnessing Data-Driven Solutions* (pp. 337–361). Springer.

Croll, H. C., Ikuma, K., Ong, S. K., & Sarkar, S. (2023). Systematic performance evaluation of reinforcement learning algorithms applied to wastewater treatment control optimization. *Environmental Science & Technology*, *57*(46), 18382–18390.

Du, T., Xu, X., Guo, D., Jiang, X., Zeng, M., Wu, N., ... & Zhang, Z. (2021). Unraveling the role of salinity in anammox-based nitrogen removal processes via data analysis from the literature and experimental validation. *Environmental Science: Water Research & Technology*, *7*(12), 2295–2306.

Elbeltagi, A., Nagy, A., Mohammed, S., Pande, C. B., Kumar, M., Bhat, S. A., ... & Kovács, E. (2022). Combination of limited meteorological data for predicting reference crop evapotranspiration using artificial neural network method. *Agronomy*, *12*(2), 516.

Esteves, F. A., Cardoso, J., Leitão, S., & Pires, E. (2023). Impact of artificial intelligence in the reduction of electrical consumption in wastewater treatment plants: a review. *Journal of Information Systems Engineering and Management*, *8*(3), 21855.

Fawzy, M., Nasr, M., Adel, S., Nagy, H., & Helmi, S. (2016). Environmental approach and artificial intelligence for Ni (II) and Cd (II) biosorption from aqueous solution using Typha domingensis biomass. *Ecological Engineering*, *95*, 743–752.

Filipe, J., Bessa, R. J., Reis, M., Alves, R., & Póvoa, P. (2019). Data-driven predictive energy optimization in a wastewater pumping station. *Applied Energy*, *252*, 113423.

Gautam, K., Sharma, P., Dwivedi, S., Singh, A., Gaur, V. K., Varjani, S., ... & Ngo, H. H. (2023). A review on control and abatement of soil pollution by heavy metals: Emphasis on artificial intelligence in recovery of contaminated soil. *Environmental Research*, *225*, 115592.

Gawde, S., Patil, S., Kumar, S., Kamat, P., Kotecha, K., & Abraham, A. (2023). Multi-fault diagnosis of industrial rotating machines using data-driven approach: A review of two decades of research. *Engineering Applications of Artificial Intelligence*, *123*, 106139.

Ghosh, D., Adhikary, S., & Sau, S. (2024). AI-based smart water quality monitoring and wastewater management: An integrated bio-computational approach. In Kumar, A., Srivastav, A. L., Dubey, A. K., Dutt, V., & Vyas, N. (Eds.) *Innovations in Machine Learning and IoT for Water Management* (pp. 127–151). IGI Global.

Goriparthi, R. G. (2024). AI-driven predictive analytics for autonomous systems: A machine learning approach. *Revista de Inteligencia Artificial en Medicina, 15*(1), 843–879.

Gouvêa, R. A. (2024). Exploring Cs3Sb2X9-type perovskites (X= Cl, Br, I) for optoelectronic applications: A theoretical investigation using density functional theory (DFT) and machine learning. Doctorate thesis.

Gu, Y., Li, Y., Yuan, F., & Yang, Q. (2023). Optimization and control strategies of aeration in WWTPs: A review. *Journal of Cleaner Production, 418*, 138008.

Guo, Y., Wang, H., Wang, B., Deng, S., Huang, J., Yu, G., & Wang, Y. (2018). Prediction of micropollutant abatement during homogeneous catalytic ozonation by a chemical kinetic model. *Water Research, 142*, 383–395.

Gupta, P., Bhardwaj, G., Dubey, S., Tayal, T., Sengupta, A., & Narad, P. (2024). AI-enabled process optimization for sustainable wastewater treatment solutions. In Garg, M. C. (Ed.) *The AI Cleanse: Transforming Wastewater Treatment Through Artificial Intelligence: Harnessing Data-Driven Solutions* (pp. 141–164). Springer.

Hassanien, A. E., Abouelmagd, L. M., Mahmoud, A. S., & Darwish, A. (2023). An optimized backpropagation neural network models for the prediction of nanomaterials concentration for purification industrial wastewater. *Engineering Applications of Artificial Intelligence, 126*, 107010.

Himeur, Y., Elnour, M., Fadli, F., Meskin, N., Petri, I., Rezgui, Y., ... & Amira, A. (2023). AI-big data analytics for building automation and management systems: A survey, actual challenges and future perspectives. *Artificial Intelligence Review, 56*(6), 4929–5021.

Hu, H., Tong, Y. W., & He, Y. (2023). Current insight into enhanced strategies and interaction mechanisms of hydrogel materials for phosphate removal and recovery from wastewater. *Science of the Total Environment, 892*, 164514.

Huang, R., Zhang, Z., Zhang, W., Mou, J., Zhou, P., & Wang, Y. (2020). Energy performance prediction of the centrifugal pumps by using a hybrid neural network. *Energy, 213*, 119005.

Jagaba, A. H., Lawal, I. M., Birniwa, A. H., Affam, A. C., Usman, A. K., Soja, U. B., ... & Yaro, N. S. A. (2024). Sources of water contamination by heavy metals. In Jaafar, J., Zaidi, A. A., & Naseer, M. (Eds.) *Membrane Technologies for Heavy Metal Removal from Water* (pp. 3–27). CRC Press.

Jiang, Y., Li, C., Song, H., & Wang, W. (2022). Deep learning model based on urban multi-source data for predicting heavy metals (Cu, Zn, Ni, Cr) in industrial sewer networks. *Journal of Hazardous Materials, 432*, 128732.

Jomova, K., Makova, M., Alomar, S. Y., Alwasel, S. H., Nepovimova, E., Kuca, K., ... & Valko, M. (2022). Essential metals in health and disease. *Chemico-Biological Interactions, 367*, 110173.

Kasa, A., & Mohd, S. F. (2024). Performance prediction evaluation of machine learning models for slope stability analysis: A comparison between ANN, ANN-ICA and ANFIS. *Journal of Electrical Systems, 20*, 4364–4374.

Khurshid, A., & Pani, A. K. (2023). Machine learning approaches for data-driven process monitoring of biological wastewater treatment plant: A review of research works on benchmark simulation model no. 1 (BSM1). *Environmental Monitoring and Assessment, 195*(8), 916.

Lan, J., Liu, P., Hu, X., & Zhu, S. (2024). Harmful algal blooms in eutrophic marine environments: Causes, monitoring, and treatment. *Water, 16*(17), 2525.

Leburu, E., Qiao, Y., Wang, Y., Yang, J., Liang, S., Yu, W., ... & Hu, J. (2024). Flexible electronics for heavy metal ion detection in water: A comprehensive review. *Biomedical Microdevices, 26*(3), 30.

Maleki, S., Karimi-Jashni, A., & Mousavifard, M. (2024). Removal of Ni (II) ions from wastewater by ion exchange resin: Process optimization using response surface methodology and ensemble machine learning techniques. *Journal of Environmental Chemical Engineering, 12*(2), 112417.

Mannina, G., Rebouças, T. F., Cosenza, A., Sànchez-Marrè, M., & Gibert, K. (2019). Decision support systems (DSS) for wastewater treatment plants: A review of the state of the art. *Bioresource Technology, 290*, 121814.

Manu, D., & Thalla, A. K. (2017). Artificial intelligence models for predicting the performance of biological wastewater treatment plant in the removal of Kjeldahl Nitrogen from wastewater. *Applied Water Science, 7*, 3783–3791.

Mazhar, T., Irfan, H. M., Haq, I., Ullah, I., Ashraf, M., Shloul, T. A., ... & Elkamchouchi, D. H. (2023). Analysis of challenges and solutions of IoT in smart grids using AI and machine learning techniques: A review. *Electronics, 12*(1), 242.

Mojiri, A., Ohashi, A., Ozaki, N., Aoi, Y., & Kindaichi, T. (2020). Integrated anammox-biochar in synthetic wastewater treatment: Performance and optimization by artificial neural network. *Journal of Cleaner Production, 243*, 118638.

Morris, J., Georgiou, I., Guenther, E., & Caucci, S. (2021). Barriers in implementation of wastewater reuse: Identifying the way forward in closing the loop. *Circular Economy and Sustainability, 1*, 413–433.

Murmu, S., Sinha, D., Chaurasia, H., Sharma, S., Das, R., Jha, G. K., & Archak, S. (2024). A review of artificial intelligence-assisted omics techniques in plant defense: Current trends and future directions. *Frontiers in Plant Science, 15*, 1292054.

Nagpal, M., Siddique, M. A., Sharma, K., Sharma, N., & Mittal, A. (2024). Optimizing wastewater treatment through artificial intelligence: Recent advances and future prospects. *Water Science & Technology*, *90*(3), 731–757.

Narvarte, B. C. V., Hinaloc, L. A. R., Gonzaga, S. M. C., Crisostomo, B. A., Genovia, T. G. T., & Roleda, M. Y. (2023). Nitrate and phosphate uptake of morphologically distinct calcified macroalgae. *Phycologia*, *62*(6), 551–560.

Olmo Bau, J. (2024). *Early Detection of Pollution Episodes in WWTPs Through the Application of Artificial Intelligence Systems*. Universitat Politècnica de Catalunya.

Qambar, A. S., & Al Khalidy, M. M. (2022). Optimizing dissolved oxygen requirement and energy consumption in wastewater treatment plant aeration tanks using machine learning. *Journal of Water Process Engineering*, *50*, 103237.

Rai, S., Jain, S., Rallapalli, S., Magner, J., Singh, A. P., & Goonetilleke, A. (2024). Effect of varying hydrologic regime on seasonal total maximum daily loads (TDML) in an agricultural watershed. *Water Research*, *249*, 120998.

Rajaee, T., Khani, S., & Ravansalar, M. (2020). Artificial intelligence-based single and hybrid models for prediction of water quality in rivers: A review. *Chemometrics and Intelligent Laboratory Systems*, *200*, 103978.

Safeer, S., Pandey, R. P., Rehman, B., Safdar, T., Ahmad, I., Hasan, S. W., & Ullah, A. (2022). A review of artificial intelligence in water purification and wastewater treatment: Recent advancements. *Journal of Water Process Engineering*, *49*, 102974.

Sahu, S., Kaur, A., Singh, G., & Arya, S. K. (2023). Harnessing the potential of microalgae-bacteria interaction for eco-friendly wastewater treatment: A review on new strategies involving machine learning and artificial intelligence. *Journal of Environmental Management*, *346*, 119004.

Salim, R., Oukhrib, R., & Zaidi, K. (2024). Protection of copper in contact with an aggressive medium using various inhibitors: Review. *Journal of Materials and Environmental Science*, *15*(10), 1383–1402.

Sanjeevi, R., Sathvara, P. B., Tripathi, S., & Anuradha, J. (2024). The AI cleanse: Revolutionizing the future of wastewater treatment with AI and machine learning. In Garg, M. C. (Ed.) *The AI Cleanse: Transforming Wastewater Treatment Through Artificial Intelligence: Harnessing Data-Driven Solutions* (pp. 245–263). Springer.

Sekulić, Z., Antanasijević, D., Stevanović, S., & Trivunac, K. (2017). Application of artificial neural networks for estimating Cd, Zn, Pb removal efficiency from wastewater using complexation-microfiltration process. *International Journal of Environmental Science and Technology*, *14*, 1383–1396.

Singh, N. K., Yadav, M., Singh, V., Padhiyar, H., Kumar, V., Bhatia, S. K., & Show, P.-L. (2023). Artificial intelligence and machine learning-based monitoring and design of biological wastewater treatment systems. *Bioresource Technology*, *369*, 128486.

Smith, J. P. (2022). *Evaluating the Sustainability Performance of US Biofuel in 2017 with an Integrated Techno-Economic and Life Cycle Assessment Framework*. Colorado State University.

Soori, M., Jough, F. K. G., Dastres, R., & Arezoo, B. (2024). AI-based decision support systems in Industry 4.0, a review. *Journal of Economy and Technology*. https://doi.org/10.1016/j.ject.2024.08.005

Thakur, A., & Dharmendra. (2024). Electrocoagulation process modelling and optimization using RSM and ANN-GA for simultaneous removal of arsenic and fluoride. *Multiscale and Multidisciplinary Modeling, Experiments and Design*, *7*(6), 5899–5914.

Therrien, J.-D., Nicolaï, N., & Vanrolleghem, P. A. (2020). A critical review of the data pipeline: How wastewater system operation flows from data to intelligence. *Water Science and Technology*, *82*(12), 2613–2634.

Upadhyay, S. K., Rani, N., Kumar, V., Mythili, R., & Jain, D. (2023). A review on simultaneous heavy metal removal and organo-contaminants degradation by potential microbes: Current findings and future outlook. *Microbiological Research*, *273*, 127419.

Vakili, M., Rafatullah, M., Yuan, J., Zwain, H. M., Mojiri, A., Gholami, Z., … & Yu, Y. (2021). Nickel ion removal from aqueous solutions through the adsorption process: A review. *Reviews in Chemical Engineering*, *37*(6), 755–778.

Wang, Z., & Man, Y. (2021). Artificial intelligence algorithm application in wastewater treatment plants: Case study for COD load prediction. In Ren, J. (Ed.) *Applications of Artificial Intelligence in Process Systems Engineering* (pp. 143–164). Elsevier.

World Health Organization. (2023). *Exposure to Lead: A Major Public Health Concern. Preventing Disease Through Healthy Environments*. World Health Organization.

Wu, J., & Wang, Z. (2022). A hybrid model for water quality prediction based on an artificial neural network, wavelet transform, and long short-term memory. *Water*, *14*(4), 610.

Wu, M., Xiong, J., Li, R., Dong, A., Lv, C., Sun, D., Abdelghany, A. E., Zhang, Q., Wang, Y., & Siddique, K. H. (2024). Precision forecasting of fertilizer components' concentrations in mixed variable-rate fertigation through machine learning. *Agricultural Water Management*, *298*, 108859.

Wurtsbaugh, W. A., Paerl, H. W., & Dodds, W. K. (2019). Nutrients, eutrophication and harmful algal blooms along the freshwater to marine continuum. *Wiley Interdisciplinary Reviews: Water*, *6*(5), e1373.

Yaqub, M., Eren, B., & Eyupoglu, V. (2021). Prediction of heavy metals removal by polymer inclusion membranes using machine learning techniques. *Water and Environment Journal*, *35*(3), 1073–1084.

Yuan, X., Li, J., Lim, J. Y., Zolfaghari, A., Alessi, D. S., Wang, Y., … & Ok, Y. S. (2023). Machine learning for heavy metal removal from water: Recent advances and challenges. *ACS ES&T Water*, *4*(3), 820–836.

Zafar, M. H., Khan, N. M., Mansoor, M., Mirza, A. F., Moosavi, S. K. R., & Sanfilippo, F. (2022). Adaptive ML-based technique for renewable energy system power forecasting in hybrid PV-wind farms power conversion systems. *Energy Conversion and Management*, *258*, 115564.

Zaferani, S. P. G., Emami, M. R. S., Amiri, M. K., & Binaeian, E. (2019). Optimization of the removal Pb (II) and its Gibbs free energy by thiosemicarbazide modified chitosan using RSM and ANN modeling. *International Journal of Biological Macromolecules*, *139*, 307–319.

Zaghloul, M. S., & Achari, G. (2022). Application of machine learning techniques to model a full-scale wastewater treatment plant with biological nutrient removal. *Journal of Environmental Chemical Engineering*, *10*(3), 107430.

Zhang, F., Li, J., Wei, Y., Dong, L., Wang, J., & Liu, J. (2024). Sustainable reduction of heavy metal pollution in zinc electrowinning: Insights from a cleaner production perspective. *Journal of Cleaner Production*, *470*, 143268.

Zhang, Y., Luo, P., Zhao, S., Kang, S., Wang, P., Zhou, M., & Lyu, J. (2020). Control and remediation methods for eutrophic lakes in the past 30 years. *Water Science and Technology*, *81*(6), 1099–1113.

Zhao, L., Deng, J., Sun, P., Liu, J., Ji, Y., Nakada, N., … & Yang, Y. (2018). Nanomaterials for treating emerging contaminants in water by adsorption and photocatalysis: Systematic review and bibliometric analysis. *Science of the Total Environment*, *627*, 1253–1263.

Zhu, X., Wang, X., & Ok, Y. S. (2019). The application of machine learning methods for prediction of metal sorption onto biochars. *Journal of Hazardous Materials*, *378*, 120727.

PART VI

AI and Desalination

A Path to Water Security

AI-Driven Optimization in Seawater Desalination

14

Sumeda Bandara and Thilini Bandara

14.1 INTRODUCTION

Access to freshwater is essential for human health and eco-economic stability (Gleick & Cooley, 2021). Climate change, global population growth, and increasing living standards are placing tremendous pressure on natural resources, leading to uncertainties in water availability, a resource vital for survival (Shemer, Wald, & Semiat, 2023). However, the increasing gap between human demand and the natural supply of freshwater drives water scarcity, impacting industrial and agricultural productivity (Gleick & Cooley, 2021). The demand for freshwater now surpasses the available supply, making it critical to tap into alternative water sources, such as treated seawater, brackish water, and wastewater (Shemer, Wald, & Semiat, 2023). Desalination involves removing salt and impurities from brine to produce potable water (Elimelech & Phillip, 2011).

Desalination technology occupies a large space in the water industry. Those emerging technologies offer numerous opportunities for expanding the water purification industry, which is now dominated by thermal desalination, membrane-based methods, and emerging desalination processes (Ahmed, Khalil, & Hilal, 2021). There are several desalination techniques: thermal methods: flashing/boiling, multi-stage flash (MSF), multiple effect distillation (MED), mechanical vapor compression (MVC), membrane distillation, humidification–de-humidification of carrier gas; crystallization: freezing, hydration; membrane methods: reverse osmosis (RO), forward osmosis, electro dialysis (ED); and other methods: air dehydration, ion exchange, dew vaporization, solvent extraction, and hybrid methods (Ghalavand, Hatamipour, & Rahimi, 2014). In the last decade, RO and MSF desalination processes have primarily dominated the desalination industry. Other thermal desalination methods, for instance, MED and MVC processes, are utilized on a much smaller scale (Al-Sahali & Ettouney, 2007). Currently, Membrane desalination with RO technology occupies roughly 69% of worldwide desalination capacity and 85% of total desalination facilities. Thermal desalination with MSF and MED occupies approximately 18% and 7% of global desalination capacity, respectively (Elsaid et al., 2020).

High energy consumption (Ghalavand, Hatamipour, & Rahimi, 2014; Zarzo & Prats, 2017), environmental concerns (Miller, Shemer, & Semiat, 2014), and operational costs (Ziolkowska, 2014) are among the most significant challenges facing desalination technologies. Both thermal and membrane filtration

DOI: 10.1201/9781003600718-20

processes are energy-intensive (Ghalavand, Hatamipour, & Rahimi, 2014), often relying on fossil fuels, which not only drive up operational costs but also contribute to global warming and climate change (Miller, Shemer, & Semiat, 2014). Additionally, the environmental impact extends to brine disposal, as the concentrated saline byproduct can harm ecosystems when released without proper treatment (Ahamed & Anwar, 2012). Maintenance and operational expenses further add to the financial burden, particularly with issues such as membrane fouling in RO systems or the wear and tear of components in thermal processes (Kabeel, El-Said, & Dafea, 2019). Addressing these challenges is critical to making desalination more cost-effective, energy-efficient, and environmentally sustainable.

Artificial Intelligence (AI), as an advancing technology, has become prevalent in engineering science and technology. It is increasingly applied to optimize the water treatment processes. At present, AI has become a prominent technology, extensively utilized to enhance the effectiveness of the seawater desalination process (Ray et al., 2022) by boosting efficiency, lowering costs (He et al., 2021), and fostering sustainability. It has evolved into an erudite tool for data processing and optimizing smart water services, effectively tackling challenges related to monitoring, management, and labor costs (Ray et al., 2022). Applications of data-driven techniques in the desalination industry can be divided into five groups in advance: optimization and correlation development, performance prediction using operational parameters, performance prediction using design parameters, and desalination system maintenance and control (Behnam, Faegh, & Khiadani, 2022). Moreover, AI facilitates demand forecasting and plant design, enabling desalination facilities to operate more efficiently and adapt to fluctuating water needs (He et al., 2021). By integrating renewable energy (Sayed et al., 2023) and addressing brine management challenges, AI promotes sustainable practices in operation (Abba et al., 2023).

14.2 AI TECHNIQUES IN DESALINATION OPTIMIZATION

AI is becoming a pivotal tool in optimizing desalination processes, addressing challenges related to water scarcity, and enhancing the efficacy and sustainability of water purification systems. Due to the well-known challenge of modeling and simulating the desalination process using traditional model-based techniques, which are likewise very limited, previous work has suggested and addressed data-driven methodologies and AI tools to accomplish so. The ability of AI-based models to transform massive data volumes into relevant information and reliable behavioral models is the primary benefit of using these techniques (Sayed et al., 2023). This synthesis explores the various types of AI techniques applied in desalination optimization, their benefits, and the challenges in the implementation process.

14.2.1 AI Techniques and Applications

14.2.1.1 Artificial neural networks (ANNs) and genetic algorithms (GAs)

Desalination processes are widely modeled and optimized using ANNs and GAs. Regression models are served by ANNs, and performance parameters are predicted using them, while global optimization is employed by GAs. In desalination systems, multi-objective optimization is provided by ANNs and GAs, balancing variables including water output, cost, and energy efficiency (Di Martino, Avraamidou, & Pistikopoulos, 2022; Mahadeva et al., 2021, 2022).

14.2.1.2 Support vector machines (SVMs) and decision trees (DTs)

Support vector machines (SVMs) are a popular machine learning (ML) technique used in various optimization processes, including the optimization of desalination processes. A seawater reverse osmosis (SWRO) desalination plant's efficiency has been maximized using SVMs (Sadi et al., 2019).

The performance of SWRO desalination plants is optimized using DTs. Actuators in the plants are adjusted by these techniques to achieve optimal performance, with higher accuracy being shown by DTs in some cases (Salem et al., 2022).

14.2.1.3 Adaptive neuro-fuzzy inference system (ANFIS)

The desalination efficiency of produced water in hydrate-based desalination processes has been effectively predicted using ANFIS. Desalination efficiency in hydrate-based desalination processes has been predicted using ANFIS models, which have been combined with SVMs and optimized using GAs. High accuracy and robustness in predicting desalination outcomes are demonstrated by these models (Mahadeva et al., 2021).

14.2.1.4 Deep learning and metaheuristic algorithms

High accuracy in predicting desalination plant performance has been shown by deep learning models, such as long short-term memory (LSTM) networks, integrated with metaheuristic algorithms like the crow search algorithm (CSA). Energy usage is optimized, and sustainable operating strategies are suggested by these models (Sayed et al., 2023).

14.3 OPTIMIZING CORE DESALINATION PROCESSES WITH AI

The current commercial technologies like RO, MSF, MED, and ED (Curto, Franzitta, & Guercio, 2021) in AI can optimize the above-mentioned commonly used desalination processes. However, as discussed, the desalination industry has been primarily dominated by RO and MSF desalination processes. In this section, we aim to discuss optimizing core desalination in RO and MSF technologies.

14.3.1 Artificial Intelligence for Reverse Osmosis (RO) in Seawater Desalination

There is great potential for improving the efficacy and efficiency of seawater desalination by the incorporation of AI into RO desalination processes. This synthesis explores the various applications and benefits of AI in RO desalination, focusing on predictive maintenance, real-time process control, and energy optimization. AI algorithms can analyze large amounts of sensor data to anticipate equipment failures, fine-tune operational parameters, and minimize energy consumption, leading to improved water quality, reduced operational costs, and a more sustainable desalination process.

14.3.1.1 AI modeling in RO desalination

AI modeling techniques such as machine-learning sequence modeling are effectively applied to control water quality and seawater desalination dosage (Zhang et al., 2024). Modeling techniques are also performed to suggest sustainable operating strategies in RO plants contributing to a more sustainable future (Abba et al., 2023). RO desalination systems have been extensively modeled using ANNs. By exposing hidden patterns and relationships in the data gathered from the RO process, these models are able to forecast the purity of water. With strong correlation coefficients between actual and anticipated outputs, ANNs have demonstrated excellent performance in this setting, suggesting accurate modeling potential.

SVMs can use RO desalination process modeling to yield substantial improvements in predicting the process performance. The gray wolf optimizer-based ANN (GWO-ANN) approach accurately predicts RO desalination plant performance, improving stability and efficiency (Mahadeva et al., 2022).

14.3.1.2 Optimization of RO systems

RO desalination plant design and operation have been optimized by the use of AI techniques, specifically ANNs and GAs. For instance, membrane behavior was properly modeled using a feed-forward ANN with rectified linear units, and the results were converted into a mixed-integer linear programming formulation that maximized water utilization and minimized energy consumption. This method helps industrial-scale RO plants operate as efficiently as possible by visualizing trade-offs between energy use and water recovery (Di Martino, Avraamidou, & Pistikopoulos, 2022).

14.3.1.3 Multi-stage RO desalination

The use of AI in multi-stage RO desalination processes has also been explored. Studies have shown that applying ANNs and SVMs to model these processes can significantly reduce the mean square error (MSE) in predictions, particularly in the second stage of the process. This indicates substantial improvements in the accuracy and reliability of the models when using AI techniques.

14.3.1.4 Pretreatment optimization

AI has also been used to improve RO desalination's pretreatment procedures. For example, response surface methodology (RSM) and ML models were used to optimize a hybrid ZnO-immobilized/photo-Fenton process. To attain the best organic degradation performance, these models were further improved using a nondominated sorting genetic algorithm (NSGA-III). Total organic carbon (TOC) and chemical oxygen demand (COD) were significantly reduced as a result of the adjusted settings, indicating the potential for AI to improve pretreatment efficiency in desalination facilities (Joy, Feroz, & Dutta, 2022).

14.3.2 Artificial Intelligence for Multi-Stage Flash (MSF) Distillation in Seawater Desalination

The application of AI in MSF distillation for seawater desalination is a growing area of research. AI techniques, particularly neural networks, are being explored to enhance the efficiency and performance of MSF plants.

14.3.2.1 Estimation of pressure drop

The pressure drop across demisters in MSF desalination processes has been estimated using LSTM neural networks. The effectiveness of the flashed-off vapor's separation from the brine droplets is largely determined by the pressure drop, which also has an impact on the MSF plant's overall performance. Prior research shows that stacked LSTM algorithms outperform conventional computational fluid dynamics (CFD) models in accurately estimating the pressure drop using both experimental and actual plant data.

14.3.2.2 Performance improvement

Accuracy and Efficiency: The stacked LSTM algorithm shows a significant improvement in estimation accuracy, with a root mean square error (RMSE) that is 40% lower than that of the CFD model. This indicates that AI models, specifically LSTM networks, can provide more precise and reliable predictions, which are crucial for optimizing the operation and performance of MSF desalination plants (Dawkrajai et al., 2021).

14.4 AI FOR ENERGY AND COST EFFICIENCY

Intensive energy requirements, inefficient energy usage, and high operational costs in seawater desalination are critically important due to their implications for environmental sustainability, economic viability, and sustaining global water security. Intensive energy requirements for processes like RO and MSF underscore a requirement for a sustainable approach. In addition, dependence on fossil fuels causes environmental impacts, and energy inefficiencies such as inefficient energy recovery systems in older desalination plants and energy losses in heat-based processes like MED are common as energy and cost efficiency issues in seawater desalination. However, energy costs can account for up to 36% of the total desalination costs (LiVecchi et al., 2019), and maintenance costs, equipment replacement (e.g., membranes, pumps) costs, and brine management costs are added as other operational expenses in seawater desalination (Genesis Water Tech, 2024).

As a solution for the above concerns, smart algorithms can be used to reduce energy consumption through ML approaches to predict and reduce peak energy usage and integration with renewable energy sources for sustainable desalination (Genesis Water Tech, 2024). In addition, AI's role in predicting chemical, energy, and equipment needs and cost minimization models for sustainable operation underscores the dynamic cost management and supply chain optimization in seawater desalination. ANNs are employed to model complex relationships between operational parameters and performance metrics in desalination plants. GAs are utilized to optimize these parameters for improved efficiency (Mjalli et al., 2007).

In process optimization, DL models analyze large datasets to identify patterns and optimize desalination processes. MAs, such as the modified whale optimization algorithm (MWOA), are combined with ANNs to enhance system performance. For example, in RO desalination, the hybridization of MWOA with ANNs has been applied to optimize RO desalination plants, resulting in improved energy efficiency. In addition, fault detection is another major concern while SVMs and DTs are utilized to detect anomalies and classify operational states in desalination systems, enabling proactive maintenance and reducing downtime (Plasencia et al., 2021). These AI techniques assist in identifying inefficient operational modes, allowing for timely interventions that enhance energy efficiency. ANFIS combines neural networks and fuzzy logic to create intelligent control systems that optimize desalination processes in real time. In SWRO desalination, ANFIS-based control systems have been developed to enhance the performance of SWRO plants, focusing on minimizing energy consumption (Krzywanski et al., 2024).

In summary, through process optimization and cost reduction, AI techniques optimize the desalination process by predicting the performance of different adsorbents and reducing the cost of chemicals used in water treatment. This leads to significant operational expenditure savings (Alam et al., 2021; Mahadeva et al., 2021). Accurate modeling of desalination plants using AI helps design engineers operate plants under stable conditions, enhancing plant efficiency and safety. This reduces production time, energy consumption, and operational risks (Mahadeva et al., 2021). AI also helps create sophisticated brine treatment methods that make it easier to recover valuable minerals from brine, reducing waste and optimizing resource use. This helps create water resource management dynamics (CFD) models that are more sustainable (Ray et al., 2022).

14.5 ENHANCING WATER QUALITY WITH AI

Optimization of membrane performance, real-time water quality monitoring, predictive maintenance, and material design for enhanced desalination are vital in enhancing water quality in seawater desalination. AI is revolutionizing seawater desalination by enhancing water quality through various innovative applications. AI models, particularly MLAs, are employed to predict and optimize the behavior of desalination membranes. By analyzing extensive datasets, AI can forecast membrane fouling and scaling, enabling proactive maintenance and operational adjustments to maintain optimal water quality (Niu et al., 2022). Additionally, a variety of AI platforms, such as ML and ANNs, are used to estimate water flux, forecast fouling behavior, and optimize operating parameters in the context of water purification and desalination operations. In these applications, models like SVMs, convolutional neural networks (CNNs), and recurrent neural networks (RNNs) have shown excellent prediction ability (Shahouni et al., 2024).

AI systems analyze data from sensors monitoring parameters such as salinity, pH, and turbidity in real time. This continuous analysis allows for immediate detection of anomalies, ensuring consistent water quality and prompt corrective actions when necessary. The research focuses on real-time monitoring and detection of contaminants, employing advanced ML algorithms, including random forest, SVMs, and neural networks, to process and classify spectral data. These models demonstrate remarkable accuracy in real-time classification, highlighting the synergy between AI and environmental science (Durgun, 2024). By optimizing operating parameters, AI is being used more and more to increase desalination process efficiency, which improves water quality and lowers energy usage. AI-powered solutions help make desalination processes more economical and sustainable by continuously monitoring and optimizing several parameters, including energy usage, feed water quality, and product water output. Furthermore, by anticipating possible problems and enabling preventative maintenance, AI can evaluate the long-term performance of membrane systems used in desalination to guarantee sustainable freshwater output (Habieeb et al., 2023). AI's predictive maintenance greatly increases the desalination systems' operational efficiency. To guarantee continuous operation within ideal parameters and reliable production of high-quality water, AI algorithms can predict any failures before they happen by evaluating equipment performance data. Additionally, AI-driven predictive maintenance may keep an eye on membrane conditions by evaluating sensor data and continuously evaluating variables like salinity, flow rates, and pressure. By enabling the early detection of anomalies, this method reduces downtime during proactive maintenance and facilitates timely maintenance (Habieeb et al., 2023).

Materials like graphene nanopores are designed and optimized for desalination using AI, especially DRL. Better water quality and more effective desalination procedures result from these AI-designed materials' increased water flux and enhanced ion rejection rates. A noteworthy study by Wang, Cao, and Farimani (2021) presented a framework for optimizing the geometry of graphene nanopores for water desalination by integrating DRL with CNNs. In this method, the CNN forecasts desalination performance indicators, including water flux and ion rejection, while the DRL agent regulates the formation of nanopores by choosing which atoms to remove. Comparing the AI-designed nanopores to conventional circular nanopores, molecular dynamics simulations showed that the former maintained competitive ion rejection rates while achieving higher water flux. The study identified that semi-oval shapes with rough edges were key factors contributing to the superior performance of these nanopores. Further research by) Wang et al. (2021) expanded on this work, utilizing a DRL framework to rapidly create and screen thousands of graphene nanopores with varying geometries. The AI-designed nanopores demonstrated higher water flux and maintained rival ion rejection rates compared to standard circular nanopores. The irregular shapes with rough edges were again identified as critical for enhanced desalination performance.

14.6 AI-DRIVEN BRINE MANAGEMENT AND ENVIRONMENTAL SUSTAINABILITY

AI-driven brine management has emerged as a transformative approach to address both the environmental challenges associated with brine disposal and the opportunities for sustainable resource recovery. Brine, a high-salinity wastewater byproduct generated from processes like desalination, industrial activities, and oil extraction, poses significant ecological threats when improperly managed. Discharging untreated brine into aquatic ecosystems can lead to increased salinity, altered water chemistry, and toxic element accumulation, harming biodiversity and water quality (Panagopoulos & Haralambous, 2020). Moreover, conventional disposal methods, such as deep-well injection and ocean discharge, can contribute to long-term soil degradation and groundwater contamination.

AI addresses these challenges by enhancing the monitoring and control of brine discharge. Advanced AI models can assess brine composition and volume in real time, allowing for data-driven decision-making and regulatory compliance (Gaudio et al., 2021; El-Chaghaby et al., 2023). MLAs can predict brine's impact on local ecosystems and help design optimal discharge strategies by integrating historical data, hydrodynamic modeling, and satellite imagery analysis. These AI tools empower industries to implement proactive measures, such as dynamic flow adjustments and dilution strategies, which minimize environmental harm.

Beyond disposal, AI-driven technologies have shown remarkable potential in transforming brine from a waste product into a valuable resource. The conversion of brine into useful byproducts involves the extraction of minerals such as lithium, magnesium, and potassium, which are critical for industries like battery manufacturing and agriculture (ScienceDaily, 2024; Kia et al., 2024). Predictive modeling and process optimization are two AI-assisted techniques that simplify the process of determining the best extraction conditions while lowering energy usage and operating expenses. Businesses may optimize the production of valuable minerals while reducing waste by utilizing AI's capacity to examine vast information and model chemical reactions. The economic potential of brine byproduct utilization is significant, as recovered minerals can be sold in global markets, contributing to resource circularity and reducing dependence on virgin mineral extraction. For example, lithium extracted from geothermal brine has become sustainable, and for the electric vehicle market, it creates new revenue streams for desalination plants (Kong et al., 2025). Moreover, the environmental benefits of brine valorization extend beyond waste reduction. Repurposing brine for mineral recovery reduces the need for harmful disposal practices, thus lowering the ecological footprint of desalination and other industrial processes.

However, the successful implementation of AI-driven brine management requires collaboration between policymakers, researchers, and industry stakeholders. Clear regulatory frameworks and technological standardization are essential to safeguard the safe and efficient application of AI tools. Furthermore, continued research and development in ML models and brine treatment technologies will be crucial for expanding the scalability and effectiveness of these approaches.

In conclusion, AI-driven brine management offers a dual advantage of mitigating environmental risks associated with brine discharge while unlocking new opportunities for sustainable resource recovery. By leveraging AI technologies for real-time monitoring and mineral extraction, industries can achieve economic benefits and reduce their environmental impact simultaneously. As global water scarcity and pollution concerns intensify, the integration of AI in brine management will be pivotal for advancing environmental sustainability and circular economy principles.

14.7 CASE STUDIES AND REAL-WORLD APPLICATIONS

14.7.1 Application of Artificial Intelligence in Desalination Plants in the Gaza Strip

14.7.1.1 Background on water scarcity in Gaza

The Gaza Strip, a densely populated region with limited freshwater resources, faces a severe water crisis. Over 95% of the coastal aquifer, the primary water source for Gaza, is deemed unfit for human consumption due to over-extraction and contamination (Oxfam International, 2022). In response, seawater desalination has emerged as a vital solution to address water scarcity. Among the desalination facilities in Gaza, AI has begun playing a pivotal role in optimizing operations, reducing costs, and ensuring sustainability (Aish, Zaqoot, & Abdeljawad, 2015). This chapter explores the application of AI in seawater desalination plants within the Gaza Strip, focusing on its transformative potential to enhance water security in this challenging environment.

The Gaza Strip, home to over 2 million people, relies heavily on desalination to meet its growing water demand. The United Nations has repeatedly highlighted the urgent need for sustainable water management in the region, emphasizing the role of advanced technologies like AI in mitigating water scarcity. Desalination plants, such as the Deir Al-Balah and the Southern Gaza Desalination Plant, have been equipped with cutting-edge technologies to address these challenges. However, operating desalination plants in Gaza comes with unique obstacles, including intermittent electricity supply, limited financial resources, and environmental concerns related to brine disposal (Mogheir & Bohissi, 2015).

14.7.1.2 Key applications of AI in the Gaza strip desalination plant

A study by Aish, Zaqoot, and Abdeljawad (2015) demonstrated the effective use of ANN models to forecast the operational performance of RO desalination facilities in the Gaza Strip. The models were designed to predict key parameters such as total dissolved solids (TDS) and the permeate flow rate one week in advance. For this purpose, five brackish water desalination plants of varying scales were selected from different regions across the Gaza Strip, including Rafah, Khanyounis, Dier Al Balah, Gaza City, and Beit Lahia. These facilities included the Al-Salam, Al-Sharqia, Al-Balad, Hanneaf, and Al-Radwan plants.

Water quality data obtained from these locations served as the foundation for developing ANN models tailored to predict desalination performance. Two specific models were created, for estimating TDS levels and for forecasting permeate flow rates. After training and validation, the models' accuracy was tested using independent datasets not previously used in the training phase. Two types of neural networks were applied: the radial basis function (RBF) and the multilayered perceptron (MLP). Among the two, the MLP model showed slightly superior predictive accuracy. When benchmarked against traditional statistical models, both ANN models yielded better performance in forecasting TDS concentrations and permeate flow rates in the studied plants.

14.7.2 Applications for AI in the Gold Coast Desalination Plant (GCDP), Seawater Desalination, in Queensland, Australia

14.7.2.1 Background on water scarcity in South East Queensland, Australia

South East Queensland (SEQ), Australia, has faced significant water scarcity challenges, particularly during the Millennium Drought (1997–2009), which severely impacted water supplies across the region. In response, the Gold Coast Desalination Plant (GCDP) was constructed to provide a climate-resilient water source, independent of rainfall. The GCDP can supply more than 10% of SEQ's daily water demand, enhancing the region's water security. The SEQ Water Grid, a network of dams, treatment plants, and pipelines, integrates the GCDP to distribute water efficiently across the region. This infrastructure allows for the strategic use of desalinated water, especially during periods of low rainfall or high demand. For instance, ahead of a forecasted dry summer, SEQ water increased production from the GCDP to preserve dam levels and ensure a reliable water supply (Seqwater, n.d.).

14.7.2.2 Key applications of AI in the Gold Coast Desalination Plant

A newly launched start-up is pioneering the application of ML and AI technologies to enhance the operation of RO membranes at the GCDP facility, to improve efficiency while minimizing energy use, emissions, and overall operational expenses (Jo, 2023). The system utilizes RO performance data, which is assessed and fed into the ML algorithm by the operations team. Based on this input, the algorithm suggests optimized operational parameters, known as setpoints, such as RO recovery rate (%) and permeate flow (m^3/h). These setpoints directly influence feed flow and pressure requirements, which in turn affect the energy used by pumps.

Three separate performance trials at GCDP uncovered a range of operational insights and obstacles. The first test showed energy savings, even though the differential pressure across the cartridge filters rose sharply during the testing phase. In the second test, issues with an energy recovery device (ERD) led to higher feed pressure needs, increasing power consumption. The third test highlighted how variations in feedwater temperature significantly impacted energy efficiency. Despite these challenges, the AI model demonstrated its capability to recommend operational setpoints that stayed within system limits and consistently delivered the desired permeate output while reducing energy demands, even in less-than-ideal conditions (Jo, 2023).

14.8 CHALLENGES AND LIMITATIONS OF AI IN DESALINATION

Optimizing desalination processes is essential for ensuring the sustainable performance of seawater desalination plants. The application of AI in seawater desalination has shown significant promise in optimizing processes, reducing costs, and improving efficiency. However, given the complexity of engineering optimization problems, some circumstances could result in less-than-ideal performance, especially when addressing the nonlinearity and unpredictability present in actual desalination processes (Aani et al., 2019). In practice, the integration of AI in desalination faces core challenges and limitations despite its transformative potential.

AI relies heavily on high-quality and extensive datasets for training and operation. Obtaining high-quality, real-time data for training AI models presents several challenges. Currently, many desalination plants lack the necessary infrastructure to collect data, significantly limiting their ability to produce robust datasets for advanced AI techniques (Krishnan et al., 2024). Apart from that, AI models are often tailored to specific processes or conditions, such as site-specific constraints. For example, desalination plants operate under varying environmental conditions, such as salinity levels, water temperatures, and energy sources. Models trained on one dataset may not perform well when applied to different locations (Sanan et al., 2023). There is often a lack of sufficient data related to specific desalination processes, particularly in unique or localized conditions. This scarcity can hinder the development of robust AI models. In contrast, data collected from desalination plants can include noise, inconsistencies, or errors, which may affect the reliability of AI predictions and decisions. Due to that, poor data preprocessing or integration methods can further exacerbate this issue in general (Ejigu, Tuo, & Liu, 2024).

Integrating AI with the systems presents several barriers, including compatibility issues, as older systems may not be designed to support modern AI technologies (Chekifi, Benmoussa, & Boukraa, 2024). Ensuring compatibility between AI algorithms and various control systems is complex and may require substantial investments in hardware and software upgrades. And purchasing advanced sensors, computational tools, and AI platforms can be prohibitively expensive, particularly for smaller desalination plants. For that reason, integrating AI into existing systems is a highly cost-intensive process and poses significant challenges, particularly in developing regions, where upgrading technology and infrastructure remains difficult.

Additionally, ensuring AI solutions are user-friendly and accessible, even for individuals without extensive technical expertise, will be crucial for creating more reliable and widely adopted systems. The skill gap in managing both legacy systems and advanced AI technologies can make integration difficult, requiring specialized expertise (Wang et al., 2021; Krishnan et al., 2024). Concern has also been raised regarding the potential over-reliance on AI systems in the absence of adequate human supervision. As AI becomes more involved in decision-making processes, it may lead to reduced human intervention, potentially overlooking ethical considerations or unintended consequences. A careful balance must be maintained, ensuring AI enhances the desalination process without compromising safety, transparency, or accountability.

Training and deploying complex models can be energy-intensive, which may offset some of the environmental benefits of optimizing desalination processes (Strubell, Ganesh, A., & McCallum, 2019). Meanwhile, technically, AI faces challenges, such as deviating from expected outcomes when unexpected changes occur. For instance, a limitation of ML is its tendency to deviate from expected results when unexpected changes occur. For example, genetic models for assessing membrane deterioration during membrane integrity tests have been developed using programming. However, it was noted that throughout these tests, a distinct phenomenon appeared in the event of a significant membrane breach (Aani et al., 2019).

14.9 FUTURE DIRECTIONS AND EMERGING TRENDS

AI's effects on desalination are growing more significant as it develops. By increasing productivity and cutting expenses, recent advancements in ML, deep learning (DL), and reinforcement learning (RL) are revolutionizing desalination procedures. At the same time, completely automated, sustainable desalination systems are becoming possible due to the combination of AI and renewable energy sources like solar and wind (Shahouni et al., 2024).

14.9.1 Advances in AI Algorithms for Desalination

One of the key areas of innovation in desalination is the use of AI to improve process efficiency. ML algorithms, for example, are being used to analyze large datasets in real time, enabling desalination plants to predict system behavior and make data-driven decisions to enhance performance (Shahouni et al., 2024). This is particularly useful in membrane-based desalination, where AI models can forecast membrane fouling and degradation, helping to optimize maintenance schedules and reduce energy consumption (Wang, Cao, & Farimani, 2021).

DL and reinforcement learning are also contributing to advancements in desalination technology. A study by Wang, Cao, and Farimani (2021) demonstrated that DRL has been successfully applied to optimize the design of graphene nanopores, an emerging material that could significantly improve water purification efficiency. By continuously learning and adapting, these AI-driven approaches help create more reliable and effective desalination systems, reducing energy demands and operational costs (Shahouni et al., 2024).

14.9.2 Integration of Renewable Energy and AI-Driven Systems

The future of desalination is not only dependent on AI-driven process optimization but also on the seamless integration of AI with renewable energy sources. As desalination plants increasingly rely on solar and wind power, AI-based control systems are being developed to manage the fluctuating energy supply efficiently (Chekifi, Benmoussa, & Boukraa, 2024). AI algorithms can analyze weather patterns and predict energy availability, allowing desalination plants to adjust their operations dynamically based on energy supply and water demand (Shahouni et al., 2024).

According to research by (Chekifi, Benmoussa, & Boukraa, 2024), AI-driven forecasting models and optimization algorithms have been instrumental in enhancing the stability and reliability of renewable-powered desalination systems. By intelligently managing power distribution and system operations, these AI models can minimize energy waste while ensuring continuous water production. This is particularly important in regions where access to electricity is limited, making sustainable desalination solutions even more critical (Shahouni et al., 2024). Even though there are still issues, such as the high upfront costs and the difficulty of integrating AI with renewable energy systems, research indicates that these problems may be solved with further developments in AI technology. The future of desalination is probably going to be defined by more intelligent, sustainable solutions that can handle the problems associated with global water scarcity as AI-powered desalination systems get more self-sufficient and energy-efficient (Shahouni et al., 2024; Chekifi, Benmoussa, & Boukraa, 2024).

14.10 CONCLUSION

In summary, the application of AI in seawater desalination represents a major step forward in addressing the escalating global water scarcity challenge. The incorporation of AI and ML technologies improves operational efficiency, promotes sustainability, and reduces the overall costs associated with desalination. Through the use of real-time data analysis, AI systems offer more precise control over key operational parameters, leading to optimized energy usage and enhanced membrane performance. These intelligent systems can anticipate equipment malfunctions and maintenance needs, helping to prevent breakdowns and extend the service life of essential components, ultimately decreasing downtime and operational expenses.

Furthermore, AI technologies enable desalination systems to respond dynamically to fluctuating environmental conditions, such as changes in seawater composition, thereby maintaining high performance levels. By minimizing energy consumption and reducing greenhouse gas emissions, AI also contributes to making desalination a more environmentally friendly method for producing freshwater. With the rising demand for clean water, AI-enhanced desalination systems offer a scalable and cost-efficient solution that is both practical and sustainable. Ultimately, AI is set to play a vital role in ensuring long-term access to reliable and eco-conscious freshwater supplies for communities around the world.

REFERENCES

Aani, S. A., Bonny, T., Hasan, S. W., & Hilal, N. (2019). Can machine language and artificial intelligence revolutionize process automation for water treatment and desalination? *Desalination*, *458*, 84–96. https://doi.org/10.1016/j.desal.2019.02.005

Abba, S. I., Usman, J., Abdulazeez, I., Lawal, D. U., Baig, N., Usman, A. G., & Aljundi, I. H. (2023). Integrated modeling of hybrid nanofiltration/reverse osmosis desalination plant using deep learning-based crow search optimization algorithm. *Water*, *15*(19), 3515. https://doi.org/10.3390/w15193515

Ahamed, M., & Anwar, R. (2012). An assessment of the environmental impact of brine disposal in marine environment. *International Journal of Modern Engineering Research*, *2*(4), 2756–2761. Retrieved from https://citeseerx.ist.psu.edu/document?repid=rep1&type=pdf&doi=03fd4aa617d99934be3e1a729a1cb6f178feb4bc

Ahmed, F. E., Khalil, A., & Hilal, N. (2021). Emerging desalination technologies: Current status, challenges and future trends. *Desalination*, *517*, 115183. https://doi.org/10.1016/j.desal.2021.115183

Aish, A. M., Zaqoot, H. A., & Abdeljawad, S. M. (2015). Artificial neural network approach for predicting reverse osmosis desalination plants performance in the Gaza Strip. *Desalination*, *367*, 240–247. https://doi.org/10.1016/j.desal.2015.04.008

Alam, G., Ihsanullah, I., Naushad, M., & Sillanpää, M. (2021). Applications of artificial intelligence in water treatment for optimization and automation of adsorption processes: Recent advances and prospects. *Chemical Engineering Journal, 427*, 130011. https://doi.org/10.1016/j.cej.2021.130011

Al-Sahali, M., & Ettouney, H. (2007). Developments in thermal desalination processes: Design, energy, and costing aspects. *Desalination, 214*(1–3), 227–240. https://doi.org/10.1016/j.desal.2006.08.020

Behnam, P., Faegh, M., & Khiadani, M. (2022). A review on state-of-the-art applications of data-driven methods in desalination systems. *Desalination, 532*, 115744. https://doi.org/10.1016/j.desal.2022.115744

Chekifi, T., Benmoussa, A., & Boukraa, M. (2024). Desalination powered by renewables: A challenge and an AI opportunity. *Water Resources Management, 38*(14), 5419–5461. https://doi.org/10.1007/s11269-024-03935-4

Curto, D., Franzitta, V., & Guercio, A. (2021). A review of the water desalination technologies. *Applied Sciences, 11*(2), 670. https://doi.org/10.3390/app11020670

Dawkrajai, J., Weerachapichasgul, W., Daosud, W., & Kittisupakorn, P. (2021). Estimation of pressure drop in a demister for multi-stage flash desalination process based on stacked long short-term memory (LSTM) neural networks. *Journal of Physics Communications, 5*(10), 105019. https://doi.org/10.1088/2399-6528/ac304c

Di Martino, M., Avraamidou, S., & Pistikopoulos, E. N. (2022). A neural network based superstructure optimization approach to reverse osmosis desalination plants. *Membranes, 12*(2), 199. https://doi.org/10.3390/membranes12020199

Durgun, Y. (2024). Real-time water quality monitoring using AI-enabled sensors: Detection of contaminants and UV disinfection analysis in smart urban water systems. *Journal of King Saud University - Science, 36*(9), 103409. https://doi.org/10.1016/j.jksus.2024.103409

Ejigu, D. A., Tuo, Y., & Liu, X. (2024). Application of artificial intelligence technologies and big data computing for nuclear power plants control: A review. *Frontiers in Nuclear Engineering, 3*. https://doi.org/10.3389/fnuen.2024.1355630

El-Chaghaby, G. A., Rashad, S., Chavali, M., Rawash, E. A., Moneem, M. A., & Radwan, S. R. (2023). Water quality and water pollution monitoring using artificial intelligence. In Mahmoud, A. E. D., Fawzy, M., & Khan, N. A. (Eds.) *Artificial Intelligence and Modeling for Water Sustainability* (pp. 71–95). CRC Press. https://doi.org/10.1201/9781003260455-5

Elimelech, M., & Phillip, W. A. (2011). The future of seawater desalination: Energy, technology, and the environment. *Science, 333*(6043), 712–717. https://doi.org/10.1126/science.1200488

Elsaid, K., Kamil, M., Sayed, E. T., Abdelkareem, M. A., Wilberforce, T., & Olabi, A. (2020). Environmental impact of desalination technologies: A review. *Science of the Total Environment, 748*, 141528. https://doi.org/10.1016/j.scitotenv.2020.141528

Gaudio, M. T., Coppola, G., Zangari, L., Curcio, S., Greco, S., & Chakraborty, S. (2021). Artificial intelligence-based optimization of industrial membrane processes. *Earth Systems and Environment, 5*(2), 385–398. https://doi.org/10.1007/s41748-021-00220-x

Genesis Water Tech. (2024, August 28). Seawater RO operating costs analysis: A comprehensive guide. Retrieved January 28, 2025, from https://genesiswatertech.com/blog-post/seawater-ro-operating-costs-analysis/

Ghalavand, Y., Hatamipour, M. S., & Rahimi, A. (2014). A review on energy consumption of desalination processes. *Desalination and Water Treatment, 54*(6), 1526–1541. https://doi.org/10.1080/19443994.2014.892837

Gleick, P. H., & Cooley, H. (2021). Freshwater scarcity. *Annual Review of Environment and Resources, 46*(1), 319–348. https://doi.org/10.1146/annurev-environ-012220-101319

Habieeb, A., Kabeel, A. E., Sultan, G., & Abdelsalam, M. M. (2023). Advancements in water desalination through artificial intelligence: A comprehensive review of AI-based methods for reverse osmosis membrane processes. *Water Conservation Science and Engineering, 8*(1). https://doi.org/10.1007/s41101-023-00227-7

He, Q., Zheng, H., Ma, X., Wang, L., Kong, H., & Zhu, Z. (2021). Artificial intelligence application in a renewable energy-driven desalination system: A critical review. *Energy and AI, 7*, 100123. https://doi.org/10.1016/j.egyai.2021.100123

Jo, J. (2023). Seawater desalination energy efficiency and emissions reduction at Gold Coast Desalination Plant. *Water e-Journal, 9*(3), 1–8. https://doi.org/10.21139/wej.2023.011

Joy, V. M., Feroz, S., & Dutta, S. (2022). Artificial intelligence-based multiobjective optimization of reverse osmosis desalination pretreatment using a hybrid ZnO-immobilized/photo-Fenton process. *Journal of Chemometrics, 36*(8). https://doi.org/10.1002/cem.3434

Kabeel, A. E., El-Said, E. M. S., & Dafea, S. A. (2019). Design considerations and their effects on the operation and maintenance cost in solar-powered desalination plants. *Heat Transfer-Asian Research, 48*(5), 1722–1736. https://doi.org/10.1002/htj.21454

Kia, A. K., Mortaheb, H. R., Salehi, M. B., & Nozaeim, A. A. (2024). Solvent extraction of lithium from brines with high magnesium/lithium ratios: Investigation on parameter interactions. *Environmental Science and Pollution Research, 31*(39), 52523–52539. https://doi.org/10.1007/s11356-024-34617-8

Kong, L., Yan, G., Hu, K., Yu, Y., Conte, N., Mckenzie, K. R., Jr, ... & Liu, X. (2025). Electro-driven direct lithium extraction from geothermal brines to generate battery-grade lithium hydroxide. *Nature Communications*, *16*(1). https://doi.org/10.1038/s41467-025-56071-x

Krishnan, A., Sundaram, T., Nagappan, B., Devarajan, Y., & Bhumika, N. (2024). Integrating artificial intelligence in nanomembrane systems for advanced water desalination. *Results in Engineering*, *24*, 103321. https://doi.org/10.1016/j.rineng.2024.103321

Krzywanski, J., Sztekler, K., Skrobek, D., Grabowska, K., Ashraf, W. M., Sosnowski, M., ... & Mika, L. (2024). AutoML-based predictive framework for predictive analysis in adsorption cooling and desalination systems. *Energy Science & Engineering*, *12*(5), 1969–1986. https://doi.org/10.1002/ese3.1725

LiVecchi, A., Copping, A., Jenne, S., Gorton, A., Preus, R., Gill, G., ... & Spence, H. (2019, April). Powering the blue economy: Exploring opportunities for marine renewable energy in various maritime. https://doi.org/10.2172/1525367

Mahadeva, R., Kumar, M., Manik, G., & Patole, S. P. (2021). Modeling, simulation, and optimization of the membrane performance of sea-water reverse osmosis desalination plant using neural network and fuzzy based soft computing techniques. *Desalination and Water Treatment*, *229*, 17–30. https://doi.org/10.5004/dwt.2021.27386

Mahadeva, R., Kumar, M., Patole, S. P., & Manik, G. (2022). Desalination plant performance prediction model using grey wolf optimizer based ANN approach. *IEEE Access*, *10*, 34550–34561. https://doi.org/10.1109/access.2022.3162932

Miller, S., Shemer, H., & Semiat, R. (2014). Energy and environmental issues in desalination. *Desalination*, *366*, 2–8. https://doi.org/10.1016/j.desal.2014.11.034

Mjalli, F. S., Abdel-Jabbar, N., Ettouney, H., & Qiblawey, H. A. (2007). Advanced computational techniques for solving desalination plant models using neural and genetic based methods. *Chemical Product and Process Modeling*, *2*(3). https://doi.org/10.2202/1934-2659.1113

Mogheir, Y., & Bohissi, N. A. (2015). Optimal management of brine from seawater desalination plants in Gaza strip: Deir AL Balah STLV plant as case study. *Journal of Environmental Protection*, *6*(6), 599–608. https://doi.org/10.4236/jep.2015.66054

Niu, C., Li, X., Dai, R., & Wang, Z. (2022). Artificial intelligence-incorporated membrane fouling prediction for membrane-based processes in the past 20 years: A critical review. *Water Research*, *216*, 118299. https://doi.org/10.1016/j.watres.2022.118299

Oxfam International. (2022, May 25). *Failing Gaza: Undrinkable Water, No Access to Toilets and Little Hope on the Horizon*. Oxfam International. Retrieved from https://www.oxfam.org/en/failing-gaza-undrinkable-water-no-access-toilets-and-little-hope-horizon

Panagopoulos, A., & Haralambous, K. (2020). Environmental impacts of desalination and brine treatment: Challenges and mitigation measures. *Marine Pollution Bulletin*, *161*, 111773. https://doi.org/10.1016/j.marpolbul.2020.111773

Plasencia, G. N. M., Camacho-Espino, J., Prats, D. Á., & Suárez, B. P. (2021). Machine learning models applied to manage the operation of a simple SWRO desalination plant and its application in marine vessels. *Water*, *13*(18), 2547. https://doi.org/10.3390/w13182547

Ray, S. S., Verma, R. K., Singh, A., Ganesapillai, M., & Kwon, Y. (2022). A holistic review on how artificial intelligence has redefined water treatment and seawater desalination processes. *Desalination*, *546*, 116221. https://doi.org/10.1016/j.desal.2022.116221

Sadi, M., Fakharian, H., Ganji, H., & Kakavand, M. (2019). Evolving artificial intelligence techniques to model the hydrate-based desalination process of produced water. *Journal of Water Reuse and Desalination*, *9*(4), 372–384. https://doi.org/10.2166/wrd.2019.024

Salem, H., El-Hasnony, I. M., Kabeel, A., El-Said, E. M., & Elzeki, O. M. (2022). Deep learning model and classification explainability of renewable energy-driven membrane desalination system using evaporative cooler. *Alexandria Engineering Journal*, *61*(12), 10007–10024. https://doi.org/10.1016/j.aej.2022.03.050

Sanan, O., Sperling, J., Greene, D., & Greer, R. (2023). Towards data-driven methods for decarbonizing reverse osmosis desalination. *2023 IEEE MIT Undergraduate Research Technology Conference (URTC)*, Cambridge, MA (pp. 1–5). https://doi.org/10.1109/urtc60662.2023.10534998

Sayed, E. T., Olabi, A., Elsaid, K., Radi, M. A., Semeraro, C., Doranehgard, M. H., ... & Abdelkareem, M. A. (2023). Application of artificial intelligence techniques for modeling, optimizing, and controlling desalination systems powered by renewable energy resources. *Journal of Cleaner Production*, *413*, 137486. https://doi.org/10.1016/j.jclepro.2023.137486

ScienceDaily. (2024, October 24). New simultaneous lithium and magnesium extraction technology. Retrieved from https://www.sciencedaily.com/releases/2024/10/241022132829.htm

Seqwater. (n.d.). Desalinated drinking water production increases, to top up SEQ Water Grid ahead of forecast dry summer. Retrieved from https://www.seqwater.com.au/news/desalinated-drinking-water-production-increases-top-seq-water-grid-ahead-forecast-dry-summer

Shahouni, R., Abbasi, M., Dibaj, M., & Akrami, M. (2024). Utilising artificial intelligence to predict membrane behaviour in water purification and desalination. *Water*, *16*(20), 2940. https://doi.org/10.3390/w16202940

Shemer, H., Wald, S., & Semiat, R. (2023). Challenges and solutions for global water scarcity. *Membranes*, *13*(6), 612. https://doi.org/10.3390/membranes13060612

Strubell, E., Ganesh, A., & McCallum, A. (2019). Energy and policy considerations for deep learning in NLP. *Proceedings of the 57th Annual Meeting of the Association for Computational Linguistics*, Florence, Italy. https://doi.org/10.18653/v1/p19-1355

Wang, Y., Cao, Z., & Farimani, A. B. (2021). Efficient water desalination with graphene nanopores obtained using artificial intelligence. *NPJ 2D Materials and Applications*, *5*(1). https://doi.org/10.1038/s41699-021-00246-9

Zarzo, D., & Prats, D. (2017). Desalination and energy consumption. What can we expect in the near future? *Desalination*, *427*, 1–9. https://doi.org/10.1016/j.desal.2017.10.046

Zhang, X., Jiang, Y., Li, K., & Sun, Y. (2024). Energy-saving analysis of desalination equipment based on a machine-learning sequence modeling. *Energy Harvesting and Systems*, *11*(1). https://doi.org/10.1515/ehs-2024-0002

Ziolkowska, J. R. (2014). Is desalination Affordable? Regional cost and price analysis. *Water Resources Management*, *29*(5), 1385–1397. https://doi.org/10.1007/s11269-014-0901-y

PART VII

AI-Integrated Water Policy and Governance

AI Enabled Water Policy of Formulation and Decision Making

15

Kavitha Manikandan and Satthiyaraju Mani

15.1 INTRODUCTION

Earth is one of the beautiful planets by nature, and it is also the hub of nature's creativity. It is easy to say but the work that has been undertaken starting from mountains, sea, desert, huge living beings such as whales, elephants, and camels to microorganisms such as viruses and subatomic particles is tremendous and unimaginable. This wonderful world will not be complete if we do not point out the creation of mankind and the creativity wouldn't have succeeded without him. He is one of the elements in nature's creation and who has the ability to challenge nature's creativity.

There are a lot of elements that exist in this world such as air, water, fire, and clouds. In those elements, water has a unique character of changing its form from one state to the other. Moreover, water is one of the primary requirements for all living beings to exist in the world. However, due to mismanagement by human beings, this abundant water is becoming useless, and there is a need to use this resource in a more effective manner. This calls for the formulation of water policy across nations.

15.2 WATER STATISTICS

The quantum of water that is available in this world is around 1.4 million km^3 and in which it is said that only 2.5% is fresh water and this makes up around 35 million km^3. Out of 35 million km^3 of fresh water, only 0.6% is readily available for human consumption, and it is in the form of lakes, rivers, and ground water. Manmade reservoirs help in storing water during abundant times, and the same is used during times of need. This manmade reservoir stores around 6,500 km^3 of water. It is said that around 50,000 dams have been built across the world as of today, and these are the major reservoirs. However, with the growth of technology, engineers and architects are working to find new ways to store the water and to use it appropriately. There are also multiple policies such as integration of rivers, enriching ground water table, and rerouting rivers to make them more effective. Adding the traditional method, artificial intelligence (AI) has its role to play and contributes significantly to society.

DOI: 10.1201/9781003600718-22

15.3 AI-DRIVEN FRAMEWORK FOR WATER MANAGEMENT

The water management topic has its roots back to civilization such as the Egyptians, the Indus Valley Civilization, which existed 8300–6500 BC. From then on, the water management policies have been evolving to accommodate growth in culture, usage of water, and the quantum of water available. However, there are no situations where mankind has succeeded the water requirement through water management. It is a continuously evolving and it will continue to evolve until mankind exists or until he needs water for his existence.

Water management is the process or method of collecting water, storing it, and using the same in a more systematic manner which helps in sustenance and growth. There have been a lot of methods or frameworks which have been considered and deployed based on human invention and growth. Technology growth has been acting as a double-edged sword for both water management and water usage. Growth in technology has helped mankind to find water 100s of feet below the soil with more accuracy, and this technology has also helped him to bring that water above the soil for his usage. Thus, more wells and bore wells have resulted in faster depletion of ground water and thereby led to a drop in water table levels. On the other hand, the usage of computers, the Internet, and technology driven by artificial technology in water management is significant, and this has led the water policy and water management framework to a new era.

Earlier, there was no information on how much rain is expected in a year and in which month there will be more rain. There is also no information on how much water will flow in the streams following the rain, the requirement of water for people usage, water required for agriculture usage, and industrial purposes. The total water management was on a crude method, and the water was stored to the possible extent. The central, state, and local bodies have been educating the public about the need for saving water and to use the same judiciously. However, the method that they adopted was a crude method. They were not aware of the growing needs for water, water leakages, quantum of water evaporated or the drop in water level due to failure of rain, or higher water availability, etc. The government and NGOs advocated to save water, and they were not aware of the future situation that might pop up. This has led to situations like unrest in the society, crop failure, etc. Thus, all these impacts have called for a more effective and successful framework to manage water.

Data are one of the great resources in the 21st century. AI and machine learning help to bring out the hidden information from the data that have been collected. The data that are collected start from the quantum of rain that comes in a place, the quantum of water absorbed in the land, water that flows in a stream, water that is wasted to quantum of water that is consumed in households, agriculture, and for industrial purposes. AI also takes into account the details such as the fluctuating temperature, growing population, and GDP growth for industrial production, and the demand for water is computed with greater accuracy. AI also helps in collecting information about the water leakages and the quantity of water that has been wasted due to the leakages. All these are precious information to create a framework for water management.

The framework that is under discussion needs to be more dynamic, and it has to consider real-time data. The framework has to provide a solution or it has to adjust the constraints based on the input that has been provided. AI-driven framework would be more real time and dynamic. It takes into account all inputs that have been provided by the user in creating water policy. This framework will consider all demand and demand fluctuations, supply, and in case of any shortages foreseen, the same would be highlighted to the user and others concerned well in advance.

The framework helps to find the average consumption of water by a user in a particular locality. In case of any higher usage by a family or in a residence, AI will help to find those outliers, and the same can be sent for audit to the concerned authority. On identification of any deviation, the government body or the local bodies can take appropriate action. Thus, the framework not only manages the water requirement in a more precise manner but it also helps in increasing revenue for the corporations.

The AI framework helps in creating a circulation list to distribute water in a more logical manner. This means that there is no need for the local body to plan and execute. The plan provided by AI would be more accurate and efficient enough in the distribution of water. One of the major factors when using AI is that there is no bias in the framework. All localities are considered with equal importance against the human preference for certain areas due to the presence of politicians or higher officials. Thus, AI-driven framework would be the best water management framework in the 21st century.

15.4 COMPARISON WITH TRADITIONAL WATER MANAGEMENT METHODS

Water management, as pointed out earlier, has been in existence for a long time. The water management methods vary from place to place based on the geographical condition of the locality, people's mindset, economic condition, etc. The methods might vary based on the preferences but water management has been happening across the world for a long time. Then, if the question is why is there still an issue of water shortage? The answer is twofold: first, the effectiveness of water management practices, and second growing needs of water coupled with fluctuating climatic conditions. Let us dive deep to understand each of the reasons and the traditional water management methods.

The only source of water from historical times to the present day is the rain. The rainwater which comes down from the clouds percolates the soil to the possible extent. The excess rainwater flows down as a stream. The stream that flows to the lake and the pond moves forward after having filled these lakes and ponds. These storage reservoirs act as a source of water for the local community. The water which overflows from lakes and ponds joins the river. The river which has collected a huge quantum of excess water flows directly to the sea.

The government bodies both central and state have built huge dams across the rivers to store the excess water that comes down during the rainy season. These dams not only act as a storage reservoir but they also protect the people from flooding, hydroelectric power generation unit, water navigation, and other sources. However, the growth in population and urbanization has led many people to build their houses very closely. These close structures have also made them cover all the free soil in and around their houses with cement and concrete, leaving no space for water to percolate. Moreover, they have also levelled up the ponds and lakes in the city to accommodate more people, and this has made the rainwater directly join the river or the drainage without percolating the soil or getting stored in a lake or pond. There is no chance for the water table to rise during any quantum of rainfall.

The second topic that is taken for the discussion is the growing need for water through the traditional water management system. As we all know, the water that is evaporated from the sea on back of sunlight gets collected in the clouds. The trees that are present on the land help to cool these clouds, and the water from these clouds comes down as rain. Deforestation, industrial pollution, and the growing need for land for agriculture and for living have led to a lower number of trees and thereby resulting in lower rainfall. This has pushed people to rely more on these dams and irrigation systems. The growing population in India has led to cover the lakes and the ponds with soil, and they have used the land mass to construct buildings and other complexes. This also had led the water table to drop down and pushed people to rely more on the traditional water management methods.

The absence of trees, global warming, and changes in climatic conditions have led to changes in the rainfall pattern. This has resulted in sometimes total drought in the country and sometimes huge rainfall, which has resulted in floods, evacuation, and huge destruction to the country's infrastructure. Thus, the traditional water management practices have been challenged in all possible ways by humans themselves and by nature.

15.5 PREDICTIVE ANALYTICS AND MACHINE LEARNING IN WATER POLICY

The water management method's whole aim is to ensure that the future demands are met. However, in the traditional method, the future demand is not computed or there were not enough room or resources to compute the same. The water is stored to the possible extent, and they are used to the maximum possible level. Predictive Analytics helps to determine the demand for water in the future. It considers all the variables which has been given as input. This means that both variable factors, which result in demand as well as supply, are taken into consideration, and the final demand due to fluctuation of these variables is computed. For instance, let us say that the population is expected to grow in an area by 5% and the demand for water may be at 3% or 2.5% considering other demands for industry and commercial usage remain the same. This is considered by predictive analytics. In similar lines, the drop in rainfall or failure of rain may result in higher demand for water. The rainfall variable, which has been considered in predictive analytics, helps to consider this, and the demand for water is computed.

The predictive analytics models help in arriving at the final demand for water to a more accurate level with less room for deviation. This helps the town planners and the government to understand what the demand for water will be going forward. They are also forced to look for other alternative plans in case of any rain failure or in case of insufficient water in the dams. Predictive analytics also helps to identify the factor which has a higher impact and the factor which has a lower impact. The town planners concentrate on the variables which has a higher impact and they try to fix the issue. They can plan the priority of the water requirement for usage, and the available water is shared based on the priority. Thus, predictive analytics has a huge role in framing water policy.

Machine learning is a wider spectrum in which predictive analytics is a subset or part of it. Machine learning helps to read the given data and tries to find the hidden information from the data. It helps to find the underlying theme or the story behind the impact. Thus, machine learning plays a crucial role in creating the water policy in the 21st century. Machine learning not only considers the water management requirement, but it also takes into account the quality of water that is being distributed. It helps to keep track of pH levels, minerals, vitamins, purity, etc. Thus, Machine learning goes a step further and explains the requirement of quality water for human consumption.

The water policy that is to be created has to be dynamic. It has to consider the change in rain pattern, the flow of water in the river, the quantum of water that comes to the dam, the change in table level, etc. The water policy also has to consider the affordability for different people to get water. The policy has to ensure that people from all walks of life get their basic needs.

15.6 ENHANCING DECISION-MAKING AND ADAPTABILITY

The water policy that has to be made is the one which would educate people on how to use water and how to save the same. The usage of water will vary based on the availability at different times. The policy also has to consider different parameters in different locations. There could be regions where the environment would be rich with water, but there is no major requirement for water, as there is no agriculture being carried out or other water-related activities. There could be areas where there is less water, but the requirement is very high.

This is one of the critical situations where water from the available location has to be moved to the place where it is required. The water policy has to pave the way for the same by educating people in the abundant area to sacrifice the excess availability. Similarly, the people in the required areas have to use the same more judiciously. It is easy to say, but difficult to be followed. It is a very delicate subject that has to be handled with more caution and importance.

AI helps to know the growing requirement of water for both human consumption and allied activities. For instance, water is one of the vital sources in vaccine manufacturing. The growing requirement for vaccines and the need for scientists to invent new vaccines is very important. From that front, there is a chance for us to miss out on the requirement for vaccine manufacturing. However, AI considers requirements from each and every part in the state or the country. This means all data are considered before arriving at the solution.

The model also helps in predicting the rainfall in the coming months. It also considers global warming, melting of snow, and El Nino impact on a particular region. This is a great achievement in technology. There is a limitation for human beings to consider the number of factors and to build a model. The AI-driven technology helps to achieve this. Thus, AI-driven water policy is more efficient and effective in decision-making. The decisions made are more accurate and have higher benefits when compared to the framework that has been formed by using traditional water management techniques.

Adaptability is one of the key features when it comes to AI-driven framework. The model is built based on the inputs that have been provided by the user. As the new data come in, the data are fed into the model for updates and new changes in the variables. This helps to consider all possible changes in the environment, changes in the supply and demand of water, and other factors. If the model is not getting adopted for the changes, then it will become useless and people will stop using the model. Users can also use different models and train the model by using the data. Once the model is trained, then the user can test the model with a portion of unused or new data to verify the results. This is called as test data. If the results that come out from the test data turn out to be satisfactory, then the model can be implemented. In case if the results are not satisfactory, the model is retrained to the expected levels.

15.7 REAL-TIME DATA AND SCALABLE SOLUTIONS

In traditional days, any policy which is being created will be continued for quite some time. These policies will not be reviewed or altered as they felt that there is no change in the underlying data or the scenario. It is believed that the situation remains the same and the policies are held back for some point in time.

The current AI-driven water policy framework has been made up of historical data fed with real-time data. This means that the model predicts the future based on the historical data. It doesn't stop there, it keeps on learning from the new data that come in. This means that the model incorporates the new data into its database to adjust the policy. The system continues to learn and updates the system to frame water policy. Moreover, the point that has to be highlighted is the intervention of human beings, which was mandated in the traditional water management policies, has been totally replaced with AI. The AI tweaks or adjusts the policy based on the change in the availability of water, and the same is implemented immediately or in the future based on the requirement.

(For instance) Let us say that there was more rainfall yesterday night and the quantum of water that has flowed into the dam is higher than usual, the model takes this immediately into its policy computation and derives the distribution of the excess water to a more needed location. If the basic amenities are fulfilled, then the model will look for the best opportunity to use this excess water so that there is a higher benefit when compared to other alternatives.

On the other hand, let us say that there is a major pipe breakage in one of the distribution lines and it is expected to take a couple of days to fix the same. The model will adjust its policy of distribution of water to the specific region and will reallocate the water to the right location. Once the breakage has been fixed, then the model will realign and it will try to allocate more water to the earlier affected area to cover up the shortfall. All these changes in the policy happen without human intervention, and because of the same, the policy that has been formed is unbiased and accurate to the best extent. AI-driven framework also helps in getting one-time inputs or special requests to fine-tune the policy and match the additional requirement.

The next feature of the AI-driven water policy is that scalable solution. This means that in the traditional method, as there are more inputs or factors that have to be considered, this solution is likely to be inaccurate or the efficiency will come down. This means that in traditional water policy, if there is a new request or there is more requirement, then the outcome quality is sacrificed due to the limitation of resources or data or lack of computers to synthesize the available data. Human intervention or any shortfalls as discussed above may lead to uneven policy distribution, which might lead to unrest as well. However, in the AI-driven water policy, the increase in the number of variables or criteria or the constraints will not have much impact on the quality of the outcome. This is one of the greatest benefits to mankind. The reason being there is a huge number of factors which have multiple impacts on the management of water. Along with these factors, there are also noises which will affect the outcome negatively. AI-driven framework identifies the right impacting factors from the noises and computes the results. The noises are removed from the system.

Moreover, the model can all be extended to new areas with the addition of new inputs. This helps the model to be applied across different geographies.

15.8 ADVANCEMENTS IN WATER RESOURCE MANAGEMENT

Water resource management is one of the areas which keeps on evolving. Each and every country uses all possible ways to manage water and their sources. China is one of the countries that has taken huge steps in water resource management. China's Three Gorges Dam, which is located on the Yangtze River, is said to have not only controlled the flooding in the country. It also has stored a huge quantum of water through which China is able to generate a maximum quantum of clean energy and at a low cost. This has helped the country to boost its economy and compete in the world market. The water from Three Gorges Dam is also used for drinking and irrigation purposes. The construction of dam requires huge capital, and it requires the government to invest in it.

In India, there are a good number of projects that have been completed in water management. The country's well-known and famous Kaleshwaram Lift Irrigation Project, which has been built on the Godavari, is a monument. It is considered to be the World's largest lift irrigation project. This means that the water which flows through the Godavari River is undisturbed till it reaches near the sea shore. Somewhere around 300–400 km before the shore, huge constructions have been made and the channels, pipelines, and other required infrastructure have been built to take the water from there through a new way. This not only helps to save the excess water, but it is also routed to a new location where there is water scarcity.

The lift irrigation project is also supported by some power generation units where the power is generated for local use. This is one of the significant moves by two state governments where the original path of the river is unaltered. The new route helps in enriching the land and also helps in bringing water to the people where earlier it was not possible.

The other important water resource management is the one which has been taken by different states and in which Karnataka leads a role. The government has identified water bodies such as ponds, lakes, and tanks, which are very crucial in the state. All these water bodies have been connected with the pipeline from the rivers. This means that when there is water flow in the river, the water is pumped through the pumping station and it is taken to all the water bodies in the state. This ensures that the water is stored in all possible locations. This process of water management, though it helps in cutting down the cost of constructing a dam, involves some costs which have been incurred in laying the pipes and canals. Moreover, the water which has been stored in different locations percolates the soil and seeps through it. This helps in improving the

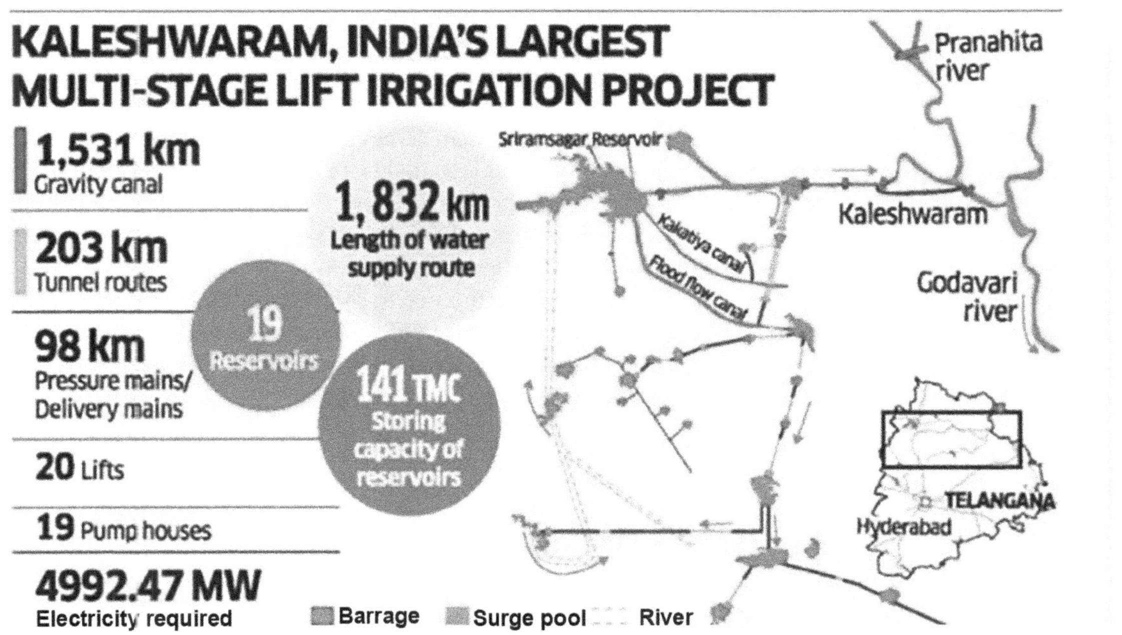

FIGURE 15.1 Map of Kaleshwaram Lift Irrigation Project.

water table of the state. The wells which have been dug by the farmers have experienced higher water levels following the implementation of the above-mentioned method. This method is also helping to have green vegetation across the state, and thereby people are also experiencing more distributed rainfall in the state.

The next important advancement in water resource management is to improve ground water level. People have been drilling bore holes and they have taken a huge quantum of water for drinking and agricultural purposes. There was no restriction in most of the states in the country. This liberty has led people to go for a higher number of bore holes to take water as much as they can for individual benefit. Too much water that has been taken out from the soil has resulted in lowering of the water table, and thereby the vegetation has also started to come down. There is also an opinion that taking water from the soil is one of the major reasons for earthquakes as they create a vacuum or empty space, which results in much easier movement of plates beneath the soil.

To control the taking of water from the soil, now the governments are coming up with restrictions in laying bore wells. The government has also found that laying roads, concrete pathways, and covering the soil surface with other materials have led to a lower area for water to percolate. This has led them find a new way to ensure that the rainwater is saved. Governments in some states have mandated the deployment of water harvesting tanks in each and every home. This means that the water which is collected on the roof of each house has to be gathered and then have to be routed to the soil post appropriate filtering and passing it on. This is one of the successful methods in the 21st century. The cost that has been incurred by each household is very low when compared to the cost incurred by the country in constructing a dam. More than the cost, the quantum of water that can be collected beneath the soil is very huge when compared to the water that is collected in the dam. Hence, the water harvesting by each household has been mandated, and it has been brought into the water policy.

There are certain countries and governments which are trying to convert seawater into potable water. To do the same, there has been a huge amount of money invested in research, and scientists across the world are working on the same. The conversion of seawater to drinking water is possible, but till now the cost to convert the same has been found to be very high. This limits the commercialization of the invented method. If the researcher is able to identify a new method that is cost-optimized and effective, then people will consider using the same.

15.9 CONCLUSION

Water is one of the most important resources not only for human beings but also for all other living beings in this world. It is our responsibility to ensure that this water is available for all without any discretion. The AI-driven water policy is one of the models that will help to ensure water is available for all. It is based on the inputs that have been provided by us, and there is no other bias. Moreover, the framework updates continuously, leaving no room or less room for any deviation or for any error.

There is also a saying that the third world war could arise due to shortage of water, and the country that has more water may be the target of the countries which has high scarcity. From that perspective, a successful AI-driven water management policy will help in avoiding a situation of this sort and help in creating a more balanced and integrated growth across countries in the world.

FURTHER READING

Ahmad, T., Ali, L., Alshamsi, D., Aldahan, A., El-Askary, H., & Ahmed, A. (2024). AI-powered water quality index prediction: Unveiling machine learning precision in hyper-arid regions. *Earth Systems and Environment*, 9, 1–18.

Alzaabi, M. S. A., & Mezher, T. (2021). Analyzing existing UAE national water, energy and food nexus related strategies. *Renewable and Sustainable Energy Reviews*, 144, 111031.

Banerjee, C., Bhaduri, A., & Saraswat, C. (2022). Digitalization in urban water governance: Case study of Bengaluru and Singapore. *Frontiers in Environmental Science*, 10, 816824.

Biswas, A. K. (1991). Water for sustainable development in the 21st century: A global perspective. *GeoJournal*, 24(4), 341–345.

Biswas, A. K. (2001). Water policies in the developing world. *International Journal of Water Resources Development*, 17(4), 489–499.

Chen, Z., Ngo, H. H., Guo, W., & Eslamian, S. (2016). Water shortages. In Eslamian, S. (Ed.) *Urban Water Reuse Handbook* (pp. 1–12). Boca Raton: CRC Press.

Di Baldassarre, G., Wanders, N., AghaKouchak, A., Kuil, L., Rangecroft, S., Veldkamp, T. I., ... & Van Loon, A. F. (2018). Water shortages worsened by reservoir effects. *Nature Sustainability*, 1(11), 617–622.

Egbemhenghe, A. U., Ojeyemi, T., Iwuozor, K. O., Emenike, E. C., Ogunsanya, T. I., Anidiobi, S. U., & Adeniyi, A. G. (2023). Revolutionizing water treatment, conservation, and management: Harnessing the power of AI-driven ChatGPT solutions. *Environmental Challenges*, 13, 100782.

Falkenmark, M., Da Cunha, L., & David, L. (1987). New water management strategies needed for the 21st century. *Water International*, 12(3), 94–101.

Farhaoui, Y., & El Allaoui, A. (2024, April). Sustainability in the Internet of Things: Insights, scope, and AI-driven optimized water management with big data integration. In *The International Workshop on Big Data and Business Intelligence* (pp. 468–475). Cham: Springer Nature Switzerland.

Garcia, E. (n.d.). AI-Driven Water Management Systems for Sustainable Smart cities.

Gohari, A., Eslamian, S., Mirchi, A., Abedi-Koupaei, J., Bavani, A. M., & Madani, K. (2013). Water transfer as a solution to water shortage: A fix that can backfire. *Journal of Hydrology*, 491, 23–39.

Goyal, M. K., Kumar, S., & Gupta, A. (2024a). *AI Innovation for Water Policy and Sustainability*. Cham: Springer Nature.

Goyal, M. K., Kumar, S., & Gupta, A. (Eds.) (2024b). AI for water policy. In *AI Innovation for Water Policy and Sustainability* (pp. 41–53). Cham: Springer Nature Switzerland.

Griffin, R. C. (2016). *Water Resource Economics: The Analysis of Scarcity, Policies, and Projects*. Cambridge, MA: MIT Press.

Kaluri, A., Enoch, J., & Magdalene, M. (2025). A survey on AI-driven strategies and solutions for environmental conservation. *Journal of Energy Technology and Environment*, 7(1), 30–37.

Liu, J., Zang, C., Tian, S., Liu, J., Yang, H., Jia, S., ... & Zhang, M. (2013). Water conservancy projects in China: Achievements, challenges and way forward. *Global Environmental Change*, 23(3), 633–643.

Obiorah, C. A., & Ndubuisi, O. G. (2025). Innovative Approach to Water Resource Management: A Focus on Efficiency and Conservation.

Pasumarti, S. S. R., & El Hajj, C. (2025). AI for sustainable development: MENA strategies. In Azoury, N., & Yahchouchi, G. (Eds.) *AI in the Middle East for Growth and Business: A Transformative Force* (pp. 43–60). Cham: Springer Nature Switzerland.

Rajput, H., & Tanwar, R. (2024). AI and water resource. In Whig, P., Sharma, P., Aneja, N., Elngar, A. A., & Silva, N. (Eds.) *Artificial Intelligence and Machine Learning for Sustainable Development: Innovations, Challenges, and Applications* (p. 75). Milton: Taylor & Francis Group.

Ratnam, M. (2025). Water supply mapping for a sustainable future: Data-driven efforts in decision making. *Smart Internet of Things*, 2(1), 1–10.

Sakkaravarthy, S., Jano, N. A., & Vijayakumar, A. (2024). Overcoming challenges in traditional waste water treatment through ai-driven innovation. In Garg, M. C. (Ed.) *The AI Cleanse: Transforming Wastewater Treatment Through Artificial Intelligence: Harnessing Data-Driven Solutions* (pp. 53–81). Cham: Springer Nature Switzerland.

Suchetana, B., Srivastava, B., Gupta, H. P., & Saharia, M. (2023, January). Promoting sustainable water usage and management with water data, AI and policy. In *Proceedings of the 6th Joint International Conference on Data Science & Management of Data (10th ACM IKDD CODS and 28th COMAD)*, Mumbai, India (pp. 308–311).

Tandon, N., & Shaheen, M. (2024). AI-driven innovations for environmental sustainability across sectors. In Singh, B. (Ed.) *Maintaining a Sustainable World in the Nexus of Environmental Science and AI* (pp. 429–454). Hershey, PA: IGI Global.

Tarhule, A. (2017). The future of water: Prospects and challenges for water management in the 21st century. *Competition for Water Resources*, pp.442–454.

Varis, O., & Vakkilainen, P. (2001). China's 8 challenges to water resources management in the first quarter of the 21st Century. *Geomorphology*, 41(2–3), 93–104.

Wang, G., Kumar, S., Huang, Z., & Liu, R. (2024). Water resource management and policy evaluation in Middle Eastern countries: Achieving sustainable development goal 6. *Desalination and Water Treatment*, 320, 100829.

Wang, Z., & Li, D. (2025). Big data and artificial intelligence-driven risk assessment and response for water resources: Model applications and future perspectives. *Advances in Resources Research*, 5(1), 414–434.

Ye, Z., Yin, S., Cao, Y., & Wang, Y. (2024). AI-driven optimization of agricultural water management for enhanced sustainability. *Scientific Reports*, 14(1), 25721.

Zhao, Y. (n.d.). Evaluating the Achievement of Singapore's Environmental Policy Goals for Enhanced Sustainability and Livability through Big Data Analysis.

16 Investigation on Some Aspects of Data-Driven Insights for Effective Water Governance

Abhijit Bora, Lima Hazarika, and Bamdeb Dey

16.1 INTRODUCTION

Clean and polluted water is the cornerstone of all life forms, serving as the primary medium for biological processes across ecosystems. From a biological standpoint, clean water is critical for sustaining aquatic and terrestrial organisms, promoting biodiversity, and maintaining healthy ecosystems. Clean water is free from contaminants, supporting the life cycles of various species without causing harm. Conversely, polluted water is characterized by the presence of harmful chemicals, heavy metals, and pathogens, posing significant risks to wildlife and humans alike (Cabral, 2010; Gerba, 2012). Pollutants can originate from industrial discharge, agricultural runoff, and domestic waste, directly affecting aquatic species and indirectly impacting terrestrial animals that rely on these water sources. Polluted water leads to bioaccumulation and biomagnification, where toxic substances accumulate in the tissues of organisms and increase in concentration up the food chain (Roets-Dlamini et al., 2022). This not only threatens aquatic biodiversity but also affects apex predators, including humans. The health of aquatic life, such as amphibians, fish, and macroinvertebrates, often reflects the overall quality of the water (Hlordzi et al., 2020). As such, maintaining clean water in an effective water governance system is integral to preserving ecological balance and preventing the collapse of aquatic ecosystems. Here, deployment of data-driven system across different stakeholders of water governance can play a vital role for the ecosystems.

DOI: 10.1201/9781003600718-23

16.2 THE BACKGROUND AND CONTEXT

Many researchers discussed that the presence of pathogenic microbes poses severe threats to both wildlife and human populations. In poor water governance, it is found that water bodies naturally harbor a diverse array of microorganisms, many of which play crucial roles in nutrient cycling and ecosystem functioning. Waterborne diseases such as yellow fever, dengue, filariasis, malaria, onchocerciasis, and sleeping sickness are transmitted by insects that breed in or near water. Common bacterial pathogens found in water include *Salmonella, Escherichia coli, Shigella, Campylobacter, Yersinia, Vibrio, Helicobacter*, and *Legionella*. Opportunistic pathogens, with *Pseudomonas aeruginosa* being the most significant, are also abundant in the environment. Viruses like norovirus, hepatitis A, coxsackievirus, echovirus, and adenoviruses are leading causes of gastroenteritis and waterborne outbreaks. Respiratory diseases are linked to viruses such as rhinoviruses, coronaviruses, parainfluenza, respiratory syncytial virus (RSV), influenza, and adenoviruses. Microbial water analysis through data-driven system, therefore, plays a critical role in monitoring the health of aquatic ecosystems and identifying potential threats to biodiversity. Studies have reported a range of microorganisms that play important roles in water treatment and monitoring, leveraging both natural processes and cutting-edge technology. *Escherichia coli (E. coli)* is generally considered a harmful pathogen, yet it serves as an indicator microorganism for monitoring water quality and detecting fecal contamination in water (Khan & Gupta, 2020). A study on aquaculture effluent reveals that it contributes to water pollution by increasing organic carbon, nitrogen, phosphorus, and other harmful substances, which can damage aquatic ecosystems and affect plant health. Bacillus species can help improve water quality in aquaculture by breaking down pollutants, but their effectiveness depends on factors like pH, temperature, and nutrient availability. Bacteria such as Nitrosomonas and Nitrobacter are key in the nitrification process, converting ammonia into nitrates, essential for the detoxification of water (Grunditz & Dalhammar, 2001). Additionally, *Pseudomonas* species are involved in the degradation of hydrocarbons and other contaminants. Studies reported that *Pseudomonas* adapts to toxic organic compounds by adjusting its carbon metabolism, and understanding how it breaks down these compounds can help develop new methods for treating environmental pollutants (Medić & Karadžić, 2022). Microalgae, like chlorella and spirulina, contribute to nutrient cycling by removing excess nitrogen and phosphorus, while also improving water quality through photosynthesis. In the industrial sector, chlorella and spirulina are used to make biopolymers, fuel cells, and solar technologies. These innovative uses could increase the value of microalgae, helping to overcome challenges like high production costs (Abreu et al., 2023). It is well known in Life Sciences that white rot fungi, basidiomycetes with the ability to degrade lignocelluloses using enzymes like peroxidase, laccase, and oxidase, play a key role in recycling carbon in lignocelluloses, which are abundant, renewable organic materials (Echezonachi, 2022). Fungi like *Aspergillus niger*, known for enzymes like cellulase, amylase, and ligninase, help break down plant waste and produce biofuels, while *Pleurotus ostreatus* (oyster mushroom) is effective in bioremediation, making it valuable for waste management and environmental cleanup (Dinakarkumar et al., 2024). *Trametes versicolor*, a typical white rot fungus, has a strong ability to degrade various pollutants due to its complex enzymes that break down different substances both inside and outside its cells (Zhuo & Fan, 2021). Fecal indicator bacteria, such as *E. coli*, are commonly used to assess water quality. These indicators signal the presence of fecal contamination and potential pathogens, guiding conservationists and public health officials in identifying polluted water sources.

Inadequate water treatment, poor sanitation, and lax enforcement of regulations are just a few of the ways that poor water governance can greatly aid in the spread of microbial diseases.

16.3 CONCEPTUAL FRAMEWORK

The inclusion of Artificial Intelligence (AI) can aid in optimizing the roles of these microorganisms by enabling real-time monitoring of their activity, and environmental parameters such as pH, turbidity, dissolved oxygen, and temperature, and adjusting treatment processes accordingly. AI-driven sensors and algorithms can predict microbial growth patterns, enhancing the efficiency of biological treatments and ensuring cleaner water with minimal human intervention.

AI is revolutionizing the water industry by enhancing the efficiency, accuracy, and sustainability of water treatment processes. AI-driven systems leverage machine learning and data analysis to optimize water purification, predict contamination events, and monitor ecosystem health in real time. By integrating AI with zoological studies, conservationists can gain deeper insights into the interactions between aquatic species at both micro and macro levels and water quality, enabling proactive measures to protect vulnerable species. Regarding drinking water, while nanofiber filters are commonly used for air purification and pollution control (Leung, 2021), this raises the question of whether they can be effectively applied for water filtration, or if AI should be used to monitor their performance in water treatment systems.

AI algorithms can analyze vast datasets, identifying patterns that may elude human observation. For instance, predictive models can forecast pollution outbreaks, allowing for timely intervention to prevent the spread of disease. Furthermore, AI-powered monitoring systems can continuously track water parameters, providing invaluable data on microbial content, chemical composition, and overall ecosystem health.

Monitoring indicator microorganisms is essential for assessing water quality and public health. While current treatment methods address contaminants, eco-friendly microbial technologies are emerging to tackle pollutants like pharmaceuticals and hormones and recover valuable resources from wastewater. However, challenges persist in data interpretation and integrating sewage analysis into public health strategies.

The fluid mechanism and monitoring system through data-driven methodologies are also essential for good water governance in order to provide stakeholders with high-quality services. Fluids are commonly utilized as thermal transporters in energy transfer apparatus. Heat-transferring fluids find significant applications in transportation industries such as aerospace and vehicle cooling systems, mechanical systems for cooling and heating in infrastructure, and cooling mechanisms in fossil fuel, cellulose and paper, and other production facilities. The thermal efficiency of heat-transferring fluids is crucial in the advancement of sustainable heat transmission apparatus across all these industries. Industries require the development of improved means of transferring heat fluids with substantially greater thermal capacity than currently available, particularly in the context of escalating global competitiveness (Jawad & Nisar, 2023; Dey et al., 2024; Choudhury & Dey, 2018). It must be acknowledged that the poor thermal efficiency of traditional heat transmission fluids has limited significant advancements in cooling capacities despite significant prior studies and experiments on heat transferring augmentation. Nonetheless, it is commonly known that at ambient temperature, the heat conductivity of rigid metals is orders substantially higher than that of fluids. For instance, at ambient temperature, copper's capacity for heating is 700 times higher than that of water and 3,000 times higher than that of motor oil. Furthermore, metallic fluids have a far higher heat conductivity than non-metallic fluids. Consequently, it is anticipated that fluids containing embedded metallic components would have substantially enhanced thermal conductivity than traditional heat transferring fluids (Malleswari et al., 2024; Verma & Meher, 2024; Mirza et al., 2024).

Nanofluid, a novel type of thermal transfer fluid that was created by steadily floating nanoparticles in traditional heat transmission fluids—typically liquids—was launched 10 years ago. Oxide nanostructures such as aluminum (Al), silicon (Si), titanium (Ti), copper (Cu), zinc (Zn), iron (Fe), magnesium (Mg), and silver (Ag) are frequently employed; base fluids include kerosene, water, motor oil, ethanol, methanol, ethylene glycol, and mono ethylene glycol. It has been demonstrated that fundamental fluids' heat transmission by convection and thermal efficiency are improved by nanofluids made of these nanoparticles floating in fluids. Therefore, studies are underway to utilize nanofluids in a variety of heating processes where traditional fluids are unable to further increase the rate of heat transmission. Using nanoparticles embedded in a traditional heat-transferring fluid for the first time, Choi and Eastman suggested that adding nanoparticles of nanometer size to the initial fluid increased heat transfer and, consequently, the nanofluid's efficiency in transferring heat (Choi & Eastman, 1995).

Since then, nanofluids have been employed in many different fields, such as photosynthesis (Dilnawaz et al., 2023), transportation, microbial fuel cells for power generation (Naha et al., 2023), radiological systems cooling (Noreen et al., 2023), pollution reduction (Jubair et al., 2025), nanomedicine delivery (Mandal et al., 2023), amplifying microreactors (Wang et al., 2023), and technical cooling (Thesiya et al., 2023). Compared to microparticles, nanoparticles have a substantially larger extent and remain suspended for a significantly longer period of time. Additionally, nanoparticles may absorb and transmit energy effectively since 20% of their atoms are close to the outer layer. Furthermore, the dispersed particles expand the mixture's overall area and its thermal capacity. In other words, utilizing smaller particles (nanometer-sized particles) instead of bigger ones (micrometer-sized particles) results in a notable increase in optimal energy transfer. This characteristic significantly increases the fluid's thermal conduction involvement since radiation takes place on a solid's interface. Significant practical challenges include problems like reduced pressure and mechanical power, particularly when working with greater nanoparticle concentrations (Shiravi et al., 2020). For broad industrial use, the economics of producing and disposing of nanofluids must also be taken into account.

Since fluid mechanics offers the scientific and engineering foundations required for efficient water resource management, it is essential for water governance. Flood control, wastewater treatment, pollution management, and water supply system design and optimization are all aided by it.

Rapid advancements in cloud computing and the Internet of Things (IoT) enable Smart Cities by combining heterogeneous computing devices to cooperatively deliver a range of widely implemented smart services (Nikouei et al., 2019; Shi & Dustdar, 2016), where the ability to gather and process data at the edge is found crucial. Even though cloud computing has improved people's lives by offering services like infrastructure as a service (IaaS), platform as a service (PaaS), and service-oriented architecture (SOA) as a service (SaaS) (Fox et al., 2009), the rise of edge computing, AI, and IoT ushers in the post-cloud computing era. An extension of SOA is a microservices architecture (MSA) (Wilde et al., 2016; Kakati & Bora, 2024).

MSAs are more decentralized than SOAs, distributing all logic, including routing and message parsing, among intelligent endpoints (Zaki et al., 2022). This is the primary distinction between the two. To manage services, MSA also uses a lightweight application programming interface (API) gateway rather than more complex and heavy ones (Blinowski et al., 2022). Each microservice manages a separate process and uses lightweight communication protocols (SOAP or REST API) to interact with peers (Bora & Bezboruah, 2020). The services are independently deployed using fully automated deployment tools and are constructed around business capabilities (Pathania, 2017).

Two important characteristics of MSA are fine granularity and loose coupling (Yu et al., 2018). There is little centralized management of these services due to fine granularity. Furthermore, the service may not always be micro (Sheng et al., 2019). Each microservices component has minimal dependencies on other independent components because there is loose coupling, which increases the independence of microservices development and deployment (Pautasso et al., 2017; Bora & Bezboruah, 2023).

On the other hand, artificial intelligence markup language (AIML) can be used to create and develop the service's intelligence. The microservice (MS) divides various business logic into several separate services that interact with one another via an application programming interface. It facilitates communication between loosely coupled services using lightweight protocols like simple object access protocol. MS can be used to implement the SOA concept. A separate team of developers can manage the MS on their own (Lyu et al., 2020). Therefore, the extra complexity that arises during deployment can be minimized. The data transfer, internet, and computing industries are undergoing tremendous change as a result of the last few decades' technological breakthroughs. The cloud computing paradigm is replacing the traditional client-server setup. Cloud computing is a system that allows users worldwide to access online resources whenever they need them. Applications, servers, storage space, networking tools, and tools for developing mobile applications are among the resources that are accessible in the cloud. Vendors provide the resources and services, and any end user can use them by purchasing the features for a fixed period of time. Cloud service providers managed and virtualized the apps or services that the vendors hosted in various data centers. Because of their scalability, modularity, and deployment ease, MS has grown in popularity across a variety of software engineering domains. Their application in automation presents an opportunity to circumvent many of the issues with traditional monolithic robotic systems, particularly with regard to swarm intelligence.

A benchmark for MS evaluation can be set for deployment. For actual deployment, Docker containers are popular and well-known in the container community, which is why researchers use them for microservices development. To enable easy execution of the application on any Linux server, Docker bundles the application and all of its dependencies into a container. For container management, it makes use of layered file system images in addition to the other Linux kernel functionalities. With the help of this feature, Docker can generate an infinite number of containers—basic images that have been enhanced with new features—from a base image. Additionally, this lowers the total amount of memory and storage needed, which facilitates quick startup.

Based on common computing paradigms utilized in the edge smart public safety system, a number of popular benchmarks are taken into consideration when designing the microservices benchmark in order to assess the performance of the containerized microservices deployment policy. The benchmark's primary metrics are as follows:

a. **CPU Performance:** The Apache Jmeter is selected to assess the CPU's performance. How well a computer's central processing unit (CPU) can carry out commands and finish tasks in a specific amount of time is referred to as CPU performance. It gauges the CPU's processing speed and program execution ability.
b. **Memory Performance:** How well a computer's memory is utilized efficiently to process the request is called as memory performance, unit usually measured in MB/s.
c. **Disk Performance:** A computer's storage device's ability to read and write data quickly and effectively is referred to as disk performance.

The data-driven strategy can be developed while following different MS patterns. The API gateway pattern can be followed to deploy a scenario like a central router where the specific data-driven request is redirected to a specific MS for the necessary execution of business logic. The central router itself can be an MS that can even validate the incoming and outgoing data-driven request response. It can also monitor the traffic log for better service in web traffic. The MS pattern of service registry and discovery pattern can be followed for the scenario where there will be a central hub that will keep tracking or updating the list of MS-based applications that are functioning for data-driven-based activity. Additionally, when a new MS-based data-driven application wants to participate in the network, that application needs to be registered in the central hub along with its information of IP address and port number. The central hub can also update itself automatically once an MS-based data-driven system stops handling requests in the network. In circuit breaker pattern, when repeated service calls fail, the pattern activates a circuit to protect the system's dependability and reroutes requests to fallback methods or default responses. It reduces the impact of service outages or slowdowns by preserving system stability and user satisfaction. It also serves as a safety mechanism in distributed systems, preventing cascading failures by isolating problematic services and automatically validating and resetting the service, allowing it to gradually recover from the problems that need to be fixed. In the saga design pattern, the transactions are divided into local transactions managed by distinct services. This pattern type serves as a fundamental idea in microservices architecture, addressing data consistency across services. The saga pattern offers a fault-tolerant and scalable approach to managing distributed transactions. It supports independence and flexibility in the architecture, guarantees data consistency, and helps the system remain flexible and reliable even in the event of partial failures by enabling individual services to independently manage their portion of the transaction (Nikouei et al., 2018).

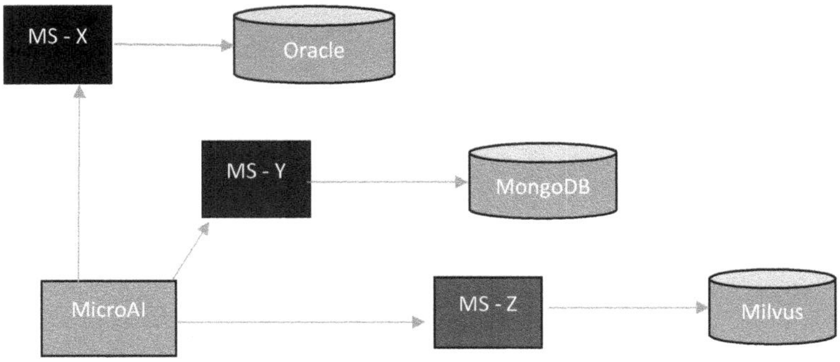

FIGURE 16.1 Block diagram of MicroAI for database per service pattern.

In the database per service pattern, a dedicated database service is assigned for a dedicated MS of business logic. In a scenario where a specific MS may need a specific datastore for the operation of relational queries, NoSQL solution, and vector database operations. In that case, managing the MS individually could be easier than managing them collaboratively. In this scenario, one can obtain the features of loose coupling. Services must use published interfaces to communicate with one another. The drawback is that dedicated databases need a way to protect against communication failures. Figure 16.1 shows a basic scenario for a database per service pattern.

In the aggregator design pattern, an aggregator service will be there, where the aggregator, on receiving the request, will forward it to different MS for necessary processing at their end. Once data-driven operations are completed, the response will be collected by the aggregator, and it will be forwarded to the initial request that was made to the aggregator.

16.4 TESTING METHOD FOR MICROAI

The audit testing of MicroAI is an important concern for an effective data-driven water governance system. The testing methodology for MicroAI is shown in Figure 16.2. Here, the MicroAI's evaluation is contrasted with that of other authors. Each author's observation is listed in Table 16.1. The following are the evaluation criteria that were used: (a) Methodology for evaluating data samples; (b) use of test beds; (c) mean value evaluation; (d) failure assessment; (e) fault count population mean and rate; (f) reliability assessment; (g) high user load behavior; (h) population mean assessment; (i) comparison of time complexity; (j) evaluation of time complexity; and (k) correlation of performance metrics. Figure 16.3 displays each study's percentage of criteria satisfaction.

16.5 OVERALL EVALUATION OF MICROAI

From Figure 16.3, it is observed that a different criteria meeting percentage exists for every observation in comparison to MicroAI. Rahmani et al. (2014) and Saddik (2006) found that it satisfies 9.09% of the total observed assessment. In contrast, Singh and Bezboruah (2015) found that it satisfies 54.55% and 18.18% of the total observed assessments in the evaluations of Kalita and Bezboruah (2012), Medhi and Bezboruah (2014), and Ba et al. (2016). Conversely, the suggested MicroAI satisfies the evaluation criteria and meets 90.91%. $O(n^2)$, or polynomial time complexity, will be the probable time complexity for MicroAI It can be said that the number of operations needed to assess the quality aspects of water

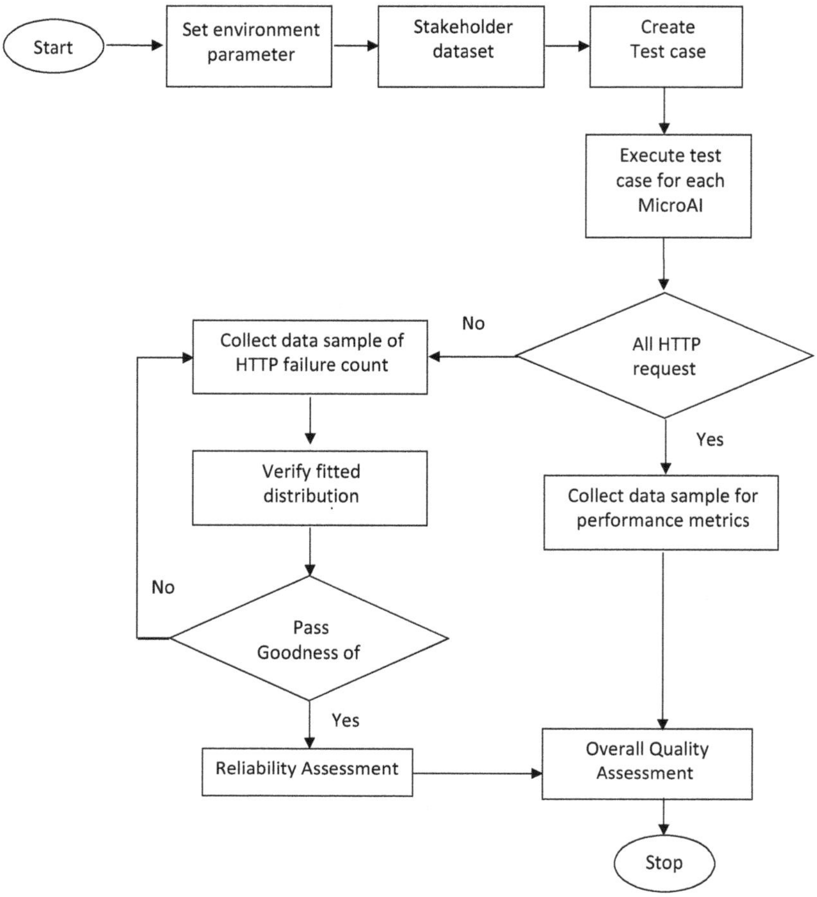

FIGURE 16.2 Testing methodology for MicroAI.

TABLE 16.1 Evaluation of MicroAI in comparison to other frameworks

		OTHER AUTHORS WORK					
CRITERIA	*MICROAI*	*KALITA AND BEZBORUAH (2012)*	*BA ET AL. (2016)*	*RAHMANI ET AL. (2014)*	*SINGH AND BEZBORUAH (2015)*	*SADDIK (2006)*	*MEDHI AND BEZBORUAH (2014)*
a	Y	Y	N	N	N	N	Y
b	Y	Y	N	Y	Y	Y	Y
c	Y	Y	N	N	N	N	Y
d	Y	N	N	N	N	N	N
e	Y	N	N	N	N	N	N
f	Y	N	N	N	N	N	N
g	Y	Y	N	N	Y	N	Y
h	Y	Y	N	N	N	N	Y
i	Y	N	Y	N	N	N	N
j	Polynomial	Not performed in the study	NP Complete	Not performed in the study	Not performed in the study	Not performed in the study	Not performed in the study
k	N	Y	N	N	N	N	Y

Y, Yes; N, No.

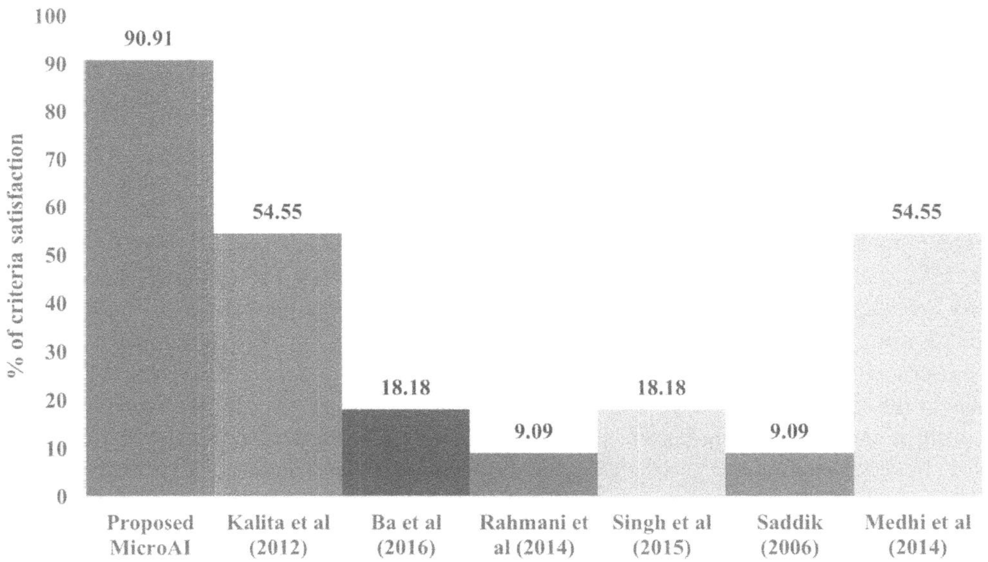

FIGURE 16.3 Criteria meeting % for every observation in comparison to MicroAI.

TABLE 16.2 Comparative analysis of the proposed framework with existing AI-based frameworks for water governance

FRAMEWORK	*DESCRIPTION*	*APPLICATION AREA*	*INTEGRATION FEATURE*	*LIMITATION*
Explainable AI (Saghya et al., 2025)	A tool to enhance sustainable water management practices	Urban water system	IoT	No feature of node-to-node communication for collaborative execution
Big data analysis and intelligent decision support system (Qi et al., 2024)	A big data analysis and intelligent decision support system for environmental water quality monitoring. It explores how AI is applied to protect and manage water environments effectively.	Water body in society	Big data and decision support system	No observation framework for assessing the performance metrics
AI for water conservancy project management (Shi & Dustdar, 2016)	Water conservancy projects through resource optimization	Flood control, irrigation, and water resource allocation	AI-based predictive modeling	Integration issues with traditional water management systems
Proposed MicroAI	Microservice-oriented swarm intelligence	Universal approach for all areas	Microservice, Swarm intelligence	Depends on valid datasets.

governance is directly correlated with the creation of new stakeholder entities. Therefore, it is possible to solve the issue that comes up when assessing the quality aspects of data-driven water governance services in polynomial time. Table 16.2 presents a comparative analysis of the proposed MicroAI with existing AI-based frameworks for water governance proposed by different authors. While evaluating the quality aspects of service delivery, it can be concluded that MicroAI can provide better service than other deployment methodologies.

16.6 CONCLUSION

When assessing the time complexity analysis, the suggested MicroAI is found to be valid. Its execution is polynomial, and the issue of assessing the data-driven-based water governance system's quality aspects can be resolved in the allotted time. When compared to other authors, the MicroAI meets all requirements. The proposed work demonstrates a MS and AI-based deployment methodology for providing better water governance services. When compared to other authors' work, the architecture and deployment feasibility are found to be applicable. From a zoological perspective, innovations in water treatment are essential for preserving biodiversity and ensuring the long-term survival of aquatic species. Healthy water bodies are crucial for sustaining fish populations, amphibians, and aquatic invertebrates, all of which contribute to the stability of larger ecosystems. By harnessing AI and other technological advancements, we can mitigate the impacts of pollution, safeguard endangered species, and promote environmental sustainability. In conclusion, investing in innovative water treatment and monitoring technologies will pave the way for a healthier planet.

REFERENCES

Abreu, A. P., Martins, R., & Nunes, J. (2023). Emerging applications of *Chlorella* sp. and *Spirulina (Arthrospira)* sp. *Bioengineering, 10*(8), 955. https://doi.org/10.3390/BIOENGINEERING10080955

Ba, S., Kawabata, A., Chatterjee, B. C., & Oki, E. (2016). Computational time complexity of allocation problem for distributed servers in real-time applications. In *Proceedings of the 18th Asia-Pacific Network Operations and Management Symposium (APNOMS)*. https://doi.org/10.1109/APNOMS.2016.7737236

Blinowski, G., Ojdowska, A., & Przybyłek, A. (2022). Monolithic vs. microservice architecture: A performance and scalability evaluation. *IEEE Access, 10*, 20357–20374. https://doi.org/10.1109/ACCESS.2022.3152803

Bora, A., & Bezboruah, T. (2020). Investigation on reliability estimation of loosely coupled software as a service execution using clustered and non-clustered web server. *International Journal of Engineering, Transactions A: Basics, 33*(1), 75–81. https://doi.org/10.5829/ije.2020.33.01a.09

Bora, A., & Bezboruah, T. (2023). Performance investigation of software agents in artificial intelligence and document object model domain. *Indonesian Journal of Electrical Engineering and Informatics, 11*(1), 97–104. https://doi.org/10.52549/ijeei.v11i1.3984

Cabral, J. P. S. (2010). Water microbiology: Bacterial pathogens and water. *International Journal of Environmental Research and Public Health, 7*(10), 3657. https://doi.org/10.3390/IJERPH7103657

Choi, S. U., & Eastman, J. A. (1995). Enhancing thermal conductivity of fluids with nanoparticles (No. ANL/MSD/CP-84938; CONF-951135-29). *Argonne National Laboratory (ANL)*, Argonne, IL, United States.

Choudhury, R., & Dey, B. (2018). Unsteady thermal radiation effects on MHD convective slip flow of visco-elastic fluid past a porous plate embedded in porous medium. *International Journal of Applied Mathematics and Statistics, 57*(2), 215–226.

Dey, B., Dukru, D., Das, T. K., & Nath, J. M. (2024). Modelling and simulating the heat transference in Casson EMHD fluid motion exacerbated by a flat plate with radiant heat and Ohmic heating. *East European Journal of Physics, (2)*, 172–180. https://doi.org/10.26565/2312-4334-2024-2-16

Dilnawaz, F., Kalaji, M. H., & Misra, A. N. (2023). Nanotechnology in improving photosynthesis under adverse climatic conditions: Cell to canopy action. *Plant Nano Biology, 4*, 100035.

Dinakarkumar, Y., Ramakrishnan, G., Gujjula, K. R., Vasu, V., Balamurugan, P., & Murali, G. (2024). Fungal bioremediation: An overview of the mechanisms, applications, and future perspectives. *Environmental Chemistry and Ecotoxicology, 6*, 293–302. https://doi.org/10.1016/J.ENCECO.2024.07.002

Echezonachi, S. O. (2022). The role of white rot fungi in bioremediation. In Malik, J. A. (Ed.) *Microbes and Microbial Biotechnology for Green Remediation* (pp. 305–320). Elsevier, Amsterdam. https://doi.org/10.1016/B978-0-323-90452-0.00034-7

Fox, A., Griffith, R., Joseph, A., Katz, R., Konwinski, A., Lee, G., Patterson, D., Rabkin, A., & Stoica, I. (2009). Above the clouds: A Berkeley view of cloud computing (Technical Report No. UCB/EECS-2009-28). University of California, Berkeley, Department of Electrical Engineering and Computer Sciences.

Gerba, C. P. (2012). Environmentally transmitted pathogens. In *Environmental Microbiology* (p. 445). https://doi.org/10.1016/B978-0-12-370519-8.00022-5

Grunditz, C., & Dalhammar, G. (2001). Development of nitrification inhibition assays using pure cultures of *Nitrosomonas* and *Nitrobacter*. *Water Research*, *35*(2), 433–440. https://doi.org/10.1016/S0043-1354(00)00312-2

Hlordzi, V., Kuebutornye, F. K. A., Afriyie, G., Abarike, E. D., Lu, Y., Chi, S., & Anokyewaa, M. A. (2020). The use of *Bacillus* species in maintenance of water quality in aquaculture: A review. *Aquaculture Reports*, *18*, 100503. https://doi.org/10.1016/J.AQREP.2020.100503

Jawad, M., & Nisar, K. S. (2023). Upper-convected flow of Maxwell fluid near stagnation point through porous surface using Cattaneo-Christov heat flux model. *Case Studies in Thermal Engineering*, *48*, 103155.

Jubair, S., Yang, J., Ali, B., Bin-Mohsin, B., & Khalifa, H. A. E. W. (2025). Analyzing the impact of non-Newtonian nanofluid flow on pollutant discharge concentration in wastewater management using an artificial computing approach. *Applied Water Science*, *15*(1), 1–13.

Kakati, P., & Bora, A. (2024). Some aspects of data engineering for edge computing using microservice design pattern. In Bora, A., Changmai, P., & Maharana, M. (Eds.) *Critical Approaches to Data Engineering Systems and Analysis* (pp. 284–298). IGI Global, Hershey, PA. https://doi.org/10.4018/979-8-3693-2260-4.ch014

Kalita, M., & Bezboruah, T. (2012). Investigations on implementation of web applications with different techniques. *IET Software*, *6*(6), 474–478. https://doi.org/10.1049/iet-sen.2011.0136

Khan, F. M., & Gupta, R. (2020). *Escherichia coli* (*E. coli*) as an indicator of fecal contamination in groundwater: A review. In *Environmental Science and Engineering* (pp. 225–235). https://doi.org/10.1007/978-3-030-45263-6_21

Leung, W. W.-F. (2021). *Nanofiber Filter Technologies for Filtration of Submicron Aerosols and Nanoaerosols* (1st ed.). Elsevier. Retrieved from https://shop.elsevier.com/books/nanofiber-filter-technologies-for-filtration-of-submicron-aerosols-and-nanoaerosols/woon-fong-leung/978-0-12-824468-5

Lyu, Z., Wei, H., Bai, X., & Lian, C. (2020). Microservice-based architecture for an energy management system. *IEEE Systems Journal*, *14*(4), 5061–5072. https://doi.org/10.1109/JSYST.2020.2981095

Malleswari, K., Nath, J. M., Reddy, M. V., Mala, M. S., & Dey, B. (2024). Thermal performance of Cattaneo-Christov heat flux in MHD radiative flow of Williamson nanofluid containing motile microorganisms and Arrhenius activation energy. *Journal of Computational and Theoretical Transport*, *54*, 1–23.

Mandal, D., Anand, K., & Prabhakar, P. K. (2023). Biomedical applications of nanofluids in drug delivery. In Kulkarni, S., Haghi, A. K., & Manwatkar, S. (Eds.) *Novel Technologies in Biosystems, Biomedical & Drug Delivery* (pp. 105–124). Springer Nature, Singapore.

Medhi, S., & Bezboruah, T. (2014). Investigations on implementation of e-ATM web services based on .NET technique. *International Journal of Information Retrieval Research*, *4*(2), 42–51.

Medić, A. B., & Karadžić, I. M. (2022). Pseudomonas in environmental bioremediation of hydrocarbons and phenolic compounds: Key catabolic degradation enzymes and new analytical platforms for comprehensive investigation. *World Journal of Microbiology and Biotechnology*, *38*(10), 1–28. https://doi.org/10.1007/S11274-022-03349-7

Mirza, A. H., Dey, B., & Choudhury, R. (2024). The detrimental effect of thermal exposure and thermophoresis on MHD flow with combined mass and heat transmission employing permeability. *International Journal of Applied Mechanics and Engineering, 29*(1), 90.

Naha, A., Debroy, R., Sharma, D., Shah, M. P., & Nath, S. (2023). Microbial fuel cell: A state-of-the-art and revolutionizing technology for efficient energy recovery. *Cleaner and Circular Bioeconomy*, *5*, 100050. https://doi.org/10.1016/j.clcb.2023.100050

Nikouei, S. Y., Chen, Y., Song, S., & Faughnan, T. R. (2019). Kerman: A hybrid lightweight tracking algorithm to enable smart surveillance as an edge service. In *2019 16th IEEE Annual Consumer Communications & Networking Conference (CCNC)*, Las Vegas, NV (pp. 1–6). IEEE. https://doi.org/10.1109/CCNC.2019.8651706

Nikouei, S. Y., Chen, Y., Song, S., Xu, R., Choi, B.-Y., & Faughnan, T. (2018). Smart surveillance as an edge network service: From Harr-cascade, SVM to a lightweight CNN. In *2018 IEEE 4th International Conference on Collaboration and Internet Computing (CIC)*, Philadelphia, PA (pp. 256–265). IEEE. https://doi.org/10.1109/CIC.2018.00044

Noreen, S., Farooq, U., Waqas, H., Fatima, N., Alqurashi, M. S., Imran, M., & Bariq, A. (2023). Comparative study of ternary hybrid nanofluids with role of thermal radiation and Cattaneo-Christov heat flux between double rotating disks. *Scientific Reports*, *13*(1), 7795. https://doi.org/10.1038/s41598-023-34783-8

Pathania, N. (Ed.) (2017). Setting up Jenkins on Docker and cloud. In *Pro Continuous Delivery* (pp. 115–143). Springer, New York. https://doi.org/10.1007/978-1-4842-2913-2_4

Pautasso, C., Zimmermann, O., Amundsen, M., Lewis, J., & Josuttis, N. M. (2017). Microservices in practice, part 1: Reality check and service design. *IEEE Software*, *34*(1), 91–98. https://doi.org/10.1109/MS.2017.24

Qi, D., Chang, B., & Li, W. (2024). Big data analysis and intelligent decision support system for environmental water quality: Application of artificial intelligence in water environmental protection. In *Proceedings of the 2024 3rd International Conference on Artificial Intelligence and Autonomous Robot Systems (AIARS)*, Bristol, United Kingdom (pp. 169–174). IEEE. https://doi.org/10.1109/AIARS63200.2024.00037

Rahmani, M., Azadmanesh, A., & Siy, H. (2014). Architectural reliability analysis of framework-intensive applications: A web service case study. *Journal of Systems and Software*, *94*, 186–201. https://doi.org/10.1016/j.jss.2014.03.070

Roets-Dlamini, Y., Moonsamy, G., Lalloo, R., & Ramchuran, S. (2022). Use of *Bacillus* spp. in the bioremediation of fats, oils and greases (FOGs), and other waste substrates in food processing effluents. *Biocatalysis and Agricultural Biotechnology*, *42*, 102351. https://doi.org/10.1016/j.bcab.2022.102351

Saddik, E. A. (2006). Performance measurement of web service-based application. *IEEE Transactions on Instrumentation and Measurement*, *55*(5), 1599–1605. https://doi.org/10.1109/TIM.2006.880288

Saghya Infant, S., Vickram, S., Saravanan, A., Mathan Muthu, C. M., & Yuarajan, D. (2025). Explainable artificial intelligence for sustainable urban water systems engineering. *Results in Engineering*, *25*, 104349. https://doi.org/10.1016/j.rineng.2025.104349

Sheng, X., Hu, S., & Lu, Y. (2019). The microservice architecture design research of financial trading system based on domain engineering. In *2018 International Symposium on Social Science and Management Innovation (SSMI 2018)*, Xi'an City, China. Atlantis Press. https://doi.org/10.2991/ssmi-18.2019.22

Shi, T., & Wu, J. (2021). Application of artificial intelligence in water conservancy project management. In *2021 2nd International Conference on Big Data & Artificial Intelligence & Software Engineering (ICBASE)*, Zhuhai, China (pp. 556–559). https://doi.org/10.1109/ICBASE53849.2021.00109

Shi, W., & Dustdar, S. (2016). The promise of edge computing. *Computer*, *49*(5), 78–81. https://doi.org/10.1109/MC.2016.145

Shiravi, A. H., Firoozzadeh, M., Bostani, H., Shafiee, M., & Bozorgmehrian, M. (2020). Experimental study on carbon nanofluid pressure drop and pumping power. *Advances in Nanochemistry*, *2*(1), 27–31.

Singh, H. K., & Bezboruah, T. (2015). Performance metrics of a customized web application developed for monitoring sensor data. In *Proceedings of the IEEE 2nd International Conference on Recent Trends in Information Systems (ReTIS)*, Kolkata, India. https://doi.org/10.1109/ReTIS.2015.7232870

Thesiya, D., Patel, H., & Patange, G. S. (2023). A comprehensive review electronic cooling: A nanomaterial perspective. *International Journal of Thermofluids*, *19*, 100382. https://doi.org/10.1016/j.ijft.2023.100382

Verma, L., & Meher, R. (2024). Study on hybrid nanofluid flow with heat transfer embedded in a porous medium with uncertain volume fraction. In *Numerical Heat Transfer*, Part A*: Applications* (pp. 1–27). https://doi.org/10.1080/10407782.2024.2352655

Wang, Y., Fei, Y., Yang, T., Luo, Z., Xu, Y., Su, B., & Lin, X. (2023). Nanotechnology for ultrafast nucleic acid amplification. *Nano Today*, *48*, 101749. https://doi.org/10.1016/j.nantod.2022.101749

Wilde, N., Gonen, B., El-Sheikh, E., & Zimmermann, A. (2016). Approaches to the evolution of SOA systems. In El-Sheikh, E., Zimmermann, A., & Jain, L. C. (Eds.) *Emerging Trends in the Evolution of Service-Oriented and Enterprise Architectures* (pp. 5–21). Springer, Cham. https://doi.org/10.1007/978-3-319-40564-3_2

Yu, D., Jin, Y., Zhang, Y., & Zheng, X. (2018). A survey on security issues in services communication of microservices-enabled fog applications. *Concurrency and Computation: Practice and Experience*, e4436. https://doi.org/10.1002/cpe.4436

Zaki, J., Islam, S. M. R., Alghamdi, N. S., Abdullah-Al-Wadud, M., & Kwak, K.-S. (2022). Introducing cloud-assisted micro-service-based software development framework for healthcare systems. *IEEE Access*, *10*, 33332–33348. https://doi.org/10.1109/ACCESS.2022.3161455

Zhuo, R., & Fan, F. (2021). A comprehensive insight into the application of white rot fungi and their lignocellulolytic enzymes in the removal of organic pollutants. *Science of the Total Environment*, *778*, 146132. https://doi.org/10.1016/J.SCITOTENV.2021.146132

17 Leveraging Data-Driven Approaches

A Case Study on Enhancing Water Governance in Africa

Basil Osayin Daudu, Salisu Ameh Amodu, Peace Ojochenemi Yakubu, and Esther Besty Anaiye

17.1 INTRODUCTION

The 21st century has so many concerns, of which the water crisis or governance in Africa is one such concern. Proper management of water resources is important for sustaining ecosystems and strengthening industrial growth, agricultural productivity, and, at the same time, attending to the needs of people. Across Africa, nations have been struggling with traditional water governance systems due to lack of data and poor infrastructure, among others. However, with the availability of advanced artificial intelligence (AI) technologies in the 21st century, African leaders have started making progress in revolutionizing water systems in their respective countries (Daudu, Osimen, & Anaiye, 2024). These AI-driven technologies provide decision-makers with real-time data, predictive analytics, and transparent systems that enhance the distribution and control of water resources. That is, improving water quality monitoring, detecting inefficiencies, and ensuring greater accountability.

According to Olatunde, Adelani, and Sikhakhane (2024), smart water management systems, such as Internet of Things (IoT), AI, and remote sensing, have revolutionized water conservation efforts, enhanced access to clean water, and promoted sustainable usage. These innovations enable the real-time monitoring and management of water resources, yielding significant environmental, social, and economic benefits. In Africa, challenges like water scarcity, inadequate infrastructure, financial limitations, and climate variability further complicate water management (Onukogu et al., 2023; Water, 2020). Africa's method of smart water control often emphasizes technology adoption (mobile technology for water quality monitoring) to enhance water accessibility, both in urban and rural regions.

DOI: 10.1201/9781003600718-24

Digital technologies have the potential to revolutionize global water systems, enabling utilities to become more resilient, innovative, and efficient. This, in turn, lays a stronger, more sustainable economic foundation for the future. Waters, Sarni, White, Cross, and Glotzbach (2022) noted that in the face of growing water challenges, water and wastewater utilities have been driven to explore new, innovative solutions—digital technologies being one of the most promising. Umgeni Water, a utility based in Durban, South Africa, has leveraged digital tools to improve water resource management and safeguard its customers from facing the same water crises experienced by Cape Town. By combining hydrologic models with monitoring devices, Umgeni Water has been able to optimize dam and reservoir storage levels.

Water management is crucial for achieving sustainable development goals such as SDG 6—quality water, SDG 13—climate change, and SDG 11—sustainable cities. The Africa Water Vision 2025, along with various declarations, aims to accelerate its implementation, offering a policy framework for the African Ministers' Council on Water (AMCOW). These declarations represent a strong blueprint for taking action on water-related issues at different levels across the continent. Three key declarations from 2008—the eThekwini Declaration, Tunis Declaration, and Sharm-el-Sheikh Declaration—highlight the importance of addressing water challenges (AMCOW, 2012).

In Sub-Saharan Africa, water resource control is faced with political unrest, population problems, and climate change. Pollard, Riddell, du Toit, Retief, and Ison (2023) assert that South Africa's integrated water resources management, implemented in 1998, serves as a leveraging strategy to promote fair, sustainable, and decentralized water governance that involves all stakeholders. Mutschinskia and Coles (2021) argue that quality water accessibility remains a key problem for rural communities in Africa, particularly over the past two decades. This issue is especially critical for agricultural farmers.

ClimDev-Africa (2013) highlights that Africa's water management is vulnerable to long-term climate variability. The effects of climate change on water resources vary across the continent, with some areas experiencing worsened water stress, while others may see a stress reduction. Key issues for Africa's water sector include limited access to water, including groundwater, and weak governance systems. This lack of access stems from inadequate infrastructure to ensure reliable water supplies for drinking, agriculture, and other needs. Afuye, Zhou, Majahana, Mazinyo, and Kalumba (2025) submit that in South Africa, water scarcity results not only from climate change but also from political inaction and ineffective institutional coordination of drought relief efforts.

Therefore, water quality monitoring is crucial for protecting the environment, safeguarding public health, and ensuring the sustainable use of water resources (Ponnuru et al., 2025). As such, this chapter explores water governance in Africa, its evolution over time, the challenges hindering effective water governance, and data-driven (AI) approaches in improving water governance across the continent.

17.2 CONCEPTUALIZING WATER GOVERNANCE

The global water crisis is fundamentally a governance issue (GWP, 2000). In response, the Hague Ministerial Declaration in 2000 emphasized the importance of water governance to promote good governance, which includes public participation and the consideration of all stakeholders of water control. Enhancing water governance involves service delivery among state actors, individuals, and the government.

Water governance is solely the government's responsibility in partnership with the citizens to ensure improvements, efficiency, and effectiveness (Stefanoa et al., 2014). On the one hand, water governance is defined in terms of the transition from policy-making to service execution (UNDP, 2013), while on the other hand, water governance is seen as a set of higher-level decisions that shape day-to-day water management (WWAP, 2003). According to the United Nations (2010), in recent years, access to clean water has been recognized as a fundamental human right.

According to Olagunju et al. (2019), improving water governance has become a top priority in both international water discussions and the national decision-making processes of many countries. Despite numerous efforts over the past years to ensure and enhance water accessibility, developing countries still battled with water security challenges (Tekken & Kropp, 2012). Eisma and Merwade (2021) point out that sand dams, a type of small-scale water harvesting structure, have gained recognition in East Africa thanks to the initiatives of private organizations.

For many years, water policy reform and governance design were largely shaped by hitch-free solutions like privatization (Blatter & Ingram, 2001). However, this approach has evolved into an advanced insight into the need for sustainable water governance that can effectively address the challenges posed by global and climate change (Pahl-Wostl, 2015). The governance of water consists of the political, social, economic, and administrative systems that influence the use, management, development of water resources, and the delivery of water services across the society.

According to OECD (2015), water governance is the set of political, institutional, and administrative rules, practices, and processes—both formal and informal—through which decisions are made and executed, stakeholders can express their interests and have their concerns addressed, and decision-makers are held accountable for managing water resources. The OECD strategies are organized into three key segments:

1. Water governance's efficiency depends on its ability to establish sustainable water policy visions at various government stages, execute, and achieve the desired outcomes.
2. The sustainability of water governance is about maximizing sustainable water management and societal well-being at minimal cost, relying on effective use of data, information, financing, regulations, and innovation.
3. Trust and engagement in water governance focus on creating public trust and inclusive stakeholder participation through democratic principles. It involves monitoring and evaluation, transparency, stakeholder involvement, and balancing interests among actors.

The Stockholm International Water Institute (2016) emphasizes that water governance is a crucial factor in advancing water control sustainability and its impact on people's livelihoods. For many, water availability is essential for survival and can be key to breaking the cycle of poverty. Water governance is less about the decisions themselves and more about how those decisions are made—specifically, who makes them, how they are made, and under what conditions (Moench, Dixit, Janakarajan, Rathore, & Mudrakartha, 2003).

Water governance refers to how allocation and regulatory decisions are made in water control and utilization of other natural resources, encompassing both government and private institutions through which order is maintained. Bilalova, Newig, and Villamayor-Tomas (2024) argue that governance is vital for ensuring water systemic long-term sustainability, and recognizing the complexities of water resource management, new and diverse governance models are emerging and being adapted to different contexts.

Tortajada (2010) pointed out that water resource implementation and control contribute to the development of other vital segments of society. Makridis (2020) acknowledges that while water governance faces numerous challenges, advances in AI can improve water governance in at least three key ways:

First, AI can enhance analytics and resource optimization, particularly by helping households reduce water waste through improved management and individual optimization of water resources. Second, AI can improve property rights and ownership delineation by using predictive models to reduce risk and enhance contracts between buyers and sellers. Since water is mobile and spatially dispersed, these models help clarify ownership and facilitate transactions. Third, AI can help balance water supply and demand by improving forecasting accuracy. Through better predictions and continuous data updates, AI supports the development of functional water markets by ensuring more reliable supply and demand management. Basically, water governance has three units, viz. Water law (water rights, inter-source links, inter-resource links, and conflict resolution), water policy (cost recovery, water transfers, project selection, and use priority), and water organization (finance, regulations, government layers, and technical capacity).

Beleyur, Revi, and Sami (2024) explain that water governance is a sustainable set of procedures involving stakeholders from various areas. Therefore, the perception of water governance is based on the "Data–Information–Knowledge–Wisdom" theory, introduced by Russell Ackoff. This theory is

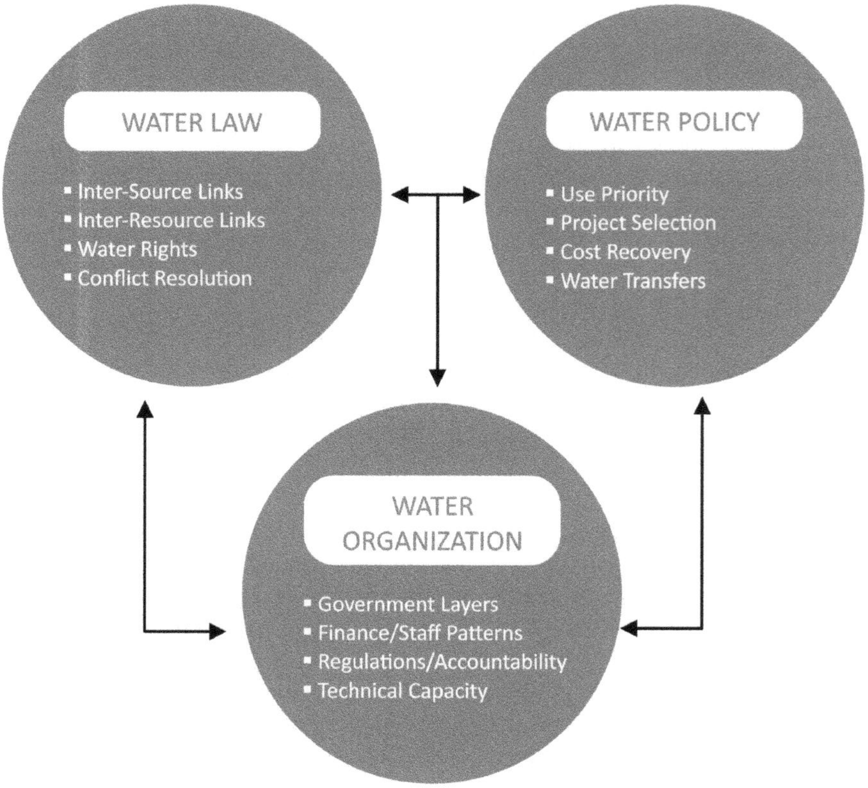

FIGURE 17.1 Units of water governance.

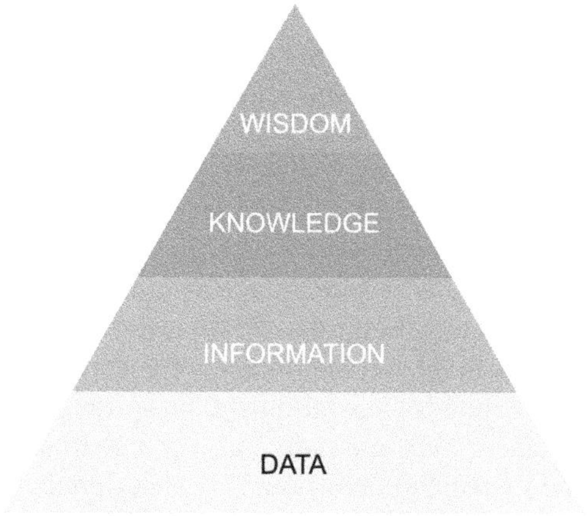

FIGURE 17.2 Ackoff's "Data–Information–Knowledge–Wisdom" pyramid.

represented as a pyramid and serves as a conceptual framework that illustrates how data can be transformed into actionable information, knowledge, and ultimately wisdom to inform decision-making.

Ackoff (1989) expanded on the original framework by adding a fifth dimension, "understanding," which connects knowledge and wisdom. In this model, wisdom involves knowledge, which involves information, consisting of data. Data represent the most basic level, with information providing context, knowledge offering guidance on its use, and wisdom determining when and why to apply it

(Kaminski, 2021). Consequently, water governance is built on key principles of good governance, including equity, efficiency, participation, decentralization, integration, transparency, and accountability. If any of these principles are absent or misapplied, achieving the desired outcomes becomes difficult, potentially leading to a water crisis or insecurity.

17.3 WATER GOVERNANCE TRENDS ACROSS AFRICA

In Africa, water governance has made remarkable changes over time, with the pre-colonial and colonial periods (before the 1900s) marking the first phase. In pre-colonial Africa, water management was largely decentralized and handled at the local or community level. Traditional practices such as rainwater harvesting, communal wells, and river-based irrigation were common, particularly in rural areas. Local knowledge and customs were essential in managing water resources. During the colonial era, European powers established centralized water management systems that primarily focused on the needs of colonial settlements, agriculture, and mining, often disregarding the needs of indigenous communities.

The Post-Independence Era (1960s–1980s) saw the nationalization of water resources, large infrastructure projects, and limited integration. Following independence, water governance shifted to state-led management. This period saw the construction of major dams, reservoirs, and irrigation systems aimed at increasing agricultural productivity, but there was little effort to involve local communities in decision-making. From the 1990s to the 2000s, integrated and participatory governance became more prominent. African nations adopted the principles of Integrated Water Resource Management, introduced by the International Conference on Water and the Environment and the Rio Earth Summit, as part of their water governance reforms. This included decentralizing water management by transferring responsibility to local and regional authorities, as well as introducing public–private partnerships(PPP) to enhance water service delivery.

From 2010 to the present, the focus has shifted toward sustainability, climate change, and SDG 6 on clean water. Efforts have been directed at improving quality water accessibility, managing water resources sustainably, and addressing inequalities. There has also been a growing emphasis on regional cooperation to control communal water resources, with initiatives such as the Nile Basin Initiative (NBI) and the Southern African Development Community (SADC) fostering collaboration among countries on transboundary water management.

According to the WHO/UNICEF Joint Monitoring Programme (2023), in 2020, only 60% of the population in the Sahel had clean water accessibility, compared to 96% in Northern Africa. While around 77% of people in the Sahel had access to clean water. The report also states that in 2020, 28% of people in the Sahel had sanitation accessibility, while 56% basic sanitation accessibility. In contrast, 85% of the Northern African population had access to safe sanitation accessibility.

Africa is home to 63 transboundary rivers, with 60% of them being governed by agreements. By 2020, the Nile Basin Initiative reported that 60 million people were benefiting from shared projects. Climate Change Adaptation emerged as a critical issue across the continent. The Intergovernmental Panel on Climate Change (2022) forecasts that over 70% of African countries will face increased water stress by 2030 due to climate change. As a result, investment in water infrastructure became a priority for African leaders, leading to the adoption of water pricing and economic instruments. The Organization for Economic Co-operation and Development (2015) reported that more than 20% of African countries now use pricing and economic tools for water management, making these strategies a key component of water governance reforms.

There is no doubt that technology has enhanced water governance across Africa. By 2020, 45% of African countries had implemented smart water metering and remote sensing technologies to improve water management and reduce losses. A notable example is Kenya's use of mobile payment platforms such as M-Pesa for water service billing, which has increased access to water payments and improved revenue

TABLE 17.1 Highlights of data-driven water governance in selected African countries

COUNTRY	*DATA-DRIVEN STRATEGY*	*ISSUES RESOLVED*	*DESIRED RESULTS*	*SOURCES*
Ethiopia	Spatial data visualization of rural water resources	Ineffective water distribution systems	40% enhancement in rural water access strategy	UNICEF (2023)
Ghana	Smart leak detection system	Excessive pipeline leakage rates	Monthly water savings of 2 million liters	Ministry of Water Resources and Sanitation (2023)
South Africa	ML-based flood forecasting	Delayed disaster relief	50% improvement in disaster response speed	Peden, Mayhew, Baker, Mayedwa, and Saunders (2023)
Nigeria	Community-based water quality tracking	Data gaps in drinking water safety	25% increase in portable water availability	Federal Ministry of Water Resources (2021)
Kenya	Real-time water monitoring	Inefficient billing systems and rampant water theft	30% drop in water pilferage	JICA (2022)

collection efficiency. By 2020, more than 20 African countries had engaged in PPP in water and sanitation, with South Africa, Morocco, and Gabon being the leaders in these initiatives.

Water quality remains a significant concern. The United Nations Environment Programme (UNEP) reported that over 60% of Africa's rivers are contaminated by industrial, agricultural, and domestic waste. Rivers like the Congo and the Nile face severe pollution, impacting water quality for millions. By 2020, 54% of African countries had aligned their national water policies with SDG 6, aiming to provide access to clean water and sanitation. However, progress has been uneven, with North Africa closer to attaining SDG 6(clean water), while Sub-Saharan Africa continues to face major challenges.

Indubitably, water is crucial for Africa's socio-economic recovery and development, yet natural and human-made challenges hinder its effective use. However, a sustainable water future requires strengthening governance, improving water knowledge, addressing urgent needs, and securing financial support (UN Water/Africa, 2025). As such, several notable initiatives across Africa are working to enhance water governance, foster cooperation, and promote sustainable practices. Some of these include the Nile Basin Initiative, the African Ministers Council on Water, WaterNet, the Zambezi River Authority (ZRA), the Lake Victoria Basin Commission (LVBC), the African Water Facility (AWF), the African Development Bank's Water Supply and Sanitation Initiative (AWSSI), Water and Sanitation for the Urban Poor (WSUP), the African Union's 2063 Agenda: Water and Sanitation Goals, and The Blue Deal—a Dutch Government Initiative.

17.4 WATER CRISIS AND ITS IMPACT ON AFRICA'S ECONOMY

The water crisis in Africa is driven by several factors, including water scarcity and climate change, limited access to clean water and sanitation, transboundary water management challenges, inadequate infrastructure and investment, political instability and poor governance, environmental degradation, unequal access to water, and financial constraints. Henry, Baldwin, and Quathamer (2015) argue that tackling water insecurity is the world's key challenge, particularly in developing countries due to infrastructural deficit and poor connectivity.

Elke (2010) highlights that climate change presents significant threats to South Africa, a country that relies heavily on surface water, which is already under strain. In response, one potential adaptation strategy is to adopt a comprehensive approach to water control based on the basin blueprint. Makayaa et al. (2020) emphasize the impact of droughts on the livelihoods of rural regions in South Africa. As such, effective coordination in managing water for agriculture and water supply is crucial for improving drought resilience. Despite the interconnectedness of South African water regulatory institutions, there are difficulties in how these institutions collaborate in tackling the drought crisis.

The governance of water accessibility is important for efficient hazard reduction during droughts, as access to water in such times is influenced by the existing governance systems (Davies, 2000). The water crisis in Africa has significant economic impacts, as water scarcity and poor water management reduce agricultural productivity, potentially leading to lower crop yields, threats to livestock, and higher food prices. Additionally, when clean water is scarce, the cost of water collection can rise significantly. The water crisis also leads to environmental degradation, including water scarcity and pollution, which harms water quality and aquatic ecosystems in Africa.

There are also instances of conflict and political instability linked to water scarcity, particularly in transboundary river basins. Examples like the Nile, Lake Chad, and the Zambezi highlight water sources that have either sparked or could potentially spark disputes between countries or communities over water rights and usage. Politically, internal displacement occurs as people move in search of water and food. Corruption and mismanagement often arise from poor water governance, leading to the misallocation of funds and resources. In some situations, inconsistent or inadequately enforced water policies can create confusion and obstruct effective management.

17.5 AI ADOPTION AND WATER GOVERNANCE IN AFRICA

17.5.1 Strengthening Water Governance in Africa with Artificial Intelligence

The first and most important question is, what exactly is AI? The concept of AI has various interpretations. It is seen as a rational agent (Russell & Norvig, 2016), an intelligent agent (Poole, Mackworth, & Goebel, 1998), and a machine with its mind (Haugeland, 1989; Bellman, 1978). AI is a field of computing that is focused on simulating human thought in machines. The rapid development of AI in the past decade has been extraordinary, and it is expected to be increasingly adopted across different fields in the future. This growth could result in AI either replacing humans or assisting them in performing tasks more efficiently (Adithiyaa & Alamelu, 2024).

McDonald (2024) likens AI to a highly adaptable technology, much like electricity, that can be applied to various situations. AI consists of advanced algorithms (the "brain" of AI), data (the "fuel" powering AI), and computing power (the machines enabling AI), all of which work together to allow machines to exhibit intelligent behavior. Algorithms guide the computer's actions, data provide the material for learning, and computing power allows machines to process information and make decisions. Training AI systems typically requires vast amounts of data that are both large and complex, and these data must be processed quickly, demanding substantial data processing ability. For instance, advanced conversational AI like Google Bard and Microsoft Copilot have been trained on datasets that include hundreds of billions of words (McDonald, 2024).

At its broadest level, AI is essentially synonymous with algorithms. The term "algorithm" comes from the name of the 9th-century Persian mathematician Muhammed ibn Musa al-Kharizmi and refers to a set of instructions aimed at solving a problem or performing a calculation. However, in a more specific sense, AI is about replicating human intelligence in computers (Sheikh, Prins, & Schrijvers, 2023). AI also refers to systems that display intelligent behavior by analyzing their environment and taking actions autonomously to achieve particular goals (AIHLEG, 2019).

Saini (2023) highlights that fundamental AI's principles include logic, understanding, strategy, assimilation, dialogue, perception, and manual dexterity. AI is a fusion of both scientific inquiry and technical expertise in creating automated systems, particularly computer programs with cognitive abilities. AI technology has a wide range of applications, including the learning community (Ahmad et al., 2023). In education, various neural networks are being used, such as content verification, maintaining academic honesty (Ade-Ibijola et al., 2022), and intellectual research support (Nakitare & Otike, 2022).

Additionally, AI is closely linked to advanced technologies like "machine learning" (ML), which has paved the way for the development of "deep learning" (DL). These innovations have led to significant breakthroughs, such as facial recognition and computers successfully playing complex games like Go. Unlike traditional approaches where computer systems adhere to rigid instructions, data science and cognitive computing algorithms can detect correlations, often referred to as "self-learning algorithms" (Sheikh, Prins, & Schrijvers, 2023). Therefore, the concept of AI encompasses the integration of ML, DL, algorithm development, and natural language processing (Akgun & Greenhow, 2022).

AI offers significant benefits to both organizations and individuals by improving efficiency, boosting productivity, saving time and effort, and enhancing overall performance (Ali et al., 2023). However, despite these advantages, AI also brings challenges, including concerns over data security, privacy, and the potential for job displacement (Bécue, Praça, & Gama, 2021). Still, AI continues to infiltrate various aspects of daily life and is increasingly applied in professional sectors like education (Chen, Chen, & Lin, 2020), healthcare (Matheny, Whicher, & Thadaney, 2020), and marketing (Vlačić, Corbo, Silva, & Dabić, 2021). It has become a pivotal technological force and trend of the 21st century.

Sustainable water resource management is crucial in tackling the increasing challenges of water scarcity, climate change, and population growth in Africa (Daudu, Osimen, & Abubakar, 2025). Advanced technologies such as cloud data, AI, and mapping software have become essential tools for promoting the sustainable use of hydro resources across the continent. These technologies offer valuable insights that help policymakers, water managers, and stakeholders in Africa make strategic judgments regarding water distribution, usage trends, and conservation efforts.

Programme, Connor, and Miletto (2022) argue that a key use of analytics in water control is its ability to monitor and forecast water accessibility. Instantaneous data from sensors, satellite imagery, and weather predictions can be analyzed to optimize water distribution, identify leaks, and prevent overconsumption. This improves transparency and accountability in water governance. By providing lawmakers with accurate and reliable data, it also builds trust and supports collaborative decision-making. This is particularly important for ensuring fair water allocation across competing sectors like agriculture, industry, and households.

Sensor-based irrigation systems can adjust watering schedules according to soil moisture levels and weather forecasts, significantly cutting down on water usage in agriculture (Ogunjobi et al., 2023). These smart irrigation systems are gaining popularity across Africa for their ability to optimize water use, conserve resources, and boost agricultural productivity. Notable examples of smart irrigation systems in Africa include Kenya's M-KOPA, Morocco's drip irrigation systems, and Nigeria's i-Farm. Additionally, by improving the efficiency of water treatment and distribution, smart water management helps reduce pollution. Advanced monitoring systems can detect and report contaminants in real time, enabling prompt corrective actions. This capability is essential for preventing pollutants from reaching natural water sources and ensuring clean drinking water, thus safeguarding aquatic ecosystems and public health (Singh & Ahmed, 2021).

Predictive analytics and ML models can provide accurate forecasts of water demand, enabling utilities to adjust supply in real time and minimize resource waste. This efficiency also benefits agricultural practices, where precision irrigation systems ensure that water is used more effectively, boosting crop yields while conserving water (Dogo, Salami, Nwulu, & Aigbavboa, 2019). Additionally, data analytics can optimize treatment processes, reducing energy consumption and maintenance costs. These savings can then be passed on to consumers through lower water bills and to governments and organizations through reduced spending on water infrastructure (Aderibigbe, Ani, Ohenhen, Ohalete, & Daraojimba, 2023).

While the benefits of analytics are clear, fully utilizing them necessitates investments in infrastructure, technical expertise, and data-sharing frameworks. As a result, it is the responsibility of governments and organizations to prioritize capacity building and foster partnerships, ensuring that analytics are effectively integrated into water resource management throughout Africa.

17.5.2 Combatting Water Crisis in Africa with Artificial Intelligence

Water security in Africa continues to be a significant challenge, with many areas struggling with scarcity, unequal access, and susceptibility to climate change. Yonazi, Kelly, Halewood, and Blackman (2012) assert that technological solutions are becoming increasingly essential to address the unique water management issues posed by Africa's diverse geographic and socio-economic conditions. For instance, mobile technology has enhanced access to water services and promoted more efficient water usage in agriculture.

Gleick (2020) argues that technologies like remote sensing, geographic information systems, and IoT sensors are instrumental in gathering important data on rainfall, groundwater levels, and river flows. These data can be analyzed to forecast droughts, track water usage, and pinpoint areas at risk of water stress. Such predictive capabilities enable policymakers to plan ahead, allocate resources more efficiently, and implement conservation strategies that are tailored to specific local conditions (Ponnuru et al., 2025). Additionally, data are essential for addressing challenges in water infrastructure. By examining information from leak detection systems and usage meters, authorities can identify inefficiencies in water supply networks and reduce losses caused by wastage or outdated infrastructure (Okello, Githiora, Sithole, & Owuor, 2024).

Kalogiannidis, Kalfas, Giannarakis, and Paschalidou (2023) suggest that technologies like blockchain and digital platforms enhance transparency and accountability by allowing real-time monitoring of water usage and resource allocations, promoting fair distribution, and helping to reduce corruption. These technologies also encourage community involvement in water governance. Mobile apps and social media platforms enable citizens to report hydrological problems, provide feedback, and engage in selection procedures. This engagement-based model strengthens governance by addressing the priorities of rural areas. Additionally, AI-driven platforms aggregate data like sensors, satellite imagery, and forecasting centers, using AI models to predict, analyze, and optimize water resource management. Examples of AI-powered platforms for water governance in Africa include WaterNet in South Africa, the Water Sector Trust Fund in Kenya, and WaterAid in Nigeria.

TABLE 17.2 Data sources for improving water governance

DATA SOURCE	*OVERVIEW*	*TECHNOLOGIES*	*APPLICATIONS*
Remote sensing	Space-based technology for monitoring water cycles	NASA Landsat, Sentinel-2, Google Earth Engine	Water resource tracking and extreme event prediction
Citizen engagement	Collaborative water quality tracking	Applets, SMS polls, digital social hubs	Identifying service gaps, policy advocacy
Government and Utility Records	Policy briefs on water infrastructure development	National water agency databases, municipal records	Policy planning, governance monitoring, and regulatory compliance
IoT and Smart sensors	Smart water network monitoring	Smart water meters, flow sensors, Supervisory Control and Data Acquisition systems	Leak detection, consumption analysis, and infrastructure maintenance
Climate and Weather Data	Climate forecasting datasets	European Centre for Medium-Range Weather Forecasts, National Meteorological Agencies	Predictive modeling for floods, droughts, and climate adaptation

According to the UN-Water report (2021), a major benefit of cloud data lies in its ability to provide real-time monitoring of water resources and usage patterns. This allows policymakers to identify inequalities in water distribution, locate underserved areas, and take prompt action (UN-Water, 2021). In rural Africa, for example, big data can highlight communities without access to clean water, guiding the installation of boreholes or water treatment facilities. Additionally, satellite data is increasingly being used in Africa to tackle water-related challenges, including resource management, environmental monitoring, enhancing agricultural practices, and handling natural disasters like floods and droughts. A notable example is NASA's GRACE satellites, which have been used to monitor groundwater depletion in regions like the Sahel and Southern Africa. These data are crucial for managing water resources in areas that rely heavily on groundwater, especially during dry seasons.

In countries such as Sudan and Nigeria, data from the European Space Agency's (ESA) Sentinel satellites have been utilized to monitor flooding events along the Nile and other major rivers. These data are combined with flood forecasting models to enhance early warning systems. In nations like Kenya and Ethiopia, MODIS satellite data have been employed to track vegetation health and assess agricultural productivity, particularly during dry spells. This information aids both farmers and governments in planning for crop management and ensuring food security. In the Nile River basin, Sentinel-2 imagery has been used to monitor water quality, detecting pollution from agricultural runoff and urban areas. Similar monitoring efforts have been implemented around Lake Victoria, where harmful algal blooms are a growing concern. In Southern Africa, Landsat data have been valuable for tracking changes in river systems and understanding the effect of global warming on regional water security, particularly concerning shrinking lakes like Lake Chad. In Egypt, satellite altimetry is employed to monitor the Mediterranean Sea and predict sea-level rise, which is crucial for managing the Nile Delta, one of the most densely populated regions in Africa.

As a result, the implementation of smart water technologies has greatly enhanced water governance throughout Africa. The influence of these AI-driven innovations has transformed multiple areas of human interaction, and water governance plays a key role in attaining societal progress across the continent.

17.6 CONCLUSION

Water governance in Africa is evolving, but it faces numerous challenges that continue to threaten water security across the continent. While traditional approaches have been attempted to address these issues, they have often fallen short. The prevalence of weak infrastructure and policy gaps in water governance across Africa is worrisome, but can be conveniently resolved through policy reforms. However, the emergence of AI-driven solutions presents new opportunities to enhance water governance by optimizing resource management, boosting efficiency, minimizing waste, and improving decision-making. By leveraging technologies such as intelligent systems, satellite data, data security, and the web of things, African nations are making meaningful progress toward establishing more efficient, transparent, and sustainable water governance systems.

17.7 RECOMMENDATIONS

This chapter provided the following insightful suggestions to enhance water governance in Africa:

1. African governments should prioritize greater investment in data infrastructure and capacity development.
2. Governments and stakeholders should establish platforms for cross-border water data sharing to facilitate more coordinated management of shared water resources.

3. Governments and water utilities should enhance the use of AI and predictive analytics to better manage water demand and optimize distribution systems.
4. Policymakers should create inclusive and sustainable data governance policies that encourage open access to water data, ensuring its ethical and transparent use.
5. Community-based monitoring systems should be enhanced to gather localized water data. By equipping local communities with tools like mobile apps or basic sensors, data accuracy can be improved, ensuring that water management practices are better aligned with local needs.

REFERENCES

Ackoff, R. L. (1989). From data to wisdom, *Journal of Applied Systems Analysis*, 16, 3–9. Retrieved from https://oro.open.ac.uk/72526/3/RA.pdf

Ade-Ibijola, A., Young, K., Sivparsad, N., Seforo, M., Ally, S., Olowolafe, A., & Frahm-Arp, M. (2022). Teaching students about plagiarism using a serious game (Plagi-Warfare): Design and evaluation study. *JMIR Serious Games*, 10(1), e33459. https://doi.org/10.2196/33459

Aderibigbe, A. O., Ani, E. C., Ohenhen, P. E., Ohalete, N. C., & Daraojimba, D. O. (2023). Enhancing energy efficiency with AI: A review of machine learning models in electricity demand forecasting. *Engineering Science & Technology Journal* 4(6), 341–356. https://doi.org/10.51594/estj.v4i6.636

Adithiyaa, D., & Alamelu, C. (2024). A review of applications and uses of artificial intelligence. *World Journal of Advanced Engineering Technology and Sciences*, 12(1), 477–484. https://doi.org/10.30574/wjaets.2024.12.1.0260

African Ministers' Council on Water (AMCOW). (2012). *Status Report on the Application of Integrated Approaches to Water Resources Management in Africa.* Retrieved from https://unepdhi.org/wp-content/uploads/sites/2/2020/06/2012_Africa_Status_Report_UK.pdf

Afuye, G. A., Zhou, L., Majahana, J. M., Mazinyo, S. P., & Kalumba, A. M. (2025). Drought disasters, vulnerability, severity, preparedness, and response to the water sector: A comprehensive model approach for water governance and policy in South Africa. *Progress in Disaster Science*, 26, 100417. https://doi.org/10.1016/j.pdisas.2025.100417

Ahmad, S. F., Han, H., Alam, M. M., Rehmat, M. K., Irshad, M., Arraño-Muñoz, M., & Montes, A. A. (2023). Impact of artificial intelligence on human loss in decision making, laziness and safety in education. *Humanities and Social Sciences Communications*, 10, 311. https://doi.org/10.1057/s41599-023-01787-8

Akgun, S., & Greenhow, C. (2022). Artificial intelligence in education: Addressing ethical challenges in K-12 settings. *AI and Ethics*, 2(3), 431–440. https://doi.org/10.1007/s43681-021-00096-7

Ali, O., Abdelbaki, W., Shrestha, A., Elbasi, E., Ali, M. A., & Dwivedi, Y. K. (2023). A systematic literature review of artificial intelligence in the healthcare sector: Benefits, challenges, methodologies, and functionalities. *Journal of Innovation and Knowledge*, 8(1), 100333. https://doi.org/10.1016/j.jik.2023.100333

Bécue, A., Praça, I., & Gama, J. (2021). Artificial intelligence, cyber-threats and Industry 4.0: Challenges and opportunities. *Artificial Intelligence Review*, 54(5), 3849–3886. https://doi.org/10.1007/s10462-020-09942-2

Beleyur, P., Revi, A., & Sami, N. (2024). *Reimagining Sub-National Water Governance.* Bangalore: The Indian Institute for Human Settlements and the Global Commission for the Economics of Water. https://doi.org/10.24943/9788198256805

Bellman, R. (1978). *An Introduction to Artificial Intelligence: Can Computers Think?* Belmont: Boyd & Fraser.

Bilalova, S., Newig, J., & Villamayor-Tomas, S. (2024). Toward sustainable water governance? Taking stock of paradigms, practices, and sustainability outcomes. *Wiley Interdisciplinary Reviews: Water*, 12, e1762. https://doi.org/10.1002/wat2.1762

Blatter, J. & Ingram, H. (2001). *Reflections on Water: New Approaches to Transboundary Conflict and Cooperation.* Cambridge, MA: MIT Press.

Chen, L., Chen, P., & Lin, Z. (2020). Artificial intelligence in education: A review. *IEEE Access*, 8, 75264–75278. https://doi.org/10.1109/access.2020.2988510

ClimDev-Africa. (2013). *Climate Change and Water in Africa: Challenges, Opportunities and Recommendations.* Policy Brief 4. Retrieved from https://www.climdev-africa.org

Daudu, B. O., Osimen, G. U., & Abubakar, A. T. (2025). Artificial intelligence, fintech and financial inclusion in African digital space. In: V. Sharma, M. Gupta, N. Arora, & A. A. Shaikh (eds.), *Fintech and Financial*

Inclusion: Leveraging Digital Finance for Economic Empowerment and Sustainable Growth (pp. 268–282). London: Routledge. https://doi.org/10.4324/9781003514114-19
Daudu, B. O., Osimen, G. U., & Anaiye, E. B. (2024). Sustainable smart cities in African digital space. In: P. Whig, P. Sharma, N. Aneja, A. A. Elngar, & N. Silva (eds.), *Artificial Intelligence and Machine Learning for Sustainable Development: Innovations, Challenges, and Applications* (pp. 100–111). Boca Raton, FL: CRC Press. https://doi.org/10.1201/9781003497189
Davies, S. (2000). Effective drought mitigation. In: D. A. Wilhite (ed.), *Drought: A Global Assessment* (pp. 3–16). London: Routledge.
Dogo, E. M., Salami, A. F., Nwulu, N. I., & Aigbavboa, C. O. (2019). Blockchain and Internet of Things-based technologies for intelligent water management system. In *Artificial Intelligence in IoT* (pp. 129–150). Cham: Springer. https://doi.org/10.1007/978-3-030-04110-6_7
Eisma, J., & Merwade, V. (2021). A data-driven approach to assessing the impact of water harvesting structures on regional water storage in East Africa. *Journal of Hydroinformatics*, 23(2), 352–247. https://doi.org/10.2166/hydro.2021.115
Elke, H.-P. (2010). South African water governance between administrative and hydrological boundaries. *Climate and Development*, 2, 111–127. https://doi.org/10.3763/cdev.2010.0038
Federal Ministry of Water Resources. (2021). *Water Sanitation and Hygiene National Outcome Routine Mapping 2021: Summary of Survey Findings*. Retrieved from https://www.unicef.org/nigeria/media/5956/file/2021WASHNORM%20Infographics.pdf
Gleick, P. H. (2020). Climate change and water policy in Africa. *Environmental Research Letters*, 15(8), 1–12.
Global Water Partnership. (2000). *Towards Water Security A Framework for Action*. Stockholm, Sweden: Global Water Partnership. Retrieved from https://www.ircwash.org/resources/towards-water-security-framework-action
Haugeland, J. (1989). *Artificial Intelligence: The Very Idea*. Cambridge, UK: MIT Press.
Henry, M., Baldwin, G., & Quathamer, G. (2015). Designing a community-based water harvesting system: Understanding water use in Endallah, *Tanzania. The Journal of Purdue Undergraduate Research*, 5(1), 38–47. https://doi.org/10.5703/jpur.05.1.05
High-Level Expert Group on Artificial Intelligence. (2019). *A Definition of AI: Main Capabilities and Scientific Disciplines*. European Commission. Retrieved from https://ec.europa.eu/newsroom/dae/document.cfm?doc_id=56341
Intergovernmental Panel on Climate Change (IPCC). (2022). *Climate Change and Water*. Cambridge, UK: Cambridge University Press.
JICA. (2022). *Kenya: Project for Strengthening Capacity in NRW Reduction*. Retrieved from https://www.jica.go.jp/Resource/project/kenya/012/materials/ku57pq00003smyw8-att/briefnote_en_202208.pdf
Kalogiannidis, S., Kalfas, D., Giannarakis, G., & Paschalidou, M. (2023). Integration of water resources management strategies in land use planning towards environmental conservation. *Sustainability*, 15(21), 15242. https://doi.org/10.3390/su152115242
Kaminski, J. (2021). Theory applied to informatics: DIKW theory editorial. *Canadian Journal of Nursing Informatics*, 16, 3–4. Retrieved from https://cjni.net/journal/?p=9374
Makayaa, E., Rohseb, M., Dayc, R., Vogeld, C., Mehtae, L., McEwenf, L., Rangecroftg, S., & Loon, A. F. V. (2020). Water governance challenges in rural South Africa: Exploring institutional coordination in drought management. *Water Policy*, 22, 519–540. https://doi.org/10.2166/wp.2020.234
Makridis, C. A. (2020). *Three Ways Artificial Intelligence can Enhance Water Governance*. Global Water Forum. Retrieved from https://www.globalwaterforum.org/2020/11/19/three-ways-artificial-intelligence-can-enhance-water-governance/
Matheny, M. E., Whicher, D., & Thadaney, I. S. (2020). Artificial intelligence in health care: A report from the national academy of medicine. *JAMA*, 323(6), 509–510. https://doi.org/10.1001/jama.2019.21579
McDonald, J. (2024). *Introduction to Artificial Intelligence (AI): Guide for Travel and Tourism Leaders*. World Travel and Tourism Council. Microsoft. Retrieved from www.microsoft.com/en-us/ai
Ministry of Water Resources and Sanitation. (2023). *Restoring and Sustaining Macroeconomic Stability and Resilience through Inclusive Growth & Value Addition*. Retrieved from https://www.mofep.gov.gh/sites/default/files/pbb-estimates/2023/2023-PBB-MWRS.pdf
Moench, M., Dixit, A., Janakarajan, M., Rathore, S., & Mudrakartha, S. (2003). *The Fluid Mosaic, Water Governance in the Context of Variability, Uncertainty, and Change*. Boulder, CO: Nepal Water Conservation Foundation, Kathmandu, and the Institute for Social and Environmental Transition.
Mutschinskia, K., & Coles, N. A. (2021). The African Water Vision 2025: Its influence on water governance in the development of Africa's water sector, with an emphasis on rural communities in Kenya: A review. *Water Policy*, 23(4), 838–861. https://doi.org/10.2166/wp.2021.032

Nakitare, J., & Otike, F. (2022). Plagiarism conundrum in Kenyan universities: An impediment to quality research. *Digital Library Perspectives*, 39(2), 145–165. https://doi.org/10.1108/dlp-08-2022-0058

Ogunjobi, O. A , Eyo-Udo, N. L., Egbokhaebho, B. A., Daraojimba, C., Ikwue, U., & Banso, A. A. (2023). Analyzing historical trade dynamics and contemporary impacts of emerging materials technologies on international exchange and the US strategy. *Engineering Science & Technology Journal*, 4(3), 101–119.

Okello, C., Githiora, Y. G., Sithole, S., & Owuor, M. A. (2024). Nature-based solutions for water resource management in Africa's arid and semi-arid lands (ASALs): A systematic review of existing interventions. *Nature-Based Solutions*, 6, 100172. https://doi.org/10.1016/j.nbsj.2024.100172

Olagunju, A., Thondhlana, G., Chilima, J. S., SèneHarper, A., Compaoré, W. R. N., & Ohiozebau, E. (2019). Water governance research in Africa: Progress, challenges and an agenda for research and action. *Water International*, 44(9), 1–26. https://doi.org/10.1080/02508060.2019.1594576

Olatunde, T. M., Adelani, F. A., & Sikhakhane, Z. Q. (2024). A review of smart water management systems from Africa and the United States. *Engineering Science & Technology Journal*, 5(4), 1231–1242. https://doi.org/10.51594/estj.v5i4.1014

Onukogu, O. A., Onyebuchi, C. N., Scott, T. O., Babawarun, T., Neye-Akogo, C., Olagunju, O. A., & Uzougbo, C. G. (2023). Impacts of industrial wastewater effluent on Ekerekana Creek and policy recommendations for mitigation. *The Journal of Engineering and Exact Sciences*, 9(4), 15890-01e.

Organization for Economic Co-operation and Development. (2015). *OECD Principles on Water Governance*. Retrieved from www.oecd.org/governance/oecd-principles-onwater-governance.htm

Pahl-Wostl, C. (2015). *Water Governance in the Face of Global Change: From Understanding to Transformation*. Cham, Switzerland: Springer. https://doi.org/10.1007/978-3-319-21855-7

Peden, A. E., Mayhew, A., Baker, S. D., Mayedwa, M., & Saunders, C. J. (2023). Exploring flood response challenges, training needs, and the impact of online flood training for lifeguards and water safety professionals in South Africa. *International Journal of Environmental Research and Public Health*, 20(16):6573. https://doi.org/10.3390/ijerph20166573

Pollard, S. R., Riddell, E., du Toit, D. R., Retief, D. C., & Ison, R. L. (2023). Toward adaptive water governance: The role of systemic feedback for learning and adaptation in the eastern transboundary rivers of South Africa. *Ecology and Society*, 28(1), 47. https://doi.org/10.5751/ES-13726-280147

Ponnuru, A., Madhuri, J. V., Saravanan, S., Vijayakumar, T., Manimegalai, V., & Das, A. (2025). Data-driven approaches to water quality monitoring: Leveraging AI, machine learning, and management strategies for environmental protection. *Journal of Neonatal Surgery*, *14*(5s), 664–675. https://doi.org/10.52783/jns.v14.2107

Poole, D. L., Mackworth, A., & Goebel, R. G. (1998). *Computational Intelligence and Knowledge*. Oxford, England: Oxford University Press.

Programme, U., Connor, R., & Miletto, M. (2022). *The United Nations World Water Development Report 2022: Groundwater Making the Invisible Visible*. Paris, France: UNESCO. Retrieved from https://coilink.org/20.500.12592/3b6g0b

Russell, S. J. & Norvig, P. (2016). *Artificial Intelligence: A Modern Approach*, 3rd ed. London, UK: Pearson.

Saini, N. (2023). Artificial intelligence & its applications. *International Journal for Research Trends and Innovation*, 8(4), 356–360.

Sheikh, H., Prins, C., & Schrijvers, E. (2023). Artificial intelligence: Definition and bnackground. In: *Mission AI, Research for Policy* (pp. 15–41), Cham, Switzerland: Springer. https://doi.org/10.1007/978-3-031-21448-6_2

Singh, M., & Ahmed, S. (2021). IoT-based smart water management systems: A systematic review. *Materials Today: Proceedings*, 46, 5211–5218. https:doi.org/10.1016/j.matpr.2020.08.588

Stefanoa, L. D., Svendsenb, M., Giordanoc, M., Steeld, B. S., Browne, B., & Wolf, A. T. (2014). Water governance benchmarking: Concepts and approach framework as applied to Middle East and North Africa countries. *Water Policy*, 16, 1121–1139. https://doi.org/10.2166/wp.2014.305

Stockholm International Water Institute (SIWI). (2016). *Water Governance. Issue Sheet*. Stockholm, Sweden. Retrieved from www.watergovernance.org

Tekken, V., & Kropp, J. P. (2012). Climate-driven or human-induced: Indicating severe water scarcity in the Moulouya River Basin (Morocco). *Water*, 4, 959–982. https://doi.org/10.3390/w4040959

Tortajada, C. (2010). Water governance: Some critical issues. *International Journal of Water Resources Development*, 26(2), 297–307. https://doi.org/10.1080/07900621003683298

UNDP. (2013). *User's Guide on Assessment of Water Governance*. Retrieved from https://www.undp.org/content/undp/en/home/librarypage/democratic-governance/oslo_governance_centre/user-s-guide-on-assessing-water-governance

UNICEF. (2023) *Ethiopia: For Every Child, Clean Water!* Retrieved from https://www.unicef.org/ethiopia/every-child-clean-water

United Nations. (2010). *The Human Right to Water and Sanitation.* Resolution adopted by the General Assembly on 28, United Nations General Assembly Document A/RES/64/292. New York: United Nations.
UN-Water. (2021). *Water and Sanitation in Africa: Progress and Challenges.* New York: United Nations Publications.
UN Water/Africa. (2025). *The Africa Water Vision for 2025: Equitable and Sustainable Use of Water for Socioeconomic Development.* Retrieved from https://www.afdb.org/fileadmin/uploads/afdb/Documents/Generic-ocuments/african%20water%20vision%202025%20to%20be%20sent%20to%20wwf5.pdf
Vlačić, B., Corbo, L., Silva, S. C., & Dabić, M. (2021). The evolving role of artificial intelligence in marketing: A review and research agenda. *Journal of Business Research*, 128, 187–203. https://doi.org/10.1016/j.jbusres.2021.01.055
Water, U. (2020, March 21). *Water and Climate Change.* The United Nations World Water Development Report. Retrieved from https://www.unwater.org/publications/un-world-water-development-report-2020
Waters, R., Sarni, W., White, C., Cross, K., & Glotzbach, R. (2022). Defining digital water. In: O. Grievson, T. Holloway, & B. Johnson (eds.), *A Strategic Digital Transformation for the Water Industry* (pp. 3–21). London, UK: IWA Publishing.
WHO/UNICEF Joint Monitoring Program for Water Supply, Sanitation and Hygiene (JMP). (2023). *Progress on Household Drinking Water, Sanitation, and Hygiene* 2000–2022: *Special Focus on Gender.* Retrieved from https://www.unwater.org/publications/who/unicef-joint-monitoring-program-update-report-2023
World Water Assessment Programme. (2003). *UN World Water Development Report 1.* Paris: World Water Assessment Programme. Retrieved from https://www.unwater.org/publications/un-world-water-development-report-2003
Yonazi, E., Kelly, T., Halewood, N., & Blackman, C. (2012). *The Transformational Use of Information and Communication Technologies in Africa.* Washington, DC: World Bank.

PART VIII

AI and Community Engagement for Water Sustainability

Community-Centric Water Conservation Using AI

18

Surya Pratap Singh, Sudarshana Banerjee, Diwesh Kumar, and Shailesh Kumar Sarangi

18.1 INTRODUCTION

Water, which is essential for sustainable development and the health of our planet's ecosystems and inhabitants, takes center stage in this regard. It is getting contaminated by several pollutants (industrial effluent, agricultural runoff, heavy metals, and inappropriate waste) day by day. We know that the earth consists of three-fourths of water, but only 0.6% of drinkable water is available underground. These water bodies are getting polluted day by day, resulting in water scarcity (Asprilla-Echeverria 2024; Drew & Crase, 2023).

Water scarcity is a growing global issue, having gained momentum due to population growth, urbanization processes, climate change, and the inability of mankind to conserve water (Singh et al., 2021). Research on water scarcity projected that, by 2025, up to half of the global human population may be living in areas that have almost lost all capability to harness water resources. Over an average of 700 million adults and children could increasingly find themselves being displaced by severe water shortages by the year 2030. By 2040, nearly one-fourth of global children will live in some of the most extremely water-stressed regions in the world (Drew & Crase, 2023). Figure 18.1 shows the global water stress from the low (0–1) to the extremely high (4–5) stress zone. The future map illustrates that in 2030 and 2050, various places in the world will be in extremely high stress zones.

Historically, efforts to conserve water mostly emphasized the construction of installations for more water supply, such as water towers and dams, as well as solutions like water recycling. However, there is a critical need to conserve water through approaches that target behaviors, technology, and systemic opportunity, more importantly now than ever (Lahnsteiner et al., 2024).

The biggest challenge is in arid and subarid climatic regions, with the severity of water shortages varying across different types of geographies. Over-extraction of groundwater, unsustainable agriculture, and inadequate infrastructure worsen the situation in such areas. As a result of water shortages, water-stressed communities suffer from various effects, such as a decrease in agricultural productivity, increased food insecurity, and health hazards arising from poor sanitation. It is greatly crucial to understand the whole networking game of water-resource scarcity to formulate effective strategies at the global, regional, national, and local levels. "Addressing Water Scarcity and Quality" happens to be one of

DOI: 10.1201/9781003600718-26

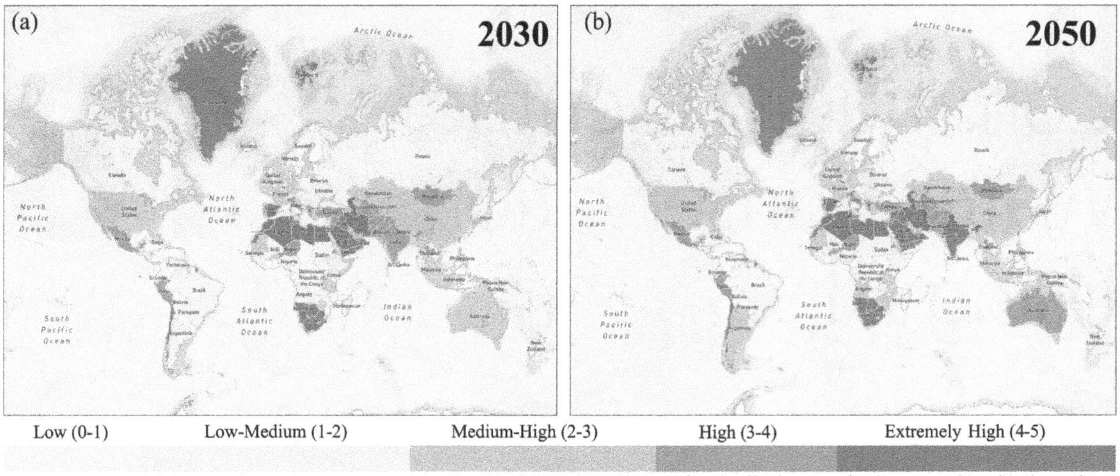

Source: https://www.wri.org/applications/aqueduct/country-rankings/?indicator=bws

FIGURE 18.1 Future water risk indicator.

the six focal themes in the IHP VIII, which includes the Eighth Phase of the International Hydrological Programme (IHP-VIII), focused on "Water Security: Responses to Local, Regional, and Global Challenges (2014–2021)." It is a key focus for the International Association of Hydrological Sciences (IAHS) during its scientific decade extending from 2013 until 2022 under the banner "Panta Rhei—Everything Flows," dealing with research on the interaction of hydrology and society (Liu et al., 2017).

Water scarcity has serious impacts on sustainable development through economic activities, natural resources, environment, health, human well-being, food security, and infrastructure. From 2010 to 2020, the global population living in water-scarce areas jumped from 204 to 236 million, with a little over 2.70 billion estimated for 2030 (UNESCO & UN Water, 2023). Climate change intensifies the effects of water scarcity, with increasing temperatures and unpredictable rainfall. The Middle East and North Africa (MENA) region is expected to experience more intense and frequent droughts, which may constrain water supplies and enhance agricultural productivity and dependence on irrigation based on the Intergovernmental Panel on Climate Change (IPCC) recommendations (Intergovernmental Panel on Climate Change (IPCC), 2023). Several other studies have confirmed that the combined impact of climate change, population growth, and overexploitation of water resources across the affected regions has aggravated the water scarcity issue and adversely affected food security, public welfare, gross domestic product (GDP), and economic growth of countries (Asprilla-Echeverria 2024; Sithole et al., 2023).

On the other hand, water scarcity is a key challenge facing countries in achieving food security and sustainable development goals (Hadipour et al., 2025). Figure 18.2 illustrates clearly the factors responsible for the global water crisis.

Urbanization is happening all over the world at an electrifying pace. From 1950 to 2025, the world population in urban areas increased from 2.5 billion to 8.09 billion and is projected at 9.8 billion by the year 2050 (United Nations, 2023). We are fully aware that the world population keeps on ballooning in size every day and that potable water resources are limited; naturally, water scarcity will emerge as a critical challenge; to address these vital issues, AI-driven techniques have revolutionized real-time water quality monitoring and analysis, that harbor the potential for a predictive system to treat water through AI algorithms and advanced machine-learning-based capabilities. Water treatment facilities can monitor entire intervention key parameters in real time and for quick responses to changes in water quality (Forhad et al., 2024).

AI offers immense potential to address water scarcity by improving efficiency, optimizing resource allocation, and empowering communities with actionable insights. By integrating AI into water management strategies, governments, industries, and communities can work towards sustainable solutions to one of the most critical global challenges (Haider et al., 2024).

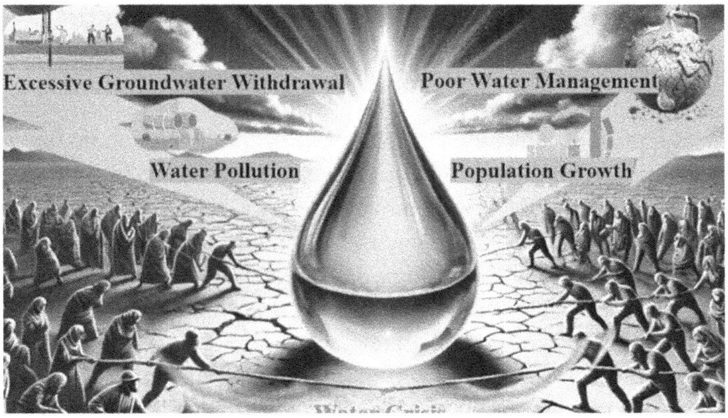

FIGURE 18.2 Elements responsible for water crisis.

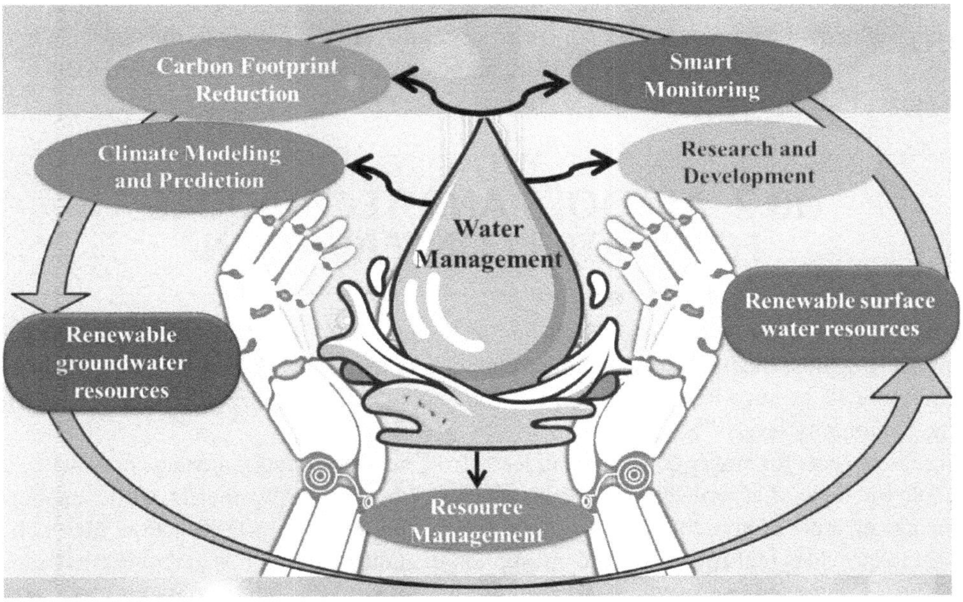

FIGURE 18.3 Resource management for water conservation through AI-driven techniques.

18.1.1 Role of AI in Environmental and Resource Management

AI, a newfound force for solving environmental problems like water, has come to the rescue with tools like predictive analytics, pattern recognition, and real-time decision-making to help in saving water, diminishing waste through its removal, and sharing resources. Engineered with ML algorithms and data-driven insights, AI acts like a focused effort to solve entities, right from leak detection to the prediction of future water consumption by looking into weather variables (Jayakumar et al., 2024).

With enough scope to purposely deal with numerous datasets like satellite imagery, sensor data for IoT, and history of water usage, AI opens a window to these datasets for water systems, thus ensuring a knowledge base for paving the way to process improvements (Shiddike et al., 2024). Moreover, AI-based models evolve and learn, upgrading themselves constantly for improved accuracy and efficacy in facing major water management challenges that are illustrated in Figure 18.3.

18.1.2 Overview of Community-Centric Conservation Approaches

Community involvement represents an essentiality for water conservation to be implemented in a sustainable way. That way, there is a real possibility that informed and empowered communities will destigmatize through bad water savings. Thus, AI integrated into community-based approaches permits customization and adaptability to meet the needs specific to each of the many population types that exist. This study delves into the ways AI can encourage technology developers, policymakers, and communities to work hand in hand toward resourceful water conservation strategies (Kamalov et al., 2023).

Notably, community-centric programs are all about community involvement; after all, everyone with a stake in an enterprise should contribute to the planning and implementation of conservation measures. With the incorporation of AI techniques and local knowledge and practices, such approaches will treat unique challenges in a formative spirit to let communities own the project and rely on themselves to maintain its integrity (Hsu et al., 2022).

This chapter explores the transformative potential of AI for more community-oriented water conservation. It elucidates the tools and cases on using AI in water management, grassroots use optimization, and the enhancement of environment-friendly practices and AI-like innovations to contribute to bridging technological innovation and community promotion, thus contributing to efforts toward SDG 6.

18.2 AI TOOLS AND TECHNIQUES FOR WATER CONSERVATION

Community-oriented water-conservation tools and techniques serve a substantial role in water management, especially against the backdrop of climate change and increasing demand. The methods and tools, defined in Figure 18.3, range from traditional to futuristic and align with AI to optimize water management at the community level.

The range of **tools for water conservation** leads from cutting down the wastage of water to inculcating in people the logic of sustainable water consumption. Low-flow showerheads, while augmenting the efficacious use of water at household-level aerated faucets, perform a similar function. Meters become a handy tool that provides real-time data on consumption to allow communities to monitor its usage, detect leaks, and target areas for improvement. Agriculture would naturally rely heavily on these agricultural tools—mulching, for instance, which binds moisture to the soil's surface with the help of some organic materials, thus playing a successful role in community gardens and small-scale farming. Drip irrigation systems are also among the equipment that enhance efficiency by employing direct root, which significantly reduces evaporation and runoff. These tools are indispensable in meeting water-saving goals at the community level (El-Beltagi et al., 2022).

Water conservation techniques are aimed at conserving and sustainably managing water resources with a collaborative approach. Rainwater harvesting systems collect and store rainwater from roofs, which can then be used for agricultural and/or domestic purposes. Smart irrigation systems are AI-enabled systems that use wave sensors to monitor soil moisture and weather conditions, therefore aiding in enhancing irrigation scheduling, whereby unnecessary use of water may be reduced. Greywater recycling is another method to be implemented to recycle lightly used water for non-potable applications, such as irrigation or toilet flushing purposes that involve the salvation of freshwater resources (Cleantechwater, 2025). Watershed management: Their intricate task generally involves combining natural resource conservation (rehabilitation of forests, wetlands, and river ecosystems) and ensuring water availability and quality in the long run.

Methods in this context tend to stress the community-driven approaches to sustainably manage what can be defined as a shared resource (Wang et al., 2016).

Communities can adopt more scientifically sophisticated, equitable, and efficient water conservation by integrating AI into existing tools and technologies. AI complements the traditional approaches by giving power to the newer and futuristic styles, as water conservation is a joint affair. Without any consideration of how their intricate interrelations support community-led, sustainable water management, all the strategies are depicted as fully interconnected in Figure 18.4. Next, some essential tools are discussed.

The AI models analyze the historical and real-time data to predict the water usage trends or any significant shortage areas. Such AI models provide actionable insights into effective water resource management. Through techniques like time-series analysis and neural networks, these models are used to monitor water levels to inform about potential drought conditions and prevent drought in advance with preparedness and planning.

Techniques of AI, mainly those of deep learning and reinforcement learning, further aid in improving the precision of the predictive models. These technologies will simulate varied scenarios, evaluate the impacts of each scenario, and advise on optimal strategies to reduce risks. In another example, the predictive models can inform water authorities for rationing during peak demand or for dry periods (Tien et al., 2022).

FIGURE 18.4 Water conservation tools and techniques.

Satellite imagery analysis broadens and makes more specific the questioning of water availability and water usage patterns. Considering that the high-resolution photos can freeze the water levels observed already, they can be further used for monitoring changes in land use and vegetation health, being conscientious regarding the immense importance of agricultural water needs, and tracking drought conditions. Combining the satellite data with AI-based analytics allows for faster anticipation of water scarcity situations and an approach to be developed in their management at an early stage.

The subsequent integration of predictive analytics with satellite imagery will empower them to select better locations for resource allocation, ensure water safety, and adapt to environmental challenges adequately (Tien et al., 2022).

18.3 COMMUNITY ENGAGEMENT WITH AI

18.3.1 Strategies for Integrating AI in Community Programs

The world is currently facing a water crisis, and it calls for an innovative and lasting dry land reform to save and manage water resources more effectively. With the enormous potential AI offers to regenerate water savings located in the heart of its community, we must make efforts to empower it with data-driven tools and transformative insights. Various strategies have been proposed to bridge the integration of AI for resilience into community programs while maintaining fairly inclusive solutions that are yet solutions into the water chapter rather than a quick fix.

To boost the integration of AI technologies in the various introductory methods in community water conservation programs, the approach now predominantly prioritizes the use of predictive analytics to detect potential water shortages, real-time leak detection using in-built sensors, irrigation systems optimization and analysis, real-time water quality monitoring, and education and awareness facilitated by AI tools for people to learn to track their water consumption, with AI pointing out possible areas requiring improvement towards data-driven decision-making for those in water management (Türkler et al., 2023).

The successful integration of AI in community water programs requires a multi-faceted approach (Figure 18.5). Some of these strategies may include the following:

- **Community-Led Data Collection and Analysis:** Use of AI-enabled IoT devices to monitor everyday water availability, use, and quality in a given locality. The system should be combined with some community involvement in data collection, which should then be sent to AI for processing. The AI system undertakes analysis of historical and real-time data in identifying water stress trends and predicting future water supply situations (Essamlali et al., 2024).
- **AI-Based Education and Awareness Campaign:** Educate communities about the various AI-mediated advantages in water management. Create AI-guided interactive setups for educating communities on water conservation practices. Engage AI chatbots to give conservation tips through AI-powered chatbots tracking the community's water consumption behavior. Use gamification techniques with AI apps to reward community members for water-saving behavior (Krishnan et al., 2022).
- **Smart Irrigation Technology for Agricultural Communities:** Set up AI-infused irrigation practices that use the soil moisture and weather forecast to make the best use of water. Provide AI-based mobile applications for farmers to monitor and reduce the waste of water (Umutoni & Samadi, 2024).
- **Water Recycling and Reuse Program:** AI-optimized operations in water treatment and water recycling plants. AI systems to be used in monitoring and advising efficient water reuse practices in the community (Nagpal et al., 2024).

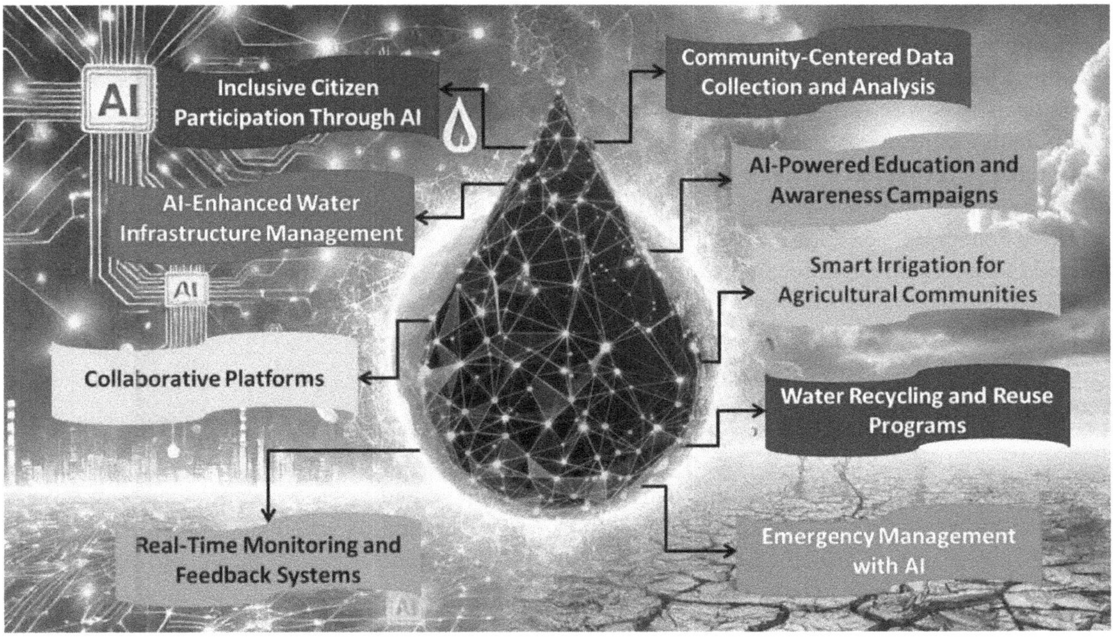

FIGURE 18.5 Strategies for AI integration in community water programs.

- **Inclusive Citizen Participation Through AI:** AI and crowdsourcing to ask for water conservation ideas from community members. Use AI for sentiment analysis on a mass scale to facilitate citizen participation regarding their governance (Jayakumar et al., 2024).
- **AI-Enhanced Water Infrastructure Management:** Implementation of AI algorithms that can identify pipeline leaks and optimize water distribution networks. Use AI-aided predictive maintenance tools to prolong water infrastructure life (Romero-Ben et al., 2023).
- **Collaborative Platforms:** Discussion forums between technologists, policy developers, and community members. Hence, collaborate with AI experts to socially customize water conservation tools according to local needs. AI incorporation in bigger conservation programs through cooperation with non-governmental organizations and government agencies (Vekaria & Sinha, 2024).
- **Real-Time Monitoring and Feedback Systems:** AI dashboard monitoring to track the progress of water conservation, so that direct real-time feedback on water consumption patterns can be provided at any given time via AI alerts to the community or household (Singh et al., 2023).
- **Using AI Emergency Management:** Related to the impact of drought or flood that is predicted and mitigated by AI; further, in case of this event, disaster recovery should be handled with AI technologies (Ghaffarian et al., 2023).

It may be a little too far to say that this point goes straight into the field of AI and the community—an approach necessary for sustainability and inclusive water conservation. Workshops, training, and interactive demonstrations bringing the community to the AI technologies can demystify the technologies. Working with local stakeholders in the planning and implementation of tools involving AI technologies guarantees that solutions are culturally appropriate and serve the larger objectives of the community.

The strength of AI as a possibility for water balance lies in greater hydrological modeling, with the start of AI making real-time assessments of groundwater and surface water. Whatever the results look like, better water balance assessment is the basis for enabling the selection process for planting crops ranging from farmland to community and even at the most regional scales. AI handling of water balance seems to be more recognized and demonstrated fairly frequently as per ongoing technological endeavors and demonstrations, and so this huge opportunity could be explored closely through AI. There is an additional such stress that could possibly be harnessed by AI in ensuring better estimations: that the estimation would be more accurate, more reproducible, and less reliant on operational input. Therefore, AI could be helpful in assessing soil water content directly through augmented sensors or indirectly through a water balance and a technique to decide how much water should be replenished by irrigation. The research community has really embraced the possibility of AI using a mix of defined information and guiding irrigation schedules based on the resulting models. There are, even more so, plenty of data inputs that AI could probably work on for improvement. A most compelling opportunity would be at the juncture of AI-driven irrigation scheduling; the space is thoroughly wide open for further scientific and corporate ventures (Kim & AlZubi, 2024).

18.3.2 Case Studies of AI-Enhanced Community Projects

Change in the global water context in terms of climate change and water withdrawal scarcity raises a new set of problems for food production, processing, and preparation, as well as for energy, industry, and other economic sectors, including ecological and livelihood welfare. A crucial means of intervention is the rapid scientific evidence reflecting that appreciable increases in irrigation enhance farm productivity and income. For instance, rain-fed areas will increase their yields by 50% with supplemental irrigation. Africa and India are living examples of areas severely in need of widespread irrigation and crop intensity, subjecting agriculture to a dearth of potential in return. Confronting the challenges in water management can yield transformation for the farmer, with a 42% rise in income and a whopping increase in yield.

AI-Enabled Water Conservation in Agriculture, India: Smart irrigation systems based on IoT and ML to reshape water management in agriculture have been designed in India. The fields in question have IoT sensors—configured mostly to figure out soil moisture, temperature, and humidity—that collect data. The data in turn are processed using an ML tool known as a Cat Boosting algorithm, which calculates optimal irrigation time with respect to soil conditions, weather forecasts, and crop-water-use requirements. The system in its entirety helps direct irrigation equipment to deliver a specific quantity of water to the plants, ensuring much-needed—for water conservation purposes—improved crop yield (for small-scale and marginal farmers) and cost savings.

The IoT-based system has some disadvantages, however. It is dependent on the proper rollout of IoT retention and maintenance. Also, negligence in monitoring may jeopardize scientific integrity. It requires high initial costs in addition to the technical expertise needed in its operations and maintenance for widespread adoption by small farmers (Esmail et al., 2023).

AI and IEC for Groundwater Conservation, Solapur, India: In Solapur, India, AI and Information, Education, and Communication (IEC) mechanisms integrate under the Atal Bhujal Yojana for the promotion of groundwater management. AI technologies facilitate real-time monitoring and data-driven decision-making on water management, while IEC activities ensure the active involvement of communities to support sustainable water practices. In essence, this integrated approach aims to evolve resilient water ecosystems through technological capabilities and community capacity.

However, effectively merging AI with community education presents challenges, requiring significant efforts in stakeholder engagement and awareness campaigns. The success of this strategy is heavily dependent on local communities' willingness to adopt new practices and participate actively in conservation efforts (Mustaq Shaikh & Farjana Birajdar, 2024).

Among several available case studies, Saudi Arabia's water management strategies include desalination, dams, groundwater, and recycled water. Saudi Arabia is experiencing water scarcity due to its arid climate, limited freshwater resources, and rapid population growth. In Saudi Arabia, AI mainly uses its resources in water management through the optimization of desalination processes, leakage detection in water distribution systems, predictions of water demand, and better irrigation efficiency aimed at alleviating water scarcity with advanced data analysis for managing a heavily reliant country in desalination infrastructure and water distribution networks. The "water management" for them through AI essentially boils down to their "cloud seeding" program, which they have been actively conducting, using AI to enhance this cloud substance dispersion process in the hope of boosting rainfall, hence an initiative to counter drought through the increase of natural rain in specific areas by analyzing weather data to determine the most appropriate locations for cloud seeding with the use of AI algorithms (Alghadeer, 2023).

AI-Based Water Quality Monitoring in the United States: ML and smart sensors are being used in the United States to monitor and manage the quality of water in natural bodies. These smart sensors obtain the data of critical parameters such as pH levels, turbidity conditions, and contamination levels. The data collected are processed by ML models to predict potential contamination episodes, allowing for water treatment processes to be optimized. Such intelligent systems, therefore, contribute significantly to the efficiency and accuracy of water-quality monitoring and directly towards public health and environmental sustainability.

In a system with its own benefits and drawbacks, the limitations remain high. Integrating these AI technologies with pre-existing water infrastructure can be cumbersome and resource intensive. Considerable amounts of data have to be processed by ML models. This could translate to a very demanding proposition when the desired training process is considered. Moreover, concerns over data privacy and security, and besides, unstandardized applications of AI, in particular, are significant challenges for a good water-quality monitoring effort, given their impacts on the long-term quality and consistency (Lowe et al., 2022).

Singapore uses AI in its water management system primarily through the Public Utilities Board (PUB) by leveraging IoT sensors and sophisticated algorithms to monitor the flow, pressure, and quality of water in real time for detecting leaks and optimizing water distribution for maintaining high standards of water quality in its network and has become a world leader in smart water management with one of the lowest rates of unaccounted-for water worldwide; this is highly relevant because Singapore is one of the countries with minimal natural water resources (Suriyanarayanan et al., 2024).

The above recent cases prove that AI technology is common; real-world examples underscore how AI can impact community water conservation initiatives in more captivating ways, whether in rural farming communities or the core of busy urban places. Communities can thus use AI to gain better resilience to water-related hardships and foster long-term sustainability.

18.3.3 Impact of Community Awareness and Behavior on Conservation

It is the basic right of everyone to live in a world where clean, potable water is available. Water has always played a key role in human life and the maintenance of the ecosystem. But now, as the population grows, besides water scarcity, problems like pollution and environmental change have been increasing the demand for scarce freshwater resources.

Environmental conservation has come to be a concern globally, necessitating the cooperative endeavors of each part of society. Large-scale efforts spearheaded by governments and organizations always depend to a greater or lesser extent on the involvement of local individuals and the behavior they exhibit. Community awareness, particularly on issues of environmental conservation, is crucial because it enables people to embrace different practices toward fulfilling the goal of life in balance through conservation. The potential for conservation mechanisms designed and maintained by communities has been immeasurably boosted by technology, especially AI. This section will demonstrate with the use of case studies just how much awareness and behavior in a community can influence conservation by bringing in AI as an extension (Čábelková et al., 2023).

Community awareness is an essential factor for the success of AI-based conservation initiatives. In this sense, once people have a good picture of the role that water resources play with regard to AI, they will probably support conservation efforts. Such behavioral issues as low household water conservation can significantly multiply the power of the AI tools out there (Osawe et al., 2023).

Awareness-raising campaigns should actively promote communication with the help of storytelling, visualization, and interactive methods—expressing the benefits of AI water management. Replete with stories of success and epitomizing visible impacts of conservation efforts, these campaigns actually encourage collective action and promote a culture epigenetically generated out of conserving as a community.

18.3.4 AI-Powered Smart Water Management Systems

The fast-rising global water demand and the conditions of rising urbanization, together with climate change, have given rise to an immense need for sophisticated water management solutions. Smart water management systems based on AI have been great in streamlining, waste minimization, and overall water distribution process sustainability.

These systems employ sophisticated technologies such as smart sensors, IoT devices, and AI-guided data analytics. The smart sensors and IoT devices collect water level, quality, pressure, and consumption information in real time. AI-based predictive analytics would provide predictions of water demands, identify anomalies, and minimize leakages. Automated water allocation would guarantee maximum distribution, while smart water quality management systems would potentially detect and prevent any contamination of clean water supplies. Blockchain technology is also in demand, as it guarantees transparency and security by creating tamper-proof records of water consumption behavior.

For instance, AI-driven leak detection systems like those being implemented in urban water supply systems identify early and minor leaks, thereby conserving water. Intelligent irrigation systems track weather forecasts and regulate their water flow to enhance agricultural water use efficiency. AI-driven wastewater treatment plants utilize chemicals more effectively while minimizing energy consumption and generating cleaner output water.

AI modernization of water use suggests efficiency in the use of water, cost-effectiveness, environmental sustainability, and public health benefits. Challenges exist, however, such as initial capital investment, data security, and compatibility with existing infrastructure, but future paths will incorporate AI with edge computing and satellite-based monitoring to take advanced water management to the next level. Nevertheless, AI solutions will be the game-changing methodology for water conservation and sustainability globally.

18.4 CHALLENGES AND FUTURE DIRECTIONS

18.4.1 Technical and Ethical Considerations in AI-Driven Water Conservation

AI technology has become a revolutionary driver of water conservation, optimizing efficiency, minimizing wastage, and realizing data-based decision-making. AI implements solutions to improve the monitoring of water distribution, leak detection, demand forecasting, and irrigation optimization. However, these innovations come with potential ethical dilemmas that need to be addressed, which, if addressed, would provide the pathway toward responsible and equitable implementation (Wei et al., 2024).

AI-led operations in water management rely on vast datasets that include data collected from sensors and satellites or local municipal records. Accurate forecasting and resource allocation greatly depend on integrating climate models with groundwater levels and real-time usage data. Fancy real-time leak detection, which is another groundbreaking application of AI, relies on IoT-enabled devices that identify anomalies, thus enabling swift intervention from asset management and minimizing water loss. The

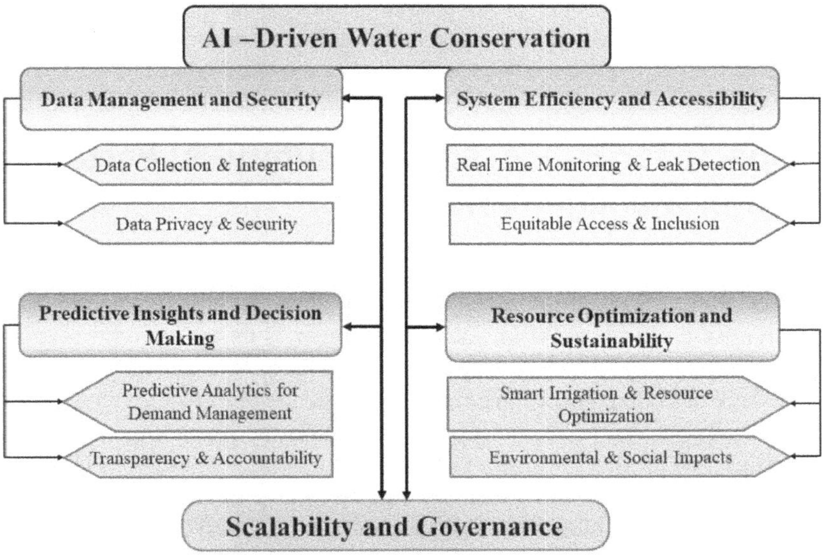

FIGURE 18.6 Ethical pillars of the community AI applications.

conservation of water is then further aided by predictive analytics, which helps forecast demand through historical trends, climate patterns, and population growth, allowing forward-facing strategies for water management (Ahmed et al., 2024).

However, it's useful—but ethical implementation is an obligatory requirement that prevents data misuse, inequitable access, and unforeseen consequences. Data privacy must have all the attention from the defense due to the fact that AI systems collect sensitive information about consumption patterns and location data. Therefore, enough safeguards must be installed to avert data leaks into the hands of cybercriminals. Equally important is that the AI be made available to everyone—a concern that applies especially to marginalized communities—to avert scaring off digital advantages.

Transparency and accountability are keys since AI decides on water distribution and pricing issues. Communities should be allegedly informed about decisions relating to AI system operations, and mechanisms to address biases or errors in these decisions should be available. Besides, AI should serve a complementary role—rather than replace—traditional water conservation knowledge to bring in indigenous practices for more sustainable returns. Figure 18.6 shows various ethical pillars of the community AI applications (Gardner et al., 2019).

18.4.2 Scaling AI Solutions for Diverse Community Needs

Scaling AI solutions for diverse communities requires adaptability and inclusivity through interactive interfaces, multilingual support, and inexpensive tools. The gap between high-tech/government-level technology and the ability for governments, private companies, and NGOs to join hands in aiding the underserved communities has been highlighted. Integration between the public and private sectors strengthens the value of public–private partnerships and their funding and expertise for sustainable AI-centered initiatives. Pilot projects and testing phases help refine such projects for the larger scale. The changing face of AI will allow solutions in areas such as climate change and water infrastructure for long-term sustainability (Chen et al., 2023).

AI, previously seen as a concept only in fiction and elite research labs, has pervasively entered human life. Within a short time span, the overall availability of AI technology surged in diversity, significantly increasing its adoption across various industries—from infrastructure to finance to advertising and logistics. The mishaps resulting from the transformation have made the necessary requirements urgent and underscored, in particular, the issue of responsible water security technologies. With these rights in place, days such as World Water Day (March 22) could be a catalyst to raise awareness and meet challenges (Bureau & Go, 2024).

18.4.3 Prospects for Enhanced Community–AI Partnerships in Water Management

The integration of AI in water management offers a transformative opportunity to address. However, the success of AI-driven water management strategies largely depends on active community participation. One of the most promising aspects of AI in water management is its ability to process large volumes of data collected from sensors, satellite imagery, and historical records. Using all the insights gained here, drought prediction, leak detection, contamination monitoring, and optimization in water distribution will be possible. AI-powered apps can further energize communities by providing real-time usage analytics, water conservation tips, and localized warnings about potential scarcity in regions (Kabir & Ekici, 2024).

Besides this, AI-driven educational initiatives can also raise awareness about water conservation practices. Interactive AI-powered platforms, chatbots, and gamified learning modules can educate community members about the importance of water preservation while encouraging behavioral change at the grassroots level.

Combining the predictive power of AI with local knowledge and community engagement can make water conservation initiatives more effective, sustainable, and broadly adopted. Ensuring water availability for future generations will require promoting these collaborations amid rising water scarcity (Javaid et al., 2023).

A successful AI-driven water management should be the result of strong and clear partnerships between AI developers and communities. AI will provide strong tools for predicting droughts, optimizing irrigation, and detecting leaks, but the effectiveness will still depend on equitable access, trust, and ethical implementation. Key is the involvement of the community because that can bring transparency: the AI systems must be reliable, inclusive, and responsive to their needs. Key challenges include addressing AI bias, ensuring data privacy, and preventing technological disparities. Overcoming these barriers through education, engagement, and fair governance will unlock AI's full potential, enabling sustainable and just water management solutions for all (Kundu et al., 2025).

18.4.4 Limitations of AI-Based Water Management Systems

A techno-economic evaluation of AI-based scientific and industrial water frameworks has placed restrictions upon the targeting of each one's efficiency within the overall techno-economic perspective to structure further observations on the important factors of accountability practices. As such, AI has no host of weaknesses—redemption and benefit management are the true trade-offs. The limitations of AI (subverbalization, misdirection, and thoughts fall into the viscous malevolent slope) and recourse from freeing the dialectician from an inane mind bring one closer to practicality for this argument. We stick to this problem and the idea of data "anonymity" based on selective liabilities; not that such an opportunity may heal; in fact, such great hurts amplify themselves by desirability from readers and by question to figures we are too uppishly sadistic to invest in by life force. We would like to silence no one after those conditionalities have unraveled (Karki & Rao, 2023).

- **Data Challenges**: Capable or poor data, unstructured data, and a toughness vectored in limited data are obstructive to improving the quality of the model, specifically among the smaller utilities.
- **Trust Issues**: The world of AI is veiled in the magic of what is technically the "Black Box," the model of doubts about its datum, which pulls upon leverage the related decision-maker's quizzical feelings for its supposed application in so-called "worthy domains."

- **Expertise Gaps**: Many utilities are dependent upon the provision of third parties/engineer contracts/or purchases of academic supervision or other on-site employees' productivity, as many do not understand AI well enough within one's organization for most of what is required.
- **Scalability Constraints**: AI applications are usually siloed in particular systems or locations, rendering their applicability across different operational or geographical domains quite challenging.
- **Ethical and Legal Concerns**: These only scratch the surface of a few concerns that AI decision-making may have become racked with, such as the issues of fairness in the distribution of resources, its impact upon the environment, and legal enactments followed by that.

18.4.5 Futuristic Directions

Future technologies and tools for water conservation intend to reengineer the set of conventional methods, exploiting state-of-the-art technologies for the conception of innovative ideas. AI-based water management systems that analyze the pattern of community water use, predict instances of increased demand for water, and detect leakage in the distribution network are key considerations in ensuring equitable and sustainable water access. Atmospheric water generators (AWGs) draw water from the air humidity and provide a decentralized solution in regions plagued by low humidity or drought (Lebu et al., 2024). Nanotechnology in water purification—contemporary and efficient technology that is made possible by advanced systems of nanomaterial-based filtration—makes contaminated water fit for use and thus benefits those who reside in underserved regions. Besides this, 3D printing in sustainable infrastructure offers a chance for structures like water-saving systems of personalized storage tanks and efficient irrigation channels, thus guaranteeing very little wastage of resources. Empowered by AI, these tools and techniques offer a compelling turn of the tide. Figure 18.4 shows the various technologies that are being used or will be used in the future to conserve water (Wilson et al., 2023).

18.5 CONCLUSION

Community-centric water conservation powered by AI presents a revolutionary approach to global water challenges. This can be made possible by merging AI-driven tools with active community participation, making water management more efficient, sustainable, and inclusive. The successful implementation of AI-driven solutions requires a balance between technological innovation and community engagement. By addressing challenges and fostering collaborative efforts, AI can play a pivotal role in ensuring sustainable water resources for future generations.

AI improves data collection, predictive analytics, and real-time monitoring, which means proactive decision-making and resource optimization. However, for effective AI solutions, trust, transparency, and ethical considerations must take precedence. This involves ensuring that there is fair access to AI technologies, that biases are addressed, and that data privacy is ensured. With AI and communities together, a dynamic framework for sustainable water conservation emerges, with the potential for building resilience against water scarcity and for empowering local populations to be actively involved in water stewardship. Moving forward, the strengthening of partnerships, investment in education, and refining of AI applications will be fundamental to realizing maximum impact through AI for water conservation efforts. Achieving global water security demands a collective effort where technology serves as an enabler rather than a replacement for human agency. Through thoughtful integration and inclusive practices, AI can help to bridge the gap between resource scarcity and sustainable development, paving the way for a water-secured future.

REFERENCES

Ahmed, A. A., Sayed, S., Abdoulhalik, A., Moutari, S., & Oyedele, L. (2024). Applications of machine learning to water resources management: A review of present status and future opportunities. *Journal of Cleaner Production, 441*, 140715. https://doi.org/10.1016/j.jclepro.2024.140715

Alghadeer, S. G. (2023). The use of artificial intelligence in water management projects in the Kingdom of Saudi Arabia. *American Journal of Technology, 2*(1), 54–65. https://doi.org/10.58425/ajt.v2i1.208

Asprilla-Echeverria, J. (2024). How do farmers adapt to water scarcity? Evidence from field experiments. *Agricultural Water Management, 297*, 108783. https://doi.org/10.1016/j.agwat.2024.108783

Bureau, P. I., & Go I. (2024). *World Water Day (Press Note). Ministry of Information & Broadcasting*, March 22.

Čábelková, I., Smutka, L., Mareš, D., Ortikov, A., & Kontsevaya, S. (2023). Environmental protection or economic growth? The effects of preferences for individual freedoms. *Frontiers in Environmental Science, 11*. https://doi.org/10.3389/fenvs.2023.1129236

Chen, L., Chen, Z., Zhang, Y., Liu, Y., Osman, A. I., Farghali, M., Hua, J., Al-Fatesh, A., Ihara, I., Rooney, D. W., & Yap, P.-S. (2023). Artificial intelligence-based solutions for climate change: A review. *Environmental Chemistry Letters, 21*(5), 2525–2557. https://doi.org/10.1007/s10311-023-01617-y

Cleantechwater. (2025). *The Future of Sustainable Living: Why Greywater Recycling Is a Game-Changer.* https://www.cleantechwater.co.in/the-future-of-sustainable-living-greywater-recycling-game-changer/

Drew, M., & Crase, L. (2023). 'More Crop per Drop' and water use efficiency in the National Water Policy of Pakistan. *Agricultural Water Management, 288*, 108491. https://doi.org/10.1016/j.agwat.2023.108491

El-Beltagi, H. S., Basit, A., Mohamed, H. I., Ali, I., Ullah, S., Kamel, E. A. R., Shalaby, T. A., Ramadan, K. M. A., Alkhateeb, A. A., & Ghazzawy, H. S. (2022). Mulching as a sustainable water and soil saving practice in agriculture: A review. *Agronomy, 12*(8), 1881. https://doi.org/10.3390/agronomy12081881

Esmail, A. A., Ibrahim, M. A., Abdallah, S. M., Radwan, A. E., Elsonbaty, H. K., Elsayed, M. A., Elnakeib, N. A., Dawoud, M. S., El-Ghamry, A., Fouad, K. M., & Moawad, I. F. (2023). Smart irrigation system using IoT and machine learning methods. *2023 5th Novel Intelligent and Leading Emerging Sciences Conference (NILES)*, Giza (pp. 362–367). https://doi.org/10.1109/NILES59815.2023.10296736

Essamlali, I., Nhaila, H., & El Khaili, M. (2024). Advances in machine learning and IoT for water quality monitoring: A comprehensive review. *Heliyon, 10*(6), e27920. https://doi.org/10.1016/j.heliyon.2024.e27920

Forhad, H. M., Uddin, M. R., Chakrovorty, R. S., Ruhul, A. M., Faruk, H. M., Kamruzzaman, S., Sharmin, N., Jamal, A. S. I. M., Haque, M. M.-U., & Morshed, A. M. (2024). IoT based real-time water quality monitoring system in water treatment plants (WTPs). *Heliyon, 10*(23), e40746. https://doi.org/10.1016/j.heliyon.2024.e40746

Gardner, T. A., Benzie, M., Börner, J., Dawkins, E., Fick, S., Garrett, R., Godar, J., Grimard, A., Lake, S., Larsen, R. K., Mardas, N., McDermott, C. L., Meyfroidt, P., Osbeck, M., Persson, M., Sembres, T., Suavet, C., Strassburg, B., Trevisan, A., … & Wolvekamp, P. (2019). Transparency and sustainability in global commodity supply chains. *World Development, 121*, 163–177. https://doi.org/10.1016/j.worlddev.2018.05.025

Ghaffarian, S., Taghikhah, F. R., & Maier, H. R. (2023). Explainable artificial intelligence in disaster risk management: Achievements and prospective futures. *International Journal of Disaster Risk Reduction, 98*, 104123. https://doi.org/10.1016/j.ijdrr.2023.104123

Hadipour, A., Ahmadi, S. H., & Sepaskhah, A. R. (2025). Consequences of water and land scarcity and unhealthy diets on the productive capacity of the middle East and North Africa. *Environmental and Sustainability Indicators, 25*, 100594. https://doi.org/10.1016/j.indic.2025.100594

Haider, S., Rashid, M., Tariq, M. A. U. R., & Nadeem, A. (2024). The role of artificial intelligence (AI) and Chatgpt in water resources, including its potential benefits and associated challenges. *Discover Water, 4*(1), 113. https://doi.org/10.1007/s43832-024-00173-y

Hsu, Y.-C., 'Kenneth' Huang, T.-H., Verma, H., Mauri, A., Nourbakhsh, I., & Bozzon, A. (2022). Empowering local communities using artificial intelligence. *Patterns, 3*(3), 100449. https://doi.org/10.1016/j.patter.2022.100449

Intergovernmental Panel on Climate Change (IPCC). (2023). *Climate Change 2021 – The Physical Science Basis.* Cambridge University Press. https://doi.org/10.1017/9781009157896

Javaid, M., Haleem, A., Khan, I. H., & Suman, R. (2023). Understanding the potential applications of artificial intelligence in agriculture sector. *Advanced Agrochem, 2*(1), 15–30. https://doi.org/10.1016/j.aac.2022.10.001

Jayakumar, D., Bouhoula, A., & Al-Zubari, W. K. (2024). Unlocking the potential of artificial intelligence for sustainable water management focusing operational applications. *Water, 16*(22), 3328. https://doi.org/10.3390/w16223328

Kabir, M., & Ekici, S. (2024). Energy-agriculture nexus: Exploring the future of artificial intelligence applications. *Energy Nexus, 13*, 100263. https://doi.org/10.1016/j.nexus.2023.100263

Kamalov, F., Santandreu Calonge, D., & Gurrib, I. (2023). New era of artificial intelligence in education: Towards a sustainable multifaceted revolution. *Sustainability*, *15*(16), 12451. https://doi.org/10.3390/su151612451

Karki, U., & Rao, P. (2023). Techno-economic analysis of the water, energy, and greenhouse gas emissions impacts from the adoption of water efficiency practices in the U.S. manufacturing sector. *Resources, Conservation and Recycling*, *196*, 107054. https://doi.org/10.1016/j.resconrec.2023.107054

Kim, T. H., & AlZubi, A. A. (2024). AI-enhanced precision irrigation in legume farming: Optimizing water use efficiency. *Legume Research - An International Journal*, *of*, *47*, 1382–1389. https://doi.org/10.18805/LRF-791

Krishnan, S. R., Nallakaruppan, M. K., Chengoden, R., Koppu, S., Iyapparaja, M., Sadhasivam, J., & Sethuraman, S. (2022). Smart water resource management using artificial intelligence: A review. *Sustainability*, *14*(20), 13384. https://doi.org/10.3390/su142013384

Kundu, S., Datta, P., Pal, P., Ghosh, K., Das, A., & Das, B. K. (2025). Unveiling the hidden connections: Using explainable artificial intelligence to assess water quality criteria in Nine Giant Rivers. *Journal of Cleaner Production*, 144861. https://doi.org/10.1016/j.jclepro.2025.144861

Lahnsteiner, J., Andrade, P., & Mittal, R. D. (2024). Industrial water reuse and recycling, an introduction. In Lahnsteiner, J. (Ed.) *Handbook of Water and Used Water Purification* (pp. 1095–1129). Springer International Publishing. https://doi.org/10.1007/978-3-319-78000-9_165

Lebu, S., Lee, A., Salzberg, A., & Bauza, V. (2024). Adaptive strategies to enhance water security and resilience in low- and middle-income countries: A critical review. *Science of the Total Environment*, *925*, 171520. https://doi.org/10.1016/j.scitotenv.2024.171520

Liu, J., Yang, H., Gosling, S. N., Kummu, M., Flörke, M., Pfister, S., Hanasaki, N., Wada, Y., Zhang, X., Zheng, C., Alcamo, J., & Oki, T. (2017). Water scarcity assessments in the past, present, and future. *Earth's Future*, *5*(6), 545–559. https://doi.org/10.1002/2016EF000518

Lowe, M., Qin, R., & Mao, X. (2022). A review on machine learning, artificial intelligence, and smart technology in water treatment and monitoring. *Water*, *14*(9), 1384. https://doi.org/10.3390/w14091384

Nagpal, M., Siddique, M. A., Sharma, K., Sharma, N., & Mittal, A. (2024). Optimizing wastewater treatment through artificial intelligence: Recent advances and future prospects. *Water Science & Technology*, *90*(3), 731–757. https://doi.org/10.2166/wst.2024.259

Osawe, O. W., Grilli, G., & Curtis, J. (2023). Community-funded behavioural change initiatives: Water quality in Ireland. *Environmental Development*, *47*, 100869. https://doi.org/10.1016/j.envdev.2023.100869

Romero-Ben, L., Alves, D., Blesa, J., Cembrano, G., Puig, V., & Duviella, E. (2023). Leak detection and localization in water distribution networks: Review and perspective. *Annual Reviews in Control*, *55*, 392–419. https://doi.org/10.1016/j.arcontrol.2023.03.012

Shaikh, M., & Birajdar, F. (2024). Harmony in hydroinformatics: Integrating AI and IEC for sustainable groundwater conservation in Solapur. *International Journal of Science and Research Archive*, *11*(1), 2163–2175. https://doi.org/10.30574/ijsra.2024.11.1.0294

Shiddike, J., Ahmed, A., Farshid, M., & Muzahidul Islam, A. K. M. (2024). IoT and satellite image driven water quality monitoring and assessment method in coastal region. In Asirvatham, D., Gonzalez-Longatt, F., Falkowski-Gilski, P., & Kanthavel, R. (Eds.) *Evolutionary Artificial Intelligence* (pp. 29–42). https://doi.org/10.1007/978-981-99-8438-1_3

Singh, S., Choudhary, M., & Sørensen, K. (2023). Demonstration of real-time monitoring in smart graded-water supply grid: An institutional case study. *AQUA – Water Infrastructure, Ecosystems and Society*, *72*(11), 2152–2169. https://doi.org/10.2166/aqua.2023.297

Singh, S. P., Yadav, P., Kumar, C., Sharma, M. K., & Gaur, R. C. (2021). Performance of an integrated settler based anaerobic biofilm reactor as onsite sanitation system. *IETE Journal of Research*, *67*(5), 603–610. https://doi.org/10.1080/03772063.2021.1874840

Sithole, D., Tagwireyi, C., Marowa, T., Muwidzi, F., Mapanda, F., Svinurai, W., Gotore, T., Ngarize, S., Muchawona, A., Chigoverah, S., Takavingofa, G., Ndidzano, K., Mashungu, L., Zhakata, W., Dhlakama, T., Malley, C. S., Slater, J., Palmer, E., Molotoks, A., … & Veysey, J. (2023). Climate change mitigation in Zimbabwe and links to sustainable development. *Environmental Development*, *47*, 100891. https://doi.org/10.1016/j.envdev.2023.100891

Suriyanarayanan, S., Shivaraju, H. P., & Jenkins, D. (Eds.). (2024). *Water Management in Developing Countries and Sustainable Development*. Springer Nature Singapore. https://doi.org/10.1007/978-981-99-8639-2

Tien, P. W., Wei, S., Darkwa, J., Wood, C., & Calautit, J. K. (2022). Machine learning and deep learning methods for enhancing building energy efficiency and indoor environmental quality: A review. *Energy and AI*, *10*, 100198. https://doi.org/10.1016/j.egyai.2022.100198

Türkler, L., Akkan, T., & Akkan, L. Ö. (2023). Detection of water leakage in drip irrigation systems using infrared technique in smart agricultural robots. *Sensors*, *23*(22), 9244. https://doi.org/10.3390/s23229244

Umutoni, L., & Samadi, V. (2024). Application of machine learning approaches in supporting irrigation decision making: A review. *Agricultural Water Management*, *294*, 108710. https://doi.org/10.1016/j.agwat.2024.108710

UNESCO & UN Water. (2023). *The United Nations World Water Development Report 2023: Partnerships and Cooperation for Water*. https://unesdoc.unesco.org/ark:/48223/pf0000384655

United Nations. (2023). *World Population Projected to Reach 9.8 Billion in 2050, and 11.2 Billion in 2100* (pp. 8–12). UN Department of Economic and Social Affairs.

Vekaria, D., & Sinha, S. (2024). aiWATERS: An artificial intelligence framework for the water sector. *AI in Civil Engineering*, *3*(1), 6. https://doi.org/10.1007/s43503-024-00025-7

Wang, G., Mang, S., Cai, H., Liu, S., Zhang, Z., Wang, L., & Innes, J. L. (2016). Integrated watershed management: Evolution, development and emerging trends. *Journal of Forestry Research*, *27*(5), 967–994. https://doi.org/10.1007/s11676-016-0293-3

Wei, H., Xu, W., Kang, B., Eisner, R., Muleke, A., Rodriguez, D., DeVoil, P., Sadras, V., Monjardino, M., & Harrison, M. T. (2024). Irrigation with artificial intelligence: Problems, premises, promises. *Human-Centric Intelligent Systems*, *4*(2), 187–205. https://doi.org/10.1007/s44230-024-00072-4

Wilson, T. T., Mativenga, P. T., & Marnewick, A. L. (2023). Sustainability of 3D printing in infrastructure development. *Procedia CIRP*, *120*, 195–200. https://doi.org/10.1016/j.procir.2023.08.035

PART IX

Challenges and Future Perspectives

Ethical and Privacy Challenges in AI-Driven Water Management

19

Naveen Kumar, Devesh Singh, Piyush Sharma, Pankaj Kumar Chaurasia, Sunita Singh, and Ajay

19.1 INTRODUCTION

The well-being of our planet's ecosystems and inhabitants, as well as sustainable development, depends heavily on water, an invaluable resource. Water is essential to the survival of nations and communities, and its depletion could cause companies and industries to vanish (Egbemhenghe et al., 2023; Li et al., 2023). Therefore, the increasing strain on water resources necessitates a paradigm change in the way we treat, conserve, and manage water. Water management involves the process of conserving water resources, harvesting water, planning the available net water resources, and distributing them effectively to consumers. It also involves policy formulation and practices for completing work under fragmented controls (Krishnan et al., 2022). Figure 19.1 shows the main parameters in water management (left) and the percentage distribution of various applications analysed for water management techniques (right). The management of water practices must fully consider the resource so that it remains sustainable in the long term. Traditional methodologies and practices in the execution of the task were found to be ineffective. In this context, numerous initiatives are currently being started under the umbrella of "responsible AI" to gather standards for the values and concepts that should drive the development of AI (Doorn, 2021; Pérez-Beltrán et al., 2024). It is not new in the field, but its use and effects today are unparalleled. As a result, people now have strong opinions regarding artificial intelligence (AI). Some portray autonomous systems "taking control" in a dystopian manner, while others believe AI may solve many of the societal problems of our day, one of which is water management (Crevier, 1993; Russell, 2022). AI simulates cognitive functions in the human brain, such as learning and problem-solving. It is a collection of technologies, computational methods, and techniques that allow machines to perceive, understand, act, and learn with human-like intelligence (Malviya and Jaspal, 2021). AI scientists and researchers are hoping to have a better understanding of intelligence by creating computer programs that perform symbolic inference or reasoning. The distribution, monitoring, and management of water resources could be completely transformed by AI-driven

DOI: 10.1201/9781003600718-28

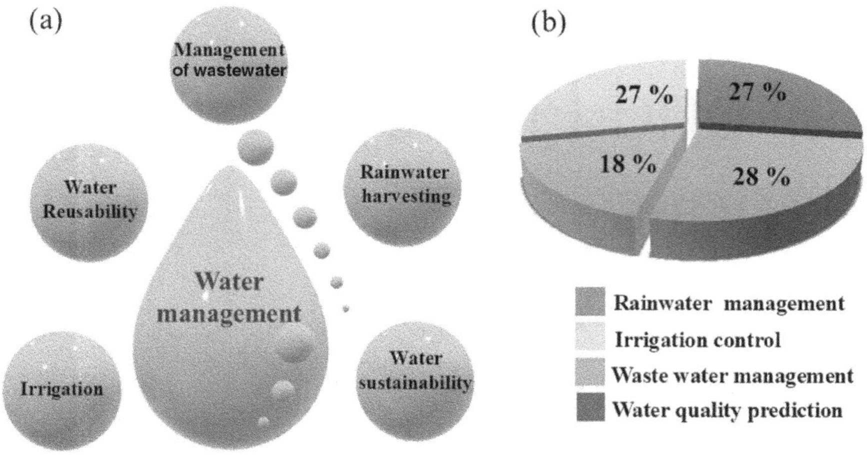

FIGURE 19.1 (a) Water management and its parameters and (b) percentage distribution of various applications analysed for water management techniques.

water management systems. These systems can forecast the trends of water use, manage water distribution, and increase efficiency by utilizing big data and sophisticated algorithms (Kamyab et al., 2023).

AI-powered water management solutions are revolutionizing resource allocation, monitoring, and optimization, ensuring sustainable usage and improving efficiency. Utilizing machine learning (ML) and Internet of Things (IoT), these systems address water scarcity, streamline distribution, and ensure sustainable use, thereby enhancing water resource management (Kamienski et al., 2019; Krishnan et al., 2022). However, there are a number of operational, privacy, and ethical issues that come with their deployment. These issues or problems include bias in data and inequity, accountability and transparency, data privacy, security and control, serious environmental impact and concerns, and serious social and cultural conflicts (Zhao et al., 2020; Doorn, 2021; Singh et al., 2024). To guarantee that AI systems in water management are fair, open, and safe while upholding people's rights, these issues must be resolved. This chapter examines the difficulties presented by AI-powered water management systems, particularly with respect to ethical and privacy challenges, and looks at possible ways to successfully deal with them. Moreover, it also explores the advantages and limitations of AI in water management.

19.2 ADVANTAGES AND LIMITATIONS OF AI IN WATER MANAGEMENT

The responsible use of AI and predictive analytics in water management can enhance environmental sustainability and boost the economy. These proactive methods ensure better resource utilization, reduce operational costs, and prevent expensive repairs. Thus, saving money is yet another significant advantage. Furthermore, the integration of IoT with AI and ML has enhanced the sufficiency of systems and optimized the use and distribution of water. One of the major benefits of AI techniques is the identification of leakages in real time. These methods use predictive algorithms to improve infrastructure and use sensors to smartly detect and fix leakages, leading to water conservation. By ensuring precise water allocation, AI methods help to reduce frequent use and needless expenses (Krishnan et al., 2022; Pérez-Beltrán et al., 2024). Moreover, AI methods play a significant role in flood management by predicting flood risks smartly by processing huge volumes of hydrological and meteorological data. Thus, these cutting-edge technologies provide early warnings, ensure timely rescue, and minimize property damage and loss of life (Krishnan et al., 2022). These systems also improve irrigation efficiency through real-time data evaluation by sensors and smart devices. Modern

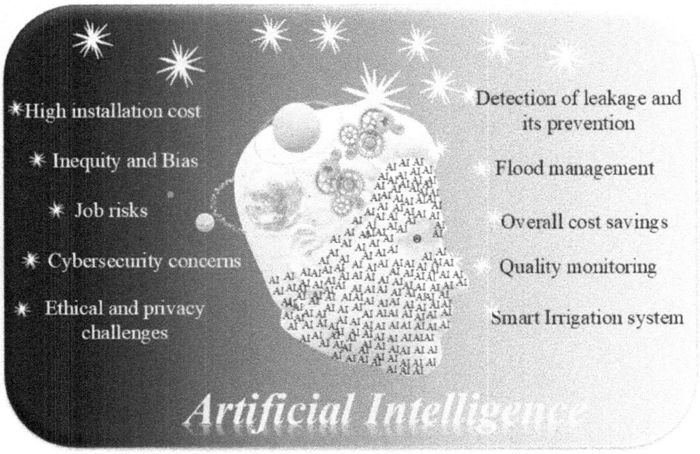

FIGURE 19.2 Some key advantages (right side) and disadvantages of AI in water management (left side).

irrigation systems distribute water according to the crop's requirement, weather predictions, and soil type, which revolutionize agriculture. These methods save water and energy with a higher yield of crops (Krishnan et al., 2022). Moreover, AI-driven devices are capable of analysing various water quality parameters such as pH, turbidity, COD, and BOD, and also, they detect pollution levels and improve water quality monitoring (Wang et al., 2023; Alprol et al., 2024). Numerous single or hybrid models such as support vector machine (SVR), linear regression (LR), random forest (RF), artificial neural network (ANN), deep neural network (DNN), convolutional neural network (CNN), adaptive network-based fuzzy inference system (ANFIS), radial basis function–artificial neural network (RBF-ANN), long short-term memory (LSTM), multi-agent reinforcement learning (MARL), and vision-based deep learning neural network (VIDLNN) are being used in water management, treatment, and quality monitoring (Malviya and Jaspal, 2021; Krishnan et al., 2022; Lowe et al., 2022; Pérez-Beltrán et al., 2024). By enabling quick pollution monitoring, these technologies promise clean drinking water and save ecosystems. Figure 19.2 shows some key advantages and limitations of AI technology in water management. Regardless of being creative and effective, AI-driven water management systems have a number of shortcomings. These systems are often inaccessible to poorly funded areas due to high installation and maintenance costs, complicating already existing disparities in water availability. As AI-driven systems automate jobs that were traditionally performed by people, there is a risk of job loss and economic disruption, particularly for workers in the water management industry. Furthermore, if the data utilized to train these systems are skewed or incomplete, algorithms may unintentionally enhance biases, which could result in an uneven distribution of water resources. For the same reason, cybersecurity issues are very real because these systems are prone to hacking or data breaches, which could jeopardize sensitive data or vital infrastructure. Their adoption is made more difficult by ethical and privacy issues, since the gathering and use of sizable datasets—which frequently contain geographical or personal data—raises concerns about data ownership and consent (Doorn, 2021; Haider et al., 2024; Singh et al., 2024).

19.3 ETHICAL AND PRIVACY CHALLENGES AND SOLUTIONS

Human civilization has undergone a new revolution thanks to AI-driven water management, but every revolution has its own set of privacy and security issues. There are many challenges involved in adopting AI-powered water management, the major ones being bias in data, inequity, accountability, transparency, data privacy, cybersecurity, environmental impacts, social and cultural conflicts, and guaranteeing fair

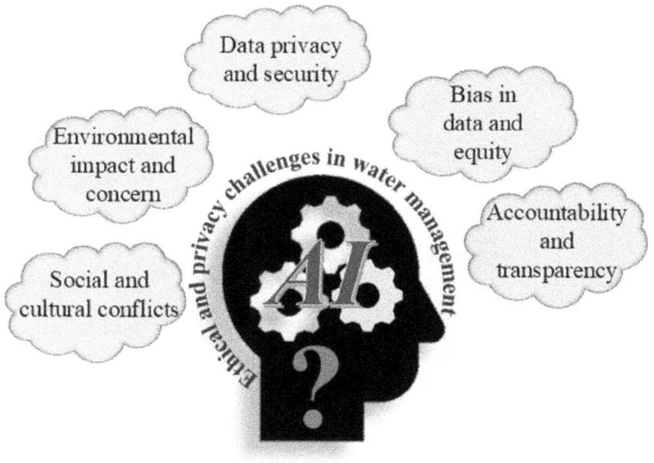

FIGURE 19.3 Ethical and privacy challenges in AI-driven water management.

access (Figure 19.3) (Zhao et al., 2020; Doorn, 2021; Egbemhenghe et al., 2023; Singh et al., 2024). This section discusses possible solutions to ethical and privacy issues, addresses previously covered issues, and provides possible practical solutions.

19.3.1 Data Privacy and Security

19.3.1.1 Data privacy: challenges, consequences, and solutions

AI systems often collect detailed information about water consumption patterns, household activities, and geographical locations to optimize resource allocation and ensure efficient usage. Sometimes these data may disclose sensitive information about the community or individuals, leading to fear about illegal access and misuse of data. Such data might be exploited for commercial purposes or used in ways that violate someone's privacy rights (Mühlhoff, 2023). Another major issue in AI-powered water management systems is how data are collected, compiled, shared, and analysed. AI-powered water management systems often involve many stakeholders from government agencies and private firms to third-party service providers, where data may be shared without consent and privacy violations may go unchecked due to inadequate regulation or oversight. For example, habits of an individual can be predicted from water consumption patterns and then misused, raising ethical concerns about surveillance and profiling (Egbemhenghe et al., 2023). If data are collected for water management purposes and someone reuses them for unrelated applications without the user's consent, then there is a strong need for regulatory guidelines to manage privacy. However, regulatory frameworks for managing privacy in AI applications are frequently incomplete or inconsistent (Folorunso et al., 2024). This gap leaves users in the lurch to further exploitation, as there is no clear legal inventory in case of invasions of privacy. Therefore, to resolve these problems, it is essential to use strong data security strategies such as encryption, anonymization, and user consent mechanisms. In this context, the promotion of international teamwork may help develop comprehensive and transparent security policies and regulations to mitigate risks regarding privacy violations, leading to the ethical use of AI in water management (Gunasekaran and Boopathi, 2023).

19.3.1.2 Data security: challenges, consequences, and solutions

Many security risks also arise due to real-time data processing, cloud infrastructure, and network systems in AI-driven water management systems. For example, the most serious fear is vulnerability to cyberattacks. Hackers may use ransomware, malware, or denial-of-service (DoS) attacks against water management, disrupting essential water supply or distribution (Aslam et al., 2023; Gulzar and Mustafa, 2024). These types of breaches pose negative effects on public health and safety, particularly for the water distributed by AI systems (Nti et al., 2023). AI algorithms generally used are reliable and accurate for forecasting the demand and quality of water. Therefore, maintaining the integrity of the data is another critical issue. International manipulation of the data by malicious adversaries, or sometimes by mistake non-intentional manipulation of data, may create very harmful conditions such as inaccurate predictions or failure to detect dangerous contaminants in water. Furthermore, the integration of IoT, sensors, and actuators in AI increases the vulnerability of the hardware to be exploited by hackers and take advantage of to weaken the whole water management system (Qiu et al., 2019; Zanfei et al., 2024). Another specific challenge in AI systems is adversarial attacks due to feeding malicious or noisy data to AI models so that they produce bad results. For example, an adversarial attack might cause an AI system to provide too much water to one area, which creates shortages in other areas. Insider threats are also a concern, which occur when someone with permissions uses their privileges incorrectly (Qiu et al., 2019). To deal with these challenges related to securities, there is a need to have a multi-layered strategy of security, such as the strongest cybersecurity infrastructure, secure authentication, encryption of transmission of data, and regular software updates. To prevent suspicious activity, a firewall and intrusion detection systems (IDSs) should be installed to monitor (Hamon et al., 2022; Mustafa et al., 2024). Strong data validation procedures and blockchain technologies produce tamper-proof transactions, ensuring reliable AI operation. Resilient AI models should be developed against adversarial attacks by using adversarial training and anomaly detection methods. To identify system vulnerabilities, regular penetration testing and security audits may be useful and essential. Regular penetration testing and security audits of entire AI systems, from cloud infrastructure to IoT devices, can be useful to identify system vulnerabilities. Furthermore, stakeholders, such as governments, local community, private firms, and researchers, can work with each other to develop comprehensive security policies (Hamon et al., 2022).

19.3.2 Bias in Data and Equity

19.3.2.1 Bias in data: challenges, consequences, and solutions

Biases in data in AI-based models may lead to poor water management in deprived areas, leading to injustices and supporting already existing inequalities. Socioeconomic status and geographical as well as historical diversities are the main sources that influence the result of AI systems and create biases in water distribution (Benjamin, 2019; Lowe et al., 2022; Alahi et al., 2023). For example, suppose an AI model is trained by using historical data. And if an area that has historically limited access to water is not well represented in the training dataset, then there is a chance of historical injustice distribution (Benjamin, 2019). In such cases, AI systems make unfair distribution of water supply and may ignore or place less or more importance on water scarcity in some areas. For example, AI system may prioritize cities over rural areas because rural water areas lack inclusive data compared to cities (Alahi et al., 2023). These types of results create worse inequality situations and marginalize the deprived communities (Sun and Scanlon, 2019). Therefore, during the implementation of AI in water management, it is essential to ensure the concerns about biases in data. To solve these problems, diverse data collection from entire regions including deprived areas, towns, and rural areas is important and should be ensured by data collection centres. In this context, synthetic data may also be helpful, which is created by simulations. Furthermore, data collection centres may bias the data in favour of rich communities or societies and exclude less wealthy areas. Because wealthier people have access to water management technologies easily, socioeconomic status, such as income inequality, also influences the results due to biases (Lowe et al., 2022). Therefore, to recognize biased data points, regular checking of AI models is essential so that the model can be made more accurate and reliable by correcting the biased data points. Additional transparency constraints and data reweighting are the most important techniques in ML algorithms for these purposes (Egbemhenghe et al., 2023).

19.3.2.2 Equity in water management: challenges, consequences, and solutions

Unfair access to technologies, cost-effectiveness, and power mobility are the factors that create equity challenges in AI-driven water distribution methods (Lutz, 2019; Mohamed et al., 2020). Some countries, especially developing countries, face challenges in accessing the technology because of restrictions, and therefore, some communities are unable to take advantage of the AI-driven water management facility (Lutz, 2019). Large infrastructural and technological funds can make it difficult for low-income communities to access AI systems. Power dynamics further complicate the matter because they can favour rich communities over poor ones (Mohamed et al., 2020; Mahmoud et al., 2023). These issues also contribute to unequal distribution of water and hinder the pervasive acceptance of AI-driven water management facilities. Biased water distribution methods can cause water shortages and social conflict in weak communities, risking their well-being and public health. Furthermore, these acts can lead to protests, disagreements, and social instability. This way, effective water distribution becomes more difficult to implement, and society may become more destabilized (Mahmoud et al., 2023; Cherniavska et al., 2024). Therefore, equal distribution of water to all irrespective of geographical locations and socioeconomic or political status is essential. In this context, to promote equity in water management, comprehensive decision-making should be ensured and targeted interventions to weak communities should be prioritized with affordable technology. Different stakeholders can train AI models with more comprehensive and relevant data. It is important to include the local population in decision-making processes so that all residents have confidence that the system will meet their demand in relation to water needs. Governments and organizations should make investments to lower the cost and increase the accessibility of AI-driven water management solutions in order to advance equity. This involves providing financial assistance to poor areas so that they can adopt this technology (Fu et al., 2023; Santos et al., 2023).

19.3.3 Accountability and Transparency: Challenges, Consequences, and Solutions

The duty of people, groups, and governments to inform stakeholders of their choices and actions is known as accountability. It includes procedures that organizations use to defend their behaviour and accept accountability for their deeds. Efficient accountability guarantees the responsible use of authority and the reduction of abuses. Transparency is the availability of information about an entity's decisions, actions, and policies. By enabling stakeholders to comprehend and examine procedures, corruption and inefficiency can be avoided. Transparency promotes public trust and well-informed decision-making (Bovens, 2007; Agrawal, 2024). Figure 19.4 shows the major challenges towards accountability and transparency. Each issue and its solutions are summarized in six subsections below.

19.3.3.1 Corruption and violation of ethics

Accountability and transparency are severely hampered by corruption. Government efficacy and trust are weakened when officials engage in fraud, bribery, or resource theft. The absence of strict anti-corruption measures erodes accountability systems by facilitating impunity. Laws protecting whistleblowers can promote the disclosure of unethical activity. Therefore, strong anti-corruption measures, such as stringent legislation, impartial investigative agencies, and severe sanctions for infractions, should be put in place by governments and organizations (Rose-Ackerman, 2002).

19.3.3.2 Absence of legal frameworks

Strong legislative structures to impose accountability and transparency are lacking in many nations and organizations. Poor oversight and little enforcement of current laws result from the absence of, or weak,

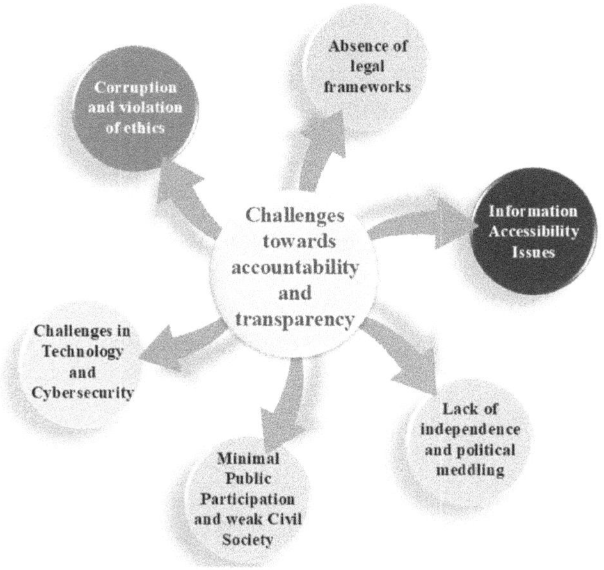

FIGURE 19.4 Accountability and transparency challenges in AI-driven water management.

legal frameworks. Sometimes, political meddling prevents laws from being implemented efficiently. Therefore, legal reforms should include open government policies and independent oversight organizations in order to strengthen accountability and openness. Political meddling must not be allowed in organizations like audit commissions, anti-corruption organizations, or ombudsman offices. Strong anti-corruption measures should be implemented by governments and organizations, including strict laws, unbiased investigative agencies, and harsh penalties for violations. Whistleblower protection laws can encourage the revelation of unethical behaviour (Rose-Ackerman, 2002; Krishnan et al., 2022).

19.3.3.3 Information accessibility issues

Transparency is severely affected due to information accessibility issues. Governments and corporate institutions often hide data in the name of security. This step further strengthens distrust and inefficiency. Information sharing is further hampered in many developing nations by weak infrastructure and digital barriers. Therefore, enacting and upholding freedom of information (FOI) laws is necessary to provide public access to data. Reports, budgets, and policy documents should be made available online by governments and organizations to encourage examination. Information shortages can be filled by digital platforms and open data initiatives (Egbemhenghe et al., 2023).

19.3.3.4 Lack of independence and political meddling

Biased decisions, less accountability, and data manipulation might result from political meddling and a lack of autonomy in AI-driven water management. Institutions cannot operate efficiently or fairly when they are dominated by political interests. Therefore, oversight organizations must be financially and operationally independent in order to fend off political meddling. These bodies should only be appointed on the basis of merit, free from political influence. External evaluations can be provided by international organizations like Transparency International (Kuziemski and Misuraca, 2020).

19.3.3.5 Minimal public participation and weak civil society

Transparency requires both a robust civil society and engaged public involvement. But in many areas, people are deterred from demanding responsibility by a lack of civic engagement and a fear of reprisals.

Suppression of whistleblowers and free press makes this problem worse. For this, governments and organizations should encourage citizens to participate in decision-making processes and educate them about their rights in order to foster civic engagement. For accountability and openness, press freedom must be strengthened, and activists must be protected (Chen, 2024).

19.3.3.6 Challenges in technology and cybersecurity

Although electronic data management and digital governance have increased transparency, there are risks associated with them as well. Data breaches, disinformation, and cybersecurity risks might harm transparency initiatives. Digital illiteracy also restricts the public's ability to participate in transparency projects. Transparency-enhancing digital tools that are safe and easy to use should be funded by governments and organizations. Financial transactions can be tracked, and corruption can be decreased with blockchain technology. To stop data breaches, cybersecurity measures should also be strengthened (Veluru, 2024).

19.3.4 Environmental Impact: Challenges, Consequences, and Solutions

Optimizing water distribution, cutting waste, and increasing efficiency are all made possible by AI-driven water management. But there are drawbacks as well, like carbon impact and climate change (Chen et al., 2023). Further, the key environmental issues in the use of AI in water management, including energy use, electronic waste (e-waste), and ecological consequences, etc., are shown in Figure 19.5 and are summarized in four subsections with possible solutions.

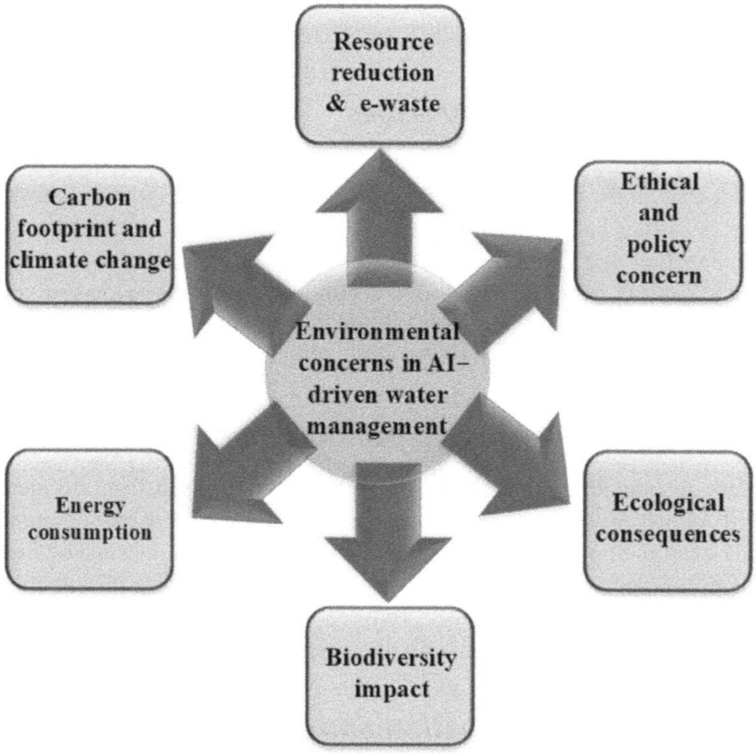

FIGURE 19.5 Major environmental concerns in AI-driven water management.

19.3.4.1 Energy consumption, carbon footprint, and climate change

AI-powered water management systems forecast demand, track quality, and optimally allocate resources. However, the high energy use and environmental impact of these systems are serious problems. AI models require huge energy for data processing, training, and real-time decisions. The sustainability of AI-driven solutions is limited by high energy consumption, which also raises operating costs and has an adverse environmental impact. The dependence on energy-intensive data centres for AI computations is a major problem, offsetting potential environmental benefits in areas like water management due to their greenhouse gases and carbon emissions. Real-time monitoring and cloud-based analytics, as well as IoT, require data transmission, which further increases energy consumption and carbon emissions (Acquaye et al., 2017; Kelechi et al., 2020; Egbemhenghe et al., 2023; Arya et al., 2024). This problem is further magnified by limited access to renewable energy, mainly in rural areas where water management by AI might be most beneficial. Therefore, making any AI system that is energy efficient is critical and challenging. Installation of energy-efficient equipment and processors such as specialist AI processors can help reduce energy consumption. The use of such renewable energy sources like solar or wind enhances the sustainability of AI systems. Additionally, varying the computation intensity according to demand may also reduce energy utilization in AI-driven systems. Sustainable AI-driven water management is further supported by improving the legislative framework and financial incentives for green AI projects (Kelechi et al., 2020; Hua et al., 2023; Ukoba et al., 2024). AI-driven water management can become more efficient, effective, and sustainable by using these discussed strategies. AI-driven water management can also become more efficient and sustainable by reducing carbon footprint and mitigating the effects of climate change.

19.3.4.2 Natural resources reduction and electronic waste (e-waste)

One of the major significant issues is "e-waste" or electronic garbage. For AI-based water management, sensors, IoT devices, and computer hardware are required, as all have limited lifespans. When they become outdated, their parts are improperly disposed, which releases harmful materials like lead and mercury into the environment, thus creating hazardous waste. Therefore, recycling and sustainable disposal are crucial challenges because rapid technological advancement also accelerates the production of e-waste. The development and implementation of AI technology are done by semiconductors and sensors, which can be obtained from rare earth elements as a significant raw material. The extraction of these materials leads to the depletion of natural resources and contributes to environmental damage—the two major consequences. The energy-intensive processes used in the production of AI-driven devices and infrastructure further worsen ecological issues (Nowakowski et al., 2020; Adnan et al., 2024). These obstacles could be overcome by a multifaceted strategy. The depleting resources can be lessened by using sustainable material procurement practices, such as using recycled and ethically mined minerals. In AI-driven water management systems, manufacturers must consider certain properties—such as longevity, modularity, and repairability—while designing devices with an increased lifespan. Strong e-waste recycling systems and advocacy for the circular economy facilitate the reuse and recovery of valuable materials. The negative effects on the environment might also be reduced by the policies that promote appropriate disposal and investments in energy-efficient or biodegradable substitutes. A crucial role in minimizing resource consumption is also played by advancements in green computing, such as energy-efficient processors and AI algorithms (Sarkhoshkalat et al., 2024). Thus, greater sustainability can be achieved in AI-driven water management by combining these methods, striking a balance between environmental responsibility and technical advancement.

19.3.4.3 Ecological impact and biodiversity

AI systems result in carbon emissions due to the huge energy consumption for data processing, training, and real-time decisions. These conditions lead to even worse climate change, which has harmful effects on ecosystems and loss of biodiversity (Reynolds et al., 2024; Zehouani et al., 2024). This problem is made worse by AI-driven solutions' dependence on non-renewable energy sources, which reduces the environmental friendliness of water management systems. Habitat disturbance brought on by the exploitation of raw materials required for AI technology, such as rare earth metals used in sensors and computer components, is another ecological effect. Mining these resources frequently results in soil erosion, water pollution, and deforestation, endangering biodiversity. Furthermore, AI-driven water management depends on massive infrastructure, such as data centres, which need water and land resources to cool, which may have an impact on nearby ecosystems and water supplies (Richards et al., 2023; Alprol et al., 2024; Reynolds et al., 2024). Poor disposal of e-waste (containing dangerous substances including lead, mercury, and cadmium) can contaminate water and soil, endangering human health and wildlife. This issue is made worse by the quick obsolescence of AI-related technologies, which raises the environmental impact of water management systems (Nowakowski et al., 2020). The disturbance of natural water ecosystems as a result of AI-controlled water extraction and distribution is a significant concern. If these systems give preference to value efficiency over ecological concerns, then they unknowingly reduce the amount of water available to rivers, wetlands, and aquatic ecosystems. Therefore, the population of some species such as fish, amphibians, and aquatic plants declines as their habitats depend on stable water levels, and this will unfortunately lead to the loss of biodiversity. This issue may be further exacerbated by the development of AI-based infrastructure, monitoring centres, and data stations. Such developments may cause habitat loss, deforestation, and alteration of land, all of which adversely affect wildlife. Thus, extreme dependence on AI systems could lead to over-reliance on data-driven decision-making that distracts the awareness towards regional ecosystems (Bousdekis et al., 2021; Zehouani et al., 2024). Therefore, it is necessary to use ecological data to train AI models for water management while ensuring that biodiversity is not sacrificed. A balanced approach is essential between AI technology and ecosystem conservation, with environmental impact assessments and sustainable infrastructure development. In order to reduce habitat loss, raw materials should be secured sustainably. The adverse effect of resource extraction on the environment can be mitigated through promoting the use of recyclable materials in AI technology with ethical mining practices. Governments and corporate entities have a responsibility to create laws that favour ecologically friendly materials. Hazardous trash can be prevented from entering landfills by incorporating take-back programmes and providing incentives for proper e-waste disposal. Moreover, ecological destruction can be reduced by exploring energy-efficient biodegradable alternatives to replace electrical parts. Thus, the above-discussed solutions can ensure a balance between technology innovation and environmental sustainability, and AI-driven management can achieve its goal with minimum ecological effects.

19.3.4.4 Ethical and policy concerns on environmental sustainability

To give preference to economic efficiency compared to ecological conservation is the main ethical concern in environmental sustainability. AI models may maximize the use of water for industries at the cost of environmental sustainability. This may lead to habitat degradation and biodiversity loss (Zehouani et al., 2024). The lack of uniform rules to govern AI in water management creates policy-related issues. The absence of well-defined guidelines on AI-driven resource distribution, monitoring of environmental issues, and data security leads to inconsistent implementation and probable environmental hazards (Ray, 2023). Therefore, strict policies should be implemented by policymakers that ensure that AI-driven water management systems give preference to ecological conservation with environmental sustainability. Further, geographical differences can also be eliminated if regulations include transparent access to AI technology that will enable morally and environmentally responsible water management techniques.

19.3.5 Social and Cultural Conflicts: Challenges, Consequences, and Solutions

AI-driven water management can enhance water efficiency and sustainability but, at the same time, may also bring social and cultural conflicts. Frequently, negligence regarding traditional water conservation methods that are deeply embedded in cultural traditions and local knowledge may create a risk of loss of cultural identity and autonomy. Some communities may be alienated if water is allocated without considering the historical and spiritual relationships with water resources (Turyasingura et al., 2024). Access imbalance to AI-based water solutions is another social challenge. For example, richer cities and companies get priority in water distribution due to their ability to afford high-tech AI infrastructure, while poor towns are left with few resources. This discrepancy fuels hostility among different social groups and promotes already existing inequalities. Generally, AI-driven systems impose unitary water management rules on every culture irrespective of local norms and traditions, which often creates cultural conflicts. In some regions, water consumption is linked to religious values, communal ceremonies, or agricultural traditions; therefore, technology-based regulation and community values may clash if AI models overlook such cultural factors. For example, automated water supply for commercial purposes may collide with religious rights or regional agricultural needs, causing a conflict situation between the local government and society (Goralski and Tan, 2020). Technologies and decision-makers need to take generous and collaborative initiatives to address these problems and interact with local communities to incorporate traditional water management expertise into AI algorithms. Digital literacy and capacity development initiatives can facilitate communities to actively participate in AI-driven water management systems. The conflict between society and technological development can be mitigated by using culturally relevant AI systems which are critical for efficient water resource delivery while respecting local customs and promoting community trust. These AI-sensitive models encourage equity by ensuring equal access to service for all, regardless of their socioeconomic status. These models incorporate underprivileged communities into AI-driven water management solutions and bridge the gap between machinery and humanity to achieve the balanced, sustainable, and ethical solution. Governments and lawmakers should also impose strict laws regarding equal access to the benefits of AI-driven water management technologies so that the marginalized communities can take advantage of innovative advanced technology.

19.4 CONCLUSION

AI with smart IoT-based emerging technologies for water management provides support to global water resources conservation. AI-powered methods have effectively optimized, predicted, and automated the most critical applications in water-based operations, such as water quality monitoring, water treatment, and water distribution across societies, communities, industries, and agriculture. These technologies improve the distribution of water resources by analysing a large amount of real-time data from water resources and users. However, due to improper data handling, sensitive personal information is being misused, which results in security vulnerabilities and loss of public trust. Therefore, concerns about data privacy, security, biases, accountability, transparency, etc., as well as environmental ethics and regulatory framework, have become more significant issues. These challenges may affect the results of AI-based methods and raise ethical concerns about long-term equity and sustainability. This chapter discussed the above challenges and solutions regarding the implementation of AI in water management. In this context, data encryption techniques, strict data protection laws, and clear data sharing guidelines will be helpful. To ensure the health of aquatic ecosystems, environmental ethics is also extremely important, and therefore, ecological impact assessment must be conducted prior to the implementation of AI technology. Finally, data management, public/legal opinion, reproducibility, and transparency in research are all important issues that need to be overcome in order to advance these intelligent applications.

REFERENCES

Acquaye, A., Feng, K., Oppon, E., Salhi, S., Ibn-Mohammed, T., Genovese, A., et al. (2017). Measuring the environmental sustainability performance of global supply chains: A multi-regional input-output analysis for carbon, sulphur oxide and water footprints. *Journal of Environmental Management*, 187, 571–585.

Adnan, M., Xiao, B., Ali, M.U., Bibi, S., Yu, H., Xiao, P., et al. (2024). Human inventions and its environmental challenges, especially artificial intelligence: New challenges require new thinking. *Environmental Challenges*, 16, 100976.

Agrawal, G. (2024). Accountability, trust, and transparency in AI systems from the perspective of public policy: Elevating ethical standards. In Kumar Shalender, K., and Singla, B. (Eds.) *AI Healthcare Applications and Security, Ethical, and Legal Considerations* (pp. 148–162). IGI Global, Hershey, PA.

Alahi, M.E.E., Sukkuea, A., Tina, F.W., Nag, A., Kurdthongmee, W., Suwannarat, K., et al. (2023). Integration of IoT-enabled technologies and artificial intelligence (AI) for smart city scenario: Recent advancements and future trends. *Sensors*, 23(11), 5206.

Alprol, A.E., Mansour, A.T., Ibrahim, M.E.E.-D., and Ashour, M. (2024). Artificial intelligence technologies revolutionizing wastewater treatment: Current trends and future prospective. *Water*, 16(2), 314.

Arya, A., Bachheti, A., Bachheti, R.K., Singh, M., and Chandel, A.K. (2024). Role of artificial intelligence in minimizing carbon footprint: A systematic review of recent insights. In Chandel, A. K. (Ed.) *Biorefinery and Industry 4.0: Empowering Sustainability* (pp. 365–386). Springer, Heidelberg.

Aslam, M.M., Tufail, A., Kim, K.-H., Apong, R.A.A.H.M., and Raza, M.T. (2023). A comprehensive study on cyber attacks in communication networks in water purification and distribution plants: Challenges, vulnerabilities, and future prospects. *Sensors*, 23(18), 7999.

Benjamin, R. (2019). *Race after Technology: Abolitionist Tools for the New Jim Code.* Cambridge and Medford: Polity Press.

Bousdekis, A., Lepenioti, K., Apostolou, D., and Mentzas, G. (2021). A review of data-driven decision-making methods for Industry 4.0 maintenance applications. *Electronics*, 10(7), 828.

Bovens, M. (2007). Analysing and assessing accountability: A conceptual framework 1. *European Law Journal*, 13(4), 447–468.

Chen, L., Chen, Z., Zhang, Y., Liu, Y., Osman, A.I., Farghali, M., et al. (2023). Artificial intelligence-based solutions for climate change: A review. *Environmental Chemistry Letters*, 21(5), 2525–2557.

Chen, S. (2024). Artificial intelligence in democracy: Unraveling the influence of social bots in brexit through cybernetics. *Transactions on Social Science, Education and Humanities Research*, 6, 324–329.

Cherniavska, T., Tanklevska, N., and Cherniavskyi, B. (2024). A decision-making system for managing the remediation of water resources in the Kherson region: Agent-oriented modeling in the context of post-war economic recovery. In Cherniavska, T. (Ed.) *Transformations of National Economies under Conditions of Instability* (p. 223). Scientific Route, Tallinn.

Crevier, D. (1993). *AI: The Tumultuous History of the Search for Artificial Intelligence.* Basic Book, New York.

Doorn, N. (2021). Artificial intelligence in the water domain: Opportunities for responsible use. *Science of the Total Environment*, 755, 142561.

Egbemhenghe, A.U., Ojeyemi, T., Iwuozor, K.O., Emenike, E.C., Ogunsanya, T.I., Anidiobi, S.U., et al. (2023). Revolutionizing water treatment, conservation, and management: Harnessing the power of AI-driven ChatGPT solutions. *Environmental Challenges*, 13, 100782.

Folorunso, A., Olanipekun, K., Adewumi, T., and Samuel, B. (2024). A policy framework on AI usage in developing countries and its impact. *Global Journal of Engineering and Technology Advances*, 21(01), 154–166.

Fu, G., Sun, S., Hoang, L., Yuan, Z., and Butler, D. (2023). Artificial intelligence underpins urban water infrastructure of the future: A holistic perspective. *Cambridge Prisms: Water*, 1, e14.

Goralski, M.A., and Tan, T.K. (2020). Artificial intelligence and sustainable development. *The International Journal of Management Education*, 18(1), 100330.

Gulzar, Q., and Mustafa, K. (2024). An analytical survey of cyber-physical systems in water treatment and distribution: Security challenges, intrusion detection, and future directions. *Security and Privacy,* 7(6), e440.

Gunasekaran, K., and Boopathi, S. (2023). Artificial intelligence in water treatments and water resource assessments. In Shikuku, V. (Ed.) *Artificial Intelligence Applications in Water Treatment and Water Resource Management* (pp. 71–98). IGI Global, Hershey, PA.

Haider, S., Rashid, M., Tariq, M.A.U.R., and Nadeem, A. (2024). The role of artificial intelligence (AI) and Chatgpt in water resources, including its potential benefits and associated challenges. *Discover Water*, 4(1), 113.

Hamon, R., Junklewitz, H., Sanchez, I., Malgieri, G., and De Hert, P. (2022). Bridging the gap between AI and explainability in the GDPR: Towards trustworthiness-by-design in automated decision-making. *IEEE Computational Intelligence Magazine*, 17(1), 72–85.

Hua, H., Li, Y., Wang, T., Dong, N., Li, W., and Cao, J. (2023). Edge computing with artificial intelligence: A machine learning perspective. *ACM Computing Surveys*, 55(9), 1–35.

Kamienski, C., Soininen, J.-P., Taumberger, M., Dantas, R., Toscano, A., Salmon Cinotti, T., et al. (2019). Smart water management platform: IoT-based precision irrigation for agriculture. *Sensors*, 19(2), 276.

Kamyab, H., Khademi, T., Chelliapan, S., SaberiKamarposhti, M., Rezania, S., Yusuf, M., et al. (2023). The latest innovative avenues for the utilization of artificial intelligence and big data analytics in water resource management. *Results in Engineering*, 20, 101566.

Kelechi, A.H., Alsharif, M.H., Bameyi, O.J., Ezra, P.J., Joseph, I.K., Atayero, A.-A., et al. (2020). Artificial intelligence: An energy efficiency tool for enhanced high performance computing. *Symmetry* 12(6), 1029.

Krishnan, S.R., Nallakaruppan, M., Chengoden, R., Koppu, S., Iyapparaja, M., Sadhasivam, J., et al. (2022). Smart water resource management using artificial intelligence: A review. *Sustainability*, 14(20), 13384.

Kuziemski, M., and Misuraca, G. (2020). AI governance in the public sector: Three tales from the frontiers of automated decision-making in democratic settings. *Telecommunications Policy*, 44(6), 101976.

Li, P., Yang, J., Islam, M.A., and Ren, S. (2023). Making ai less "thirsty": Uncovering and addressing the secret water footprint of ai models. *arXiv preprint arXiv:2304.03271.*

Lowe, M., Qin, R., and Mao, X. (2022). A review on machine learning, artificial intelligence, and smart technology in water treatment and monitoring. *Water*, 14(9), 1384.

Lutz, C. (2019). Digital inequalities in the age of artificial intelligence and big data. *Human Behavior and Emerging Technologies*, 1(2), 141–148.

Mahmoud, A.E.D., Fawzy, M., and Khan, N.A. (2023). *Artificial Intelligence and Modeling for Water Sustainability: Global Challenges.* CRC Press.

Malviya, A., and Jaspal, D. (2021). Artificial intelligence as an upcoming technology in wastewater treatment: A comprehensive review. *Environmental Technology Reviews*, 10(1), 177–187.

Mohamed, S., Png, M.-T., and Isaac, W. (2020). Decolonial AI: Decolonial theory as sociotechnical foresight in artificial intelligence. *Philosophy & Technology* 33, 659–684.

Mühlhoff, R. (2023). Predictive privacy: Collective data protection in the context of artificial intelligence and big data. *Big Data & Society*, 10(1), 20539517231166886.

Mustafa, R., Sarkar, N.I., Mohaghegh, M., and Pervez, S. (2024). A cross-layer secure and energy-efficient framework for the Internet of Things: A comprehensive survey. *Sensors*, 24(22), 7209.

Nowakowski, P., Szwarc, K., and Boryczka, U. (2020). Combining an artificial intelligence algorithm and a novel vehicle for sustainable e-waste collection. *Science of the Total Environment*, 730, 138726.

Nti, E.K., Cobbina, S.J., Attafuah, E.E., Senanu, L.D., Amenyeku, G., Gyan, M.A., et al. (2023). Water pollution control and revitalization using advanced technologies: Uncovering artificial intelligence options towards environmental health protection, sustainability and water security. *Heliyon*, 9(7), e18170.

Pérez-Beltrán, C., Robles, A., Rodriguez, N., Ortega-Gavilán, F., and Jiménez-Carvelo, A. (2024). Artificial intelligence and water quality: From drinking water to wastewater. *TrAC Trends in Analytical Chemistry*, 172, 117597.

Qiu, S., Liu, Q., Zhou, S., and Wu, C. (2019). Review of artificial intelligence adversarial attack and defense technologies. *Applied Sciences*, 9(5), 909.

Ray, P.P. (2023). Leveraging deep learning and language models in revolutionizing water resource management, research, and policy making: A case for ChatGPT. *ACS ES&T Water*, 3(8), 1984–1986.

Reynolds, S.A., Beery, S., Burgess, N., Burgman, M., Butchart, S.H., Cooke, S.J., et al. (2024). The potential for AI to revolutionize conservation: A horizon scan. *Trends in Ecology & Evolution*, 40(2), 191–207.

Richards, C.E., Tzachor, A., Avin, S., and Fenner, R. (2023). Rewards, risks and responsible deployment of artificial intelligence in water systems. *Nature Water*, 1(5), 422–432.

Rose-Ackerman, S. (2002). "Grand" corruption and the ethics of global business. *Journal of Banking & Finance*, 26(9), 1889–1918.

Russell, S. (2022). Human-Compatible Artificial Intelligence. *Human-like Machine Intelligence*, 1, 3–22. https://doi.org/10.1093/oso/9780198862536.003.0001

Santos, E., Carvalho, M., and Martins, S. (2023). Sustainable water management: Understanding the socioeconomic and cultural dimensions. *Sustainability*, 15(17), 13074.

Sarkhoshkalat, M.M., Afkham, A., Bonyadi Manesh, M., and Sarkhosh, M. (2024). Circular economy and the recycling of E-waste. In Kasinathan, K., Ladchumananandasivam, R., and Mohamed, S.B. (Eds.), *New Technologies for Energy Transition Based on Sustainable Development Goals: Factors Contributing to Global Warming* (pp. 319–354). Springer.

Singh, S., Kaur, R., and Agrawal, N.K. (2024). Ethical considerations in AI-driven wastewater treatment. In Garg, M.C. (Ed.), *The AI Cleanse: Transforming Wastewater Treatment Through Artificial Intelligence: Harnessing Data-Driven Solutions* (pp. 313–335). Springer.

Sun, A.Y., and Scanlon, B.R. (2019). How can Big Data and machine learning benefit environment and water management: A survey of methods, applications, and future directions. *Environmental Research Letters*, 14(7), 073001.

Turyasingura, B., Ayiga, N., Byamukama, W., Kayusi, F., Tumuhimbise, M., and Tumushabe, T. (2024). Application of artificial intelligence (AI) in environment and societal trends: Challenges and opportunities. *Babylonian Journal of Machine Learning*, 2024, 177–182.

Ukoba, K., Olatunji, K.O., Adeoye, E., Jen, T.-C., and Madyira, D.M. (2024). Optimizing renewable energy systems through artificial intelligence: Review and future prospects. *Energy & Environment*, 35(7). 0958305X241256293. https://doi.org/10.1177/0958305x241256293

Veluru, C.S. (2024). Responsible artificial intelligence on large scale data to prevent misuse, unethical challenges and security breaches. *Journal of Artificial Intelligence & Cloud Computing. SRC/JAICC-349.* 331(3), 2–6. http://dx.doi.org/10.47363/JAICC/2024(3)331

Wang, Y., Cheng, Y., Liu, H., Guo, Q., Dai, C., Zhao, M., et al. (2023). A review on applications of artificial intelligence in wastewater treatment. *Sustainability*, 15(18), 13557.

Zanfei, A., Lombardi, A., De Luca, A., and Menapace, A. (2024). AI-Forecast: An innovative and practical tool for short-term water demand forecasting. *Water Supply*, 24(4), 1352–1363.

Zehouani, N., Ababou, M., and Faquir, S. (2024). Harnessing artificial intelligence for environmental sustainability: A comprehensive review of its applications in biodiversity, energy, transportation, and water management. In *International Conference on Digital Technologies and Applications* (pp. 410–419), Benguerir, Morocco. Springer, Heidelberg.

Zhao, L., Dai, T., Qiao, Z., Sun, P., Hao, J., and Yang, Y. (2020). Application of artificial intelligence to wastewater treatment: A bibliometric analysis and systematic review of technology, economy, management, and wastewater reuse. *Process Safety and Environmental Protection*, 133, 169–182.

20 Environmental Impacts of Artificial Intelligence in Water Management

Surya Konappan, Cheran Easwaran, Sathya Priya Ramalingam, and Krishnan Ramanujam

20.1 INTRODUCTION

Artificial intelligence (AI) has become a transformative force in water management, providing innovative solutions to enhance efficiency, optimize resource distribution, and improve water quality monitoring (Ashoka et al., 2024). By utilizing machine learning algorithms, data analytics, and Internet of Things (IoT) devices, AI-driven systems can predict water demand, detect leaks, and promote the sustainable use of water resources (Nti et al., 2022). However, the environmental footprint of AI must not be overlooked. Integrating AI into water management necessitates significant computational power, leading to substantial energy consumption. High-performance computing systems used for training AI models and processing real-time data contribute to increased electricity demand, often relying on non-renewable energy sources (Akter, 2024). Furthermore, AI-driven systems depend on advanced digital infrastructure, which, if not managed sustainably, can result in electronic waste and environmental degradation. In addition to energy concerns, there are risks associated with an over-reliance on AI models. Incomplete or biased datasets can produce inaccurate predictions, potentially leading to ineffective water resource planning and management (Hernandez et al., 2024). Moreover, prioritizing AI-based solutions over traditional, nature-based conservation methods may disrupt sustainable water management practices. Issues like prolonged droughts, industrial wastewater pollution, and inefficient agricultural irrigation systems require a holistic approach that integrates AI advancements with ecological sustainability. To address these environmental challenges, adopting sustainable AI practices is crucial. Incorporating renewable energy sources, enhancing algorithm efficiency, and ensuring regular validation of AI models can help minimize the carbon footprint of AI-powered water management systems. Additionally, collaboration among policymakers, environmental scientists, and AI developers is essential to ensure that AI technologies support existing conservation efforts rather than replace them.

DOI: 10.1201/9781003600718-29

This chapter examines the environmental trade-offs of AI in water management, emphasizing both its potential benefits and ecological concerns. By confronting these challenges, we can pursue a balanced approach where AI serves as a tool for sustainable water management while mitigating its environmental impact.

20.2 APPLICATIONS OF AI AND ITS POSITIVE ENVIRONMENTAL IMPACTS ON WATER MANAGEMENT

Rapid advancements in AI have established it as a transformative force in water management, facilitating more efficient resource utilization, real-time monitoring, and predictive analytics (Nagpal et al., 2024). Major technology companies, such as Google, Microsoft, and Apple, are making significant investments in AI research, highlighting its growing importance across various sectors, including environmental sustainability. Water pollution, a critical global issue, arises from industrial activities, agricultural runoff, and domestic wastewater discharge. Contaminants from these sources pose serious threats to aquatic ecosystems and public health. AI-driven technologies are essential in addressing these challenges by improving wastewater management, optimizing treatment processes, and predicting pollution trends. Machine learning algorithms can analyze large datasets, identify pollution sources, and detect anomalies in water quality, enabling proactive interventions (Safeer et al., 2022).

AI-powered systems are also transforming air and water pollution control by enhancing emission treatment efficiency and automating monitoring processes. In water management, AI-driven predictive models forecast contamination events, optimize filtration systems, and improve overall water treatment efficiency (Lowe et al., 2022). Innovations such as neural networks for pollutant detection, AI automation in water treatment plants, and predictive models for algal bloom occurrences are revolutionizing water quality management. Additionally, the integration of IoT devices with AI enhances real-time water quality monitoring. Smart sensors deployed in water bodies provide continuous data streams, allowing AI models to dynamically analyze water composition. This integration facilitates early hazard detection, minimizes environmental risks, and supports better decision-making for sustainable water resource management (Kamyab et al., 2023). By leveraging AI, the water industry can significantly enhance efficiency, reduce waste, and promote long-term environmental sustainability (Table 20.1).

TABLE 20.1 AI-driven technologies in water management applications (Kumari et al., 2024)

APPLICATION AREA	*AI TECHNIQUES USED*	*KEY BENEFITS*	*EXAMPLE STUDIES*	*PROJECTS/CASE STUDIES*
Demand forecasting	Neural networks, regression analysis	Accurate demand prediction, optimized resource allocation	Singapore Smart Water Grid	Smart Water Grid Systems in Singapore
Water quality monitoring	Classification algorithms, anomaly detection	Real-time quality assessment, pollution prevention	California Water Quality Monitoring	AI-Based Water Monitoring in California
Irrigation management	Deep learning, reinforcement learning	Improved water conservation, enhanced crop yield	Indian PMKSY Program	Precision Irrigation in India
Flood prediction	Time series analysis, pattern recognition	Early warning systems, damage prevention	Japan Flood Prediction Systems	AI-Based Flood Forecasting in Japan
Leak detection	Image recognition, anomaly detection	Reduced water loss, predictive maintenance	Thames Water, UK	AI-Powered Leak Detection in the UK

20.2.1 Artificial Intelligence in Water Quality Monitoring

Water quality inspection is a vital aspect of environmental monitoring, focusing on the appropriate use of water for various purposes, including drinking, industrial applications, and ecosystem health maintenance. Assessments of water quality rely on a comprehensive set of criteria encompassing chemical, physical, and biological parameters, each offering valuable insights into the water's condition (Korostynska et al., 2012). Physical attributes, such as temperature, humidity, and color, are evaluated to determine the aesthetic and sensory qualities of water. Chemical parameters, including pH, dissolved oxygen, phosphorus, nitrogen, toxic metals, and organic pollutants, provide essential information about chemical balance and the presence of harmful substances (Omer, 2019). Biological parameters, such as microorganisms including viruses, bacteria, and phytoplankton, are crucial for assessing water's suitability for human use and its ecological impact (Korostynska et al., 2012). Traditional water quality monitoring methods have typically relied on manual processes that demand considerable human effort (Sani, 2023). These methods involve collecting water samples from various locations and sending them to laboratories for analysis. In the laboratory, techniques such as spectroscopy and microscopy are employed for chemical analysis and microorganism detection, respectively. While these methods have been standard for years, they have notable limitations. A significant drawback is the delay between sample collection and result analysis, which can postpone the identification of contamination incidents, potentially jeopardizing public health and ecosystems. Additionally, conventional methods often require expensive equipment and specialized expertise, limiting their accessibility and frequency, particularly in resource-constrained settings (Rainbow et al., 2020).

Furthermore, traditional methods often fail to provide a comprehensive view of water quality due to their irregular and static nature (Table 20.2). They cannot capture temporal variations in water quality

TABLE 20.2 Comparative chart on traditional and current monitoring techniques (Rajitha et al., 2024)

ASPECT	*TRADITIONAL MONITORING TECHNIQUES*	*CURRENT MONITORING TECHNIQUES*
Data collection method	Manual sampling and laboratory analysis, requiring human effort and time	Real-time sensor networks and IoT-enabled automated data collection
Sampling frequency	Periodic and infrequent sampling, leading to potential data gaps	Continuous, real-time, and high-frequency monitoring for comprehensive data
Parameter monitoring	Limited parameters (e.g., pH, turbidity) with minimal real-time insights	Multi-parameter monitoring (e.g., pH, dissolved oxygen, conductivity, temperature, contaminants)
Data processing and analysis	Manual data entry and analysis, prone to human errors and delays	Automated data aggregation, AI-driven analytics, and cloud-based reporting
Detection and alerting	Reactive approach, identifying contamination only after an issue occurs	Predictive modeling with AI for early detection and proactive intervention
Cost and resource requirements	Labor-intensive, high operational costs, and time-consuming processes	Cost-effective, requiring minimal manual intervention, and scalable solutions
Scalability and coverage	Limited spatial coverage with low-resolution data	Broad spatial coverage, remote monitoring, and high-resolution insights
Integration with treatment systems	Disconnected from real-time treatment processes, causing delays in response	Seamless integration with AI-enabled treatment systems for real-time adjustments
Adaptability to environmental changes	Poor adaptability to changing conditions, requiring manual recalibration	Dynamic, adaptive monitoring with automated responses to environmental variations

caused by seasonal changes, weather events, or sporadic pollution incidents, leading to data gaps that hinder effective water resource management (Younos and Heyer, 2015). These methods typically focus on specific areas, which may not accurately reflect the overall state of the aquatic environment, making it challenging to assess the health of entire ecosystems or large-scale water supply systems. Therefore, while traditional monitoring techniques have offered valuable insights, their inherent limitations underscore the need for more advanced and real-time monitoring systems to achieve a dynamic and holistic understanding of water quality.

20.2.2 Artificial Intelligence in Sewage Treatment

Artificial neural networks (ANNs), inspired by the human brain's neural system, are powerful tools for analyzing complex relationships and large datasets. These models excel at identifying intricate patterns in data that traditional methods may overlook. Specifically, backpropagation neural networks (BPNNs) have proven effective in predicting the water quality index (WQI), a key metric for assessing water quality. BPNN simplifies WQI calculations, enabling faster and more accurate predictions. By leveraging ANN's capabilities for rapid and precise data processing, researchers gain deeper insights into the factors influencing river water quality, ultimately enhancing environmental protection efforts (Frincu, 2025). AI is also revolutionizing sewage treatment by integrating real-time monitoring, predictive analytics, and automation to improve efficiency, reduce costs, and promote environmental sustainability. AI-driven systems employ IoT sensors to continuously monitor essential water parameters such as pH, chemical oxygen demand (COD), biochemical oxygen demand (BOD), and nutrient levels. These data are analyzed automatically to detect anomalies and provide real-time insights (Mustafa et al., 2021). Predictive maintenance powered by AI identifies potential issues before breakdowns occur, preventing equipment failures, reducing downtime, and lowering operational costs.

AI optimizes treatment processes by adjusting aeration, sludge removal, and chemical dosing in real time, which enhances energy efficiency and pollutant removal. It also supports biological treatment by monitoring microbial activity, preventing bacterial imbalances, and improving activated sludge processes. Smart sewer systems benefit from AI as well, enabling real-time decision-making to detect leaks, blockages, and overflow risks, thereby ensuring seamless wastewater management and reducing the need for manual intervention (Liu, 2024). By dynamically adapting to environmental conditions, AI-driven sewage treatment systems ensure compliance with regulatory standards, contribute to sustainable water management, and minimize environmental impact. Ultimately, AI enhances the efficiency and effectiveness of sewage treatment, making it a vital tool for improving water quality management and fostering environmental sustainability (Figure 20.1).

20.2.3 Artificial Intelligence in Leak Detection System

AI is transforming leak detection in water infrastructure by improving accuracy, efficiency, and proactive management. Traditional leak detection methods often depend on manual inspections and reactive maintenance, which can be time-consuming, costly, and error-prone (Abdelmageed et al., 2022). In contrast, AI-driven systems utilize machine learning algorithms to analyze extensive datasets—including pipe material, age, diameter, environmental conditions, and historical failure records—to predict leaks before they happen (Baird et al., 2019). This predictive capability allows water utilities to shift from a reactive to a proactive maintenance strategy, significantly reducing water loss, infrastructure damage, and operational costs. Additionally, AI-powered leak detection systems continuously enhance their performance as they collect more data, refining predictive accuracy and optimizing resource allocation (Mehmood et al., 2020). By integrating AI into water management systems, municipalities and utility

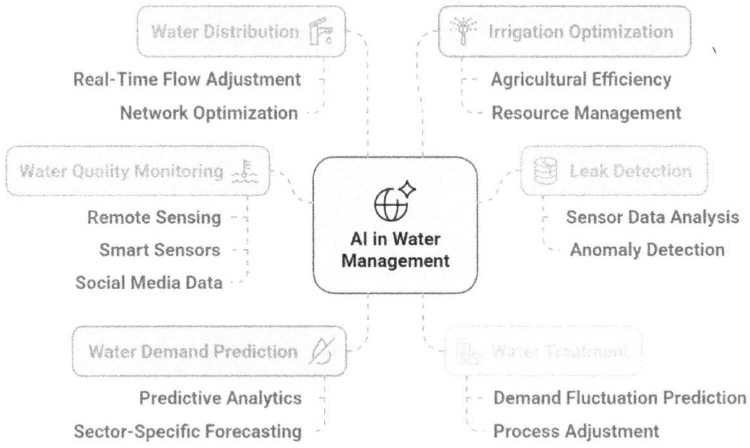

FIGURE 20.1 The various applications of AI in water management.

companies can improve the sustainability of water distribution networks, mitigate risks associated with pipeline failures, and ensure efficient water conservation. Aging water infrastructure poses serious risks, including water loss, road collapses, and increased maintenance costs. Traditional inspection methods are inefficient, necessitating AI-driven solutions for proactive monitoring. IoT-enabled sensor networks offer real-time data collection, but underground water pipes present challenges for wireless communication (Guezouli et al., 2024). Magnetic induction technology and low-frequency (LF) wake-up systems improve data transmission while conserving power. AI-powered acoustic sensing enhances leak detection by filtering out background noise, enabling precise identification. A hybrid data collection approach, combining static sensors and mobile data gathering, optimizes pipeline monitoring over large areas (Dada et al., 2024). As AI technology advances, its importance in predictive maintenance and smart infrastructure management will become even more crucial, driving efficiency, cost-effectiveness, and long-term environmental sustainability.

20.2.4 Artificial Intelligence in Optimization of Water Consumption

AI is revolutionizing water resource management by promoting efficient use, predictive analytics, and sustainability. AI-powered platforms harness automated data intelligence, integrating real-time inputs from sensors, weather forecasts, and operational parameters to provide actionable insights for optimizing water consumption (Dada et al., 2024). Machine learning algorithms analyze extensive datasets, uncovering hidden patterns and inefficiencies that exceed human capability, thereby enhancing decision-making in industrial water management. One of the key benefits of AI-driven systems is their predictive capability, which enables businesses to forecast future water usage based on historical trends and environmental conditions (Jayakumar et al., 2024). This proactive approach minimizes waste, prevents shortages, and ensures optimal resource allocation. In addition to conservation, AI-based water optimization leads to significant cost savings by reducing unnecessary water expenditures and improving operational efficiency. Furthermore, AI plays a crucial role in broader sustainability goals, assisting industries in adopting responsible water practices in response to growing global water scarcity. By incorporating AI into water management, industries can strike a balance between operational needs and environmental responsibility, paving the way for a future where technology is integral to sustainable resource utilization.

20.2.5 AI in Flood and Drought Monitoring and Management

Monitoring and assessing floods and droughts is challenging due to the diverse water usage across various disciplines, which leads to different measurement indicators. Management strategies are typically divided into structural and non-structural measures (Jayakumar et al., 2024). Structural measures consist of civil engineering projects within catchments, while non-structural approaches encompass forecasting, early warning systems, and policy interventions. AI plays a crucial role in enhancing these efforts by improving early warning systems, land-use planning, and the development of resilient infrastructure. In hydrology, AI models are widely utilized for streamflow forecasting, rainfall–runoff modeling, flood inundation mapping, and water level prediction (Brunner et al., 2021). The development of hybrid AI models aims to reduce uncertainty and enhance predictive accuracy, while transfer learning techniques are being explored to improve flood forecasting and management. For drought management, AI assists in identifying vulnerable areas, generating scenarios, quantifying hydrological responses, and evaluating mitigation strategies. The integration of AI-powered satellite data analysis with *in situ* measurements and hydrological models enhances groundwater management. Future research is focused on hybrid AI techniques for realistic flood and drought simulations, leveraging probabilistic and nature-inspired forecasting methods to achieve more effective water resource management.

20.2.6 Smart Water Grid

The smart water grid, introduced in 2007, represents a groundbreaking approach to water management by integrating real-time, two-way communication networks with advanced sensors, measurement tools, and control systems (Omar et al., 2023). This system continuously monitors, diagnoses, and optimizes water distribution. By leveraging AI, it enhances operational efficiency, improves resource allocation, and promotes sustainability while minimizing water loss and reducing maintenance costs. The smart water grid consists of two key components: an integrated data and electronic alerts system, which facilitates real-time data collection, anomaly detection, and automated alerts for issues such as leaks, pressure fluctuations, and water quality concerns; and a decision support tools model that uses AI-driven predictive analytics to optimize water distribution, forecast demand, and plan infrastructure maintenance (Dada et al., 2024). The implementation of AI-powered smart water grids (Table 20.3) is transforming water management in various countries such as Malta, Singapore, South Korea, and India, as well as the US state Florida, ensuring efficient, resilient, and sustainable water supply systems for the future.

TABLE 20.3 Role of AI in smart water grid systems (Kumari et al., 2024)

APPLICATION OF AI	*FUNCTIONS IN SMART WATER GRID*	*BENEFITS*
Predictive analysis	Forecasts water demand and identifies potential leak risks	Reduces waste and enhances efficiency
Real-time monitoring	Tracks water quality, pressure, and flow rates	Ensures water safety and reliability
Automated leak detection	Detects leaks through AI-driven pattern recognition	Minimizes water loss and reduces repair costs
Demand optimization	Adjusts distribution based on real-time usage patterns	Balances supply with consumer demand

20.3 NEGATIVE ENVIRONMENTAL IMPACTS OF AI IN WATER MANAGEMENT

20.3.1 Carbon Footprint of AI Technologies

AI is now an integral part of modern society, driving innovations from semi-autonomous vehicles to recommendation systems and advanced language models. However, the rapid growth of AI brings considerable environmental challenges, primarily due to high electricity consumption associated with training, hardware manufacturing, and maintenance (Liu and Yin, 2024). This results in substantial CO_2 emissions, particularly in areas reliant on fossil fuels. Large AI models carry a disproportionate carbon footprint, influenced by factors such as computing time, energy-intensive GPUs, electricity sources, and lifecycle emissions from material extraction to disposal. Although AI accounted for an estimated 2.1%–3.9% of global greenhouse gas (GHG) emissions in 2020—compared to 27% from transportation and 24% from industry—its expanding presence raises growing concerns among researchers and policymakers regarding sustainability (Luccioni et al., 2023).

To mitigate AI's carbon footprint, several strategies can be employed (Figure 20.2). (i) Model optimization techniques, including compression and pruning, can help eliminate redundant parameters and reduce computational power requirements. Approaches such as quantization, distillation, and transfer learning further enable the creation of smaller, more efficient models, thus lowering energy consumption. (ii) Sustainable computing choices are vital for minimizing emissions. Training AI in energy-efficient data centers powered by clean energy significantly reduces environmental impact, while optimizing cloud resources ensures that computing power aligns with actual needs, preventing unnecessary energy waste. (iii) Transparency and regulation are crucial for promoting accountability. AI companies should disclose

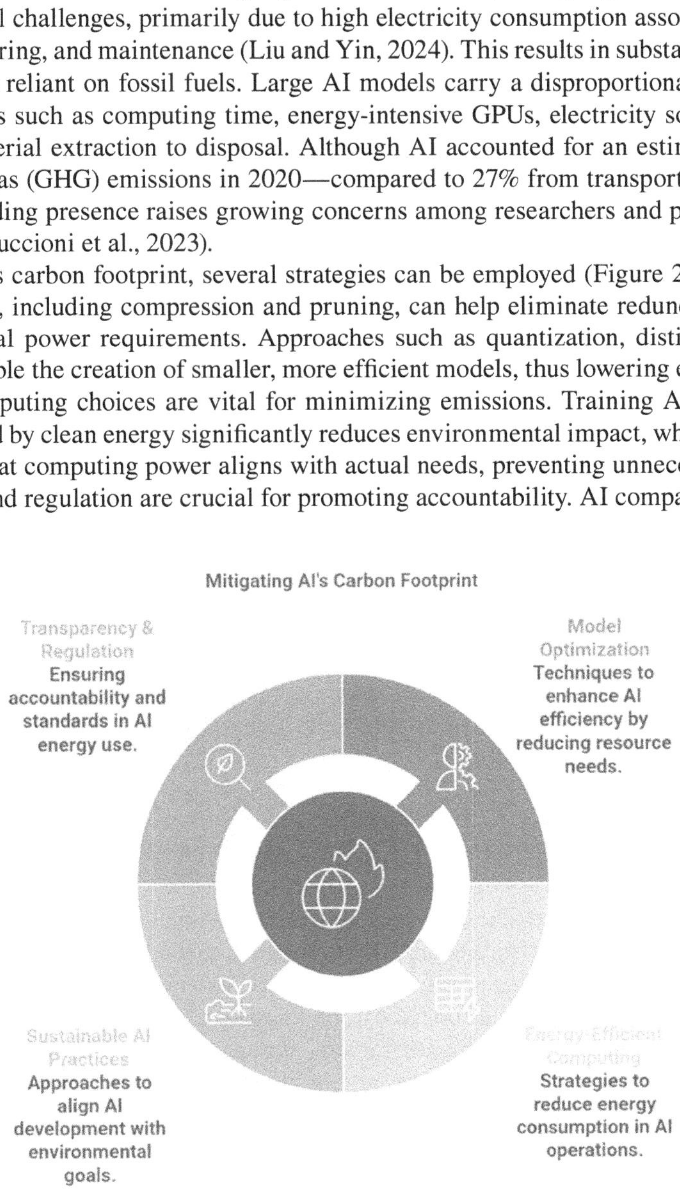

FIGURE 20.2 Mitigation strategies for AI's carbon footprint.

TABLE 20.4 Estimated CO_2 emissions of various AI models during training (Kirkpatrick, 2023)

AI MODEL	*YEAR*	*PARAMETERS (BILLION)*	*TRAINING CO_2 EMISSIONS (METRIC TONS)*	*KEY CONSIDERATIONS*
GPT-1	2018	0.12	Not available	Small-scale model
GPT-2	2019	1.5	Not available	Increased complexity
GPT-3	2020	175	~500	Trained on older, less efficient hardware
BLOOM	2022	176	25 (training only), 50 (full lifecycle)	Trained using nuclear energy in France
Meta OPT	2022	175	~75	More efficient than GPT-3

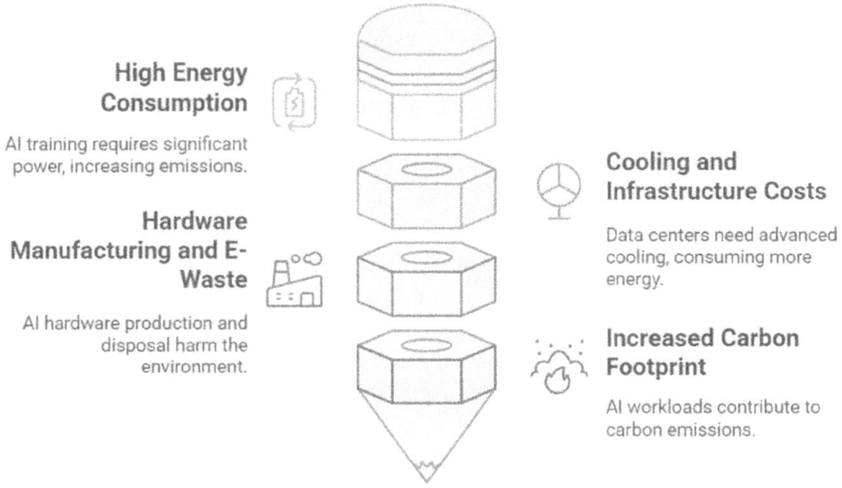

FIGURE 20.3 The major environmental impacts of AI models and their training process.

their training energy usage and emissions to encourage responsible development (Wu et al., 2024). Future policies may require CO_2 transparency and mandate reduction strategies, ensuring that AI's growth aligns with sustainability objectives (Table 20.4).

20.3.2 Resource Intensity of Data Centers and AI Training

AI is being increasingly integrated across various industries, leading to a growing demand for advanced computing power and data storage (Sala et al., 2024). However, this advancement comes with significant environmental costs. The training and operation of AI require extensive computational resources, resulting in high energy consumption and substantial carbon footprints. Training AI models necessitates large data centers equipped with high-performance servers, storage systems, and networking infrastructure (Tabbakh et al., 2024). The process of training large-scale models involves running intensive computations over prolonged periods, and its results are depicted in Figure 20.3.

20.3.3 Potential Misuse or Over-Reliance on AI Tools

Over-reliance on AI in water management is shaped by sociotechnical factors, including system design, human biases, environmental conditions, and human–AI interaction. A significant concern is automation bias, where users may trust AI-driven water resource models without fully understanding their limitations. Although researchers are focusing on explainability and transparency, these aspects alone do not guarantee improved decision-making, especially in high-stakes areas such as flood forecasting, irrigation optimization, and water quality monitoring, where human oversight is still crucial (Wei et al., 2024).

Another challenge is the lack of external validity in AI-driven water management research. Many models are developed and tested under controlled conditions that do not adequately account for real-world variables, such as unpredictable climate patterns, infrastructure limitations, and socio-economic constraints. For instance, while AI can help predict water demand, the actual implementation varies across regions due to differences in governance, weather fluctuations, and unforeseen crises (Nishant et al., 2020). Additionally, AI's probabilistic nature introduces uncertainty, which can result in misinterpretation and inconsistent trust in its recommendations. Overdependence on AI may diminish human expertise in water management, leading to passive decision-making (Nishant et al., 2020). If stakeholders rely solely on AI-generated insights for water allocation, conservation strategies, or disaster response, critical thinking and adaptability may decline, potentially resulting in suboptimal or even detrimental outcomes.

20.4 CHALLENGES AND OPEN ISSUES IN ARTIFICIAL INTELLIGENCE-BASED WATER MANAGEMENT

AI-based water management faces several challenges and open issues, including data quality, model reliability, and environmental variability. Many AI models rely on incomplete or inconsistent data, which affects prediction accuracy in areas such as flood forecasting and water distribution (Drogkoula et al., 2023). Additionally, the lack of real-world validation raises concerns about their applicability across diverse climates and infrastructure conditions. AI's dependence on high computational power also contributes to environmental costs, requiring energy-efficient solutions (Haider et al., 2024). Moreover, ethical concerns, such as equitable water allocation and algorithmic biases, need careful oversight. The major challenges associated with AI-based water management are depicted in Figure 20.4.

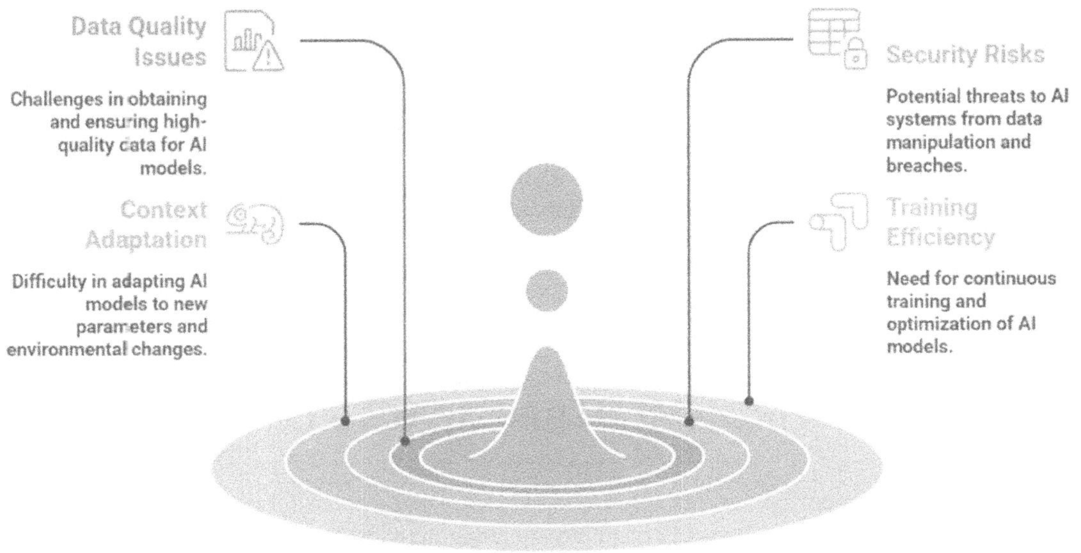

FIGURE 20.4 Challenges and open issues in AI-based water management.

20.4.1 Data Quality and Availability

Deep learning applications in water management rely on extensive datasets for precise predictions and informed decision-making (Tanchangya et al., 2024). However, accessing critical datasets from government and industry is often hindered by legal, political, and commercial constraints. Demographic factors further limit data availability. Even when large datasets are acquired, ensuring their quality poses a significant challenge, as they may contain noise, outliers, or incomplete information (Fu et al., 2022). Inadequate data preprocessing can result in unreliable models, ultimately impacting decision-making accuracy in real-time scenarios.

20.4.2 Security Challenges

AI-driven water management systems encounter substantial security risks due to their dependence on vast amounts of open-source and user-generated data. Attackers may manipulate input data, skewing decision-making processes, which could lead to excessive irrigation or disruptions in water supply operations (Richards et al., 2023). A compromised system can result in severe economic and environmental consequences. Therefore, it is essential to incorporate robust cybersecurity measures—such as encryption, anomaly detection, and strict access controls—to safeguard against data manipulation and ensure system reliability.

20.4.3 Context-Aware Data Analysis

Deep learning models often struggle to adapt to changing system requirements. Unlike traditional algorithms, neural networks do not inherently consider contextual variations, resulting in inefficiencies when new parameters or environmental factors arise (Vander Mijnsbrugge et al., 2023). For instance, smart water quality sensors trained on existing parameters would necessitate extensive retraining if new metrics are introduced. This ongoing requirement for retraining complicates the scalability and integration of AI systems with evolving technologies, particularly in IoT-based real-time applications. Developing context-aware AI models capable of dynamically adapting to changes is crucial for enhancing long-term efficiency.

20.4.4 Training Efficiency and System Optimization

To maintain accuracy in real-time applications, deep learning models must undergo continuous training and optimization. However, updates to system components—such as sensors or data sources—can alter the network's behavior, leading to performance degradation. Additionally, new algorithmic revisions may not always achieve the same level of accuracy or optimization as previous versions. Therefore, establishing efficient retraining processes and incorporating adaptive learning techniques are vital for sustaining the performance of AI-driven water management systems (Wang and Machida, 2024).

20.5 STRATEGIES TO MITIGATE ENVIRONMENTAL IMPACTS OF AI

AI presents immense opportunities, but it also carries significant responsibilities regarding its environmental impacts. The technology's supply chain and ongoing maintenance contribute to substantial ecological costs, particularly through high energy consumption during the training and operation of complex machine learning models. These impacts are evident throughout the entire lifecycle of AI systems, from development to deployment and maintenance (Amiri et al., 2024).

20.5.1 Energy Consumption

AI models, especially sophisticated ones like deep learning and natural language processing, necessitate considerable computational power, leading to high energy use. While the energy consumed during model training is substantial, it is often overshadowed by the energy required for inference. The choice of model can significantly affect the overall environmental footprint of an AI system (Li et al., 2023).

20.5.2 Optimizing Model Efficiency

Hardware Improvements: Enhancements at the hardware level can lead to increased efficiency. If advancements in microchip technology continue as predicted by Moore's Law, we may see more powerful and efficient GPUs in the future. However, some experts suggest that this trend may plateau around 2025 due to physical limitations. Innovations in quantum computing and new chip architectures could play a vital role in achieving efficiency (Liu and Yin, 2024).

Software Innovations: AI researchers are developing techniques to improve model efficiency during both training and inference:

Model Pruning: This technique involves removing unnecessary parameters from a neural network, thereby reducing its size and improving inference efficiency.

Quantization: This process maps continuous values to a smaller set of discrete values, optimizing performance.

Knowledge Distillation: This method transfers knowledge from a larger pretrained model to a smaller one, maintaining performance while reducing resource demands.

20.5.3 Utilizing Renewable Energy

Investing in renewable energy sources is crucial for mitigating the carbon footprint of AI systems in data centers. Solar panels, wind turbines, and hydropower generate electricity without carbon emissions. Major tech companies like Google, Microsoft, and Meta are leading investments in clean energy infrastructure (Liu and Yin, 2024).

20.5.4 Addressing Greenhouse Gas Emissions

The energy-intensive nature of AI often results in significant GHG emissions, particularly when powered by non-renewable sources. Additionally, large data centers require extensive cooling systems that consume large amounts of water, exacerbating water scarcity issues in vulnerable regions (Cowls et al., 2023).

20.5.5 Managing Hardware and E-Waste

The production and disposal of AI hardware—such as servers and GPUs—contribute to environmental degradation through resource extraction and electronic waste. Implementing responsible e-waste management practices is essential for minimizing these impacts.

20.5.6 Lifecycle Management Strategies

Model Selection: Opt for AI models with a lower environmental footprint based on their energy consumption metrics.

Regular Updates: Continuously improve models to enhance efficiency and reduce energy usage over time.

20.5.7 Application-Specific Strategies

Smart Grids: Utilize AI to optimize energy consumption in smart grids, reducing peak demand while integrating renewable sources.

Sustainable Transportation: Develop AI systems that optimize traffic flow and promote electric vehicle usage.

Environmental Monitoring: Employ AI for real-time tracking of environmental parameters such as air quality and pollution levels.

By adopting these strategies, stakeholders can work toward minimizing the environmental impacts associated with AI while leveraging its potential for positive change (Li et al., 2023).

20.5.8 Policy and Regulation

Without a strong, binding commitment from the AI industry, regulators and policymakers play a crucial role in promoting sustainable practices within the sector. This is particularly evident in the European Union, where the recently passed EU AI Act aims to establish voluntary industry standards that focus on minimizing energy and resource consumption associated with high-risk AI systems throughout their lifecycle (Alder et al., 2024). The Act emphasizes the importance of developing energy-efficient general-purpose AI models, such as those used by OpenAI (Bolón-Canedo et al., 2024). Additionally, it mandates that providers of these general-purpose AI models disclose information regarding their known or estimated energy consumption. This move toward transparency is essential for stakeholders to understand the environmental impact of AI technologies and to encourage more sustainable practices across the industry.

20.6 FUTURE TRENDS AND OPPORTUNITIES

The future of water management is being significantly shaped by emerging AI technologies, presenting numerous opportunities for enhancing sustainability and efficiency. AI is poised to revolutionize how we monitor and manage water resources through advanced capabilities such as real-time water quality monitoring, leak detection, and flood prediction. By utilizing data from sensors and IoT devices, AI can continuously assess water quality, detect contaminants, and predict potential infrastructure failures, thereby ensuring safer and more reliable water systems. Moreover, the integration of AI with renewable energy sources offers a promising avenue for optimizing energy use in water management. By leveraging AI algorithms to analyze energy consumption patterns, water utilities can enhance their operations while minimizing their carbon footprint. This synergy not only promotes the use of clean energy but also supports the transition toward more sustainable practices in the water sector. AI's role in achieving global water sustainability goals cannot be overstated. It enables more efficient allocation of resources, supports demand forecasting, and enhances decision-making processes through predictive analytics. By employing digital twins and advanced simulation models, stakeholders can anticipate changes in water supply and demand dynamics, allowing for proactive management strategies that align with sustainability objectives.

As we look ahead, the combination of AI with cutting-edge technologies such as blockchain and quantum computing will further enhance transparency and efficiency in water management. These advancements will facilitate better data sharing among stakeholders and optimize complex resource allocation challenges. Overall, the future trends in water management highlight a transformative shift toward smarter, more resilient systems that prioritize sustainability and effective resource utilization.

20.7 CONCLUSION

In conclusion, the integration of AI in water management represents a transformative opportunity to enhance sustainability and efficiency in addressing global water challenges. AI applications, such as smart water monitoring systems, predictive analytics for water quality, and optimization techniques for agricultural usage, are revolutionizing how we manage this vital resource. These innovations not only improve water conservation and quality management but also significantly reduce energy consumption and mitigate risks associated with floods and droughts. However, it is crucial to acknowledge the potential negative environmental impacts of AI technologies, including their carbon footprint and the resource intensity of data centers. Challenges such as data availability, ethical implications, and balancing cost-effectiveness with environmental concerns must also be addressed to ensure responsible implementation. Real-world case studies highlight successful applications of AI in various contexts, demonstrating its effectiveness in enhancing water management practices. Looking ahead, emerging AI technologies combined with renewable energy sources offer promising pathways to achieve global water sustainability goals. By developing energy-efficient AI systems and adopting green practices, stakeholders can mitigate environmental impacts while leveraging the benefits of AI. Policymakers play a critical role in establishing frameworks that support sustainable AI adoption in water management, ensuring that these advancements contribute positively to both the environment and society as a whole. The future of water management lies in harnessing AI's capabilities responsibly and innovatively to create resilient systems that safeguard our most precious resource.

REFERENCES

Abdelmageed, S., Tariq, S., Boadu, V., & Zayed, T. (2022). Criteria-based critical review of artificial intelligence applications in water-leak management. *Environmental Reviews*, *30*(2), 280–297. https://doi.org/10.1139/er-2021-0046

Akter, M. S. (2024). AI for sustainability: Leveraging technology to address global environmental challenges. *Journal of Artificial Intelligence General Science*, *3*(1), 40–48. https://doi.org/10.60087/jaigs.v3i1.64

Alder, N., Ebert, K., Herbrich, R., & Hacker, P. (2024). AI, climate, and transparency: Operationalizing and improving the AI Act. *arXiv preprint arXiv:2409.07471*. https://doi.org/10.48550/arXiv.2409.07471

Amiri, Z., Heidari, A., & Navimipour, N. J. (2024). Comprehensive survey of artificial intelligence techniques and strategies for climate change mitigation. *Energy*, *308*, 132827. https://doi.org/10.1016/j.energy.2024.132827

Ashoka, P., Devi, B. R., Sharma, N., Behera, M., Gautam, A., Jha, A., & Sinha, G. (2024). Artificial intelligence in water management for sustainable farming: A review. *Journal of Scientific Research and Reports*, *30*(6), 511–525. https://doi.org/10.9734/jsrr/2024/v30i62068

Baird, G. M., Hatler, D., & Carpenter, P. (2019). How cost effective is machine learning/AI applied to leak detection and pipe replacement prioritization? In Heidrick, J. W., & Mihm, M. S. (Eds.) *Pipelines 2019* (pp. 274–283). Reston, VA: American Society of Civil Engineers. https://doi.org/10.1061/9780784482506.029

Bolón-Canedo, V., Morán-Fernández, L., Cancela, B., & Alonso-Betanzos, A. (2024). A review of green artificial intelligence: Towards a more sustainable future. *Neurocomputing*, *599*, 128096. https://doi.org/10.1016/j.neucom.2024.128096

Brunner, M. I., Slater, L., Tallaksen, L. M., & Clark, M. (2021). Challenges in modeling and predicting floods and droughts: A review. *Wiley Interdisciplinary Reviews: Water*, *8*(3), e1520. https://doi.org/10.1002/wat2.1520

Cowls, J., Tsamados, A., Taddeo, M., & Floridi, L. (2023). The AI gambit: Leveraging artificial intelligence to combat climate change: Opportunities, challenges, and recommendations. *AI & Society*, *38*, 1–25. https://doi.org/10.1007/s00146-021-01294-x

Dada, M. A., Majemite, M. T., Obaigbena, A., Daraojimba, O. H., Oliha, J. S., & Nwokediegwu, Z. Q. S. (2024). Review of smart water management: IoT and AI in water and wastewater treatment. *World Journal of Advanced Research and Reviews*, *21*(1), 1373–1382. https://doi.org/10.30574/wjarr.2024.21.1.0171

Drogkoula, M., Kokkinos, K., & Samaras, N. (2023). A comprehensive survey of machine learning methodologies with emphasis in water resources management. *Applied Sciences*, *13*(22), 12147. https://doi.org/10.3390/app132212147

Frincu, R. M. (2025). Artificial intelligence in water quality monitoring: A review of water quality assessment applications. *Water Quality Research Journal*, *60*(1), 164–176. https://doi.org/10.2166/wqrj.2024.049

Fu, G., Jin, Y., Sun, S., Yuan, Z., & Butler, D. (2022). The role of deep learning in urban water management: A critical review. *Water Research*, *223*, 118973. https://doi.org/10.1016/j.watres.2022.118973

Guezouli, L., Guezouli, L., Djeghaba, M. B. E., & Bentahrour, A. (2024). IoT and AI for real-time water monitoring and leak detection. *Journal of Renewable Energies*, *27*(2), 243–281. https://doi.org/10.54966/jreen.v27i2.1210

Haider, S., Rashid, M., Tariq, M. A. U. R., & Nadeem, A. (2024). The role of artificial intelligence (AI) and ChatGPT in water resources, including its potential benefits and associated challenges. *Discover Water*, *4*(1), 113. https://doi.org/10.1007/s43832-024-00173-y

Hernandez, D., Pasha, L., Yusuf, D. A., Nurfaizi, R., & Julianingsih, D. (2024). The role of artificial intelligence in sustainable agriculture and waste management: Towards a green future. *International Transactions on Artificial Intelligence*, *2*(2), 150–157. https://doi.org/10.33050/italic.v2i2.552

Jayakumar, D., Bouhoula, A., & Al-Zubari, W. K. (2024). Unlocking the potential of artificial intelligence for sustainable water management focusing operational applications. *Water*, *16*(22). https://doi.org/10.3390/w16223328

Kamyab, H., Khademi, T., Chelliapan, S., SaberiKamarposhti, M., Rezania, S., Yusuf, M., & Ahn, Y. (2023). The latest innovative avenues for the utilization of artificial intelligence and big data analytics in water resource management. *Results in Engineering*, *20*, 101566. https://doi.org/10.1016/j.rineng.2023.101566

Kirkpatrick, K. (2023). The carbon footprint of artificial intelligence. *Communications of the ACM*, *66*(8), 17–19. https://doi.org/10.1145/3603746

Korostynska, O., Arshak, K., Velusamy, V., Arshak, A., & Vaseashta, A. (2012). Recent advances in point-of-access *water quality monitoring: Water quality. In Technological Innovations in Sensing and Detection of Chemical, Biological, Radiological, Nuclear Threats and Ecological Terrorism* (pp. 261–268). Springer. https://doi.org/10.1007/978-94-007-2488-4_29

Kumari, R., Ravinder, R., Kumar, H., Meena, M., Patial, B., & Sharma, R. (2024). The role of artificial intelligence and its tool in water management. *International Journal of Research Publication and Reviews*, *5*(11), 6263–6270. https://doi.org/10.55248/gengpi.5.1124.3401

Li, P., Yang, J., Islam, M. A., & Ren, S. (2023). Making AI less "thirsty": Uncovering and addressing the secret water footprint of AI models. *arXiv preprint arXiv:2304.03271*. https://doi.org/10.48550/arXiv.2304.03271

Liu, V., & Yin, Y. (2024). Green AI: Exploring carbon footprints, mitigation strategies, and trade-offs in large language model training. *Discover Artificial Intelligence*, *4*(1), 49. https://doi.org/10.1007/s44163-024-00149-w

Liu, X. (2024). Optimization of sewage treatment processes: Process control based on artificial intelligence. *Applied and Computational Engineering*, *93*, 185–190. https://doi.org/10.54254/2755-2721/93/20240981

Lowe, M., Qin, R., & Mao, X. (2022). A review on machine learning, artificial intelligence, and smart technology in water treatment and monitoring. *Water*, *14*(9), 1384. https://doi.org/10.3390/w14091384

Luccioni, A. S., Viguier, S., & Ligozat, A. L. (2023). Estimating the carbon footprint of Bloom, a 176B parameter language model. *Journal of Machine Learning Research*, *24*(253), 1–15. https://doi.org/10.48550/arXiv.2211.02001

Mehmood, H., Liao, D., & Mahadeo, K. (2020, September). A review of artificial intelligence applications to achieve water-related sustainable development goals. In *2020 IEEE/ITU International Conference on Artificial Intelligence for Good (AI4G)* (pp. 135–141). IEEE. https://doi.org/10.1109/AI4G50087.2020.9311018

Mustafa, H. M., Mustapha, A., Hayder, G., & Salisu, A. (2021, January). Applications of IoT and artificial intelligence in water quality monitoring and prediction: A review. In *2021 6th International Conference on Inventive Computation Technologies (ICICT)* (pp. 968–975). IEEE. https://doi.org/10.1109/ICICT50816.2021.9358675

Nagpal, M., Siddique, M. A., Sharma, K., Sharma, N., & Mittal, A. (2024). Optimizing wastewater treatment through artificial intelligence: Recent advances and future prospects. *Water Science & Technology*, *90*(3), 731–757. https://doi.org/10.2166/wst.2024.259

Nishant, R., Kennedy, M., & Corbett, J. (2020). Artificial intelligence for sustainability: Challenges, opportunities, and a research agenda. *International Journal of Information Management*, *53*, 102104. https://doi.org/10.1016/j.ijinfomgt.2020.102104

Nti, E. K., Cobbina, S. J., Attafuah, E. E., Opoku, E., & Gyan, M. A. (2022). Environmental sustainability technologies in biodiversity, energy, transportation and water management using artificial intelligence: A systematic review. *Sustainable Futures*, *4*, 100068. https://doi.org/10.1016/j.sftr.2022.100068

Omar, P. J., Gupta, P., & Wang, Q. (2023). Exploring the rise of AI-based smart water management systems. *AQUA—Water Infrastructure, Ecosystems and Society*, *72*(11), iii–iv. https://doi.org/10.2166/aqua.2023.005

Omer, N. H. (2019). Water quality parameters. In *Water Quality-Science, Assessments and Policy* (Vol. 18, pp. 1–34). IntechOpen. https://doi.org/10.5772/INTECHOPEN.89657

Rainbow, J., Sedlackova, E., Jiang, S., Maxted, G., Moschou, D., Richtera, L., & Estrela, P. (2020). Integrated electrochemical biosensors for detection of waterborne pathogens in low-resource settings. *Biosensors*, *10*(4), 36. https://doi.org/10.3390/bios10040036

Rajitha, A., Aravinda, K., Nagpal, A., Kalra, R., Maan, P., Kumar, A., & Abdul-Zahra, D. S. (2024). Machine learning and AI-driven water quality monitoring and treatment. In *E3S Web of Conferences* (Vol. 505, p. 03012). https://doi.org/10.1051/e3sconf/202450503012

Richards, C. E., Tzachor, A., Avin, S., & Fenner, R. (2023). Rewards, risks and responsible deployment of artificial intelligence in water systems. *Nature Water*, *1*(5), 422–432. https://doi.org/10.1038/s44221-023-00069-6

Safeer, S., Pandey, R. P., Rehman, B., Safdar, T., Ahmad, I., Hasan, S. W., & Ullah, A. (2022). A review of artificial intelligence in water purification and wastewater treatment: Recent advancements. *Journal of Water Process Engineering*, *49*, 102974. https://doi.org/10.1016/j.jwpe.2022.102974

Sala, R., Filippini, F., Ardagna, D., Lezzi, D., Lordan, F., & Thiem, P. (2024, April). Greening AI: A framework for energy-aware resource allocation of ML training jobs with performance guarantees. In *International Conference on Advanced Information Networking and Applications* (pp. 110–121). Cham: Springer Nature Switzerland. https://doi.org/10.1007/978-3-031-57931-8_11

Sani, S. A. (2023). Drawbacks of traditional environmental monitoring systems. *TMP Universal Journal of Research and Review Archives*, *2*(2). https://doi.org/10.5539/cis.v16n3p30

Tabbakh, A., Al Amin, L., Islam, M., Mahmud, G. I., Chowdhury, I. K., & Mukta, M. S. H. (2024). Towards sustainable AI: A comprehensive framework for Green AI. *Discover Sustainability*, *5*(1), 408. https://doi.org/10.1007/s43621-024-00641-4

Tanchangya, T., Raihan, A., Rahman, J., & Ridwan, M. (2024). A review of deep learning applications for sustainable water resource management. *Global Sustainability Research*, *3*(4), 48–73. https://doi.org/10.56556/gssr.v3i4.1043

Vander Mijnsbrugge, D., Ongenae, F., & Van Hoecke, S. (2023). Context-aware deep learning with dynamically assembled weight matrices. *Information Fusion*, *100*, 101908. https://doi.org/10.1016/j.inffus.2023.101908

Wang, Z., & Machida, F. (2024, July). Maintaining performance of a machine learning system against imperfect retraining. In *2024 IEEE 48th Annual Computers, Software, and Applications Conference (COMPSAC)* (pp. 1782–1787). IEEE. https://doi.org/10.1109/COMPSAC61105.2024.00281

Wei, H., Xu, W., Kang, B., Eisner, R., Muleke, A., Rodriguez, D., de Voil, P., Sadras, V., Monjardino, M., & Harrison, M. T. (2024). Irrigation with artificial intelligence: Problems, premises, promises. *Human-Centric Intelligent Systems*, *4*, 187–205. https://doi.org/10.1007/s44230-024-00072-4

Wu, C. J., Acun, B., Raghavendra, R., & Hazelwood, K. (2024). Beyond efficiency: Scaling AI sustainably. *IEEE Micro*, *44*(5), 37–46.

Younos, T., & Heyer, C. J. (2015). Advances in water sensor technologies and real-time water monitoring. In *Advances in Watershed Science and Assessment* (pp. 171–203). Springer. https://doi.org/10.1007/978-3-319-14212-8_7

Understanding Water Scarcity

The Role of AI in Sustainable Solutions

21

Elif Esra Altuner, Halil Burak Kaybal, Hasan Ulus, and Fatih Sen

21.1 INTRODUCTION

Today, over 40% of the global population experiences water scarcity, and more than 700 million people lack access to clean drinking water (Shemer, Wald, & Semiat, 2023). This pressing issue arises when regional water demand surpasses the available resources, driven by factors such as population growth, climate change, and inadequate water management. The consequences are far-reaching, affecting communities, industries, and ecosystems, and often leading to competition, conflicts, and challenges in meeting water demands (Rozhok, Sushchev, Tatarinov, & Revetria, 2019). Over the next 30 years, urbanization, population growth, and socio-economic development are projected to drive a 50%–80% increase in urban, industrial, and domestic water demand (Flörke, Schneider, & McDonald, 2018; Garrick et al., 2019). Concurrently, climate change will disrupt the spatial and temporal distribution of water resources (Greve et al., 2018), posing significant threats to food security, economic growth, and social stability (Wu et al., 2024; Darabi, Ghorbani-HasanSaraei, et al., 2023). Prolonged droughts, intensified by climate change, endanger human lives, the environment, and economic systems, causing declines in river flows, falling water levels in reservoirs, and reduced groundwater recharge (Darabi, Ghorbani-HasanSaraei, et al., 2023; Darabi, Karimi-Maleh, et al., 2023; Kocak et al., 2024). These changes hinder access to water for agriculture, industry, and domestic use (Hisdal et al., 2024; Rozhok et al., 2021; Schilstra, Wang, van Oel, Wang, & Cheng, 2024). To build resilience, existing drought management approaches need adaptation to counteract climate change impacts (Otieno & Ochieng, 2004). Failure to do so may exacerbate the water scarcity crisis, threatening the achievement of the United Nations Sustainable Development Goals (SDGs), including SDG11 (Sustainable Cities and Communities) and SDG6 (Clean Water and Sanitation) (He et al., 2021; Larsen, Hoffmann, Lüthi, Truffer, & Maurer, 2016; Schewe et al., 2014).

DOI: 10.1201/9781003600718-30

Droughts emerge from intricate interactions between meteorological, hydrological, and ecological processes, which complicate accurate modeling. Effective water resource management requires integrating spatial and temporal data at local, regional, and global levels while accounting for changing demands from population growth, urbanization, and socio-economic shifts (Dada et al., 2024; Y. Wang & Razmjooy, 2024). Groundwater, a key resource for drought resilience and sustainability, necessitates precise predictions of future water demands (Zhan et al., 2023). However, limited understanding of groundwater recharge rates, dynamics, and interactions with surface water complicates the development of accurate models. Socio-economic variables, such as land-use patterns, water policies, and agricultural practices, further influence drought outcomes (Inman & Ribarova, 2009).

Climate change intensifies global water challenges, exacerbating extreme events like droughts and floods (Y. Wang & Razmjooy, 2024). Fluctuating temperatures and altered precipitation patterns amplify these phenomena, reducing water availability and reliability by shifting the volume, timing, and location of rainfall (Yavarian, Jalalian, & Tahmasebi, 2023). Predicting long-term drought patterns is further hampered by uncertainties in climate and water resource modeling. Accurate forecasts require reliable, high-quality historical data on climate, hydrology, and land conditions. However, inconsistent or limited data networks often undermine prediction efforts (J. Wang & Azam, 2024).

Incorporating the underlying dynamics of water scarcity into models is critical for developing actionable drought forecasts and effective management strategies. Translating scientific findings into practical insights is essential for policymakers, water managers, and impacted communities. Effective communication systems and decision-support tools are vital to bridge the gap between knowledge and practical action (Zhang et al., 2024). Such tools enable efficient water management, enhance food security, and mitigate the effects of drought. Models and forecasts of drought patterns play an integral role in sustainable water planning, addressing uncertainties linked to water stress, and averting future crises (Luo, Xu, & Chen, 2019). These tools facilitate the development of more efficient and effective water management strategies, ensuring the long-term availability of water resources (Trogadas et al., 2024).

The rise of "smart" and data-driven solutions has revolutionized how resources are managed, enabling innovative strategies and preventive measures. Among the most advanced methods for water resource management are artificial intelligence (AI)–based systems. Early decision-support technologies relied on static, rule-based algorithms that became ineffective under dynamic conditions. AI overcomes these limitations by employing adaptive pattern recognition, which evolves with new data inputs. This technology mimics human learning processes: during a "learning phase," input data are matched with known outcomes to train the system, followed by an "operational phase," where the system identifies patterns in new datasets (Goralski & Tan, 2020).

AI has proven to be a versatile and efficient tool for addressing complex real-world challenges. Its speed, accessibility, and independence from domain-specific expertise make it a favored solution across disciplines (Rajaee, Ebrahimi, & Nourani, 2019). In medicine, AI aids in disease prevention, diagnosis, and treatment, while in finance, it predicts market movements. Additionally, it serves as an effective resource for managing supply chain risks by identifying disruptions, strategizing responses, and minimizing losses.

By leveraging AI algorithms and big data analytics, water service providers can make more efficient decisions, improve water quality, enhance customer service, and reduce costs (Jenny, Alonso, Wang, & Minguez, 2020). In recent years, new AI approaches aligned with human and ethical principles have emerged, leading to the development of "responsible artificial intelligence" (RAI). Although AI-based methods are not as widely applied in the water sector as in other industries like energy, healthcare, or transportation, three key insights stand out. These include the use of AI techniques in multi-objective optimization, the development of data-driven models that avoid the pitfalls of purely data-centric approaches, and participatory decision-making processes informed by prior experiences in the water sector (Tiza, 2023). In conclusion, AI is recognized as a valuable tool for water resource management in an ever-changing environment.

21.2 UNDERSTANDING WATER SCARCITY

Water is a critical resource for human health, ecosystems, agriculture, industry, and environmental sustainability. It also holds significant cultural and social values (Eslamian, 2016). However, climate change, population growth, rapid urbanization, and deteriorating water quality are increasingly turning water scarcity into a major threat. With economic growth and population expansion, water is expected to become one of the most strategic resources in the coming years, particularly in arid and semi-arid regions (Gohari et al., 2013; Young, Shah, & Kimaite, 2008).

Water scarcity is defined as the lack of access to safe drinking water or water shortage and typically occurs when demand exceeds supply (UNICEF, 2019). Reaching or approaching the sustainable limits of existing water resources is among the primary causes of water scarcity. Inadequate per capita water availability also characterizes this condition (Kummu et al., 2016). Factors such as climate change, the state of water resources at the basin level, and the adaptive capacity of water management systems combine to make water scarcity a complex and dynamic issue. Water scarcity arises when the water supply in a particular area is inadequate to satisfy the needs of its population (Kummu et al., 2016). Moreover, the lack of rainwater retained in the soil (green water) and the limited availability of directly usable liquid water (blue water) contribute to the depletion of water resources (Falkenmark, 2013).

The effects of water scarcity extend across various sectors, impacting agriculture, industry, trade, tourism, biodiversity preservation, and households. These challenges pose significant risks to global economic stability, food security, sustainable development, and overall human well-being (Salehi, 2022). In Australia, water scarcity has led to population losses in rural inland areas (e.g., northeastern, southeastern, and western Australia), further reductions in irrigation services, failures of local businesses, and rising unemployment (Hogan & Young, 2013). In the Middle East, water resources have caused severe disputes among countries in the region, with similar situations predicted in Central and Southeast Asia. Increasing tensions over water access are projected to lead to armed conflicts over the control of rivers and groundwater in the future (Chartres & Williams, 2006; Jury & Vaux Jr, 2007; Seckler, 1998). The United Nations has stated that water shortages (e.g., declining water supply, desertification, recurring floods, and increased salinity in coastal areas) may indirectly lead to internal instability in many regions. Such disruptions can trigger migration, displacement, and population movements, fueling social and ethnic tensions in cities and local areas (McAdam, 2011). Additionally, an Australian report identifies drought as a significant cause of suicide and mental illness among Australian farmers and rural communities (Hogan & Young, 2013).

Water scarcity is a major global issue. The number of "water refugees" worldwide rose from 25 million in 1995 to 50 million in 2010 and is expected to continue growing rapidly in the coming decades (Eslamian, 2016). Since the 1900s, water scarcity has increased by 58%, rising from 0.24 billion to 3.8 billion affected individuals (Adetoro et al., 2020; Kummu et al., 2016). By 2050, at least one-quarter of the world's population is expected to live in regions with regular or recurrent freshwater scarcity (Hou, Wen, Liu, & Dong, 2021). Currently, 1.42 billion people, including 450 million children, live in areas of severe water scarcity (Diop, 2002). Each year, approximately 4 billion people—representing 75% of the global population—experience extreme water shortages (Mekonnen & Hoekstra, 2016). Moreover, 3.2 billion individuals reside in agricultural areas with significant water constraints (FAO, 2020).

In response to potential water scarcity, long-term strategies such as water conservation and demand management are gaining importance. However, in some cases, these measures are insufficient to bridge the supply–demand gap, necessitating the adoption of sustainable management solutions and technological innovations. Utilizing alternative water sources, improving aquifer storage, and treating otherwise unusable water are essential approaches in this context (Smith, 2011). Furthermore, AI tools significantly contribute by reducing water losses, lowering costs, and enhancing the performance of water services

(Doorn, 2021). This digital transformation supports both SDG6 and SDG13 goals, promoting investments to mitigate the impacts of climate change (Jenny et al., 2020). Efforts to preserve and sustainably manage water resources are becoming essential for the long-term growth of societies.

21.3 CAUSES AND EFFECTS OF WATER SCARCITY

Water scarcity is an increasingly severe global problem impacting billions of individuals annually. This challenge stems from a mix of natural and human-driven causes, such as inadequate water management, pollution, excessive consumption, climate change, population expansion, and industrial processes. Its effects are extensive, influencing agriculture, public health, economic systems, and ecosystems (Agnes et al., 2021).

Water scarcity arises from both natural processes and human activities. Climate change and global warming are significant contributors, altering precipitation patterns, raising temperatures, and intensifying droughts and floods. Higher temperatures accelerate evaporation from water bodies such as rivers and lakes, reducing freshwater availability (Kummu et al., 2016). Rising sea levels also exacerbate the crisis by contaminating freshwater sources with saltwater (Šulyová, Vodák, & Kubina, 2021). Human activities such as agriculture and industrialization play a crucial role as well. Agriculture, particularly irrigation, is the largest consumer of freshwater. For instance, over 60% of South Africa's freshwater is used for agriculture, with half allocated to irrigation (Adetoro et al., 2020). The energy sector also demands significant water resources for cooling power plants and processing fuels, further straining water supplies (Burek et al., 2016). Urbanization and population growth worsen the problem. Cities, which now host more than half the global population, face water pollution and depletion due to inadequate infrastructure and excessive use (WHO, 2019). By 2050, cities are expected to house 68% of the global population, placing immense pressure on limited water resources (Mitlin, Beard, Satterthwaite, & Du, 2019).

The effects of water scarcity are profound and multifaceted. In agriculture, irregular rainfall patterns reduce crop yields, increasing reliance on irrigation and worsening water shortages. Global water demand is projected to grow by 20%–30% by 2050 due to population growth and rising agricultural needs, potentially leading to food security challenges, particularly in arid and semi-arid regions (Burek et al., 2016). Health impacts are another major concern. Water contaminated with chemicals, oils, and waste poses serious risks to human health (Kerstens, 2013). Waterborne diseases and inadequate sanitation disproportionately affect vulnerable populations in developing countries suffering from water scarcity. Economic and social consequences include conflicts over water resources. Disputes over access to water can escalate into violent conflicts or regional wars, particularly in politically unstable areas (Eslamian, 2016). In rural areas, lack of access to clean water drives migration, further straining resources in urban areas. Environmental impacts are equally alarming. Overuse and pollution of water bodies threaten biodiversity and disrupt ecosystems. Natural disasters such as floods exacerbate water pollution and infrastructure damage, making water unsafe and inaccessible for extended periods (Kerstens, 2013).

Addressing water scarcity requires a combination of policy interventions, technological innovations, and sustainable practices. Effective water management demands stricter regulations to prevent pollution and advanced technologies like AI to optimize resource use. Governments must invest in water infrastructure and enforce laws to minimize waste. Public awareness and education on responsible water use are equally critical. In agriculture, adopting water-efficient practices and technologies can significantly reduce consumption. Similarly, recycling and reusing water in industrial processes can alleviate pressure on freshwater resources. Ultimately, water scarcity is a critical issue of the 21st century, driven by factors such as climate change, overuse, and pollution. Its effects on agriculture, health, and ecosystems highlight the urgent need for coordinated global efforts to mitigate the crisis. By implementing sustainable water

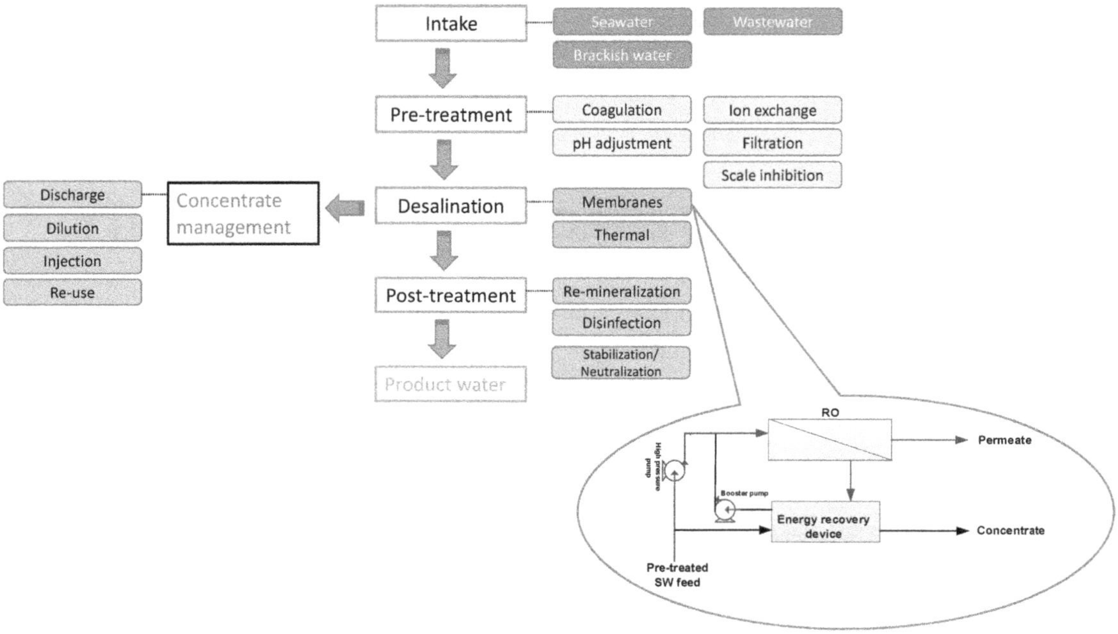

FIGURE 21.1 Impact areas of water scarcity (Shemer et al., 2023). (Reprinted (adapted) with permission from Shemer et al. (2023). Copyright (2024) MDPI-Membranes.)

management practices and leveraging technological advancements, we can meet growing water demands and secure this vital resource for future generations.

The impacts of water scarcity extend far beyond the direct lack of water, deeply affecting societal and ecological systems (Figure 21.1). Lack of access to drinking water directly impacts lives, making it harder for people to obtain fresh, clean water. This leads to health problems, as individuals cannot maintain essential bodily functions (Kerstens, 2013). Additionally, insufficient water supply hampers the growth of crops, causing food shortages and creating severe scarcity for both people and animals in the affected regions. Children are particularly vulnerable, as water scarcity disrupts their growth and education, leading to a lack of schooling. Water scarcity also facilitates the spread of microorganisms, increasing health issues. This contributes to sanitation deficiencies and psychological problems, particularly depression and anxiety. Moreover, water scarcity results in the destruction of habitats and loss of biodiversity. Decreasing food sources for animals and plants leads to the extinction of species (Kerstens, 2013). Poverty is directly impacted by water scarcity, as individuals without access to water remain impoverished, trapped in a difficult cycle. Migration becomes inevitable as people search for arable land, exposing them to various dangers during displacement. Ultimately, water scarcity is not merely a resource issue but a global crisis threatening all life systems.

21.4 TRADITIONAL MEASURES TO ADDRESS WATER SCARCITY

The global water crisis is one of the most pressing challenges of the 21st century, affecting millions of people, ecosystems, and economies. The following essential measures can help address the growing issue of water scarcity, ensuring a more sustainable future for all (Shemer et al., 2023).

21.4.1 Education and Awareness

Raising awareness about water scarcity is critical for encouraging individuals and organizations to take proactive steps in water conservation. Educational programs should emphasize the value of water and promote sustainable practices at all levels, from households to industries. Simple household measures, such as using water-efficient faucets, harvesting rainwater for irrigation, and choosing drought-tolerant plants, can reduce personal water consumption (Sachidananda, Webb, & Rahimifard, 2016). For industries, implementing water-saving technologies, such as closed-loop cooling systems, can significantly decrease water usage (Bauer & Wagner, 2022). Furthermore, educational initiatives can help drive behavioral change, fostering a culture of water conservation globally.

21.4.2 Water Catchment and Harvesting Technologies

Water harvesting technologies play a crucial role in supplementing water supply, especially in areas facing frequent water shortages. Traditional methods like rainwater harvesting (RWH) are highly effective for reducing stormwater runoff and providing an alternative water source for non-potable uses (Alim et al., 2020; Baba et al., 2018). However, the feasibility of RWH systems depends on water quality and treatment processes to remove contaminants. Aquifer recharge, where water is injected back into underground reservoirs, is another promising solution for increasing groundwater levels (GWLs) and mitigating the impacts of hydrometeorological events (Basel et al., 2021; Malik, Giordano, & Sharma, 2014). Despite the high costs and technical challenges, these technologies can play a significant role in sustainable water management, especially in arid and semi-arid regions.

21.4.3 Pollution Control

Water pollution, caused by both organic and inorganic pollutants, poses a significant threat to human health and ecosystems (Figure 21.2). Emerging contaminants (ECs), including pharmaceuticals, plastics, and pesticides, have raised alarms due to their detrimental effects on water quality (Sauvé &

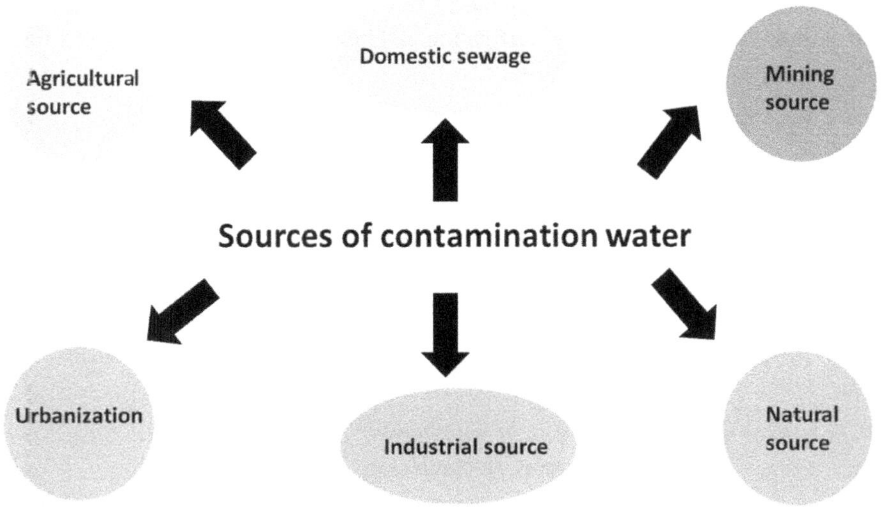

FIGURE 21.2 Sources of contamination of water (Singh et al., 2022). (Reprinted (adapted) with permission from Singh et al. (2022). Copyright (2024) IntechOpen-The Toxicity of Environmental Pollutants.)

Desrosiers, 2014; Wu et al., 2023). Addressing water pollution requires a multi-faceted approach that includes the development of advanced treatment technologies and strict regulations on industrial discharges. The implementation of physical, chemical, and biological treatments can help reduce the impact of these pollutants on water bodies (Rathi, Kumar, & Show, 2021; Zhao et al., 2018). Moreover, promoting the use of eco-friendly materials and reducing plastic waste can prevent contamination from reaching water sources.

21.4.4 Centralized Governance

Effective water governance is fundamental for ensuring equitable and efficient water management. A centralized governance system, led by a water authority, is responsible for regulating water-related activities, such as licensing wells, desalination, wastewater treatment, and infrastructure development. These authorities must establish clear guidelines for water pricing and allocation to optimize usage (Bar-Nahum, Reznik, Finkelshtain, & Kan, 2022; Domènech, 2011). However, governments alone cannot address the complexities of water management; thus, collaborative governance is essential. This approach involves the active participation of multiple stakeholders, including government agencies, nongovernmental organizations (NGOs), community groups, and the private sector, which brings together expertise and resources to manage water resources more effectively. This collaborative effort is crucial for understanding regional and global water needs, planning for extreme events like droughts or floods, and ensuring water availability across all sectors (Galvez, Rojas, Bennison, Prats, & Claro, 2020; Pugel et al., 2022).

21.4.5 Irrigation and Agricultural Practices

Agriculture is the largest consumer of water, accounting for about 70% of global freshwater withdrawals (Figure 21.3) (FAO, 2020). To optimize water use in agriculture, precision irrigation technologies such as drip irrigation and smart irrigation systems are essential. These technologies monitor soil moisture, weather conditions, and plant needs, ensuring water is applied efficiently and at the right time. Additionally, strategies like crop rotation and conservation tillage can reduce water waste and improve soil health (Bwambale, Abagale, & Anornu, 2022; Levidow et al., 2014). Although drip irrigation systems offer significant water savings, they face challenges like clogging and salinization. Therefore, innovative solutions such as subsurface irrigation and fertigation can help address these issues and further enhance water-use efficiency in agriculture (Tal, 2016).

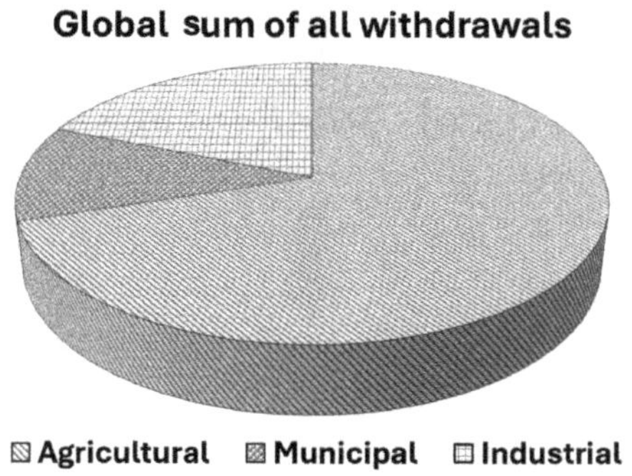

FIGURE 21.3 Global water consumption rates (Panhwar et al., 2022). (Reprinted (adapted) with permission from Panhwar et al. (2022). Copyright (2024) IntechOpen-Water Quality-New Perspectives.)

21.4.6 Water Infrastructure

A robust water infrastructure system is essential for the efficient management and distribution of water. Aging infrastructure, improper maintenance, and cyber-physical threats are some of the challenges facing water systems today (Pamidimukkala, Kermanshachi, Adepu, & Safapour, 2021). Investing in the repair and modernization of infrastructure is crucial to ensuring the continuous delivery of clean water. Furthermore, the integration of cyber-physical systems (CPSs) can improve the monitoring and management of water systems, enabling real-time detection of leaks, contamination, and inefficiencies (Ameen et al., 2023; Liu, Yang, & Yang, 2021; Ozdemir, Turkan, Kilinc, Altuner, & Sen, 2022). However, these advancements must be complemented by strong cybersecurity measures to protect the water infrastructure from potential cyber-attacks, which could compromise water safety and availability (Taormina, Galelli, Tippenhauer, Salomons, & Ostfeld, 2017).

21.4.7 Transboundary Water Cooperation

Water resources often span multiple countries, requiring international cooperation to ensure their sustainable use. Transboundary water cooperation involves shared management of river basins, aquifers, and lakes to optimize water use and minimize conflicts (Petersen-Perlman, Veilleux, & Wolf, 2017). A prominent example is the Jordan–Israel energy and water agreement, which demonstrates how countries can collaborate on water and energy resources to benefit both parties. Effective transboundary cooperation requires transparent information sharing, equitable access, and political will from all involved countries (Shemer et al., 2023). Such collaboration not only enhances regional stability but also ensures that water resources are managed in a way that benefits all nations involved.

21.4.8 Novel Technologies

Incorporating novel, cost-effective, and robust technologies is crucial for improving water and wastewater treatment processes (Erduran, Altuner, & Şen, 2023; Tatli, Gunduz, Sahin, Altuner, & Dar, 2024). However, the adoption of new technologies in the water sector faces challenges such as high upfront costs, infrastructure limitations, and regulatory barriers (Ahmed, Johnson, Hashaikeh, & Hilal, 2023). To overcome these hurdles, governments and private sectors should invest in research and development, fostering innovation and creating a regulatory environment that supports new technological advancements. Additionally, integrating emerging technologies with existing systems will help increase efficiency and scalability, making them accessible to a broader range of communities and industries (Süsser, 2020).

21.5 THE USE OF ARTIFICIAL INTELLIGENCE IN WATER RESOURCE MANAGEMENT

The management of global water resources has become increasingly complex due to factors such as population growth, rapid urbanization, and climate change. In response to these challenges, innovative solutions offered by AI and machine learning (ML) technologies significantly contribute to the efficient and sustainable use of water. AI-based water technology solutions enable the development of hybrid approaches that integrate centralized and decentralized water systems. This integration allows for overcoming the limitations of current centralized water systems and building more resilient infrastructure (Sarni et al., 2018).

AI and ML are widely utilized to optimize wastewater treatment processes. Traditional methods like membrane filtration, coagulation, and chlorination can be made more efficient with AI models. Specifically, approaches that reduce chemical consumption or increase the use of waste-derived adsorbents lower costs while aligning with sustainability principles (Lowe, Qin, & Mao, 2022). Additionally, studies on removing heavy metals and organic pollutants highlight the predictive and optimization capabilities of AI (Che, Wei, & Wen, 2022; El Bilali, Taleb, & Brouziyne, 2021). In the agricultural sector, which consumes over 70% of global water resources (El Bilali et al., 2021), increasing population and food demand necessitate more efficient irrigation management. AI and ML assist in predicting the water needs of plant species, determining when, where, and how much irrigation is required. These approaches aim to enhance both water use efficiency and crop yields (Glória, Cardoso, & Sebastião, 2021; Sidhu, Kumar, & Rana, 2020). Urban water distribution systems suffer from leakage, causing daily losses of up to 45 million m^3 in developing countries. Sensor-based monitoring systems supported by AI and ML techniques provide effective solutions to reduce these losses. These systems not only detect leaks but also optimize overall water consumption control, preventing economic losses (Alves Coelho, Glória, & Sebastião, 2020; Mounce, Pedraza, Jackson, Linford, & Boxall, 2015). In hydrology, AI and ML are applied in critical areas such as rainfall forecasting, GWL monitoring, and disaster prediction. These applications, including early warning systems for floods and droughts, contribute to preventing both human and economic losses (Mosaffa, Sadeghi, Mallakpour, Jahromi, & Pourghasemi, 2022; Schmidt, Heße, Attinger, & Kumar, 2020). Thus, AI and ML offer revolutionary opportunities in water resource management. Spanning a broad range of uses, from wastewater treatment to agricultural irrigation, and urban infrastructure to hydrology, these technologies not only ensure the efficient use of water but also support sustainability and circular economy goals. As highlighted in the literature, these multidisciplinary approaches hold promise for addressing global water issues (Doorn, 2021; Lowe et al., 2022).

21.5.1 Reducing Water Losses with AI

Globally, approximately 126 billion m^3 of water, valued at around US Dollar (USD) 39 billion, is lost annually due to non-revenue water (NRW). Major causes of water losses include network failures resulting from high pressure and the use of outdated pipe materials in older networks (Rojek & Studzinski, 2019). Detecting and locating leaks is crucial for reducing water losses and ensuring the sustainable use of resources. The volume of water lost due to leaks is often dependent on the time elapsed between the occurrence of a leak and its detection by the water service provider (Hu, Chen, Chen, Tan, & Shen, 2021). These losses can be controlled through optimizing pump operations and pump stations or renewing pipeline networks. However, the optimization process is computationally intensive as it requires complex algorithms, while network renewal incurs high costs (Rojek & Studzinski, 2019). To prevent water losses, various initiatives leverage data from sensors, advanced numerical techniques, and water network modeling. Two main approaches have been adopted for network analysis: data-driven and physics-based methods. Physics-based methods combine statistical tools with hydraulic modeling to provide digital simulations of water networks and allow real-time scenario testing. These methods serve as the starting point for the digital transformation of water management units. In contrast, data-driven approaches utilize ML or AI algorithms to extract information and identify patterns by analyzing large datasets without relying on network equations (Jung & Kim, 2018; Vrachimis, Eliades, & Polycarpou, 2018; Yipeng Wu & Liu, 2017). With access to big data, AI can offer solutions to minimize water losses.

21.5.2 Groundwater Management with AI

The measurement and analysis of GWLs are essential for managing groundwater resources. Information on GWL fluctuations can be used to assess groundwater availability and directly analyze the impacts of groundwater development on wells and aquifers. Continuously recorded GWL time series provide valuable insights into aquifer dynamics (Butler Jr, Stotler, Whittemore, & Reboulet, 2013). Therefore, water managers and engineers must model and predict GWL to evaluate groundwater resources and maintain the water supply–demand balance (Rajaee et al., 2019). Licensing and permitting systems play a critical role in groundwater sustainability. These systems define groundwater rights and monitor usage over time to ensure effective management. Licenses serve to protect groundwater quality, regulate extraction rates, and maintain usage within environmentally, socially, and technically acceptable limits. Economic mechanisms and the establishment of groundwater rights and licensing are vital components of sustainable permitting systems. Although traditional physical or conceptual models have been used for modeling GWLs, these models often face practical limitations due to the need for extensive data and numerous parameters. In many cases, when data are limited, achieving accurate predictions becomes a priority, making AI models a viable option. AI, in conjunction with the Internet of Things (IoT), facilitates monitoring groundwater usage. Wireless sensor networks (WSNs) enable the collection of data on physical conditions and remote communication, allowing system management without requiring extensive physical infrastructure (Sauvé & Desrosiers, 2014).

21.6 CONCLUSION

Water scarcity remains one of the most critical challenges of the 21st century, threatening global sustainability and human well-being. This chapter underscores the complexity of water scarcity as a multidimensional problem, shaped by climate change, population growth, urbanization, and mismanagement of water resources. The analysis reveals that traditional water management strategies, while valuable, are insufficient to tackle the escalating demand and challenges posed by water scarcity. Thus, a paradigm shift toward innovative and sustainable solutions is essential.

AI emerges as a transformative tool in addressing water scarcity. Its ability to process large datasets, predict hydrological trends, and optimize water usage offers a significant advantage in resource management. AI-driven applications in leak detection, irrigation efficiency, drought prediction, and groundwater monitoring provide practical solutions that enhance decision-making, reduce water loss, and improve system resilience. Moreover, the integration of AI with smart technologies, such as IoT-enabled sensors and real-time data analytics, exemplifies a forward-looking approach to water resource management, bridging the gap between science and actionable policies.

The chapter also highlights the importance of international collaboration, policy reform, and stakeholder engagement in ensuring the equitable distribution and sustainable use of water resources. Addressing water scarcity requires a holistic approach that aligns technological advancements with socio-economic considerations, public awareness, and global cooperation. Achieving this alignment will not only mitigate the immediate impacts of water scarcity but also contribute to the attainment of the United Nations SDGs, particularly those focused on clean water, sanitation, and sustainable cities. This chapter lays the groundwork for future research and practical applications in sustainable water management by identifying key directions for progress. Developing enhanced AI algorithms is essential, focusing on integrating diverse datasets—including climate, socio-economic, and ecological variables—to improve prediction accuracy and decision-making in water management. Climate-resilient strategies must also be prioritized, leveraging AI to adapt systems to the uncertainties posed by climate change, such as shifting rainfall patterns and increasing drought frequencies. Additionally, the exploration of decentralized water systems, which combine centralized and localized approaches, can bolster resilience and reduce reliance on large-scale infrastructure. Addressing ethical concerns in AI deployment is equally critical, ensuring that technological advancements benefit marginalized communities and promote social equity in water distribution. Lastly, fostering interdisciplinary collaboration among hydrologists, data scientists, policymakers, and local communities is vital for designing context-specific, regionally tailored solutions. Together, these directions emphasize the importance of innovative technology and global cooperation in securing water resources for a sustainable and equitable future.

REFERENCES

Adetoro, A. A., Abraham, S., Paraskevopoulos, A. L., Owusu-Sekyere, E., Jordaan, H., & Orimoloye, I. R. (2020). Alleviating water shortages by decreasing water footprint in sugarcane production: The impacts of different soil mulching and irrigation systems in South Africa. *Groundwater for Sustainable Development*, *11*, 100464.

Agnes, E. A. B., Matchawe, C., Nsawir, B. J., Manuela, B. B., Adjele, J. J., Mouafo, H. T., ... & Noni, L. D. (2021). The use of alternative water sources as a means of adaptation to water shortages in Nsimeyong, Yaounde city: A quality assessment. *Scientific African*, *13*, e00861.

Ahmed, F., Johnson, D., Hashaikeh, R., & Hilal, N. (2023). Barriers to innovation in water treatment. *Water*, *15*(4), 773.

Alim, M. A., Rahman, A., Tao, Z., Samali, B., Khan, M. M., & Shirin, S. (2020). Suitability of roof harvested rainwater for potential potable water production: A scoping review. *Journal of Cleaner Production*, *248*, 119226.

Alves Coelho, J., Glória, A., & Sebastião, P. (2020). Precise water leak detection using machine learning and real-time sensor data. *IoT*, *1*(2), 474–493.

Ameen, F., Altuner, E. E., Tiri, R. N. E., Gulbagca, F., Aygun, A., Sen, F., ... & Dragoi, E. N. (2023). Highly active iron (II) oxide-zinc oxide nanocomposite synthesized *Thymus vulgaris* plant as bioreduction catalyst: Characterization, hydrogen evolution and photocatalytic degradation. *International Journal of Hydrogen Energy*, *48*(55), 21139–21151.

Baba, A., Tsatsanifos, C., El Gohary, F., Palerm, J., Khan, S., Mahmoudian, S. A., ... & Angelakis, A. N. (2018). Developments in water dams and water harvesting systems throughout history in different civilizations. *International Journal of Hydrology*, *2*(2), 155–171.

Bar-Nahum, Z., Reznik, A., Finkelshtain, I., & Kan, I. (2022). Centralized water management under lobbying: Economic analysis of desalination in Israel. *Ecological Economics*, *193*, 107320.

Basel, B., Hernández Quiroz, N., Velasco Herrera, R., Santiago Alonso, C., & Hoogesteger, J. (2021). Bee mietii rak rkabni nis (The people know how to seed water): A Zapotec experience in adapting to water scarcity and drought. *Climate and Development*, *13*(9), 792–806.

Bauer, S., & Wagner, M. (2022). Possibilities and challenges of wastewater reuse: Planning aspects and realized examples. *Water*, *14*(10), 1619.

Burek, P., Satoh, Y., Fischer, G., Kahil, M. T., Scherzer, A., Tramberend, S., ... & Flörke, M. (2016). Water futures and solution-fast track initiative (Final Report). IIASA Working Paper. IIASA, Laxenburg, Austria: WP-16-006.

Butler Jr, J., Stotler, R., Whittemore, D., & Reboulet, E. (2013). Interpretation of water level changes in the high plains aquifer in western Kansas. *Groundwater*, *51*(2), 180–190.

Bwambale, E., Abagale, F. K., & Anornu, G. K. (2022). Smart irrigation monitoring and control strategies for improving water use efficiency in precision agriculture: A review. *Agricultural Water Management*, *260*, 107324.

Chartres, C., & Williams, J. (2006). Can Australia overcome its water scarcity problems? *Journal of Developments in Sustainable Agriculture*, *1*(1), 17–24.

Che, Y., Wei, Z., & Wen, J. (2022). Construction of surface water pollution prediction model based on machine learning. *Scientific Programming*, *2022*(1), 7211094.

Dada, M. A., Oliha, J. S., Majemite, M. T., Obaigbena, A., Nwokediegwu, Z. Q. S., & Daraojimba, O. H. (2024). Review of nanotechnology in water treatment: Adoption in the USA and prospects for Africa. *World Journal of Advanced Research and Reviews*, *21*(1), 1412–1421.

Darabi, R., Ghorbani-HasanSaraei, A., Masoomzadeh, S., Monadi Sefidan, A., Gulbagca, F., Elhouda Tiri, R. N., Zghair Al-Khafaji, A. H., Altuner, E. E., Sen, F., Davarnia, B., & Mortazavi, S. M. (2023). Enhanced photocatalytic performance of auto-combusted nanoparticles for photocatalytic degradation of azo dye under sunlight illumination and hydrogen fuel production. *Chemosphere*, *336*, 139266. https://doi.org/10.1016/J.CHEMOSPHERE.2023.139266

Darabi, R., Karimi-Maleh, H., Shahidi, S. A., Gulbagca, F., Nour Elhouda Tiri, R., Monadi Sefidan, A., Dehdashtian, N., Alhrasishawi, W., Aygun, A., Esra Altuner, E., & Sen, F. (2023). Hexamethylenetetramine-based nanomaterial synthesis, characterization, investigation of photocatalytic activity against Rhodamine B dye and kinetic study for hydrogen fuel production. *Fuel*, *352*, 128841. https://doi.org/10.1016/J.FUEL.2023.128841

Diop, S. (2002). *Vital Water Graphics: An Overview of the State of the World's Fresh and Marine Waters*, UNEP, Nairobi.

Domènech, L. (2011). Rethinking water management: From centralised to decentralised water supply and sanitation models. *Documents d'anàlisi geogràfica*, *57*(2), 293–310.

Doorn, N. (2021). Artificial intelligence in the water domain: Opportunities for responsible use. *Science of the Total Environment, 755*, 142561.

El Bilali, A., Taleb, A., & Brouziyne, Y. (2021). Groundwater quality forecasting using machine learning algorithms for irrigation purposes. *Agricultural Water Management, 245*, 106625.

Erduran, V., Altuner, E. E., & Şen, F. (2023). Stability and validation of bionanomaterials. In Öztürk, M. A., Roy, A., Bhat, R. A., Vardar-Sukan, F., & Tonelli, F. M. P. (Eds.) *Synthesis of Bionanomaterials for Biomedical Applications* (pp. 251–263). Amsterdam: Elsevier.

Eslamian, S. (2016). *Urban Water Reuse*. Nova Yorque: CRC Press Taylor & Francis Group.

Falkenmark, M. (2013). Growing water scarcity in agriculture: Future challenge to global water security. *Philosophical Transactions of the Royal Society A: Mathematical, Physical and Engineering Sciences, 371*(2002), 20120410.

FAO. (2020). *The State of Food and Agriculture 2020. Overcoming Water Challenges in Agriculture*. Roma: FAO.

Flörke, M., Schneider, C., & McDonald, R. I. (2018). Water competition between cities and agriculture driven by climate change and urban growth. *Nature Sustainability, 1*(1), 51–58.

Galvez, V., Rojas, R., Bennison, G., Prats, C., & Claro, E. (2020). Collaborate or perish: Water resources management under contentious water use in a semiarid basin. *International Journal of River Basin Management, 18*(4), 421–437.

Garrick, D., De Stefano, L., Yu, W., Jorgensen, I., O'Donnell, E., Turley, L., ... & Punjabi, B. (2019). Rural water for thirsty cities: A systematic review of water reallocation from rural to urban regions. *Environmental Research Letters, 14*(4), 043003.

Glória, A., Cardoso, J., & Sebastião, P. (2021). Sustainable irrigation system for farming supported by machine learning and real-time sensor data. *Sensors, 21*(9), 3079.

Gohari, A., Eslamian, S., Mirchi, A., Abedi-Koupaei, J., Bavani, A. M., & Madani, K. (2013). Water transfer as a solution to water shortage: A fix that can backfire. *Journal of Hydrology, 491*, 23–39.

Goralski, M. A., & Tan, T. K. (2020). Artificial intelligence and sustainable development. *The International Journal of Management Education, 18*(1), 100330.

Greve, P., Kahil, T., Mochizuki, J., Schinko, T., Satoh, Y., Burek, P., ... & Langan, S. (2018). Global assessment of water challenges under uncertainty in water scarcity projections. *Nature Sustainability, 1*(9), 486–494.

He, C., Liu, Z., Wu, J., Pan, X., Fang, Z., Li, J., & Bryan, B. A. (2021). Future global urban water scarcity and potential solutions. *Nature Communications, 12*(1), 4667.

Hisdal, H., Tallaksen, L. M., Gauster, T., Bloomfield, J. P., Parry, S., Prudhomme, C., & Wanders, N. (2024). Hydrological drought characteristics. In Tallaksen, L. M., & van Lanen, H. A. J. (Eds.) *Hydrological Drought* (pp. 157–231). Amsterdam: Elsevier.

Hogan, A., & Young, M. (2013). Visioning a future for rural and regional Australia. *Cambridge Journal of Regions, Economy and Society, 6*(2), 319–330.

Hou, C., Wen, Y., Liu, X., & Dong, M. (2021). Impacts of regional water shortage information disclosure on public acceptance of recycled water: Evidences from China's urban residents. *Journal of Cleaner Production, 278*, 123965.

Hu, Z., Chen, B., Chen, W., Tan, D., & Shen, D. (2021). Review of model-based and data-driven approaches for leak detection and location in water distribution systems. *Water Supply, 21*(7), 3282–3306.

Inman, D., & Ribarova, I. (2009). Modelling water saving in the upper Iskar region, In *Integrating Water Systems*, 1st edition, Taylor & Francis.

Jenny, H., Alonso, E. G., Wang, Y., & Minguez, R. (2020). Using artificial intelligence for smart water management systems. *ADB Briefs, 1*(143).

Jung, D., & Kim, J. H. (2018). State estimation network design for water distribution systems. *Journal of Water Resources Planning and Management, 144*(1), 06017006.

Jury, W. A., & Vaux Jr, H. J. (2007). The emerging global water crisis: Managing scarcity and conflict between water users. *Advances in Agronomy, 95*, 1–76.

Kerstens, S. (2013). *Downstream Impacts of Water Pollution in the Upper Citarum River, West Java, Indonesia: Economic Assessment of Interventions to Improve Water Quality*.

Kocak, Y., Aygun, A., Altuner, E. E., Gonca, S., Berikten, D., Tiri, R. N. E., & Sen, F. (2024). Eco-friendly production of platinum nanoparticles: Physicochemical properties, evaluation of biological and catalytic activities. *International Journal of Environmental Science and Technology, 21*, 51–62. https://link.springer.com/article/10.1007/S13762-023-05232-W

Kummu, M., Guillaume, J. H., de Moel, H., Eisner, S., Flörke, M., Porkka, M., ... & Ward, P. J. (2016). The world's road to water scarcity: Shortage and stress in the 20th century and pathways towards sustainability. *Scientific Reports, 6*(1), 1–16.

Larsen, T. A., Hoffmann, S., Lüthi, C., Truffer, B., & Maurer, M. (2016). Emerging solutions to the water challenges of an urbanizing world. *Science, 352*(6288), 928–933.

Levidow, L., Zaccaria, D., Maia, R., Vivas, E., Todorovic, M., & Scardigno, A. (2014). Improving water-efficient irrigation: Prospects and difficulties of innovative practices. *Agricultural Water Management, 146*, 84–94.
Liu, Q., Yang, L., & Yang, M. (2021). Digitalisation for water sustainability: Barriers to implementing circular economy in smart water management. *Sustainability, 13*(21), 11868.
Lowe, M., Qin, R., & Mao, X. (2022). A review on machine learning, artificial intelligence, and smart technology in water treatment and monitoring. *Water, 14*(9), 1384.
Luo, Z., Xu, H., & Chen, F. (2019). Audio sentiment analysis by heterogeneous signal features learned from utterance-based parallel neural network. *Paper presented at the AffCon@ AAAI.*
Malik, R. P. S., Giordano, M., & Sharma, V. (2014). Examining farm-level perceptions, costs, and benefits of small water harvesting structures in Dewas, Madhya Pradesh. *Agricultural Water Management, 131*, 204–211.
McAdam, J. (2011). *Climate Change Displacement and International Law: Complementary Protection Standards.* UNHCR, Division of International Protection, Geneva.
Mekonnen, M. M., & Hoekstra, A. Y. (2016). Four billion people facing severe water scarcity. *Science Advances, 2*(2), e1500323.
Mitlin, D., Beard, V. A., Satterthwaite, D., & Du, J. (2019). Unaffordable and undrinkable: Rethinking urban water access in the global south, World Resources Institute.
Mosaffa, H., Sadeghi, M., Mallakpour, I., Jahromi, M. N., & Pourghasemi, H. R. (2022). Application of machine learning algorithms in hydrology. In Pourghasemi, H. R. (Ed.) *Computers in Earth and Environmental Sciences* (pp. 585–591). Amsterdam: Elsevier.
Mounce, S., Pedraza, C., Jackson, T., Linford, P., & Boxall, J. (2015). Cloud based machine learning approaches for leakage assessment and management in smart water networks. *Procedia Engineering, 119*, 43–52.
Otieno, F. A., & Ochieng, G. M. (2004). Water management tools as a means of averting a possible water scarcity in South Africa by the year 2025. *Water SA, 30*(5), 120–124.
Ozdemir, S., Turkan, Z., Kilinc, E., Altuner, E. E., & Sen, F. (2022). Anoxybacillus flavithermus loaded ɣ-Fe_2O_3 magnetic nanoparticles as an efficient magnetic sorbent for the preconcentrations of Cu (II) and Mn (II). *Food and Chemical Toxicology, 168*, 113334.
Pamidimukkala, A., Kermanshachi, S., Adepu, N., & Safapour, E. (2021). Resilience in water infrastructures: A review of challenges and adoption strategies. *Sustainability, 13*(23), 12986.
Panhwar, A., Abro, R., Kandhro, A., Khaskheli, A. R., Jalbani, N., Gishkori, K. A., Mahar, A. M., Qaisar, S., Panhwar, A., Abro, R., Kandhro, A., Khaskheli, A. R., Jalbani, N., Gishkori, K. A., Mahar, A. M., & Qaisar, S. (2022). Global water mapping, requirements, and concerns over water quality shortages. In *Water Quality - New Perspectives.* IntechOpen. https://doi.org/10.5772/INTECHOPEN.108331
Petersen-Perlman, J. D., Veilleux, J. C., & Wolf, A. T. (2017). International water conflict and cooperation: Challenges and opportunities. *Water International, 42*(2), 105–120.
Pugel, K., Javernick-Will, A., Peabody, S., Nyaga, C., Mussa, M., Mekonta, L., ..., & Mulatu, T. (2022). Pathways for collaboratively strengthening water and sanitation systems. *Science of the Total Environment, 802*, 149854.
Rajaee, T., Ebrahimi, H., & Nourani, V. (2019). A review of the artificial intelligence methods in groundwater level modeling. *Journal of Hydrology, 572*, 336–351.
Rathi, B. S., Kumar, P. S., & Show, P.-L. (2021). A review on effective removal of emerging contaminants from aquatic systems: Current trends and scope for further research. *Journal of Hazardous Materials, 409*, 124413.
Rojek, I., & Studzinski, J. (2019). Detection and localization of water leaks in water nets supported by an ICT system with artificial intelligence methods as a way forward for smart cities. *Sustainability, 11*(2), 518.
Rozhok, A., Storozhenko, A., Valiaeva, A., Sushchev, S., Ugarov, A., & Revetria, R. (2021). Requirements for the automated monitoring system to reduce environmental risk during operation of trunk pipelines. In *IOP Conference Series: Earth and Environmental Science* (Vol. 864, p. 012049).
Rozhok, A., Sushchev, S., Tatarinov, V., & Revetria, R. (2019). Mathematical model for the calculation of dynamic risk analysis of emergencies at hydraulic structures. In *AIP Conference Proceedings* (Vol. 2195, p. 020048).
Sachidananda, M., Webb, D. P., & Rahimifard, S. (2016). A concept of water usage efficiency to support water reduction in manufacturing industry. *Sustainability, 8*(12), 1222.
Salehi, M. (2022). Global water shortage and potable water safety: Today's concern and tomorrow's crisis. *Environment International, 158*, 106936.
Sarni, W., Stinson, C., Mung, A., Garcia, B., Bryan, S., & Swanborough, J. (2018). Harnessing the fourth industrial revolution for water. *World Economic Forum.*
Sauvé, S., & Desrosiers, M. (2014). A review of what is an emerging contaminant. *Chemistry Central Journal, 8*, 1–7.
Schewe, J., Heinke, J., Gerten, D., Haddeland, I., Arnell, N. W., Clark, D. B., ... & Colón-González, F. J. (2014). Multimodel assessment of water scarcity under climate change. *Proceedings of the National Academy of Sciences, 111*(9), 3245–3250.

Schilstra, M., Wang, W., van Oel, P. R., Wang, J., & Cheng, H. (2024). The effects of reservoir storage and water use on the upstream–downstream drought propagation. *Journal of Hydrology*, *631*, 130668.

Schmidt, L., Heße, F., Attinger, S., & Kumar, R. (2020). Challenges in applying machine learning models for hydrological inference: A case study for flooding events across Germany. *Water Resources Research*, *56*(5), e2019WR025924.

Seckler, D. W. (1998). *World Water Demand and Supply, 1990 to 2025: Scenarios and Issues* (Vol. 19). IWMI, Basel.

Shemer, H., Wald, S., & Semiat, R. (2023). Challenges and solutions for global water scarcity. *Membranes*, *13*(6), 612.

Sidhu, R. K., Kumar, R., & Rana, P. S. (2020). Machine learning based crop water demand forecasting using minimum climatological data. *Multimedia Tools and Applications*, *79*, 13109–13124.

Singh, A., Sharma, A., Verma, R. K., Chopade, R. L., Pandit, P. P., Nagar, V., Aseri, V., Choudhary, S. K., Awasthi, G., Awasthi, K. K., Sankhla, M. S., Singh, A., Sharma, A., Verma, R. K., Chopade, R. L., Pandit, P. P., Nagar, V., Aseri, V., Choudhary, S. K., ... & Sankhla, M. S. (2022). Heavy metal contamination of water and their toxic effect on living organisms. In *The Toxicity of Environmental Pollutants*. IntechOpen. https://doi.org/10.5772/INTECHOPEN.105075

Smith, T. (2011). *Overcoming Challenges in Wastwater Reuse: A Case Study of San Antonio, Texas*. Harvard University, Cambridge.

Šulyová, D., Vodák, J., & Kubina, M. (2021). Effective management of scarce water resources: From antiquity to today and into the future. *Water*, *13*(19), 2734.

Süsser, D. (2020). Accelerating Cleantech Commercialization in Israel. Green Innovation as Catalyst for Sustainable Development. Policy Paper Series "Decarbonization Strategies in Germany and Israel". Potsdam/Tel Aviv: Institute for Advanced Sustainability Studies (IASS), Israel Public Policy Institute (IPPI), Heinrich-Böll-Stiftung Tel Aviv.

Tal, A. (2016). Rethinking the sustainability of Israel's irrigation practices in the Drylands. *Water Research*, *90*, 387–394.

Taormina, R., Galelli, S., Tippenhauer, N. O., Salomons, E., & Ostfeld, A. (2017). Characterizing cyber-physical attacks on water distribution systems. *Journal of Water Resources Planning and Management*, *143*(5), 04017009.

Tatli, H. N., Gunduz, S. B., Sahin, M., Altuner, E. E., & Dar, U. A. (2024). Fluorimetric methods for determination of aluminum in water resources utilizing newly synthesized N,N′-bis(2,5-dihydroxybenzylidene)-4,4′-diamino diphenyl ether. *Spectrochimica Acta Part A: Molecular and Biomolecular Spectroscopy*, *321*, 124726.

Tiza, M. T. (2023). An overview of the application of artificial intelligence in water engineering. *International Journal of Engineering Science and Application*, *7*(1), 22–31.

Trogadas, P., Cho, J. I., Rasha, L., Lu, X., Kardjilov, N., Markötter, H., ... & Coppens, M.-O. (2024). A nature-inspired solution for water management in flow fields for electrochemical devices. *Energy & Environmental Science*, *17*(5), 2007–2017.

UNICEF. (2019). *Progress on Drinking Water, Sanitation and Hygiene, 2000–2017*, UNICEF, New York.

Vrachimis, S. G., Eliades, D. G., & Polycarpou, M. M. (2018). Real-time hydraulic interval state estimation for water transport networks: A case study. *Drinking Water Engineering and Science*, *11*(1), 19–24.

Wang, J., & Azam, W. (2024). Natural resource scarcity, fossil fuel energy consumption, and total greenhouse gas emissions in top emitting countries. *Geoscience Frontiers*, *15*(2), 101757.

Wang, Y., & Razmjooy, S. (2024). Prediction of drought hydrological and water scarcity based on optimal artificial intelligence by developing a metaheuristic optimization algorithm. *Physics and Chemistry of the Earth, Parts A/B/C*, *135*, 103669.

World Health Organization. (2019). *Progress on Household Drinking Water, Sanitation and Hygiene 2000–2017: Special Focus on Inequalities*. World Health Organization, Geneva.

Wu, F., Yang, X., Cui, Z., Ren, L., Jiang, S., Liu, Y., & Yuan, S. (2024). The impact of human activities on blue-green water resources and quantification of water resource scarcity in the Yangtze River Basin. *Science of the Total Environment*, *909*, 168550.

Wu, Y., & Liu, S. (2017). A review of data-driven approaches for burst detection in water distribution systems. *Urban Water Journal*, *14*(9), 972–983.

Wu, Y., Tiri, R. N. E., Bekmezci, M., Altuner, E. E., Aygun, A., Mei, C., ... & Sen, F. (2023). Synthesis of novel activated carbon-supported trimetallic Pt–Ru–Ni nanoparticles using wood chips as efficient catalysts for the hydrogen generation from $NaBH_4$ and enhanced photodegradation on methylene blue. *International Journal of Hydrogen Energy*, *48*(55), 21055–21065.

Yavarian, H., Jalalian, H., & Tahmasebi, A. (2023). Assessment of farmers' attitudes towards water scarcity challenge and its adaptation strategies in villages of Bahar County (Hamadan Province, Iran). *Applied Researches in Geographical Sciences*, *23*(71), 21–38.

Young, G., Shah, B., & Kimaite, F. (2008). UN Water Report. Status report on integrated water resources management and water efficiency plans, United Nations.

Zhan, C., Zhang, X., Tian, G., Pham, D. T., Ivanov, M., Aleksandrov, A., … & Wu, Z. (2023). Environment-oriented disassembly planning for end-of-life vehicle batteries based on an improved northern goshawk optimisation algorithm. *Environmental Science and Pollution Research, 30*(16), 47956–47971.

Zhang, J., Chen, L., Xie, Y., Yang, P., Li, Z., Guo, H., … & Liu, L. (2024). Climate change mitigation in energy-dependent regions: A carbon tax-based cross-system bi-layer model with equilibrium-optimization superposition effects. *Resources, Conservation and Recycling, 200*, 107315.

Zhao, L., Deng, J., Sun, P., Liu, J., Ji, Y., Nakada, N., … & Yang, Y. (2018). Nanomaterials for treating emerging contaminants in water by adsorption and photocatalysis: Systematic review and bibliometric analysis. *Science of the Total Environment, 627*, 1253–1263.

Index

AAS (Atomic Absorption Spectroscopy) 70–71
ABM *see* agent-based model
Academic Press 37, 88, 169
accountability 34, 145, 266, 293, 297, 300–301, 319, 328–329, 332–334, 337–338, 347
accuracy 70–72, 74, 76, 84–87, 102–103, 122, 222, 225–226, 236–237, 239–240, 248, 259–260, 274
  improved 85, 311
  improving 118, 344
  level of 219, 243, 350
accuracy rates 121–122, 130, 132, 142–143
ACM Computing Surveys 137, 339
ACS Applied Nano Materials 78, 109
ACS ES&T Water 111, 136, 253, 339
activated carbon 66, 198
activated sludge 134, 193–195
activated sludge processes 14, 128
Active Server Pages (ASP) 187
adaptation 18, 54, 127, 157, 168, 220, 235, 305, 366
adaptation strategies 9, 16, 369
Adaptive-Network-based Fuzzy Inference System (ANFIS) 67, 79, 85, 118, 149, 190, 248, 251, 259, 261, 329
adaptive strategies 167, 170, 323
addition 56, 241
Addressing Nutrient Cycling 234
adsorbents 66, 109, 183, 196, 198, 209–210, 212, 261
  efficient 94, 107
adsorption 12, 66–69, 74, 77, 94, 183, 192–196, 198, 204, 212, 247–248, 253
adsorption capacity 183, 197–198, 214
adsorption efficiency 66, 206, 248
Adsorption Isotherm 196, 207
Adsorption isotherm of MB 207–208
adsorption mechanisms 78, 208, 214
adsorption performance 77, 213
adsorption processes 37, 66, 151, 183, 189, 192, 198, 200, 206, 252, 268
adsorption techniques 66, 147
adsorption thermodynamics 197, 208
adsorption time 248
adsorptive removal 183, 214
Advanced Dynamic Water Quality Assessment 93–111
advanced oxidation processes 19, 185
Advanced Research 107, 354, 366
Advanced Semiconductor Manufacturing Conference (ASMC) 110
advanced sensors 98, 179, 188, 237, 265, 346
advanced technologies 78, 83, 87, 105, 151, 242, 244, 300, 337, 339, 359
advancements 14–15, 86–87, 93–94, 96–97, 104–106, 120–122, 131–132, 135–136, 142–143, 266, 278–279, 284, 353
  continuous 129, 151, 229
Advances in Engineering Software 37–38
aeration 181, 187, 237, 244, 247, 251
aeration process 200, 213, 247
aeration rates 23, 239, 249
aeration tanks 128, 187
Africa 4–5, 17, 30, 293–294, 297–302, 304–305, 316, 366
  middle East and North 310, 322
African countries 297–98
African digital space 303–304
African leaders 293, 297
African Ministers' Council 294, 303
African Ministers' Council on Water (AMCOW) 294, 303
African nations 297, 302
African Water Facility (AWF) 298
Africa Water Vision 294, 306
agent-based model (ABM) 234, 239–240, 249
agricultural activities 72–73, 146
agricultural areas 35, 358
Agricultural Biotechnology 106, 292
agricultural fields 71–72, 143
agricultural practices 9, 13, 92, 158, 170, 177, 302, 357, 362
agricultural production 16, 73
agricultural productivity 35, 160–161, 257, 293, 297, 299–300, 310
agricultural runoff 10, 14, 25, 48, 92, 138, 164, 282, 302, 309, 342
agricultural sectors 3, 15, 53, 140, 164–165, 364
agricultural systems 9, 15, 26
agricultural wastes 24, 149, 151, 198–199
agricultural water management 16, 19, 37, 54, 160, 169, 172, 252, 281, 322, 366–368
agricultural water use efficiency 141, 318
agricultural water waste 26
agriculture, smart 77, 143, 169
AI adoption 33, 299
AI advancements 139, 341
AI algorithms 24, 26–27, 31–32, 74, 166, 199–200, 222–223, 262, 265–266, 331, 335, 337, 364
AI and ML in water management 156–157, 166
AI applications xiii, 24, 46, 85–86, 131–132, 140, 146, 150, 164, 167, 170, 236, 238–239, 244, 247, 321, 330, 353
AI applications in water management 144, 151
AI approaches 183, 357
AIARS (Artificial Intelligence and Autonomous Robot Systems) 292
AI-based algorithms 184, 224
AI-based analytics 28, 314
AI-based DSS 242, 244
AI-based methods 68, 187, 229, 268, 337, 357
AI-based models 52, 71, 83, 258, 311, 331
AI-based monitoring 52, 187
AI-based PdM 51
AI-based sensors 25–26
AI-based systems 26, 51, 139, 143
AI-based technologies 46, 83, 150
AI-based water management 335, 349

AI capabilities 23, 72
AI cleanse 135, 213–214, 250–252, 281, 340
AI deployment 145, 365
AI-designed nanopores 262
AI-driven brine management 262–263
AI-driven ChatGPT solutions 150, 280, 338
AI-driven devices 329, 335
AI-driven framework 274–275, 277–278
AI-driven models 31, 46
AI-driven optimization 116–117, 281
AI-driven PdM systems 51, 116–118, 121, 126, 132–133
AI-driven platforms 74, 301
AI-driven process optimization 129, 266
AI-driven solutions 22, 36, 111, 120, 145, 165, 302, 321, 335, 345
AI-driven systems 25, 32, 34–35, 141, 160, 165, 266, 335, 337, 341, 344–345
AI-driven technologies 22, 151, 172, 263, 277, 342
AI-driven water management 320, 329, 333–337
AI-driven water management facilities 332
AI-driven water management solutions 332, 337
AI-driven water management systems 280, 330, 335–337, 350
AI-driven water policy 277–278, 280
AI-driven water quality monitoring 61, 355
AI-enabled process optimization 213, 251
AI-enabled sensors 88, 268
AI implementation 130–131, 140, 238
AI in Civil Engineering 173, 324
AI Innovation for Water Policy and Sustainability 280
AI innovations 87, 138, 143
AI Innovations for Enhanced Water Management 139–153
AI in seawater desalination 265, 267
AI integration xiii, 237–238, 315
AI in water management 22–23, 31–34, 36, 184, 320, 328–331, 334, 336, 338, 342–343, 345, 347–349, 353
AI methods 69, 139, 183–184, 328
AIML (artificial intelligence markup language) 285
AI modeling 187, 259
AI models
  creating 36, 76
  decentralized 58
  hybrid 237, 240–242, 247
  train 332, 336
  training 265, 341, 348
  in WWT 149, 237
AIMP (Automated Intelligent Microfluidic Platform) 76
AIP Conference Proceedings 108, 368
AI platforms 262, 265
AI-Powered decision support systems 235–253
AI-powered drones 26, 77
AI-powered DSS 242–244
AI-powered systems 159, 342
AI-powered technologies 23, 26
AI-powered water management systems 328, 330
AI-powered water management systems 341
AI researchers 145, 351
AI revolutionizing industries 172, 190
AI solutions 33–34, 241, 265, 318, 321
AI systems 30–34, 51, 159–163, 167, 239, 241–242, 262, 266–267, 314, 319–320, 330–332, 335–338, 351–352
  hybrid 247
AI systems result 331, 335
AI techniques 70–71, 81, 85, 118, 128, 139–140, 183–184, 241–243, 258, 260–261, 328
  advanced 74, 86, 265
AI technologies 32–34, 62, 140, 183–184, 236–238, 241, 247–248, 264–265, 267, 314–319, 329, 336, 352–353
  application of 238, 240
  emerging 352–353
  implementation of 335, 338
AI tools 72, 141–143, 146, 150, 183–184, 230, 258, 263, 314, 317, 348, 358
AI TOOLS 23, 312
AI Training 347–348
Alexandria Engineering Journal 213, 269
algae blooms 163–164
algae treatment 193–194
algorithms 63–65, 69–71, 75, 81, 83–84, 102–103, 140–141, 156–157, 220–221, 225, 299, 316–317, 328–329
  advanced 55, 73, 299
  intelligent 23, 153
  predictive 150–151, 328
allocation 6, 59, 159, 221, 289, 295, 362
Altuner 93, 95, 106–108, 110, 363, 366–369
amalgamation 217, 222, 226–227, 229
AMCOW (African Ministers' Council on Water) 294, 303
AMP databases 226
AMPs *see* antimicrobial proteins
AMR 219, 222–225, 229
  resisting 223
AMR determinants 224, 226
AMR escalation 223
anaerobic bioreactors (AnBRs) 142, 194
analytics 25, 36, 132, 156, 276, 295, 300–301, 343
  advanced 118, 120, 129
AnBRs (anaerobic bioreactors) 142, 194
ANFIS *see* Adaptive-Network-based Fuzzy Inference System
ANFIS models 248, 259
ANL (Argonne National Laboratory) 290
ANN error parameters 210–211
ANN-GA model 249
ANN model 87, 147–148, 200–201, 208, 210–212, 243–244, 246, 264
ANN results 85, 209–210
Annual Reviews in Control 172, 323
anomaly detection 49, 97, 100, 107, 109–110, 118, 121–122, 134, 168, 342, 346, 350
antibiotic resistance 231, 233
antimicrobial activities 226
antimicrobial drugs 223–225
Antimicrobial Peptide Database (APD) 226, 233
antimicrobial proteins (AMPs) 225–226
antimicrobial resistance genes (ARGs) 224–225
APD (Antimicrobial Peptide Database) 226, 233
applications, operational 37, 170, 354
Applied Water Science 90, 251, 291
aquatic ecosystems 13, 90, 94, 263, 282–283, 299–300, 336, 338
ARGs *see* antimicrobial resistance genes
arsenic contaminants 67–68
artificial intelli- gence 339, 354
artificial intelligence applications 81–91, 190, 268, 322, 353
artificial intelligence in healthcare 231–232
Artificial Intelligence in Water Management 140, 153–154

artificial intelligence methods 154, 368
artificial intelligence models 38, 136, 251
Artificial Intelligence Review 214, 232–233, 251–252, 303, 355
artificial intelligence techniques 153, 171, 269, 353
artificial neural networks 37, 78–80, 84, 134–136, 139, 152–153, 221, 234–236, 251–252, 258, 329
artificial neural networks (ANNs) 63, 66–67, 70–71, 79–80, 83–85, 116, 121, 139–141, 146–149, 179–180, 183, 192, 199–201, 206–208, 211, 236–243, 246–248, 258–262, 344
assessment 12, 61 71, 76–77, 88–89, 91–92, 109, 134, 146, 149, 343
Atomic Absorption Spectroscopy (AAS) 70–71
automated irrigation systems 35, 133, 160
automated systems 23, 73, 178, 188, 221, 300
automation 35–37, 42, 71–73, 144–145, 150–151, 171–172, 177–181, 183–189, 222, 286, 344
automation in water treatment xiii, 184, 188
automation technologies 129, 177, 187

back-propagation neural network (BPNN) 70–71, 75, 234, 248, 344
bacteria 12, 18, 30, 82, 93, 138, 193–195, 199, 205, 218, 222–223, 283
bacterial strains 225–226, 233
biases, algorithmic 217, 227, 349
biochemical oxygen demand (BOD) 118, 128, 142, 149, 199, 201, 211, 237, 243, 329, 344
biofilm formation 5–6, 186
biological disks 193–194
biological treatments 26, 182, 194–195, 284, 344, 362
biological wastewater treatment systems 136, 214, 252
bioremediation 69, 283, 290, 292
Bioresource Technology 91, 136, 153, 214, 251–252
blockchain technology 58, 188–189, 318, 331, 334
BOD *see* biochemical oxygen demand
BPNN *see* back-propagation neural network
brackish water (BW) 11, 181, 257, 264
BW (brackish water) 11, 181, 257, 264

capabilities, predictive 24, 117, 159, 301, 344
capacity, global desalination 12, 257
carbon emissions 10, 13, 335, 351
carbon footprint 124, 335, 341, 347–348, 351–354
carbon nanotubes (CNTs) 93–94, 97, 104
CART (Classification and Regression Tree) 81, 84
case studies 65, 76–77, 89–90, 96, 101, 133, 135–136, 164, 169–171, 243–244, 263, 291, 316–317, 369
CBM (condition-based maintenance) 118–119, 135
CCD *see* central composite design
cellular networks 44, 56
central composite design (CCD) 192, 202–203, 211
central processing unit (CPU) 180, 286
CFD (computational fluid dynamics) 260–261
chemical composition 4, 74, 76, 107, 284
chemical contaminants 62, 65, 70, 74
Chemical Engineering Journal 37, 77, 151, 153, 189–190, 213–214, 268
chemical oxygen demand (COD) 23, 117–118, 142, 199, 201, 211, 234, 237, 242–243, 260, 329, 344
chemical precipitation 66, 68–69
chlorination 152, 195, 200, 364
circular economy 20, 36, 235, 251, 335, 340
cities, sustainable 36, 294, 356, 365
classification 18, 46–47, 70, 74, 81, 84, 103, 117, 179–180, 219, 224, 238
Classification and Regression Tree (CART) 81, 84
classification issues 70, 179, 221
clean drinking water 3, 22, 30, 36, 67, 141, 162, 300, 329, 356
Climate and Development 16, 304, 366
climate change 3–4, 7–8, 14–15, 18–19, 22, 34, 36–37, 167–168, 257–258, 294–295, 297–298, 300–301, 309–310, 334–335, 356–359, 365, 367–368
  effects of 167, 294, 335
climate change and water 304, 316, 319
climate change impacts 20, 116, 155, 169
climate change mitigation 323, 353, 370
climate data, historical 27, 31, 54, 160
climate-resilient water policies 28, 31
cloud computing 42, 44–45, 57, 80, 105, 109, 171–172, 178, 186, 188, 285, 291
cloud infrastructure 57, 330–331
cloud storage 56–57, 179
clustering algorithms 97, 102–103
CNNs *see* convolutional neural networks
CNTs *see* carbon nanotubes
coagulation 12, 147, 180–181, 185, 188, 194–195, 200, 364
coagulation flocculation 194, 201
COD *see* chemical oxygen demand
coefficient 183, 235, 237, 239, 241
collaboration 11–12, 95, 106, 150, 167, 291, 297, 320, 341, 363
communities 22–25, 28–31, 36, 105–106, 166–167, 302, 310, 312–321, 327, 330–332, 337, 356
  global xiv, 16, 106
  marginalized 319, 337, 365
  rural 30, 294, 304, 358
Community-Centric Water Conservation 311–323
community engagement xiii, 15, 314, 320–321
community involvement 28–29, 301, 312, 314
community level 33, 297, 312
companies, private 166, 319
Composite Structures 109–110
composition 163, 181–82, 193
comprehensive analysis 88, 125, 132, 220, 222, 226
comprehensive framework 15, 126, 355
comprehensive review 18, 88–91, 107, 110, 117, 171–172, 212–213, 231, 233, 268, 339–340
comprehensive solution 49, 101, 103
comprehensive survey 189, 339, 353–354
computational biology 220
computational fluid dynamics (CFD) 260–261
Computer Sciences 60, 291
computer systems 63, 219, 300
computing power 51, 63, 87, 299, 347–348
concentration 48, 63, 71, 128, 134, 178, 185, 205–207, 209–210, 241, 282
  dye 196, 205–206
  effluent 201, 245
  increased 82, 240
condition-based maintenance (CBM) 118–119, 135
conservation xiii, 7, 14, 22, 24, 28, 30, 36, 162, 164, 280–281, 317, 319, 345
  ecological 87, 336
  environmental 239, 280, 304, 317
conservation efforts 158, 300, 317–318, 341

conservation tillage (CT) 8–9, 14, 16–18, 362
consumption 14, 31, 34, 162, 168, 227, 236, 240, 247, 312, 359
consumption patterns 155, 159, 319
consumption trends 24, 29–30
contaminant levels 66–67, 102, 163
contaminants 5, 25, 62–63, 66–68, 70, 89, 93–94, 103–104, 183, 193–195, 282–283, 342–343
  detecting 93, 103, 105
  detection of xiii, 83, 88, 96, 262, 268
contamination 25, 41, 46, 49, 52, 59, 84–85, 87, 138, 162–163, 361–363
contamination events 97, 105, 109, 284
contamination levels 59, 74, 317
continent 294, 297, 300, 302
continuous monitoring 3, 44, 58, 83, 162
contribution xiv, 36, 97, 108, 123, 153, 223, 229
Convolutional neural networks (CNNs) 46, 59, 63, 72–73, 78, 84–85, 101–102, 109, 122, 142–143, 221–222, 262, 329
cooperation 11–12, 19, 298, 315, 368
  international 15, 168, 363
corruption 299, 301, 332, 334, 339
cost-benefit analysis 51, 127, 132, 168
cost reductions 14, 118, 120, 123–125, 133, 142, 261
costs 13, 32, 51, 178, 236, 244, 247, 261, 278–279, 332, 343–344
  computational 46, 227
  reducing 11, 13, 120, 124, 264–265
cost savings 32, 119, 125, 145, 178, 238, 244, 247, 316
countries 4, 11–12, 150–151, 167, 275, 277–280, 293–294, 297–299, 310, 358, 363
CPU (central processing unit) 180, 286
creation 28, 32–33, 63–64, 122, 182, 187, 239, 242, 246, 273, 347
Critical Reviews in Environmental Science and Technology 16, 88, 154, 190
crop health 27, 35, 52, 54, 72, 143, 166
crop rotation 9, 14, 362
crops 9–10, 26, 52–54, 65, 71–72, 74, 143, 160–161, 164, 329, 359–360, 364
crop shifting 9, 17
crop stress 27, 160
crow search algorithm (CSA) 135, 259
CSA (crow search algorithm) 135, 259
CT *see* conservation tillage
cultural conflicts 328–329, 337
cutting-edge technologies 24, 26, 28–29, 36, 178, 264, 283, 328, 353
cybersecurity measures 334, 350, 363

damage, environmental 160, 164, 335
data analytics 24, 72–73, 93, 105, 118–119, 121, 140, 186, 300, 341, 357
databases 188, 224, 226–227, 238, 250, 277, 287
  antimicrobial peptide 226, 233
data-driven approaches 95, 117, 164, 250, 304–305, 364, 367, 369
data-driven decision-making 36, 116, 157, 263, 314, 336
data-driven decisions 74, 163, 266
data integration 72, 118, 121, 130, 280
data management 56–57, 96, 167, 242, 244
data manipulation 203, 333, 350
data mining 230, 244, 247
data preprocessing 102, 142, 265, 350
data privacy 59, 97, 105, 319–321, 328–330, 337
  and security 165, 168, 317, 330
data quality 34, 36, 51, 128, 133, 230, 246–247, 349–350
data sample 287–288
data security 33, 56, 118, 179, 217, 227–228, 230, 300, 302, 330, 336
datasets 73–74, 95–96, 118–119, 183–184, 217, 220–224, 226–227, 265, 311, 344–345, 365
  complex 101, 141
  labeled 157, 221
data storage 33, 44, 186, 348
DBP formation 147–148, 152
DBPs *see* disinfection by-products
DCF *see* Deep Cascade Forest
Decision support systems (DSS) xiii, 54, 104, 153, 235–253
decision trees (DT) 46–47, 69–71, 81, 83–85, 97, 191–192, 204, 221, 241–242, 258–259, 261
Deep Cascade Forest (DCF) 84–85
Deep learning (DL) 37, 49, 63, 83, 85, 138–139, 141–143, 179–180, 220–221, 266, 300
deep neural network (DNN) 50, 75, 84–86, 89, 108, 122, 143, 221, 329
degradation
  environmental 298–299, 341, 351
  photocatalytic 106, 366
desalination 7, 11–16, 20–21, 28–29, 36, 151, 190, 257–258, 261–270, 316, 362
  thermal 12, 257
Desalination and Water Treatment 135, 171, 189, 213, 269, 281
desalination efficiency 13, 259
desalination facilities 11, 258, 260, 263
desalination industry 257–259
desalination plants 11–14, 25, 258, 261, 264–266
desalination processes 12–13, 15, 257–259, 261–262, 265–266, 268
  optimizing 258, 265–266
desalination systems 258, 261–262, 266–269
desalination technologies xiii, 14, 257, 266, 268
Detecting Chemical Contaminants 63–79
detection 63–65, 71, 108, 145, 217–231, 342–343, 354–355, 364
  leakage 55, 154, 173, 316
devices, smart 22, 44, 328
diagnostic procedure 218–219, 222
diagnostic techniques 218–219
digital simulations 123, 364
digital transformation 108, 118, 120–121, 125, 152, 359, 364
digital twins 23, 122–123, 125, 132, 134–137, 353
digital twin technology 123–124, 132–136
discharges, industrial 25, 138, 240, 282
disease detection 53–54, 72
disinfection 6, 50, 93, 115, 182, 188
  electrochemical 182–183
disinfection by-products (DBPs) 21, 50, 147–148, 152, 180
dissolved organic carbon (DOC) 148, 187
dissolved oxygen 5, 41, 48–49, 52, 59, 62, 75, 128, 149, 163–164, 211, 343
distribution systems 28, 36, 141, 158, 163, 298, 303
DL *see* Deep learning
DL algorithms 72, 87, 222
DL models 63, 122, 140, 142–143, 150–151, 221, 261
DOC (dissolved organic carbon) 148, 187

downtime
minimizing 129, 132, 136
reducing 118, 130, 133, 261, 344
Downtime Optimization 117–137
drinking water 5, 29–30, 81–82, 86, 145, 152–153, 182, 284, 339, 360, 369
drought conditions 25, 313–314
drought prediction 31, 171, 320, 365, 369
droughts 7–8, 22–23, 27–28, 31, 35, 139, 165–166, 299, 301–302, 304, 353–354, 357–358, 364
DSS *see* decision support systems
DT *see* decision trees

early detection 27, 101, 135, 146, 156, 164, 252, 262, 343
Earth Systems and Environment 89, 268, 280
ecological effects 90, 336
economic activities 115, 161, 310
economic analysis 51, 366
economic benefits 10, 131, 263, 293
economic growth 11–12, 15, 22, 80, 141, 151, 310, 322, 356, 358
economy, circular water 26, 29, 35
ecosystems 10–11, 65, 81–82, 86, 282, 327, 329, 336, 343–344, 356, 358–361
ECs *see* emerging contaminants
edge computing 58–59, 111, 128, 133, 285, 291–292, 318, 339
education 29, 57, 106, 139, 300, 303, 314, 320–321, 359–361
artificial intelligence in 303, 323
effective water governance 282–283, 285, 287, 289, 291, 362
electricity consumption 143, 244, 347
electrocoagulation 66, 187, 194–195
electronic waste 335, 341, 351
emerging contaminants (ECs) 10, 16, 18, 49–50, 89, 154, 214, 253, 361, 368, 370
emerging technologies 19, 30, 88, 133, 167, 188, 230, 250, 257, 337, 363
Enabled Water Policy 275, 277, 279, 281
energy 124, 134, 186, 197–198, 247, 251, 261, 268–270, 329, 335, 340–341, 351, 353
clean 278, 347, 352
energy consumption 14–15, 25, 187, 192, 200, 211, 213, 244, 259–261, 266, 335, 341, 351–353
high 32, 199, 257, 335, 348, 350
minimizing 201, 261, 267, 318
reducing 200, 300
Energy Conversion and Management 18, 189, 253
energy efficiency 96, 121, 134, 168, 258, 261, 344
energy-efficient 44, 58, 258, 267, 335–336
energy optimization 124, 191–213, 259
energy requirements 143, 153
energy savings 244, 247, 265
energy sector 11, 50, 133, 359
energy usage 33, 121, 259, 261–263, 267, 348, 352
Engineering Applications of Artificial Intelligence 79, 173, 250–251
Engineering Fracture Mechanics 109–111, 135
Engineering Science & Technology Journal 303, 305
environment 27–29, 55, 57–58, 69, 72, 77, 82, 109–110, 142, 182–184, 192, 221, 235–236, 238, 243, 335–336
aquatic 74, 82, 89, 239, 344
environmental analysis 70, 76, 100
environmental benefits 199, 263, 266, 335
environmental changes 49, 103, 122, 317, 343
Environmental Chemistry Letters 89, 322, 338
environmental conditions 103, 140, 158, 161, 165, 265, 344–345, 348
environmental factors 35, 52, 73, 102, 160, 163–164, 235, 350
environmental footprint 97, 165, 341, 351–352
environmental harm 25, 73, 163, 263
Environmental Impact Reduction 191–213
environmental impacts 11, 14–15, 70, 72–73, 94, 96, 149–150, 258, 261, 267–269, 328–329, 334–335, 341–345, 347–353, 359
environmental monitoring 60, 77, 81, 87–88, 100, 103, 108, 171, 302, 343, 352
Environmental Monitoring and Assessment 21, 61, 78–79, 90, 171, 251
Environmental Nanotechnology 106, 108–109
environmental parameters 24, 52, 78, 127–128, 284, 352
environmental pollution 38, 69, 81–82, 89
Environmental Research 76, 89–90, 213, 250
Environmental Research Letters 20, 304, 340, 367
environmental risks 105, 117, 136, 245, 263, 342, 368
environmental science xiii, 17–19, 77, 107, 152, 250, 252
Environmental Science and Pollution Research 61, 78, 168, 172, 212, 268, 370
environmental scientists 106, 341
environmental sustainability 12, 14, 22, 88, 115, 151, 261–263, 281, 336, 340, 342, 344–345
equipment failures 118, 120, 128, 178, 200, 237, 259, 344
equitable distribution 30, 165, 365
evaluation 20, 61, 63, 77, 91, 93, 115, 137, 142, 150, 199, 287–288
evaluation criteria 184, 287
evolution 44, 125–126, 130, 133, 217, 224, 227, 230, 292, 294, 324
experimental design 152, 204, 206
experimental results 121–122
experimental values 202, 207–210, 248

farmers 9–10, 26–27, 35, 52–54, 71–74, 85, 160, 279, 302, 314, 316, 322
farming, sustainable 9–10, 153
fault detection 121–122, 128, 132, 135, 237, 261
feature extraction 117, 127, 143, 227
Fenton process 185, 190
filters, biological 193–194
filtration 26, 29, 68, 93, 100, 115, 124, 181, 185, 188, 193–195, 200
FIS *see* fuzzy inference systems
flocculation 147, 180–181, 188, 194–195, 200
flood control 28, 285, 289
flood forecasting 34, 346, 349
flood prediction 27, 352, 354
floods 6–8, 11, 22–23, 27, 31, 35–36, 139, 301–302, 346, 353, 357, 359, 362, 364
flow sensor 42–43, 301
food security 6, 10, 15, 78, 164, 171, 302, 310, 356–358
forecast droughts 22, 27, 301
forecast water demand 30, 158, 165, 346
freshwater 3–4, 13, 29, 37, 86, 178, 252, 257, 359
freshwater resources 4, 15, 30, 171, 312, 317, 359
freshwater sources 11, 14, 93
Freundlich model 196, 207
Frontiers in Microbiology 17, 230, 233
fuzzy inference systems (FIS) 66, 70–71, 75

GAs *see* genetic algorithms
Gaza strip 263–264, 267, 269
GCDP *see* Gold Coast Desalination Plant
generalized regression neural network (GRNN) 50, 71, 191, 201
general packet radio service (GPRS) 50, 56
Genesis Water Tech 261, 268
genetic algorithms (GAs) 75, 120, 139, 146, 172, 191–192, 195, 199–201, 203, 205, 211, 235, 238, 242, 246, 258–261
geographic information systems (GIS) 30, 38, 58, 93, 301
GIS *see* geographic information systems
global water crisis xiii, 3, 20, 294, 310, 360, 367
global water issue 22, 364
global water scarcity 14, 20, 97, 263, 267, 270, 345, 369
Gold Coast Desalination Plant (GCDP) 264–265, 268
government agencies 7, 106, 315, 330, 362
GPRS (general packet radio service) 50, 56
GRNN (generalized regression neural network) 50, 71, 191, 201
groundwater 5, 8, 20, 81, 84–86, 180–181, 294, 302, 309, 315–316, 357–358, 366
groundwater levels (GWLs) 8–9, 27, 140, 301, 318, 361, 364–365, 368
groundwater management 29, 316, 346, 364
groundwater quality 85, 88–89, 170, 365
GWLs (groundwater levels) 361, 364–365, 368

healthcare domains 218, 229, 231
health risks 4–5, 21, 71, 82, 87, 90–91, 147, 164
heavy metal contamination 77–78, 90, 234, 240, 369
heavy metal pollution 77, 90, 240, 246, 250, 253
heavy metal removal 37, 94, 111, 236, 241–242, 244–245, 253
heavy metal removal efficiencies 101, 243–244
heavy metals (HM) 25, 62, 64–65, 67–71, 75–79, 82, 92–94, 149, 193–195, 199, 235, 240–241, 244–245, 248, 250–251
  removing 241, 245, 364
hidden layers 71, 207–208, 211–212, 221
high accuracy 84, 116, 121, 183, 240, 259
high costs 67, 74, 145, 199, 361, 364
high-quality data 51, 166, 243
HM *see* heavy metals
Household Drinking Water 20, 306, 369
human activities 48, 62, 82, 138, 142, 359, 369
human beings 82, 219, 221, 225, 228, 273, 277, 280
human consumption 146, 177–178, 263, 273, 276–277
human effort 139, 212, 343
human errors 150, 177, 219, 228, 343
human health 71, 82, 85–87, 89, 92, 178, 227, 240–241, 257, 336, 358–359, 361
human intelligence 63, 157, 200, 228, 230, 299
human intervention 41, 58, 104, 139, 266, 277–278
hybrid approaches 85, 102–103, 133, 170, 363
hybrid models 38, 243, 252, 329
hydrogen fuel production 107, 366
hydrological models 7, 37, 153, 346

identification of key water parameters 37, 88
IDQD 226
IEC 316, 323
IIoT 187, 188, 190
implementations 130–133, 238
improving water quality 25, 76, 86, 93, 146, 163, 172, 283, 293, 344
increment of diversity 226, 231, 232
independent variables inputs 209, 210
industrial activities 138, 163, 342
industrial application 185
industrial effluent 62, 109, 309
industrial IoT 122, 187
industrialization 41, 59, 62, 92, 146, 155, 359
infectious diseases 218, 226–229, 231, 232
information
  education 316
  financing 295
information management 169, 355
information processing 77, 151, 179
information systems 30, 169, 250, 301
infrastructure
  ecosystems 323, 355
  financial 293
  improper 363
infrastructure development 362
infrastructure maintenance 159, 301, 346
infrastructure management 168, 315, 345
innovations
  challenges 281, 304
  designed 99
innovations for enhanced water 139, 141, 143, 145, 147, 149, 151, 153
innovations for enhanced water management 139, 141, 143, 145, 147, 149, 151, 153
innovations for environmental 281
innovations in machine 89, 251
innovations in machine learning 89, 251
input data 71, 74, 200, 203, 211, 242, 350, 357
input parameters 179, 180, 202
input variables 148, 183, 187, 202
integrated pest management 72, 73, 78
integrated water resource management 297
intelligence, harnessing 135, 250–252, 281, 340
intelligence of wastewater treatment 128
intelligence water security 301
intergovernmental panel on climate change (IPCC) 304, 310, 322
international water management institute 369
ion exchange 67, 69, 94, 193, 251, 257
IoT *see* Internet of Things
IPCC *see* intergovernmental panel on climate change
IPM *see* integrated pest management
irrigation drip 26
irrigation efficiency 316, 328, 365
irrigation frequency 143
irrigation management 52, 53, 143, 168, 364
irrigation methods 26, 53, 160, 167
irrigation monitoring 366
irrigation practices 160, 166, 369
irrigation scheduling 143, 153, 172, 312, 316
irrigation system 27, 133, 151, 153, 169, 171, 322, 367
irrigation technologies 26, 362
irrigation, wastewater 36
irrigation water 37, 53, 60, 61, 84, 91, 143, 171
IWMI *see* international water management institute

Jordan–Israel 363
judgments 29, 32, 51, 139, 300
judiciously 274, 277
justification 51, 132

key aspects 102, 124–126
key domains 229
key findings 36, 85, 240
key performance indicators 127
key water parameters 37, 41, 88
kinetic 73, 78, 103, 106, 212, 214, 251, 366
KNN 75, 83, 84

Langmuir 196, 207, 208
large datasets 85, 105, 141, 150, 161, 186, 241, 261, 266, 342, 344, 350, 364, 365
leak detection 139, 184, 318, 346
learning methods 88, 100, 103, 107, 172, 231, 232, 253, 322, 323
learning models 37, 47, 60, 96, 101, 102, 106, 143, 159, 169, 251, 259, 269, 350, 351
legal frameworks 33, 332, 333
linear regression 48, 66, 71, 83, 86, 102, 143, 329
local bodies 274, 275
local communities 106, 166, 297, 303, 316, 337, 365
logistic regression 46, 81, 84, 102, 221
LOR *see* logistic regression

machine learning algorithms 46, 60, 78, 90, 91, 93, 95, 97, 99, 101–103, 105, 107, 109, 111, 157, 169, 232, 233, 341, 344, 345, 367
machine learning algorithms for *see* machine learning
machine learning models 37, 88, 106, 107, 159, 164, 165, 168, 251, 259, 303, 351, 369
machine learning techniques 50, 106, 153, 232, 251, 253
maintenance costs 51, 120, 121, 124, 127, 131, 133, 150, 261, 300, 329, 345
maintenance schedules 118, 120, 125, 266
maintenance strategies 117, 119, 125, 128, 136
management practices 82, 140, 141, 159, 163, 239, 240, 275, 289, 341, 353, 360
management strategies 72, 82, 138, 140, 141, 150, 158, 167, 239, 316, 320, 346, 352, 353, 357, 365
management system 53, 55, 56, 118, 164, 200, 275, 317, 331
marine 86, 90, 92, 240, 251, 267
marine environments 251
marine pollution 90, 91, 269
mass spectrometry 62, 76
mean square error 84, 96, 116, 183, 191, 202, 210–212, 235, 246, 260
membrane applications 182
membrane bioreactor 19, 29, 136, 193, 201
membrane desalination 257, 269
membrane distillation 21, 190
membrane filtration 30, 117, 193, 195, 200, 257, 364
membrane fouling 180, 258, 262, 269
membrane performance 93, 94, 261, 267
metaheuristic 238, 259, 369
metal concentration 69, 100, 101, 103, 153, 241, 244
metal contaminants 66, 77, 78, 90, 234, 240, 369
metal pollutants 241
metal pollution 77, 84, 89, 235, 240, 246, 250, 253
metal removal 37, 94, 101, 111, 236, 241–246, 248, 251, 253
MIC *see* minimum inhibitory concentration
microalgae 75, 76, 79, 135, 201, 212, 252, 283
micro- and nanoplastics 74, 75, 82, 86, 284
microbes 65, 69, 70, 102, 128, 138, 149, 218, 222–224, 229
microbial activity 44, 199, 222, 344
microbial threats 217–219, 221, 223, 225, 227, 229, 231, 233
microbiological 199, 217, 222, 230, 232, 252
microbiology 184, 217, 218, 290, 291
microplastic 25, 62, 74, 76–79, 82, 87, 89–91, 285
micropollutant 37, 62, 65, 86, 108, 185, 187, 190, 195, 251
minimum inhibitory concentration 224, 225, 232
MLM *see* machine learning models
MLP *see* multilayer perceptron
MLR *see* multiple linear regression
MNPs *see* micro- and nanoplastics
model performance 96, 121, 129, 249
model selection 96, 220, 352
monitoring water quality 42, 85, 102, 109, 111, 141, 156, 162, 179, 283, 343
MSE *see* mean square error
multilayer perceptron 71, 75, 86, 115, 117, 187, 191, 199, 201, 235, 248, 264
multiple linear regression 48, 50, 66, 71, 79, 84, 86

nanocomposite membrane 97, 99, 100, 107, 110
nanofiber membrane 93, 94, 96, 107–109
nanofiltration 17, 68, 147, 148, 182, 194, 267
nanomaterial 94, 96, 98, 196, 198, 205–208, 321
nanoparticle 106, 108, 110, 192, 285
nanoplastics 74, 75, 78, 82
nanosensor 100–103, 105
natural language processing 219, 300, 351
natural processes 148, 283, 359
nature communications 79, 231, 269, 367
nature medicine 233
nature sustainability 280, 367
nature water 21, 339, 355
neural network 37, 55, 63, 70, 73, 75, 78, 79, 84–86, 116, 122, 130, 142, 143, 148, 150, 154, 179, 192, 201, 222, 232, 234–236, 239–241, 244, 246, 250–252, 267–269, 329, 368
neural network algorithm 79, 143
neuro 79, 85, 118, 149, 151, 234, 259
normalized root mean square 116, 235, 246
NRMSE *see* normalized root mean square
nutrient cycling 234, 238, 239, 283
nutrient management 238–240
nutrient recycling 235, 236, 239
nutrient removal 26, 201, 237, 239, 240, 242, 245, 248, 249, 253
nutrient removal efficiency 239, 240, 245

operational cost 54, 101, 103, 118, 125, 128, 129, 133, 144, 184, 186, 239, 244, 247, 257–259, 261, 266, 328, 343, 344
operational data 50, 51, 96, 127, 237
operational efficiency 24, 35, 51, 97, 116–118, 120, 125, 126, 128, 132, 133, 145, 159, 188, 238, 244, 262, 267, 345, 346
operational optimization 130
operational parameters 66, 116, 127, 259, 261, 265, 345
optimization and automation 37, 151, 189, 268
optimization and control 187, 251

optimization and environmental 193, 195, 197, 199, 201, 203, 205, 207, 209, 211, 213
optimization in seawater 259, 261, 263, 265, 267, 269
optimization in water 115, 178, 179, 181, 183, 185, 187–189, 320
optimization water treatment 201
organic carbon 148, 187, 260, 283
organic compounds 67, 93, 183, 195, 283
organic matter 48, 89, 147, 193–195, 201, 204, 211
organic pollutants 82, 89, 90, 96, 97, 185, 193, 199, 292, 343, 364
oxidation 19, 185, 187, 193–195, 200, 214
oxygen demand 23, 101, 117, 118, 142, 184, 191, 199, 234, 237, 344
oxygen level 187

PAHs *see* polycyclic aromatic
pattern recognition 128, 226, 232, 311, 342, 346, 357
PCA *see* principal component analysis
PCR *see* polymerase chain reaction
PdM *see* predictive maintenance
PdM adoption 117, 121, 130
PdM implementation 120, 131
PdM systems 121, 126, 133
performance metrics 122, 127, 130, 248, 249, 261, 287, 288, 292
performance prediction 149, 251, 258, 269
pesticide application 72, 73
pesticide detection 71, 72, 76, 78, 79
pesticide residues 71, 72
pesticide spraying 73, 77
PGPR *see* plant growth promoting rhizobacteria
plant growth promoting rhizobacteria 69, 235
PLC *see* programmable logic controller
pollutant detection 88, 91, 102, 106, 342
pollutant removal 94, 117, 118, 183, 236, 238, 240, 243, 244, 344
pollutant removal efficiency 236, 240, 244
polluted water 67, 80, 82, 84, 195, 282, 283
pollution control 25, 78, 82, 83, 89, 102, 171, 284, 342, 361
pollution episodes 252
pollution levels 25, 46, 103, 163, 329, 352
pollution monitoring 188, 268, 329
pollution parameters 199, 211
pollution removal 184, 192, 195, 212
pollution research 38, 60, 61, 78, 168, 172, 212, 268, 370
pollution sources 49, 58, 60, 81, 342
polycyclic aromatic 82, 250
polymerase chain reaction 50, 84, 218, 224
polyvinyl chloride 75, 82
population expansion 28, 41, 62, 358, 359
population growth 25, 145, 155, 161, 162, 257, 300, 309, 310, 316, 319, 356–359, 363, 365
potential failures 117, 124, 125, 159
potential health risks 82, 146
potential risks 52, 55, 163
precipitation patterns 357, 359
predictive accuracy 95, 116, 120, 237, 240, 242, 264, 344
predictive analysis 36, 46, 106, 236, 269, 346
predictive analytics 22, 23, 25, 32, 35, 36, 54, 101, 104, 128, 132, 163, 236, 238, 276, 293, 300, 303, 311, 314, 318, 319, 321, 328, 342, 344–346, 352, 353
predictive efficiency 236, 243, 244
predictive insights 28, 54, 87, 102, 119, 132
predictive maintenance 24, 29, 49–51, 59, 115–133, 135, 150, 158, 159, 188, 237, 259, 261, 262, 315, 342, 345
predictive modeling 37, 113, 140, 141, 192, 213, 222, 241, 244, 246, 263, 301, 343
predictive performance 102, 116, 212
principal component analysis 49, 83, 84, 100, 101, 115, 148, 149, 179, 221
privacy challenges 328–331, 333, 335, 337, 339
privacy violations 56, 330
proactive approach 97, 118, 159, 162, 163, 345
proactive maintenance 125, 129, 200, 261, 262, 344
process optimization 125, 126, 148, 178, 179, 184, 186–188, 261, 263, 266
programmable logic controller 180, 190
public health 52, 58, 59, 82, 85, 93, 101, 115, 144, 162–164, 284, 300, 317, 318, 331, 332, 342, 343, 359
public health outcomes 93, 101
public utilities board 165, 317
PVC *see* polyvinyl chloride

radial basis function 71, 75, 115, 191, 264, 329
rainfall patterns 23, 27, 146, 155, 160, 162, 359, 365
rainfall prediction 143, 168
rainwater harvesting 21, 143, 312, 361
random forest 46, 69, 71, 74, 83, 84, 95, 101, 102, 117, 149, 179, 180, 221, 241, 262, 329
RBFN *see* radial basis function
real-time monitoring 99, 124, 125, 346
recurrent neural network 50, 51, 63, 101, 143, 221, 222
reducing waste 23, 24, 124, 158, 161, 167, 261, 263
regression analysis 71, 179, 180, 203, 342
regression models 48, 96, 102, 140, 141, 258
regression neural network 50, 71, 191
reinforcement learning 63, 131, 132, 149, 158, 169, 220, 241, 242, 250, 266, 313, 329
remote sensing 30, 45, 52, 58, 69, 80, 93, 141, 156, 164, 293, 297, 301
removal efficiency 93, 94, 101, 236, 239–245, 249, 252
renewable energy 30, 258, 261, 266, 267, 335, 336, 341, 351
renewable energy sources 266, 335, 336, 341, 351
resource distribution 59, 72, 158, 336, 341
resource management 150, 156, 295, 357
resource use 36, 54, 101, 151, 156–158, 178, 261, 359
resource utilization 31, 38, 54, 120, 136, 177, 328, 342, 345, 353
resource waste 35, 178, 300
response surface methodology 96, 110, 184, 191, 192, 201–203, 211, 213, 214, 235, 250–253, 260
reverse osmosis 17, 20, 25, 29, 37, 182, 257–259, 267–269
RFs *see* random forest
risk analysis 81, 87, 88, 368
risk assessment 29, 80, 81, 83, 86, 87, 126, 139
river flow 37, 140, 152, 301, 356
RNN *see* recurrent neural network
RSM *see* response surface methodology
RWH *see* rainwater harvesting

safe drinking water 30, 80, 92, 115, 146, 163, 358
sanitation 35, 36, 67, 80, 82, 132, 183, 283, 297, 298, 309, 356, 359, 360, 365
sanitation systems 19, 183, 368
satellite data 27, 30, 55, 166, 302, 314, 346

satellite imagery 26, 74, 263, 300, 301, 311, 314, 320
SCADA *see* supervisory control and data acquisition
SCAE *see* stacked convolutional autoencoder
SDG *see* sustainable development goal
seawater desalination 18, 257, 259–261, 263–265, 267–269
seawater desalination plants 264, 265, 269
sedimentation 124, 147, 181, 182, 193–195
sensor data 42, 45, 57, 97, 104, 117, 121, 129, 130, 132, 143, 158, 164, 179, 184, 186, 259, 262, 311
sensor network 52, 60, 163, 179, 241, 343, 345, 365
sensor technologies 56, 58, 122, 355
separation and purification 77, 107, 109, 111
smart cities 44, 55, 101, 124, 285
smart devices 22, 44, 328
smart irrigation 23, 24, 26, 29, 36, 141, 160, 300, 312, 314, 316, 362
smart sensors 24, 136, 142, 317, 318, 342
smart technology 28, 60, 108, 323, 339, 354, 368
smart water grid 37, 54, 165, 166, 342, 346
smart water management systems 29, 293, 305, 318, 355
smart water quality 60, 170, 178, 251, 318
socioeconomic 87, 306, 331, 332, 337
soil conditions 24, 29, 54, 164, 166, 316
soil erosion 336
soil health 26, 35, 65, 362
soil moisture 23, 24, 26, 27, 31, 35, 52, 53, 139, 140, 143, 160, 164, 166, 300, 312, 314, 316
soil moisture levels 27, 53, 140, 160, 300
soil moisture sensors 27, 160
sorption 94, 196, 197, 199, 204, 207–209, 211, 212, 253
stacked convolutional autoencoder 122, 130, 132, 134
Stockholm 295, 304, 305
storage tanks 20, 81, 321
supervised learning 46, 47, 49, 83, 101, 157, 220
supervisory control and data acquisition 61, 122, 124, 126, 127, 130, 133, 135, 136, 139, 142, 145, 180, 188, 301
support vector machine 70, 75, 81, 84, 117, 192, 203, 221, 235, 239, 329
support vector regression 50, 86, 102, 135, 143, 201, 329
surface water 81, 85, 143, 146, 147, 177, 181, 298, 315, 357
suspended particles 43, 180, 181
suspended solids 48, 181, 193–195, 235, 239
sustainability goals 236, 345, 352, 353
sustainable agriculture 26, 37, 77, 354, 366
sustainable approach 77, 150, 235, 261
sustainable development 35, 81, 149, 150, 204, 294, 309, 310, 321, 327, 356, 358
sustainable development goal 35, 36, 83, 169, 294, 297, 298, 310, 312, 356
sustainable future 87, 169, 281, 354, 360
sustainable management 55, 87, 101, 145, 235, 358
sustainable operations 118, 168
sustainable performance 240, 265
sustainable practices 258, 298, 352, 359, 361
sustainable resource 126, 156, 187, 262, 263, 345
sustainable smart cities 134, 280, 304
sustainable solutions 94, 267, 310, 356, 365
sustainable wastewater treatment 19, 184, 213, 251
sustainable water management 24, 30, 36, 82, 96, 116, 159, 166, 295, 313, 341, 342, 365
sustainable water management practices 82, 159, 341
sustainable water resource management 18, 103, 150, 154, 300
SVR *see* support vector regression
system components 125, 127, 350
system optimization 127, 183, 350
system reliability 119, 120, 127, 131–133, 161, 350

TDS *see* total dissolved solids
technological advancements 117, 290, 360, 365
technological innovation 93, 139, 312, 321, 354, 359
TOC *see* total organic carbon
total dissolved solids 50, 101, 192, 199, 201, 211, 235, 237, 242, 243, 264
total nitrogen 118, 187, 235, 239, 249
total organic carbon 63, 260
total phosphorus 118
total suspended solids 117, 128, 235, 239, 246, 248, 249
training data 57, 83, 85, 148, 184, 224, 226, 228
transfer learning 57, 72, 136, 221, 346
transformative force 341, 342
transformative potential 96, 98, 104, 105, 264, 265, 312
TSS *see* total suspended solids
turbidity 23, 25, 41, 43, 45, 46, 48, 49, 58, 96, 101, 102, 149, 162, 163, 181, 187, 211, 237, 262, 284, 317, 329, 343

UNDP *see* united nations development programme
united nations development programme 83, 294, 298, 305
united nations sustainable development goal 356
unsupervised learning 49, 83, 97, 100, 102, 157, 158
urban water systems 35, 101, 103, 158
urinary tract infections 127, 188, 199, 223
UTI *see* urinary tract infections

validation 32, 66, 72, 76, 84, 85, 95, 127, 128, 143, 200, 211, 226, 264, 331, 341, 349
valuable insights 85, 96, 117, 130, 132, 300, 343, 344
valuable resource 157, 263, 284
variables in wastewater 178
variations in water quality 25, 242, 343
vector machine 70, 75, 81, 84, 117, 192, 203, 221, 235, 239, 329

wastewater management 36, 101, 116, 187, 235, 236, 243, 244, 342, 344
wastewater recycling 29, 142
wastewater reuse 187, 251, 366
wastewater treatment 101, 141, 185, 201, 213, 251
wastewater treatment efficiency 22, 134, 250
wastewater treatment facilities 26, 129, 131, 188
wastewater treatment plants 131, 142, 150, 178, 186, 187, 235, 237, 239, 242–246, 250–252, 318
water accessibility 293–295, 297, 299, 300
water availability and quality 312
water bodies 25, 27, 44, 62, 66, 68, 82, 84, 92, 104, 138, 146, 164, 235, 240, 278, 283, 309, 342, 359, 362
water conservation efforts 293
water conservation initiatives 317, 320
water conservation strategies 53, 312
water conservation tools 313
water consumption patterns 158, 315, 330
water crisis 20, 22, 36, 150, 263, 293, 294, 297–299, 301, 310, 311, 314, 360
water demand 23, 24, 30, 54, 116, 140–142, 145, 155–159, 161, 162, 165, 166, 264, 266, 300, 303, 316, 318, 341, 346, 349, 356, 359, 369
water distribution networks 60, 141, 165, 172, 315, 317, 323, 345
water efficiency 24, 28, 35, 164, 166, 337

water flux 94, 187, 262
water governance services 289, 290
water governance system 282, 290, 293, 302
water harvesting 143, 279, 295, 304, 361, 366, 368
water infrastructure management 168, 315
water management practices 141, 289, 360
water management systems 36, 318, 335
water monitoring system 41, 42, 56, 58, 61, 105, 353
water monitoring techniques 41
water pollution 25, 60, 66, 78, 81, 141, 142, 164, 177, 283, 336, 342, 359, 361, 362
water pricing 30, 297, 362
water purification and wastewater 214, 252, 355
water purification systems 30, 258
water purification technologies 35
water quality components 85
water quality index 50, 61, 76, 86, 280, 344
water quality management 42, 106
water quality monitoring 39, 98, 103, 310, 342
water quality sensors 350
water resource management using artificial intelligence 37, 323, 339
water safety 19, 85, 101, 298, 305, 314, 346, 363, 368
water scarcity and future challenges 18
water scarcity in agriculture 38, 170, 367
water security and resilience 170, 323
water security challenges 295
water shortages 22, 27, 31, 87, 159, 280, 309, 314, 332, 358, 359, 361, 366
wavelengths 74, 98, 206
weather conditions 143, 162, 312, 362
weather data 27, 140, 170, 317
weather forecasting 27, 29, 123, 160
weather patterns 25–28, 72, 143, 161, 164–167, 266
weather stations 27, 31, 32, 143, 160
WHO *see* world health organization
world health organization 20, 21, 138, 232, 240, 252, 273, 295, 297, 306, 321, 359, 369
WQI *see* water quality index
WWTP *see* wastewater treatment plants

zoological 284, 290

For Product Safety Concerns and Information please contact our EU
representative GPSR@taylorandfrancis.com
Taylor & Francis Verlag GmbH, Kaufingerstraße 24, 80331 München, Germany

www.ingramcontent.com/pod-product-compliance
Ingram Content Group UK Ltd.
Pitfield, Milton Keynes, MK11 3LW, UK
UKHW061914111225
465993UK00010B/1280
9781032988061